APS
Advances in Pharmacological Sciences

Effects of Nicotine on Biological Systems

Edited by
Franz Adlkofer
Klaus Thurau

Birkhäuser Verlag
Basel · Boston · Berlin

Volume Editors' Addresses:

Franz Adlkofer
Forschungsrat Rauchen ,
und Gesundheit
Königswinterer Straße 550
D-W-5300 Bonn 3
F. R. G.

Klaus Thurau
Physiologisches Institut der
Universität München
Pettenkofer Straße 12
D-W-8000 München 2
F. R. G.

Library of Congress Cataloging-in-Publication Data

Effects of nicotine on biological systems / edited by Franz Adlkofer,
Klaus Thurau.
(Advances in pharmacological sciences)
Includes bibliographical references.
ISBN 3-7643-2519-4 (alk. paper) –
ISBN 0-8176-2519-4 (alk. paper)
1. Nicotine-Physiological effect.
I.   Adlkofer, Franz, 1935–.
II. Thurau, Klaus.
III. Series.
[DNLM: 1. Nicotine-pharmacology. QV 137 E27]
QP801.N48E34 1991
615.9'52379–dc20

Deutsche Bibliothek Cataloging-in-Publication Data

Effects of nicotine on biological systems /ed. by: Franz Adlkofer;
Klaus Thurau. - Basel; Boston; Berlin: Birkhäuser,
1991
(Advances in pharmacological sciences)
ISBN 3-7643-2519-4 (Basel . . .)
ISBN 0-8176-2519-4 (Boston)
NE: Adlkofer, Franz [Hrsg.]

Product Liability: The publisher can give no guarantee for information about drug
dosage and application thereof contained in this book. In every individual case the
respective user must check its accuracy by consulting other pharmaceutical literature.

PREFACE

As part of its scientific activities, the German Research Council on Smoking and Health regularly provides opportunities for scientists to discuss progress in the field of nicotine research. In this context, the Research Council sponsored a Satellite Symposium in Hamburg, June 28-30, 1990 entitled "Effects of Nicotine on Biological Systems". This meeting was held in conjunction with the XIth International Congress of Pharmacology in Amsterdam and follows the first Satellite Symposium on Nicotine which was convened in Brisbane, Australia in 1987. The aim of these conferences has been to discuss state of the art research on the pharmacology and toxicology of nicotine and its metabolites and to integrate this information to help define nicotinic actions on the central and peripheral nervous system as well as to evaluate health or behavioral effects associated with use of this alkaloid. Furthermore, at this conference, potential therapeutic benefits of nicotine for certain disease states were discussed. Smoking and the health effects of smoking were dealt with only as far as they could not be separated from the effects of nicotine.

This volume contains the lectures presented at the symposium and illustrates that knowledge of nicotine has advanced considerably in recent years with regard to mechanisms of its actions. Despite such progress however, it is apparent that a large number of questions remain unanswered, especially in the light of new insight into cellular and molecular mechanisms which can be affected by nicotine. For this reason, nicotine research constitutes a continuing challange to scientists, which is likely to sustain the need for further research well into the forseeable future.

The need for such research becomes also evident when one considers the decisive role which nicotine might play in determining smoking behavior and its importance in the development of cigarettes which could reduce a health risk to cigarette smokers. It is only on the basis of scientific results and a more complete understanding of the effects of nicotine on biological systems that the concept of a less harmful cigarette can be further advanced.

There are currently more than one billion smokers worldwide and this number is not likely to decline substantially in the near future. Cigarette consumption in Third World countries where at least two thirds of the total world population lives, still increases every year and is by no means offset by the decrease in consumption in developed countries. Therefore, it remains essential that both health education and smoking cessation efforts be complemented by modification to a less harmful product. While such a concept has been regarded as counterproductive by some antismoking organizations, it must be moved forward to assist those persons who have decided to smoke and will not stop. As formulated by G. Gori, former Director of the National Cancer Institute: "Leaving smokers to their fate is neither humane nor economic, particularly when their risk can be substantially reduced."

The Council thanks all scientists who presented their results at the conference in Hamburg and encourages them and others involved in nicotine research to continue their efforts into the elucidation of the mechanisms of action of nicotine and its biological effects. We look forward to the Third Satellite Symposium on Nicotine which is being planned for 1993 in conjunction with the XIIth International Congress of Pharmacology.

Klaus Thurau, M.D.
Professor of Physiology
Chairman of the
Research Council on
Smoking and Health

Franz Adlkofer, M.D.
Professor of Medicine
Scientific Secretary of the
Research Council on
Smoking and Health

Contents

III: General Pharmacology of Nicotine

IV: Mechanisms of Nicotine Actions in the CNS

V: Behavioral Effects of Nicotine in Animals

VI: Behavioral Effects of Nicotine in Man

VII: Nicotine and Human Disease

XII

Effects of Nicotine on Biological Systems
Advances in Pharmacological Sciences
© Birkhäuser Verlag Basel

INTRODUCTORY REMARKS

EFFECTS OF NICOTINE ON BIOLOGICAL SYSTEMS

Sir Peter Froggatt

Independent Scientific Committee on Smoking an Health, Department of Health and Social Security, Hannibal House, Elephant and Castle, London SE1 6TE, England.

I am chairman of the Independent Scientific Committee on Smoking and Health in the United Kingdom. As our name implies, we advise the UK government on scientific matters concerning smoking and health. Sometimes we also advise the tobacco companies. Naturally we are interested in <u>all</u> aspects of smoking including the biological effects of nicotine – which is the subject of this Symposium. But we are concerned with more than just learning the latest <u>scientific</u> findings; we have a duty to reduce the harmful effects of smoking. Basically, then, we <u>apply</u> scientific knowledge rather than <u>add to</u> it. To protect the health of smokers we instigated some years ago a 'low tar', or more accurately, 'product – that is 'cigarette' – modification programme'. We argued that the 'tar' component contains factors which can cause or trigger the process of lung cancer, chronic obstructive lung disease, and possibly some of the other tobacco-related diseases as well. Therefore, if we reduce the 'tar' yield we will reduce the dosage of disease determinants and <u>in time</u> will reduce the incidence of these diseases. This is a very simplistic model but is valid on a dose-response hypothesis, and we have recently shown it to be true for lung cancer and COPD in both younger men and women. Of course we don't know which of the 4000 identified components in tobacco 'tar' are harmful. We don't even know for certain if the causative agents are in the 'tar' at all. Perhaps they are in the nicotine, humectant, or purely vapour fraction: perhaps in all of them. Most investigations, though not all, believe that they are mainly in the 'tar' - certainly for lung cancer and COPD - and we accept this opinion until we have hard evidence to the contrary.

This approach is completely impiric: we change something suspect - 'tar', and observe the result - incidence of disease. It postulates no rationale or mechanism of action. It is not scientifically satisfactory but we have no choice. Anyhow, many epidemics have been reduced before the causative agent was identified or its mode of action known - plague and cholera and scurvy for example. But its not enough merely to have lower tar cigarettes available; we must ensure that people will smoke them. One of the ways of doing this is to manipulate nicotine yields in tobacco because smokers smoke mainly for the effect of nicotine. This is a pragmatic approach but one within the bounds of contemporary scientific knowledge. To be credible it must be sensitive to recent research findings and this is why a Symposium like this one here in Hamburg, the one three years ago at Gold Coast in Queensland, and the CIBA Foundation one last year on 'the biology of nicotine dependence', are so important to us.

We started our product modification programme 15 years ago. We accepted, and still accept, that nicotine is the principal habituating agent in tobacco, but at that time we did not know if it was actually harmful to health in the doses delivered during smoking. We acted with caution: we were after all dealing with the health of the public. We decided that nicotine yields of cigarettes must fall: this would, we believed, both decrease dependence on nicotine and reduce its toxic effects - if it had any. We thought that this would be easy - after all increased filtration and ventilation which were being used to reduce 'tar', would also reduce nicotine in about the same proportion. This didn't happen. What did happen was that cigarette manufacturers changed to tobacco blends with higher nicotine yields in order to maintain their markets. In the fifteen years, 1974-1989, the sales-weighted average 'tar' of cigarettes in the UK fell by 28 per cent but the sales-weighted average nicotine fell by only 13 per cent, less than half. In fact in the 12 years 1975-1987 sales-weighted nicotine yields in the UK <u>didn't fall at all</u> even though the number of low nicotine brands on the market increased from 23 to 37, the proportion of smokers smoking them increased from 5 to 16 per cent, and sales-weighted average 'tar' yields fell by 23 per cent. This suggested a consumer optimum in the UK for nicotine of at least 1.2-1.3 mg/cigarette machine yield. It also suggested that efforts to reduce nicotine yields in the short-term would be resisted by both consumers <u>and</u> manufacturers.

However, we were able to make a virtue out of necessity. If nicotine were not harmful to the health of smokers then we could use a consumer-optimum level of nicotine combined with a lower tar to tempt people to smoke lower tar brands, which would be less harmful than higher tar brands. But is nicotine in smoking doses harmless to health? This is the problem we faced ten years ago, and we still face it. If it <u>is</u> harmful then we could not use nicotine in this way. If it is <u>not</u> harmful then we might use nicotine in this way, providing we can answer the moral accusation that we are encouraging the use of an addictive drug especially in the young. We believe we <u>can</u> answer this moral case. We also believe that, in smoking doses, nicotine has not been definitely proved to be harmful to healthy smokers though it might exacerbate pre-existing cardio-vascular disease - but we will hear a lot more about this during the symposium. A major concern is tobacco-specific nitrosamines and we follow work in this area closely. As a result we have been able to follow our policy and have encouraged less reduction in nicotine than in tar. This means that the Tar/Nicotine ratio in UK cigarettes has decreased - and it is now under 11 as against 14 in 1975.

So far so good. But there is one final, vital point. What I have said would make most sense if human beings were machines. But they are not. Different people smoke different cigarettes in different ways at different times and for different reasons. This variability is important because nicotine yields can be increased if the smoker takes longer, larger, or more frequent puffs – the so-called 'compensatory smoking'. In practice it means that the smoker of lower nicotine cigarettes may increase his nicotine intake to the level associated with a higher nicotine cigarette simply by smoking the cigarette more intensively. This may or may not matter but 'tar' intake will therefore also increase – which does matter. However, this 'compensation' is never 100 per cent and therefore in compensatory smoking 'tar' intake will be less than it would have been with a higher nicotine, higher tar cigarette – not as low as a calculation based on machine yields, but always less. The effects of the Product Modification Programme are therefore

impeded but not negatived by compensatory smoking. Nicotine of course is not the only determinant of 'compensation' but it is the main one. We consider nicotine and compensatory smoking so important to our programme of action that we sponsored a Symposium three years ago in London which was published (by Oxford University Press in 1989) under the title 'Nicotine, Smoking, and the Low Tar Programme'.

Mr Chairman, I have said enough to show how important nicotine is to any action programme aimed to reduce smoking in the population, and so aimed to reduce the harmfulness of smoking to the smoker. It is the very centrepiece of policy: we must find out as much about it as we can. This meeting addresses key questions and maintains the momentum of nicotine research of the last few years. It will add to our knowledge – which is good, and make action programmes more effective – which is also good. I am indebted to the German Research Council on Smoking and Health, its Director – Professor Adlkofer, and its Chairman – Professor Thurau, for organising and sponsoring this important Symposium, and for enabling me to say a few words to you. I take great pleasure in joining in this welcome.

I.

Ethnopharmacology of Nicotine

Effects of Nicotine on Biological Systems
Advances in Pharmacological Sciences
© Birkhäuser Verlag Basel

THE ETHNOPHARMACOLOGY OF TOBACCO IN NATIVE SOUTH AMERICA*

J. Wilbert

Department of Anthropology, University of California, Los Angeles, Haines Hall 372, Los Angeles, CA 90024, U.S.A., and Botanical Museum of Harvard University

SUMMARY: The use of tobacco by South American Indians is deeply rooted in their culture and thought. From early pre-Columbian times to the present, tobacco has functioned as an important drug for magico-religious, medicinal, and recreational purposes. Data culled from about 1,800 ethnographic sources and pertaining to nearly 300 societies reveal that the Indians employ six major and several minor means of nicotine application. A comparison of the ethnographical data with experimental clinical studies of tobacco indicates that pharmacology corroborates the nicotine therapy and practice of South American shamans.

HISTORICAL ANTECEDENTS

Upon their return, in 1492, from an exploratory trip through coastal Cuba, two men of Columbus's crew recounted how they had become the first Europeans to witness the custom of smoking tobacco. Several weeks prior to this event and soon after first landfall, Columbus had been offered tobacco by the Indians, puzzling what to do with the shriveled leaves and wondering why such withered matter should be considered an appropriate gift for the momentous occasion.

Something else the discoverers could not have known at the time, was that by then the Indians had been living in the New World for some 40,000 years and that for about the last 8,000 years they had been using tobacco in one form or other. Originating in Asia and after traversing North and Central America, their ancestors had reached the southern subcontinent in bands of Paleo-Indian hunters and gatherers. By 11.000 years ago, they had occupied the open lowlands of the southern cone of South America, a region, it will be recalled, largely coterminous with the world distribution center of <u>Nicotiana</u>, from which all tobacco-producing plants are derived. However, surrounded as they were by the largest number of wild nicotianas, the early hunters of the southern lowlands actually never used tobacco until late historic times. Only with the discovery of root crop agriculture, some 8.000 years ago, did Neo-Indian horticulturalist begin to penetrate the rain forest of Amazonia, where they cultivated a dozen or so species, especially <u>Nicotiana</u> <u>rustica</u> and <u>N</u>. <u>tabacum,</u> and eventually dispersed the latter cultigens throughout the northern half of the subcontinent, the Caribbean, and beyond.

It is peculiar that from prehistoric times until roughly 1700 of the historic era, South American Indians took tobacco primarily for magico-religious and related medicinal purposes; - hedonistic indulgence of the drug is usually symptomatic of Western acculturative influence and secularization. Accordingly, tobacco practice was in the hands of shamans who, as mystical specialists, are believed to have officiated in practically every band of hunters and swidden horticulturalist and whose purpose it was-- and is -- to mediate between their fellowmen and the Supernaturals. In the pursuit of their office, tobacco shamans seek acute nicotine intoxication in order to reach altered states of consciousness and to become supernaturally empowered through spirit communication. However, what needs emphasizing at this point, is that shamanism as such is not an invention of Neo-Indian tobacco growers. Rather, shamanism figures as the oldest religion of mankind, antedating the immigration of Paleo-Indian hunters into the New World. Here as elsewhere, drug-free shamans of pre-agricultural societies reached trance states through endogenous and ascetic techniques of ecstasy, whereas tobacco-shamans use nicotine in order to achieve ad hoc states of trance and to find pre-extant shamanic tenets corroborated through the pharmacological action of the drug.*

METHODS OF NICOTINE ADMINISTRATION

Short of injection, South American tobacco shamans take nicotine via all humanly possible routs of application; be they gastrointestinal, respiratory, or percutaneous. They chew tobacco quids, drink tobacco juice and syrup, lick tobacco paste, administer tobacco suppositories and enemas, snuff rapé, inhale tobacco smoke, and apply tobacco products to the skin and to the eye.

GASTROINTESTINAL ADMINISTRATION

Tobacco Chewing: Being of wide-flung and scattered distribution in South America, chewing is probably one of the oldest methods of tobacco consumption. Sucked rather then chewed, the quid is mixed with ashes or lime and retained in the cheek or lower lip for hours at a time. Although not required for nicotine liberation, the mentioned alkalizing substances accelerate and intensify the action of the drug, in as much as nicotine is readily miscible in the increased salivary secretions and as the alkalized buccal environment is prepared for optimal absorption. In addition to the buccal cavity, absorption takes place in the stomach and the small

---

(*) For a broader discussion of the general topic and for detailed documentation consult, J. Wilbert (1987) Tobacco and Shamanism in South America. New Haven: Yale University Press, on which this essay is largely based.

intestines because shamans tend to swallow rather than expectorate the extracted juices.

<u>Tobacco Drinking and Licking</u>: The ingestion of tobacco in liquid form, either as an infusion or as a syrup, is practiced by a large number of tribes in the Guianas and the upper Amazon basin. Infusions of steeped or boiled tobacco leaves are mixed with salt or ashes and imbibed either through the mouth or nose. Shamans, in order to preserve their power, are required to drink tobacco juice periodically throughout life. A shamanic apprentice undergoing initiation imbibes ever-increasing amounts of tobacco juice until he goes into convulsions and remains comatose for hours. In other cases, a large bowlful of the infusion is poured through a funnel into the swooning novice's mouth, rendering him unconscious and leaving him literally at the brink of death. Initiates who fail to vomit most of the nicotine-laden liquid are expected to contract chronic infirmity that may end in death. Also patients suffering from malaise or nonspecific ailments are made to swallow large quantities of tobacco juice.

More commonly than in Guiana, tribes of the upper Amazon boil the tobacco leaves to produce a syrup or paste. Syrup concentrates are viscous but sufficiently liquid to be drunk like tobacco juice. Alkalizing agents are usually not added. Paste, known as <u>ambil</u>, is a tar-like substance (often mixed with ashes) which is passed with a finger or a stick across the teeth, the gums, and the tongue. <u>Ambil</u> is a potent concentrate and even small amounts will stupefy the unaccustomed user. Only the heavily habituated ones dare swallow pellets of the paste.

Both tobacco liquids and paste are commonly ingested simultaneously with other tobacco products (cigars, rape), and serially or alternately with hallucinogens of different kinds (<u>Banisteriopsis caapi</u>, <u>Brugmansia</u>, <u>Erythroxylum</u>, <u>Virola</u>). Also alcoholic beverages may be consumed while drinking tobacco juice and/or licking <u>ambil</u>.

<u>Suppositories and Enemas</u>: Suppositories of green tobacco or rapé

are probably more often employed than the ethnographic literature leads to expect. They serve mainly medicinal purposes for constipation and against helminthic infestation.

South American Indians administer enemas by means of two different types of syringe; one type is made either of a straight or funneled bone or cane through which the clyster is blown by an assistant into the rectum of a receiver. The other consists of a bulb made of a bladder, leather, or rubber and a nozzle of bone or reed. The rubber bulb syringe is an invention of South American Indians and occurs in the Amazon region. Enema syringes are employed for medicinal or intoxicating/ritual purposes with clysters prepared from ginger, pepper, tobacco and antiseptic herbs for the former, and a variety of hallucinogens (Anadenanthera colubrina, Banisteriopsis caapi, Brugmansia, Virola), and tobacco for the latter. Thus, tobacco clysters are administered for both medicinal and ritual purposes, and at least in one known instance, tobacco syrup is mixed with the hallucinogen Banisteriopsis caapi. Escaping from digestive changes along the gastrointestinal tract as well as from hepatic reaction, rectally taken nicotine is as appreciated for its cathartic effect as it is feared for its toxicity and potential lethality.

RESPIRATORY ADMINISTRATION

Tobacco Snuffing: The snuffing of tobacco powder or rape is practiced by tribes on the Orinoco, the Northwest Amazon, the Montana-Rio Purus, the Guapore, and the Andean region. Sporadic cases have also been recorded from northern Colombia, the lower Amazon, and the Gran Chaco. Tobacco rapé is snuffed singularly and blended with other intoxicating substances like Anadenanthera, Erythroxilum, Virola, and still others.

Tobacco rapé is taken through (a) self-administration by inhalation with or without a snuffing device and (b) mutual administration of two persons with the aid of an insufflator. The effects produced by these two methods may differ in intensity, but both are conducive to nicotine absorption and can be expected to

lead to substantial blood levels of the drug. Unaccustomed users
may remain unconscious for several hours. But strongly habituated
practitioners in Brazil take dozens of insufflations of tobacco
rapé, singly or mixed with _Virola_ powder during an extended
seance. Taking tobacco blended with hallucinogens aims at
affecting not only the central nervous system but also the
sympathetic and parasympathetic components of the autonomic system
for different effects; and South American shamans are known for
their ability to orchestrate a variable neurological trance
experience. To increase secretion, plant ashes are often mixed
with the snuff powders. This furthers the transport of nicotine
and intensifies its action while, at the same time, it provides
an alkalizing agent to liberate coca (_Erythroxylum_) and possibly
parica (_Virola_) in blended powders.

Tobacco Smoking: Tobacco smoking is the most widely distributed
practice of tobacco administration in South America. Of the two
methods of respiratory administration, smoking is more effective
even than snuffing, as nicotine is absorbed through the large
absorptive surface in the bronchioles and alveoli of the lungs and
reaches the bloodstream quickly and in large amounts. To satisfy
their craving, shamans smoke more or less regular-sized cigars
incessantly. However, on frequent ritual occasions they are also
known to roll gigantic cigars; some being 90 cm long and 2 cm
thick, others consisting of a 40 cm long and 6 cm thick funnel
made of a twisted tobacco leaf - or sundry other wrappers -- and
filled with pounded dry tobacco. Large-size cigars of the former
kind are smoked by a single individual. Those of the second kind
are placed in a cigar-holder about 60 cm long, like a two-pronged
fork, and passed around in communal smoking. With few exceptions,
cigar smokers inhale the smoke, often employing hyperventilation
and deep sucking gasps while vigorously working their shoulders
and lungs like bellows. In addition to the large amounts of
tobacco and the intensity of smoking, the toxicity of ritual cigar
smoking is much enhanced by rolling cigars from _Nicotiana rustica_
instead of N. tabacum, with nicotine concentrations in excess of

18 percent and potent enough to produce hallucinations and catatonia. Pipe-smoking is also an autochthonous method of South American tobacco adminstration. It is practiced by means of very ancient tubular pipes, a variety of elbow pipes, and others. Absorption occurs via the gastrointestinal and the respiratory tracts, and acute intoxication has been reported from many tribes on the subcontinent.

A very special manner of respiratory administration is by free tobacco smoke in the atmosphere. Streamers of smoke are blown from an inverted cigar, from the burning end of a cigar, or from the mouth of a partner onto the head, face, and body of a receiver. In other cases, bamboo tubes filled with dried tobacco leaves are used in mutual administration, with one partner blowing the smoke directly into the open mouth of the other. The alternating receivers aim at inhaling or swallowing the smoke in its entirety.

Percutaneous Administration: Application of tobacco to the intact and broken skin is a widely practiced method of native nicotine administration. Readily penetrating the skin, percutaneous absorption of the alkaloid can cause severe and fatal poisoning in man. Topically applied tobacco products aim at local effects. However, passage by passive diffusion of nicotine through the cornified layer to the underlying dermis well supplied with lymph and blood capillaries, leads also to systemic effects. Since the rate and speed of absorption of pharmacolocigally significant amounts depend largely on contact intensity, shamans effect prolonged and continuous delivery by applying wet tobacco leaves, paste, snuff plasters, juice ablutions, nicotine-laden sputum, and controlled fumigations. Functioning more or less as sustained-release mechanisms of application, these practices have an analgesic effect. Also, through nicotine-triggered norepinephrine release, flow in skin vessels is significantly reduced, and shamans apply tobacco as a febrifuge.

Finally, tobacco smoke and juice are applied to the eye where nicotine is absorbed from the conjunctiva of the inner surface of the eyelid and the forepart of the eyeball. The blowing of tobacco smoke into the eye causes copious lacrimation so that the liquids involved in both techniques of administration further the penetration of the alkaloid through the membranous barrier for rapid and systemic effects.

TOBACCO SHAMANISM AND THERAPY

Given the overall unpleasantness and outright health and even life threatening nature of tobacco use, why should the drug have played such an important role in the history of South American Indians? Of course, like psychotropic drugs everywhere, tobacco provides the user with a means of escape and release from stress. In traditional context, it did so for a select few of society, giving them privileged status and a window to the Otherworld. Shamans must continuously prove their spiritual empowerment to the community and to themselves. And tobacco helps them do that.

To prove their otherworldliness, shamans take advantage of the biphasic drug in tobacco and display through nicotine induced nausea, heavy breathing, vomiting, and prostration their ritual illness; manifest through tremors, convulsion or seizure, their agony and, in acute narcosis, demonstrate through transitory respiratory arrest that they have died as ordinary persons but come back to life with a supernatural body. (This status of a resurrected hero is basic to the appreciation of a tobacco shaman's role as a protector, healer, and soul guide.) Knowing how to dose themselves appropriately, he repeatedly risks these pharmacological conditions of progressive blockade of impulse transmission at autonomic ganglia and central stimulation, relying on prompt biotransformation of nicotine to negotiate the threshold of life and death.

Thanks to the various effects of nicotine on the body, the initiated shaman is a changed person. For instance, he sees best in crepuscular light and discern enemies and game animals approaching in the dark. Under the effects of tobacco amblyopia he can not only see but live in the darkness of the underworld; a condition, which has probably served as a model for the association of tobacco shamanism with underworld cosmology. The symptoms of a so-called "smoker's throat" let him communicate with the spirits in what is considered an appropriate voice. Increased perspiration and nicotine-triggered liberation of norepinephrine with resulting drops in skin temperature enable him to perform various heat-defying feats. He is believed to assimilate the magical heat of his sacred cigar and be the master of fire and fever. Effects of nicotine on the gastrointestinal tract due to parasympathetic stimulation, dulling of the taste buds, depression of gastric contractions, elevation of blood sugar levels, release of epinephrine and its excitatory action on the central nervous system, among others, depress the shaman's hunger feelings and make him pine for spirit food; i.e., tobacco, rather than for ordinary human fare. Like regular food, tobacco posses strong coercive power. Since tobacco is the food of the spirits and since spirits, lacking tobacco, are fully dependent upon the shaman for their sustenance, the latter uses the drug as a means of coercion to obtain favors from the gods.

Shamans are expected to protect their communities from a host of evil spirits, sorcerers, and other powers of disintegration. In many societies they are believed to engage in such battles not in human form but as were-jaguars, and there exists a close relationship in South American cosmology between the shaman and the jaguar. Nicotine causes a number of changes that activate his aggressiveness. First, night vision lets him see in the dark like the nocturnal jaguar, and a deep husky voice, a furred tongue, and a peculiar fusty body odor liken him to the powerful cat. Second, nicotine induces the cholinergic preganglionic fibers of the sympathetic nervous system to discharge the arousal hormones epinephrine and norepinephrine, which mobilize the shaman's body for emergency reaction. Third, the nicotine-specific reaction causes the appropriately enculturated shaman to adopt a fighting posture rather than to ignore it as a "cold alert". Thus, nicotine-mediated physiological changes and neuropsychological states combine with cultural conditioning to constitute the main ingredients of were-jaguar shamanism. Rather than just permitting the practitioner to enact characteristic jaguar behavior, they provide him with an essential feeling of "jaguarness," confirming his shamanic status and role.

The verified paranormal condition of the tobacco shaman gives him the essential authority to function as a healer, and nicotine provides his praxis with increased efficiency. From the time of his initiation, the shaman's breath is believed to have therapeutic properties and tobacco smoke makes this normally invisible pneumatic power apparent. Its effectiveness is dramatically manifested at harvest time, when seeds that had been blown upon with tobacco smoke bear richer fruit than those that had not. Some eight percent of the insecticides in tobacco are said to be transferred into the mainstream smoke, so that seeds freed of insect infestation have indeed been cured by the shaman's breath.

Shamans also blow thick clouds of smoke over patients, capturing it under their cupped hands, directing it over wounds, and massaging it over ailing body parts. In addition, smoke is blown into the face of the patient and into his/her eyes nose, and mouth. It is administered to a hollow tooth and the wound of tooth extraction. Fumigating patients in this manner is said to calm the pain, suppress fever, and cure. Other than just smoke, shamanic therapists blow nicotine-laden sputum, tobacco powder, and juice to heal wounds. Thus, nicotine administration through directed smoke blowing, spit blowing with nicotinic saliva, tobacco juice, and powder, saliva massages, juice ablutions and, one might add, rape and tobacco leaf plasters and compacts, involve the respiratory, gastrointestinal, and dermal routs, delivering the drug in amounts large enough to be locally and systemically effective.

REFERENCES

Wilbert, J., (1987) Tobacco and Shamanism in South America. Yale University Press, New Haven.

II.

Metabolism of Nicotine

Effects of Nicotine on Biological Systems
Advances in Pharmacological Sciences
© Birkhäuser Verlag Basel

IMPORTANCE OF NICOTINE METABOLISM IN UNDERSTANDING THE HUMAN BIOLOGY OF NICOTINE

Neal L. Benowitz

University of California, San Francisco, San Francisco General Hospital Medical Center, San Francisco, California 94110

SUMMARY: Consideration of pharmacokinetic and metabolic processes is essential to our understanding of the effects of nicotine and tobacco in people. Metabolic studies indicate rapid metabolism, which explains low oral bioavailability and physiological influence such as food on the rate of nicotine metabolism. Individual variability in rate of metabolism may explain in part individual differences in tobacco consumption and/or adverse effects of tobacco use. Some metabolites of nicotine, such as nicotine Δ-1'-(5')-iminium ion, beta-nicotyrine, cotinine or nornicotine, may have pharmacologic activities that could contribute to the effects of nicotine. Studies of nicotine metabolism also provide a basis for biochemical assessment of nicotine exposure from tobacco or environmental tobacco smoke.

INTRODUCTION

Many papers are presented in this book describing the effects of nicotine on membrane receptors, cells, organ systems and whole animals and people. How nicotine gets from tobacco or a medicinal formulation into the body, how much enters the circulation, where it goes, how long it remains in various organs, and what happens to it after it enters the body are essential to our understanding of the human biology of nicotine. This paper will briefly review aspects of the pharmacokinetic processes, with emphasis on the importance of metabolism in understanding the human pharmacology of nicotine.

Discussions of the pharmacokinetics of nicotine have been presented elsewhere (Benowitz, 1988; Benowitz et al., 1990). Essential factors relevant to understanding pharmacologic effects are as follows.

1. Nicotine is well absorbed through the lungs as well as across the buccal and nasal mucosa and the skin.

2. Absorption through the lungs is rapid so that nicotine moves into the arterial circulation quickly and reaches the brain within 10 seconds after a puff; absorption from smokeless tobacco, nicotine chewing gum and transdermal nicotine is slower.

3. After cigarette smoking, concentrations of nicotine in the arterial circulation may peak as high as 100 ng/ml (6×10^{-7} M), although because of extensive tissue uptake venous levels are only one-third that high. Peak concentrations in the brain may be two to three times higher. This concentration range needs to be considered in evaluating experimental studies of effects of nicotine on cell systems and on animals for their relevance to humans.

4. After the peak concentrations following smoking, nicotine rapidly distributes out of the brain into the blood stream and into body organs such as skeletal muscle. This pattern of rapid uptake and tissue distribution results in transient peaks and troughs of concentrations of nicotine in rapidly perfused organs, such as the brain. Such a pattern favors greater pharmacologic effects with less development of tolerance for any particular dose. Similar sharp peaks and troughs are not seen with slower-releasing nicotine preparations such as oral snuff or nicotine chewing gum, where there is considerable time for development of tolerance and the magnitude of pharmacologic effects tends to be less.

5. Nicotine in the body has a half-life of 2-3 hours. As a consequence, with regular smoking blood levels rise for 8-12 hours, and persist overnight even while the person sleeps. Thus, the body is exposed to nicotine 24 hours a day.

6. The level of nicotine in the body is determined by the rate of intake of nicotine (that is, the smoking rate or the consumption rate of smokeless tobacco

or other forms of nicotine) and the rate of metabolism. Insofar as cigarette smokers tend to regulate levels of nicotine in the body, the rate of metabolism of nicotine may influence how much a smoker smokes to achieve a desired blood level and effect. Individual differences in the rate of metabolism may, therefore, influence smoking behavior and/or smoking-related adverse health risks.

In considering the pharmacokinetics of nicotine, it is apparent that rate of metabolism of nicotine affects its pharmacology by determining the level of nicotine in the body with any given rate of consumption of tobacco. After entering the body, nicotine is metabolized primarily by the liver, although there is evidence indicating a certain amount of metabolism by the lung. The rate of nicotine metabolism is rapid, and appears to be limited by liver blood flow such that 50-70% of the nicotine in the blood passing through the liver is extracted for metabolism. Biological consequences of this high degree of hepatic extraction include extensive first-pass metabolism of oral nicotine and the susceptibility of nicotine metabolism to factors that affect liver blood flow. For example, a high protein meal increases liver blood flow and accelerates the metabolism of nicotine (Lee et al., 1989).

Aspects of nicotine metabolism are important for other reasons. Nicotine metabolism results in the generation of a number of metabolites, as is discussed in subsequent papers. Some of these metabolites may have pharmacologic activities that could contribute to the effects of nicotine in people. For example, as discussed by Castagnoli, nicotine Δ-1'-(5')-iminium ion, generated by microsomal oxidation of nicotine, is chemically reactive and could covalently bind to macromolecules in cells (Shigenaga et al., 1988). Such covalent binding could mediate injury to cells, and thus contribute to neoplasia. Beta-nicotyrine, a minor metabolite, is also highly reactive (Shigenaga et al., 1989). In animals, pretreatment with beta-nicotyrine reduced the toxicity of subsequent doses of nicotine (Stalhandski and Slanina, 1982). The possibility that local metabolism of nicotine in particular organs may generate active metabolites which could bind intracellularly or be trapped inside cells with profound pharmacologic effects deriving from very small quantitites of metabolites is an exciting one that deserves further study.

Cotinine has been reported to affect neurotransmitter systems in vitro and has been shown to have vasodilator and hypotensive actions in intact animals (Kim et al., 1968; Dominiak et al., 1985). Of interest is that blood pressure tends to be lower in smokers than in nonsmokers and in smokers blood pressure is inversely correlated to plasma cotinine concentrations (Benowitz and Sharp, 1989). While a number of mechanisms for blood pressure lowering have been proposed, cotinine itself may contribute to the lower blood pressure of cigarette smokers. As demonstrated in papers by Jacob and Neurath and co-workers in this volume, nornicotine is a minor metabolite of nicotine. Nornicotine has pharmacologic activity and toxicity comparable to that of nicotine (Risner et al., 1988). Nicotine-derived N-nitrosoamines are carcinogenic and may contribute to tobacco-related cancer. These nitrosoamines are present in smokeless tobacco and tobacco smoke, but, as suggested by Hoffman in this volume, could also be endogenously nitrosated in the gastrointestinal tract. "In that there is gastro-enteric recirculation of nicotine, such a metabolic process could be relevant to the long-term toxicity of nicotine." Of note, Tricker and Preussmann report in this volume that nitrosation does not occur in artificial serum and gastric juice, suggesting that nitrosation of nicotine in vivo is not a significant problem. However, nornicotine, a secondary amine, would be expected to be nitrosated much more easily than nicotine, and could be a substrate for endogenous formation of nitrosoamines.

Finally, studies of nicotine metabolism provide a basis for estimating human exposure to nicotine. Cotinine, quantitatively the major metabolite of nicotine, is itself metabolized much more slowly, and persists in the body for a much longer time, compared to nicotine. The half-life averages 19 hours as compared to 2 or 3 hours for nicotine. Because cotinine has a long half-life, the time of sampling with respect to smoking the last cigarette or the time of day is less critical than for nicotine. Also, concentrations of cotinine in plasma, saliva and urine are proportional, so there is considerable flexibility in the selection of biological fluids to monitor. Concentrations of cotinine are also technically easier for most laboratories to measure than are concentrations of nicotine.

On average, about 70% of nicotine is converted to cotinine. Using averaged data for nicotine and cotinine metabolism, one can estimate that a person with a blood cotinine level of 100 ng/ml consumes on average 9 to 12 mg nicotine per 24 hours. Thus, the average smoker with a cotinine level of 300 ng/ml will be consuming 27-36 mg nicotine per day, which is consistent with experimental studies.

Cotinine is metabolized primarily to trans-3'-hydroxycotinine, which is excreted in the urine to a greater extent than is cotinine itself. Thus, 3'-hydroxy-cotinine may as good or a better marker of nicotine consumption than cotinine. Dr. Curvall and co-workers in this volume show that nicotine, cotinine, and 3'-hydroxy-cotinine undergo a conjugation reaction and that, by measuring these metabolites and their conjugates, as much as 80-90% of a dose of nicotine can be accounted for in the urine. Recoveries of this high percentage in the urine offer the best pros-pect to date for precise quantitation of human exposure to nicotine. Development of such biochemical markers of nicotine intake will facilitate epidemiologic and experimental studies in smokers and in nonsmokers exposed to environmental tobacco smoke.

Acknowledgements. Research supported in part by USPHS grants DA02277, CA32389, and DA01696.

REFERENCES

Benowitz, N.L. (1988) N. Engl. J. Med. 319, 1318-1330.
Benowitz, N.L., Porchet, H.C., and Jacob, P. III. (1990) In: Nicotine Psycho-pharmacology. Molecular, Cellular, and Behavioral Aspects (S. Wonnacott, M.A.H. Russell, and I.P. Stolerman, Eds), Oxford Scientific Publications, Oxford, pp. 112-157.
Benowitz, N.L., and Sharp, D.S. (1989) Circulation 80, 1309-1312.
Dominiak P., Fuchs, G., von Toth, S., and Grobecker, H. (1985) Klin. Wochenschr. 62, 90-92.
Kim, K.S., Borzelleca, J.F., Bowman, E.R., and McKennis, H., Jr. (1968) J. Pharmacol. Exp. Ther. 161, 59-69.
Lee, B.L., Jacob, P. III, Jarvik, M. E., and Benowitz, N.L. (1989) Pharmacol. Biochem. Behav. 9, 621-625.
Risner, M.E., Cone, E.J., Benowitz, N.L., and Jacob, P. III. (1988) J. Pharmacol. Exp. Ther. 244, 807-813.

Shigenaga, M.K., Kim, B.H., Caldera-Munoz, P., Cairns, T., Jacob, P. III, Trevor, A.J., and Castagnoli, N., Jr. (1989) Chem. Res. Toxicol. 2, 282-287.

Shigenaga, M. K., Trevor, A.J., and Castagnoli, N. Jr. (1988) Drug Metab. Disp. 16, 397-402.

Stalhandske, T., and Slanina, P. (1982) Toxicol. Appl. Pharmacol. 65, 366-372.

Effects of Nicotine on Biological Systems
Advances in Pharmacological Sciences
© Birkhäuser Verlag Basel

THE *IN VITRO* METABOLIC FATE OF (S)-NICOTINE

N. Castagnoli, Jr.[*], M., Shigenaga,[#] T. Carlson,[&] W.F. Trager[&] and A. Trevor[#]

[*]Department of Chemistry, Virginia Polytechnic Institute and State University, Blacksburg, Virginia, 24061, USA, [#]Department of Pharmacology, University of California, San Francisco, California, 94143, USA, and [&]Department of Medicinal Chemistry, University of Washington, Seattle, Washington 98195, USA.

SUMMARY: The primary *in vitro* metabolic transformations for the tobacco alkaloid (S)-nicotine proceed by N'-oxidation, oxidative N-dealkylation and ring α-carbon oxidation. The aldehyde oxidase catalyzed oxidation of the unstable $\Delta^{1',5'}$-iminium intermediate formed by ring α-carbon oxidation generates (S)-cotinine which in turn is bio-oxidized *in vivo* to a variety of metabolites. This paper reviews these pathways and presents recent findings on the interaction of the $\Delta^{1',5'}$-iminium intermediate with monoamine oxidase B and the fate of the resulting pyrrolic metabolite β-nicotyrine.

The tobacco alkaloid (S)-nicotine (**1**) undergoes extensive metabolism in man and experimental animals with only about 10% being excreted unchanged (Benowitz, 1988). Quantitative studies have established (S)-cotinine (**2**) and *trans*-3-hydroxy-(S)-cotinine (**3**) as major urinary metabolites (Dagne & Castagnoli, 1972b; O'Doherty *et al.*, 1988; Neurath *et al.*, 1988; Jacob *et al.*, 1988). Nornicotine (presumably as the (S)-enantiomer **4**), the N'-oxide **5** [probably as a mixture of diastereomers (Jenner *et al.*, 1973)], and the pyridyl N-oxide **6** of (S)-cotinine (Dagne & Castagnoli, 1972) also have been characterized in the urine of various species. Only about 40-70% of the (S)-nicotine entering the body can be accounted for as urinary metabolites (Neurath & Plein, 1987; Kyerematen *et al.*, 1988).

Extensive *in vitro* metabolic studies employing liver preparations have established that the N'-oxidation of (S)-

nicotine is catalyzed principally by a flavin containing
monooxygenase (Damani *et al.*, 1988; Jenner *et al*, 1973). The
cytochrome P-450 monooxygenases catalyze the α-carbon oxidative
conversions that eventually lead to (S)-cotinine and nornicotine
(Nakayama *et al.*, 1982). These α-carbon oxidations involve
initial formation of the $\Delta^{1',5'}$-iminium species **7** (Murphy, 1972),
a reactive electrophilic intermediate which may be in
equilibrium with the corresponding enamine free base **8** (Obach &
Van Vunakis, 1988; Overton *et al.*, 1985), and the N-
methyleneiminium species **9** (Nguyen *et al.*, 1976). The iminium
species are readily trapped by cyanide ion to generate the
corresponding stable α-cyanoamines **10** and **11** (Murphy 1972;
Nguyen *et al.*, 1979; Nguyen *et al.*, 1976). Spontaneous
hydrolytic cleavage of intermediate **9** via carbinolamine **12**
yields nornicotine. The iminium intermediate **7** is oxidized
further to (S)-cotinine in a reaction catalyzed by the liver
cytosolic enzyme aldehyde oxidase (Brandange & Lindbloom, 1979).
(S)-cotinine also is formed to a minor extent in the presence of
hepatic microsomal preparations (Peterson *et al.*, 1987). It is
of interest to note that although the number of (S)-nicotine
derived urinary metabolites is large, only nornicotine, the N'-
oxide **5** and the C5'-oxidation products [iminium species **7** and
(S)-cotinine] have been identified as *in vitro* metabolites of
(S)-nicotine. Thus, the origins of the more extensively
oxidized metabolites of this alkaloid remain obscure.

Considerable species variation has been observed in the
extent to which liver preparations catalyze the C- *vs* N'-
oxidation of (S)-nicotine. McCoy *et al.* report that microsomes
prepared from livers isolated from uninduced hamsters catalyze
the formation of the N'-oxide over the $\Delta^{1',5'}$-iminium species by
a ratio of 1.2:1 to 2:1 (McCoy *et al.*, 1986). Similar results
have been obtained with rat and rabbit liver preparations (McCoy
et al., 1989; Kyerematen *et al.*, 1988). Consistent with the
view that a flavin monooxygenase is the principal N'-oxidase,
purified rabbit liver cytochrome P-450 isozymes yielded no N'-
oxide (McCoy *et al.*, 1989). Ratios of C5'- *vs* N'-methyl

oxidation with these purified enzymes ranged from about 14:1
(cytochromes P-450IIB1 and IIC3) to 1:2 (cytochromes P-450 1A2
and IIE1). The V_{max} values for 5'-oxidation ranged from 1.4 to
29.4 nmoles/min/nmole enzyme and for N-methyl oxidation from 1.0
to 8.6 nmole/min/nmole enzyme. Thus, in general the α-carbon
regioselectivity observed *in vivo* is reflected in the *in vitro*
metabolic profile. These results emphasize the influence that
selective isozyme induction may have on the metabolic fate of
(S)-nicotine. Species dependent differences also have been
observed in the *in vitro* metabolism of (S)- *vs* (R)-nicotine. Of
particular note is the reported stereospecific formation by
guinea pig microsomal preparations of the N-methylated
pyridinium metabolites **13** and **14** from the unnatural (R)-
enantiomer only (Nwosu & Crooks, 1988).

More recent studies have established that (S)-nicotine is
metabolized by rabbit lung (Williams *et al.*, 1990b) and nasal
(Williams *et al.*, 1990a) microsomal preparations to the $\Delta^{1',5'}$-
iminium species and, in lesser amounts, to the N'-oxide and
nornicotine. Rabbit lung cytochrome P-450IIB4 has been shown to
catalyze the C5'-oxidation of (S)-nicotine with a V_{max} of 1.5
nmoles/min/nmole enzyme and a K_m of 70 μM (Williams *et al.*,

1990). Cytochrome P-450NMa, an unusual isozyme isolated from the nasal mucosa of the rabbit (Williams *et al.*, 1990a), catalyzes this reaction with greater efficiency (V_{max} 28 nmoles/min/nmole enzyme; K_m 35 μM). A novel pathway for (S)-nicotine metabolism involving rabbit lung microsomal protaglandin H synthase to yield the enamine **8** also has been described recently (Mattammal *et al.*, 1987).

The mechanism by which cytochrome P-450 catalyzes the α-carbon oxidations of tertiary amines such as (S)-nicotine remains poorly defined. The generally accepted pathway (Ortiz de Montellano, 1998) involves initial electron transfer from the pyrrolidine nitrogen lone pair to the highly reactive cytochrome P-450 iron-oxo complex $[Fe^{3+}(O)]$ to form the aminium radical species **15** and $[Fe^{3+}(\cdot O^-)]$. Subsequent loss of H^+ (illustrated in Scheme 1 below with C5'-oxidation only) from the α-carbon atom yields the carbon centered radical **16** and $[Fe^{3+}(\cdot OH)]$. Radical recombination generates the carbinolamine **17** and $[Fe^{3+}]$, resting form of the enzyme. By analogy, abstraction of a proton from the N'-methyl group would lead to N-hydroxymethyl intermediate **12** which, upon N–C bond cleavage, would yield nornicotine.

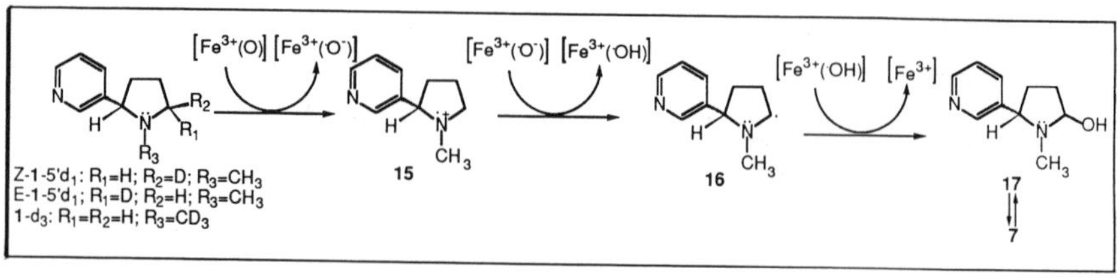

Scheme 1. Proposed reaction pathway for the cytochrome P-450 catalyzed oxidation of (S)-nicotine and its deuterated analogs.

More recent studies utilizing the diastereomeric C5' monodeutero compounds (Z)-**1**-d_1 and (E)-**1**-d_1 have established that rabbit and human liver microsomal cytochrome P-450 (Peterson *et al.*, 1987) and purified rabbit liver cytochrome P-450IIB4 (Carlson, *et al.*, 1990) catalyze the oxidation of (S)-

nicotine by a reaction pathway that involves the stereoselective loss of the C5'-proton *trans* to the pyridyl group. Although these preparations also catalyzed the oxidation of the N-methyl group, no evidence could be obtained for oxidation at the C2' position even though the aminium radical generating electrochemical and photochemical model reactions did result in C2' oxidation (Peterson & Castagnoli, 1988). Consistent with other studies on the cytochrome P-450 catalyzed oxidations of aliphatic tertiary amines, no significant deuterium isotope effect could be detected in the C5' oxidation pathway. On the otherhand, intramolecular isotope effect studies with cytochrome P-450IA1 and (S)-N'-dideuteromethylnicotine (1-CD$_2$H) have demonstrated an intramolecular kinetic isotope effect of 3.7 for the formation of the methyleneiminium species **9**, an outcome which is consistent with a hydrogen atom abstraction pathway rather than an electron transfer process as has been generally accepted (Carlson *et al.*, 1990).

MATERIALS AND METHODS

The experimental methods employed to generate the unpublished results described in this paper will be found in the Ph.D thesis titled "Studies on the Metabolism and Bioactivation of (S)-Nicotine and β-Nicotyrine" by Mark Kazuo Shigenaga, University of California, San Francisco, California (1990).

RESULTS AND DISCUSSION

Although the Δ$^{1',5'}$-iminium intermediate **7** is chemically reactive (Obach & Van Vunakis, 1988), its lifetime in the liver cell is likely to be quite transient because of the efficiency with which it oxidized to (S)-cotinine (Brandange & Lindbloom, 1979). Consistent with prediction, the covalent binding of radiolabeled (S)-nicotine to microsomal proteins is essentially abolished when the aldehyde oxidase containing cytosolic fraction is added to the incubation mixture (Shigenaga *et al.*, 1987). In view of its reactivity, the fate of the iminium intermediate **7** formed in extrahepatic tissues which are not rich in aldehyde oxidase is of some interest. The excellent MAO–B substrate properties observed for the cyclic allylamine 1-methyl–4–phenyl–1,2,3,6–tetrahydropyridine MPTP (Chiba *et al.*, 1985) prompted us to examine the possibility that the iminium species **7**, as its enamine free base **8**, also may be a substrate for this flavin oxidase. HPLC analysis of incubation mixtures containing synthetic **7** (0.5 mM) and MAO–B (0.7 units) showed the pargyline inhibitable, time dependent formation of a metabolite with retention time and diode array UV spectral characteristics (Fig. 1) identical to those of β–nicotyrine **(18)**. Comparision of the GC–EI mass spectrum of the metabolite to that of synthetic β–nicotyrine (Fig. 2) confirmed this assignment.

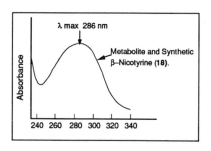

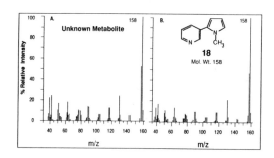

Figure 1. HPLC-diode array spectrum of (S)-nicotine iminium ion derived metabolite and β-nicotyrine.

Figure 2. GC-EIMS of (S) nicotine iminium ion derived metabolite and synthetic β-nicotyrine.

As expected, β-nicotyrine proved to be an excellent substrate for rabbit liver and lung microsomal enzymes (Shigenaga *et al.*, 1989). Lung preparations proved to be considerably more active on a per nmole cytochrome P-450 basis than liver preparations (Fig. 3). Figure 4 compares the HPLC-diode array spectrum of the principal metabolite formed in these incubation mixtures with that of the pyrrolinone **20** (see below). Unfortunately, the metabolic product proved to be too unstable to isolate. HPLC-CI mass spectral analysis of the metabolite displayed a single ion at m/z 175 (MH⁺) equivalent to the introduction of one oxygen atom into the substrate molecule. The HPLC—collision activated decomposition (CAD) mass spectrum of this ion (Fig. 5) showed several prominent fragment ions corresponding to the loss of CO (m/z 147), CO plus CH$_3$ (m/z 132) and CO plus CH$_3$N (m/z 118). These results argued for the presence of a lactam moiety and hence for the pyrrolinone structures **19** or **20**.

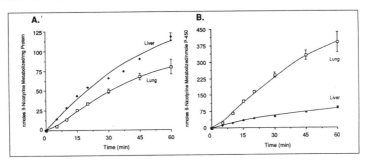

Figure 3.   Comparison between liver and lung microsomal metabolism of β-nicotyrine: (A) per mg of microsomal protein; (B) per nmole of cytochrome P-450.

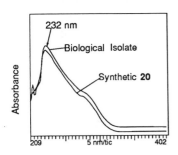

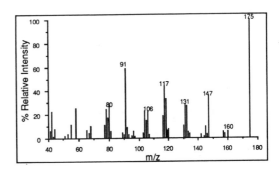

Figure 4. HPLC-diode array spectra of β-nicotyrine metabolite and **20**.

Figure 5. HPLC-CAD/MS of the m/z 175 ion derived from the β—nicotyrine metabolite.

The attempted synthesis of **19** from (S)-cotinine was approached via the 3-phenylselenyl intermediate **21** which, as the corresponding Se-oxide **22** yielded the 3-pyrrolinone **19** (Scheme 2). This compound proved to be in equilibrium with the corresponding 4-pyrrolinone **20** which eventually was isolated and

fully characterized. The metabolite generated from β-nicotyrine
proved to be identical to pyrrolinone **20**. Although relatively
stable at pH 7.6, compound **20** underwent rapid autoxidation under
acidic and basic conditions to yield two isomeric oxidation
products.   We have characterized one of these as the
hydroxypyrrolinone **21**. Current studies are underway to
determine if these products of β-nicotyrine metabolism are
present in the urine of smokers.

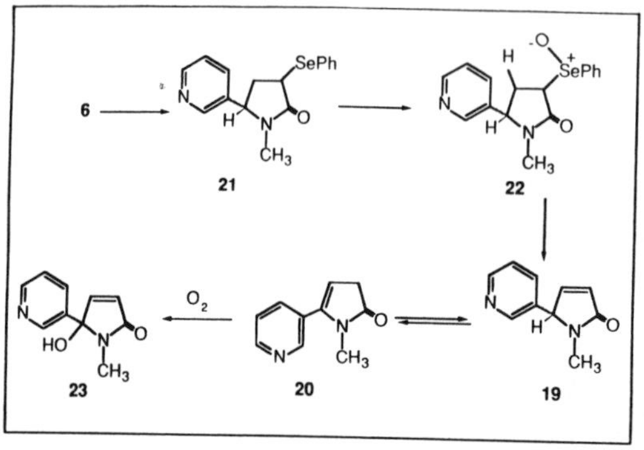

Scheme 2. Reaction pathway leading to the pyrrolinones **19,20**
and **23**.

CONCLUSIONS: Although a large number of (S)-nicotine
metabolites have been identified in the urine of smokers and
(S)-nicotine treated experimental animals, the N'oxide **5** and
(S)-cotinine represent the only major metabolic products
isolated from tissue incubations. Variations in cytochrome P-
450 isozyme composition which may be determined by the extent
and type of enzyme induction influences the metabolic profile of
(S)-nicotine *in vitro* and presumably *in vivo*. The reactive
$\Delta^{1',5'}$-iminium species formed by the cytochrome P-450 catalyzed
α-carbon oxidation of (S)-nicotine is readily converted to the
corresponding lactam (S)-cotinine in a reaction catalyzed by
liver aldehyde oxidase. The fate of this iminium ion in
extrahepatic tissues not rich in aldehyde oxidase remains to be
explored in detail.

ACKNOWLEDGEMENTS: Supported by the Peters Center for the Study of Parkinson's Disease and Central Nervous System Disorders.

REFERENCES

Benowitz, N.D. (1988) In: The Pharmacology of Nicotine. (M.J. Rand and K. Thurau, Eds), IRL Press, Oxford, pp. 3-18.

Brandange, S. and Lindbloom, L. (1979) Biochem. Biophys, Res. Comm. 91, 991-996.

Carlson, T., Trager, W.F., Peterson, L. and Castagnoli, N., Jr., ISSX Abstract, San Diego, CA, October, 1990.

Chiba, K., Peterson, L.A., Castagnoli, K.P., Trevor, A.J., and Castagnoli, N. Jr. (1985) Drug Met. Disp. 13, 342-347.

Dagne, E. and Castagnoli, N., Jr. (1972a) J. Med. Chem. 15, 356-360.

Dagne, E. and Castagnoli, N., Jr. (1972b) J. Med. Chem. 15, 840-841.

Damani, L.A., Pool, W.L., Crooks, P.A., Kaderlik, R.K., and Ziegler, D.M. (1988) Mol. Pharmacol. 33, 702-705.

Jacob, P. III, Benowitz, N.L., and Shulgin, A.T. (1988) Biochem. Behav. 30, 249-253.

Jenner, P., Gorrod, J.W., and Beckett, A.H. (1973) Xenobiotica 3, 563-572.

Kyerematen, G.A., Owens, G.F., Chattopadhyay, B., deMethizy, J.D., and Vessel, E.S. (1988) Drug Meta. Disp. 16, 823-828.

Kyerematen, G.A., Taylor, L. H., DeBethizy, J.D., and Vesell, E.S. (1988) Drug Met. Disp. 16, 125-129.

Mattammal, M.B., Lakshimi, V.M., Zenser, T.V., and Davis, B.B. (1987) J. Pharmacol. Exp. Ther. 242, 827-832.

McCoy, G.D., DeMarco, G.J., and Koop, D.R. (1989) Biochem. Pharmacol. 38, 1185-1188.

McCoy, G.D., Howard, P.C., and DeMarco, G.J. (1986) Biochem. Pharmacol. 35, 2767-2773.

Murphy, P.J. (1972) J. Biol. Chem. 218, 2796-2800.

Nakayama, H.T., Nakashima,T., and Kurogochi, Y. (1982) Biochem. Biophys. Res. Comm. 108, 200-205.

Neurath, G.B. and Plein, F.G. (1987) J. Chrom. 415, 400-406.

Neurath, G.B., Dünger, M., Krenz, O., Orth, D., and Pein, F.G. (1988) Klin. Wochenschr. 66 (Suppl XI), 2-4.

Nguyen, T.L., Gruenke, L.D. and Castagnoli, N., Jr. (1976) J. Med. Chem. 19, 1168-1169.

Nguyen, T-L., Gruenke, L.D., and Castagnoli, N., Jr. (1979) J. Med. Chem. 22, 259-263.

Nwosu, Z. and Crooks, P.A. (1988) Xenobiotica 18, 1361-1372.

O'Doherty, S., Revans, A., C.L. Smith, M. McBride, and M. Cooke (1988) J. High Res. Chrom. & Chrom Comm. 11, 723-726.

Obach, R.S. and Van Vunakis, H. (1988) Biochem. Pharmacol. 37, 4601-4604.

Ortiz de Montellano, P. R. (1989) TiPS 10, 354-359.

Overton, M., Hickman, J.A., Threadgill, M.D., Vaughn, K. and Gescher, A. (1985) Biochem. Pharmacol. 34, 2055-2061.

Peterson, L.A., Trevor, A., and Castagnoli, N. Jr. (1987) J. Med. Chem. 30, 249-254.

Peterson, L.A.. and Castagnoli, N., Jr. (1988) J. Med. Chem. 31, 637-640.

Shigenaga, M.K., Kim., B.H., Caldera-Munoz, P., Cairns, T., Jacob, P. III, Trevor, A.J., and Castagnoli, N., Jr., (1989) Chem Res. Tox. 2, 282-287.

Shigenaga, M.K., Trevor, A. J., and Castagnoli, N., Jr. (1987) Drug. Metab. Disp. 16, 397-402.

Williams, D.E., Ding, X., and Coon, M. J. (1990a) Biochem. Biophy. Res. Comm. 166, 945-952.

Williams, D.E., Shigenaga, M.K., and Castagnoli, N., Jr. (1990b) Drug. Met. Disp. 18, in press.

Effects of Nicotine on Biological Systems
Advances in Pharmacological Sciences
© Birkhäuser Verlag Basel

OXIDATIVE METABOLISM OF NICOTINE *IN VIVO*

Peyton Jacob, III, and Neal L. Benowitz

University of California, San Francisco, San Francisco General Hospital Medical Center, San Francisco, California 94110

SUMMARY: Nicotine undergoes a variety of metabolic reactions involving oxidation, most of which result in transformations of the pyrrolidine ring. The major route of metabolism in most mammalian species involves oxidation of the 5'-carbon atom to give the γ-lactam derivative cotinine. Oxidation of the N-methyl group results in demethylation to nornicotine, which is a minor metabolite in humans. Nicotine also undergoes oxidation of the pyrrolidine nitrogen atom to give a diasteriomeric mixture of N-oxides. Cotinine is extensively metabolized in humans, with only 10-20% being excreted unchanged in urine. It appears that the major urinary metabolite of cotinine is trans-3'-hydroxycotinine which results from oxidation adjacent to the carbonyl. Formation of this metabolite in humans has been shown to be highly stereoselective. Cotinine is also metabolized by N-oxidation of the pyridine nitrogen atom to give cotinine N-oxide which is a minor metabolite in humans, accounting for only a few percent of the nicotine absorbed by smokers. Other metabolites resulting from oxidative degradation of the pyrrolidine ring have been reported, but the pathways by which they are formed are not well understood, and their quantitative importance in humans is unknown. In this paper the various oxidative pathways will be discussed, with emphasis on quantitative and stereochemical aspects of nicotine metabolism in humans. Applications of metabolic data in clinical research will also be discussed.

NICOTINE METABOLIC PATHWAYS

Although nicotine is structurally a relatively simple substance, more than a dozen metabolites (Gorrod and Jenner, 1975), most of which are formed by oxidative pathways (Fig. 1), have been characterized. The major metabolite in most mammalian species is the γ-lactam derivative cotinine, which results from oxidation of the 5'-carbon atom of the pyrrolidine ring. Another important route of metabolism involves oxidation of the pyrrolidine nitrogen atom, which introduces a

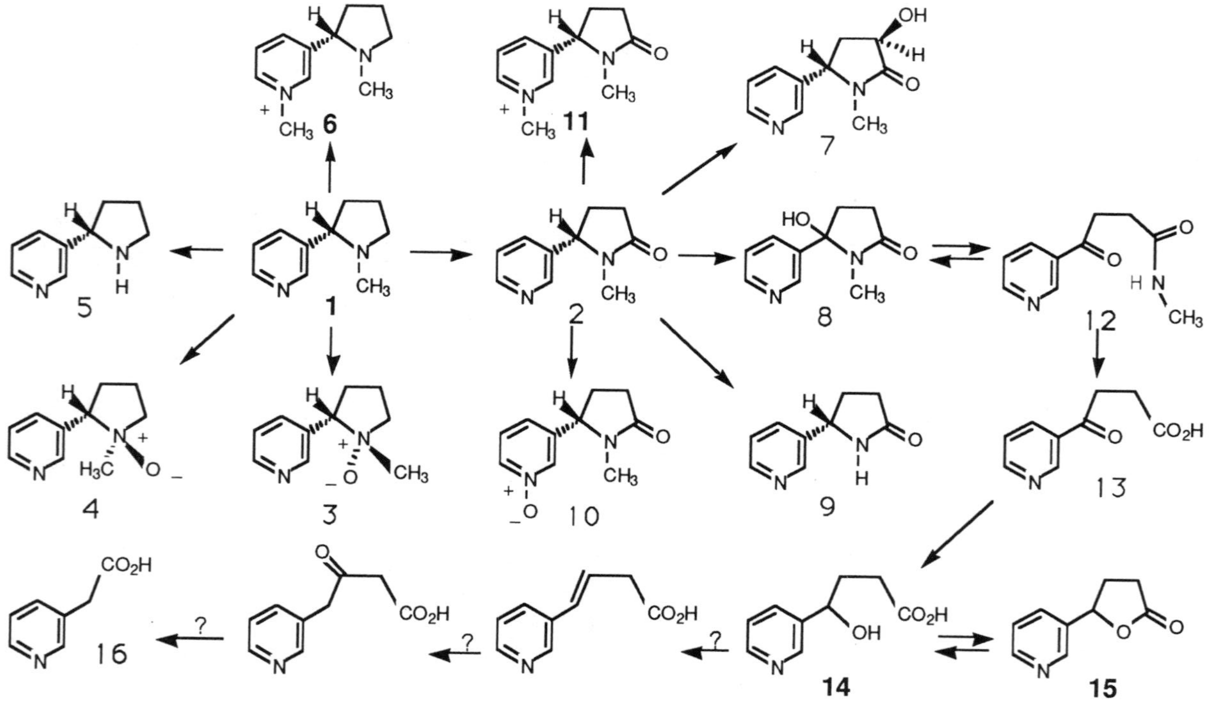

Figure 1. Nicotine Metabolic Pathways. 1 = Nicotine; 2 = Cotinine; 3 = 1'-(S),2'(S)-trans-nicotine-1'-N-oxide; 4 = 1(R),2'(S)-cis-nicotine-1'-N-oxide; 5 = Nornicotine; 6 = Nicotine isomethonium ion (N-methylnicotinium ion); 7 = Trans-3'-hydroxycotinine; 8 = 5'-Hydroxycotinine; 9 = Norcotinine (desmethylcotinine); 10 = Cotinine-N-oxide; 11 = Cotinine methonium ion (N-methylcotininium ion); 12 = γ-3-Pyridyl-γ-oxo-N-methylbutyramide; 13 = γ-3-Pyridyl-γ-oxobutyric axid; 14 = γ-3-Pyridyl-γ-hydroxybutyric acid; 15 = 5-(3-Pyridyl)-tetrahydrofuran-2-one; 16 = 3-Pyridylacetic acid.

second asymmetric center, and results in the formation of two diasteriomeric N'-oxides. Nicotine is demethylated to nornicotine which, in addition to being a metabolite, is one of the more plentiful alkaloids present in tobacco (Leete, 1983). Methylation to give nicotine isomethonium ion (N-methylnicotinum ion) is one of the few non-oxidative pathways.

The major metabolite cotinine is extensively metabolized. It is hydroxylated to 3'-hydroxycotinine and 5'-hydroxycotinine. 5'-Hydroxycotinine is in equilibrium with a ring opened keto amide. Cotinine is demethylated to norcotinine in some species, although it has not yet been shown to be a nicotine metabolite in humans. Cotinine is also oxidized on the pyridine nitrogen to give cotinine N-oxide, and methylated on the pyridine nitrogen to give cotinine methonium ion.

Several metabolites resulting from further degradation of the pyrrolidine ring have been characterized, although the pathways by which they are formed are not well understood. It is believed that oxidation of cotinine to 5'-hydroxycotinine followed by ring opening and hydrolysis of the amide bond gives rise to γ-3-pyridyl-γ-oxobutyric acid, which is reduced to the corresponding alcohol. The alcohol can cyclize to the lactone 5-(3-pyridyl)-tetrahydrofuran-2-one. Further degradation to 3-pyridylacetic acid has been reported, but the mechanism of its formation is uncertain. McKennis and co-workers, who carried out pioneering studies of nicotine metabolism in the 1960's, proposed a sequence of reactions analogous to the beta-oxidation of fatty acids (Schwartz and McKennis, 1963), but none of the intermediates have been identified.

QUANTITATIVE ASPECTS OF NICOTINE METABOLISM

To assess the quantitative importance of various metabolic pathways in humans, we determined 24-hour urinary excretion of various nicotine metabolites in 25 smokers (Benowitz et al., 1990); Shulgin et al., 1987) (Table 1). In these subjects, a mean of 9% of absorbed nicotine was excreted in urine unchanged. Nicotine N'-oxide in urine accounted for 4% of ingested nicotine. We estimated that about 70% of nicotine is metabolized to cotinine, but only 10% was excreted in urine unchanged. The major urinary metabolite is trans-3'-hydroxycotinine, accounting for 39% of ingested nicotine. Only 3% of absorbed nicotine appeared in urine as cotinine-N-oxide. These metabolites account for about 65% of the nicotine absorbed by this group of smokers.

Table 1. Urinary Excretion of Nicotine and Metabolites in 25 Cigarette Smokers

Compound	Excretion (µg per 24 hr)			Percent of Absorbed Nicotine
	Mean	SD	Range	Mean
Nicotine	2520	1830	212-5790	9%
Nicotine-N'-oxide	1080	686	135-3190	4
Cotinine	2880	1540	441-6200	10
Cotinine-N'-oxide	716	370	68-1540	3
Trans-3'-hydroxy-cotinine				39[1]

[1] Mean 48-hour urinary excretion in 5 human subjects who were given an intravenous infusion of nicotine following seven days abstinence from tobacco.

The metabolite nornicotine is of particular interest due to its pharmacologic activity (Risner et al., 1988) and because it may be an in vivo precursor for the carcinogen N-nitrosonornicotine (Hoffmann and Hecht, 1985). However, its presence as a metabolite per se cannot be determined directly in smokers since it is also a tobacco alkaloid. As part of studies of nicotine metabolism using stable isotope methodology, we were able to determine to what extent nornicotine is a metabolite in humans. This was carried out by administering deuterium-labeled nicotine intravenously to eight smokers. Urine was collected for 96 hours, and assayed for deuterium-labeled nornicotine by GC-MS. These subjects received a mean of 33 mg of deuterium-labeled nicotine, and excreted a mean of about 100 µg labeled nornicotine, or ~ 0.3% of the dose (Table 2). Thus, nornicotine is a minor urinary metabolite in humans.

Table 2. Urinary Excretion of Nornicotine in Eight Smokers Following Intravenous Infusion of Stable Isotope Labeled Nicotine

	(S)-Nicotine-3',3'-d_2 (mg)	Nornicotine-3',3'-d_2 (µg)	Recovered in Urine (% of Dose)
Mean	33.1	108	0.34
Range	17-50	54-240	0.19-0.63
SD	14	62	0.15

STEREOCHEMICAL ASPECTS OF NICOTINE METABOLISM

Nicotine in tobacco is levorotatory, and has the (S)-configuration at the asymmetric carbon. It appears that nicotine in tobacco is virtually 100% the (S)-isomer. However, the optical isomer (R)-nicotine is present in tobacco smoke, in amounts of up to 10% of the total nicotine, presumably formed by racemization during the process of smoking (Pool et al., 1985). As is often the case with drugs that exist as stereoisomers, (R)- and (S)-nicotine have distinctly different biologic effects. (S)-Nicotine is much more potent pharmacologically (Risner et al., 1988), and may be metabolized differently than (R)-nicotine (Jenner et al., 1973).

To explore the effects of stereochemistry on metabolic disposition, we carried out studies comparing the pharmacokinetics of nicotine enantiomers in rabbits (Jacob et al., 1988). On separate days, rabbits were given intravenous infusions of (R)- or (S)-nicotine. Blood concentrations of nicotine and its metabolite cotinine were determined at regular intervals during and after the infusion. We found that blood concentration-time profiles for (R)- and (S)-nicotine were similar, indicating very little effect of stereochemistry on the rate of elimination of the enantiomers. However, blood concentrations of the metabolite cotinine differed significantly; they were much higher following infusion of (S)-nicotine. Initially, we proposed that this was due to higher fractional conversion of (S)-nicotine than (R)-nicotine to cotinine. However, it turned out that there was another explanation. In another experiment we compared the elimination kinetics following intravenous infusion of (R)- and (S)-cotinine. At equivalent doses, concentrations of (S)-cotinine were higher than those of (R)-cotinine due to slower elimination of the (S)-isomer.

The results are summarized in Table 3. Total clearance for (S)- and (R)-nicotine was virtually the same, about 60 ml/min/kg. The half-life of (R)-nicotine was slightly longer and volume of distribution somewhat larger, although the difference was not statistically significant. The fractional conversion of nicotine to cotinine was about the same for the two isomers, about 50%. On the other hand, the clearance of (S)-cotinine was only about half the clearance of (R)-cotinine. Consequently, in studies of nicotine metabolism, it is important to know whether a pure stereoisomer or a racemic mixture of the two isomers is used. Previous studies in laboratory animals (Morselli et al., 1967; Kyerematen et al., 1988) and in humans (Kyerematen et al., 1982) have been carried out using racemic carbon-

14 labeled nicotine and cotinine. Metabolic data from such studies is likely to be different from data obtained using pure (S)-nicotine.

Table 3. Disposition Kinetics of Nicotine and Cotinine Enantiomers in Rabbits

	Total Clearance (ml/min/kg)	Half-Life (min)	Volume of Distribution (Steady state) (L/kg)	Fractional Conversion to Cotinine
(S)-Nicotine	57	80	2.9	0.44
(R)-Nicotine	60	107	3.4	0.48
(S)-Cotinine	6.9	153	1.0	
(R)-Cotinine	13.1	124	1.2	

STEREOCHEMICAL ASPECTS OF METABOLITE FORMATION

Conversion of (S)-nicotine to (S)-cotinine has been shown to occur stereospecifically (McKennis et al., 1963), and presumably the same is true for the analogous conversion of (R)-nicotine to (R)-cotinine. Hydroxylation of (S)-cotinine to 3'-hydroxycotinine could theoretically produce two isomers, cis-3'(S),5'(S) and trans-3'(R),5'(S) isomers. Likewise, there are two possible 3'-hydroxycotinine isomers derived from (R)-cotinine: trans-3'(S),5'(R) and cis-3'(R),5'(R (Fig. 2). Because there is some (R)-nicotine in tobacco smoke, all four isomers are potential metabolites in smokers.

Previous studies have shown that the metabolite of (S)-cotinine is largely, if not entirely, the trans-isomer. This metabolite was first isolated from smokers' urine by McKennis and co-workers in the 1960's, but the structure was not definitively established (McKennis et al., 1963). Subsequently, Dagne and Castagnoli isolated a metabolite from monkey urine following cotinine administration that had properties consistent with those of the metabolite isolated by McKennis (Dagne and Castagnoli, 1972). By total synthesis and spectroscopic methods, they were able to show that the hydroxylation had occurred in the 3'-position and that it had the trans-configuration. However, both McKennis and Castagnoli's studies involved isolation of the metabolite from urine and several purification steps including crystallization, so it is possible that the cis-isomer could have been lost if some were present.

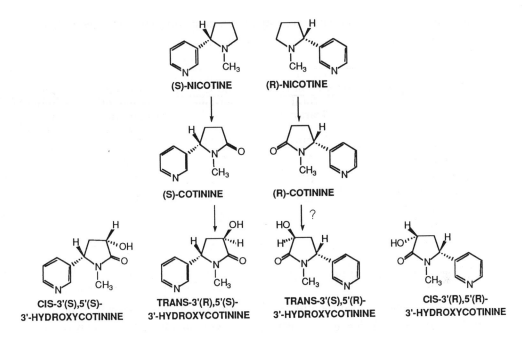

Figure 2. Stereochemical Aspects of Nicotine Metabolism

Consequently, we carried out a study to determine the stereochemical purity of 3'-hydroxycotinine produced by smokers (Jacob et al., 1990). This involved development of a chromatographic assay to distinguish the cis- and trans-isomers. Urine samples from seven smokers were extracted and carried through the assay. It was found that the stereochemical purity of metabolic 3'-hydroxycotinine ranged from 95-98% trans. Due to the possibility of a small amount of isomerization occurring during the assay procedure, these results represent the minimum stereochemical purity of the human metabolite. Since 3'-hydroxycotinine appears to be the major urinary metabolite of nicotine, there is interest in its use as a marker for exposure to tobacco. In studies of the metabolic disposition of 3'-hydroxycotinine it will be important to use the pure trans-3'(R),5'(S)-isomer.

BIOCHEMICAL MARKERS OF EXPOSURE TO TOBACCO SMOKE

Estimation of nicotine intake is useful to researchers studying smoking behavior and nicotine dependency, to epidemiologists assessing the health risks of tobacco usage or exposure to environmental tobacco smoke, and to therapists treating tobacco dependence. Since there is considerable variability in the way people smoke, it is not possible to estimate nicotine intake reliably based on the amount of tobacco consumed. Consequently, we carried out a study to examine the correlation of nicotine intake with various biochemical markers of smoke exposure present in biologic fluids of smokers (Benowitz and Jacob, 1984). In this study, afternoon blood concentrations of nicotine or carboxyhemoglobin correlated fairly well with daily intake of nicotine (Table 4). Urinary measures of nicotine and cotinine correlated less well with nicotine intake, which is unfortunate since non-invasive methods such as these are useful for studies carried out in the natural environment. Since urinary excretion of nicotine and cotinine comprise only a small percentage of nicotine intake, it is likely that the modest correlation is due to individual variability in metabolism and excretion. We propose that urinary exretion of the major metabolite trans-3'-hydroxycotinine, or total excretion of nicotine, cotinine, and trans-3'-hydroxycotinine would correlate well with nicotine intake. Further research will be necessary to test this possibility.

Table 4. Correlation of Various Markers of Tobacco Smoke with Daily Intake of Nicotine

Measure	Correlation (r)
Blood nicotine concentration at 1600 h	0.81*
Carboxyhemoglobin level at 1600 h	0.69*
Blood cotinine concentration at 1600 h	0.53**
Urinary cotinine excretion at 24 h collection	0.62*
Urinary nicotine excretion at 24 h collection	0.39

* $p < 0.01$; ** $p < 0.05$

CONCLUSIONS

To summarize, nicotine undergoes a variety of metabolic transformations, most of which involve oxidative degradation of the pyrrolidine ring. Some of these transformations are well understood, including stereochemical and quantitative aspects. But despite the fact that tobacco is consumed by millions of people, our knowledge of the biotransformation pathways of this simple substance are still incomplete. Elucidating the pathways of nicotine metabolism is not only scientifically interesting, but also of value in understanding the pharmacology and toxicology of tobacco.

Acknowledgements. The authors are grateful to Lourdes Abayan, Pat Buley, Mario Cave, James Copeland, Irving Fong, Lila Glogowski, David Lau, Clarissa Ramstead, Chin Savanapridi, Alexander Shulgin, Sandra Tinetti, Kaye Welch, Margaret Wilson, Lisa Yu, Qi Zhang and Ying Zhang for their contributions to this research. Supported in part by U.S. Public Health Services grants DA02277, CA32389, and DA01696. Human studies were conducted in the General Clinical Research Center of San Francisco General Hospital Medical Center with support by the Division of Research Resources, National Institutes of Health (RR-00083).

REFERENCES

Benowitz, N.L., and Jacob, P. III. (1984) Clin. Pharmacol. Ther. 35, 499-504.
Benowitz, N.L., Porchet, H., and Jacob, P. III. (1990) In: Nicotine Psychopharmacology, Molecular, Cellular, and Behavioral Aspects (Wonnacott, S., Russell, M.A.H., and Stolerman, I.P., Eds), Oxford University Press, London, pp. 113-157.
Dagne, E. and Castagnoli, J., Jr. (1972) J. Med. Chem. 15, 356-360.
Gorrod, J.W., and Jenner, P. (1975) In: Essays in Toxicology, Academic Press, New York, pp. 35-78.
Hoffmann, D., and Hecht, S.S. (1985) Cancer Research 45, 934-944.
Jacob, P. III, Benowitz, N.L., Copeland, J.R., Risner, M.E., and Cone, E.J. (1988) J. Pharm. Sci. 77, 396-400.
Jacob, P. III, Shulgin, A.T., and Benowitz, N.L. (1990) J. Med. Chem. 33, 1888-1891.
Jenner, P., Gorrod, J.W., and Beckett, A.H. (1973) Xenobiotica 3, 573-580.
Kyerematen, G.A., Damiano, M.D., Dvorchik, B.H., and Vesell, E.S. (1982) Clin. Pharmacol. Ther. 32, 769-780.
Kyerematen, G.A., Taylor, L.H., deBethizy, J.D., and Vesell, E.S. (1988) Drug Metab. Disposition 16, 125-129.

Leete, E. (1983) In: Alkaloids: Chemical and Biological Perspectives, vol. 1 (S.W. Pelletier, Ed), Wiley-Interscience, New York, pp. 85-152.

McKennis, H., Jr., Turnbull, L.B., Bowman, E.R., and Tamaki, E. (1963) J. Org. Chem. **28**, 383-387.

Morselli, L., Ong, H.H., Bowman, E.R., and McKennis, H., Jr. (1967) J. Med. Chem. **10**, 1033-1036.

Pool, W.F., Godin, C.S., and Crooks, P.A. (1985) The Toxicologist **5**, 232.

Risner, M.E., Cone, E.J., Benowitz, N.L., and Jacob, P. III. (1988) J. Pharmacol. Exp. Ther. **244**, 807-813.

Schwartz, S.L., and McKennis, H., Jr. (1963) J. Biol. Chem. **238**, 1807-1812.

Shulgin, A.T., Jacob, P. III, Benowitz, N.L., and Lau, D. (1987) J. Chromatogr. **423**, 365-372.

Effects of Nicotine on Biological Systems
Advances in Pharmacological Sciences
© Birkhäuser Verlag Basel

DETECTION OF NORNICOTINE IN HUMAN URINE AFTER INFUSION OF NICOTINE

G.B. Neurath, D. Orth, and F.G. Pein

Institut für Biopharmazeutische Mikroanalytik, Bötelkamp 35, D-2000 Hamburg 54, FR Germany

SUMMARY: The role of oxidative N-dealkylation as a pathway of the human metabolism of nicotine has been investigated after intravenous infusion of nicotine. Nornicotine has been found in the urine presenting evidence for the existence of this metabolic step. But, only less than 1% of the administered dose of nicotine is excreted as nornicotine possibly indicating rapid further metabolism.

Oxidative N-dealkylation is a major pathway in drug metabolism (Williams, 1971). Nornicotine would be the product of nicotine by this route (Papadopoulos and Kintzios, 1963, McKennis, Schwartz, and Bowman, 1964).

After intraarterial application of nicotine, 8.9% of the administered dose were found in the urine of rats as nornicotine by Kyerematen et al. (1987). Nornicotine was also found in recent smoker studies by Neurath et al. (1988), but the nornicotine was not in excess of 1.9%, 1.7%, and 1.4% respectively of the excreted sum of the total nicotine metabolites. Similar results have recently been reported (Zhang et al., 1990).

As nornicotine occurs in minor and varying concentrations in most of the tobaccos (Kuhn, 1964), its occurrence in the urine of smokers does not prove its formation in the human metabolism: It may be directly absorbed by the smoker and pass unchanged to the urine.

Direct evidence of its metabolic formation can best be achieved from experiments under direct application of nicotine to men.

Therefore, the nornicotine formation after infusion of nicotine was studied in the urinary excretion of six male regularly smoking volunteers.

MATERIAL AND METHODS

Materials: Nicotine was purchased from Merck, Darmstadt, distilled under reduced pressure, and stored in the dark under nitrogen. Nornicotine was synthesized according M.W. Hu et al. (1974), bp_{11} 131, purity 99% (GC).

Subjects and protocol: Six healthy male cigarette smokers - mean age 32.3 y, range 24 to 41 y - gave their written consent and refrained from smoking for five days. Ammonium chloride (1 g) was given orally after each meal on the preceding day to adjust the urine to slightly acidic conditions. Nicotine was administered by indwelling catheters to the incubital vein. Nicotine was applied in six separate infusions utilizing an automatic infusion device, Braun Melsungen, at a rate of 2 µg/min·kg with breaks of 20 min each between infusions. Physiological sodium chloride solutions were infused when the catheter was not in use. An electrocardiogram was recorded during the infusion period. Urine specimens were sampled from 15 hours before to 60 hours after the end of the infusion period. Total urinary excretion was collected in time intervals -15 to -3, -3 (begin of the infusion) to 0 (end of the infusion), 0-3, 3-6, 6-9, 9-12, 12-24, 24-30 and on in six hours intervals to 60 hours after the end of the infusion.

Extraction and derivatization of nornicotine from serum and urine: To 1 ml serum or urine in a 3-ml stoppered tube, 50 ng anabasine (5 µl of a solution of 10 µg anabasine in 1 ml water),

100 µl 1.5 n aqueous sodium hydroxide solution, and 1 ml benzene containing 10% heptanoic acid anhydride are added. The mixture is treated with a Rotarymixer for 30 min and centrifuged for 10 min at 4,200 r/min. 750 µl of the benzene phase are transferred into a 1-ml Reactivial and treated with 200 µl 1 n hydrochloric acid with a Vortex-mixer. After centrifuging (3 min at 4,200 r/min), 150 µl of the hydrochloric acid are transferred into a 1.5-ml tube. 100 µl 6 n aqueous sodium hydroxide solution and 50 µl benzene are added. After treating the mixture with a Vortex-mixer for 1 min, the mixture is centrifuged at 4,200 r/min for 3 min. 10 µl of the benzene phase are injected into the gas chromatograph.

Gaschromatography: Apparatus: Packard Instrument Model 437 with integrator Spectra-Physics 4290. Column: 2.4 m, 2 mm ID 1.5% Superox 0.1 on Chromosorb WHP 100-120 mesh. Injection onto 2% Superox 0.6 on Chromosorb WHP 100-120 mesh. Conditions: Oven 250°C isotherm, injector 280°C, detector (NP) 300°C, carrier gas: 30 ml/min nitrogen, 4 ml/min hydrogen, 50 ml/min air. Under these conditions, the following retention times were obtained: nornicotine heptanoate 6.25 min, anabasine heptanoate 7.10 min. Linear Regression: Samples of urine to which between 5 and 40 ng nornicotine/ml had been added (n = 9) were analysed resulting in a linear regression: $r^2 = 0.997$.

RESULTS AND DISCUSSION

The slope of urinary excretion of nornicotine after the intravenous infusion of a mean of 8.8 ± 1.4 mg nicotine to six male habitual smokers is shown (Fig.1).

Nornicotine is excreted with a half-life of about 12 hours. The total amount of nornicotine excreted by the subjects - mean of 46.6 µg - corresponds to only 0.5% of the dose. But, in contrary to small amounts of nornicotine generally found in the urine of smokers, who may inhale nornicotine from tobacco, the

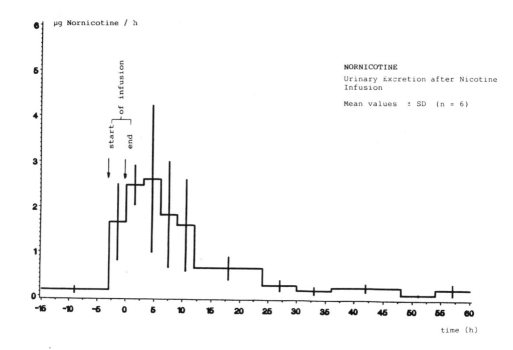

figures after intravenous infusion of pure nicotine give evidence to the formation of nornicotine by oxidative dealkylation
in the human body.. The amounts are smaller than those found after arterial administration to rats (8.9%) by Kyerematen et al.
(1987).

The serum concentrations of nornicotine exceeded the detection limit only occasionally.

CONCLUSION: Evidence has been given for the formation of nornicotine from nicotine in man. This pathway plays only a minor
role in human metabolism of the tobacco alkaloid.

Acknowledgement: We thank Dr. G. Scherer at Analytisch-biologisches Forschungslabor, Munich, for the intravenous administration of nicotine.

REFERENCES

Hu, M.W., Bondinell, W.E., and Hoffmann, D. (1974) J. Labelled
    Compounds 10, 79.
Kuhn, H. (1964) Fachl. Mitt. Österr. Tabakregie 5, 73-82.
Kyerematen, G.A., Taylor, L.H., deBethizy, J.D., and Vesell,
    E.S. (1987) J. Chromatogr. 419, 191-203.
McKennis, H., Schwartz, S.L., and Bowman, E.R. (1964) J. biol.
    Chem. 239, 3990-3996.
Neurath, G.B., Dünger, M., Orth, D., and Pein (1987) Int.
    Arch. Occup. Environ. Health 59, 199-201.
Papadopoulos, N.M., Kintzios, J.A. (1963) J. Pharmacol. 140,
    269-277.
Williams, R.T. (1971) In: Handbook of Experimental Pharmacology,
    XXVIII/2, p.229.
Zhang, Y., Jacob, P., and Benowitz, N. (1990) J. Chromatogr.
    525, 349-357.

CONCENTRATION OF NICOTINE IN SERUM AND SALIVA AFTER INTRAVENOUS
INFUSION OF THE ALKALOID AND AFTER SMOKING

L.Jarczyk[1], H.Maier[2], I.A.Born[3], G.Scherer[1], F.Adlkofer[1]

[1] Analytisch-biologisches Forschungslabor Prof. Dr. F. Adlkofer,
Goethestraße 20, 8000 München 2, F.R.G.
[2] Universitäts HNO-Klinik, Heidelberg, F.R.G.
[3] Institut für Allgemeine Pathologie und Pathologische Anatomie,
Heidelberg, F.R.G.

SUMMARY: The contribution of salivary glands to the excretion of
nicotine and tobacco specific nitrosamines (TSNA) which may derive
from nicotine was investigated in two experiments. In the first
experiment, four healthy male cigarette smokers received three
nicotine infusions. In the second experiment, four healthy male
cigarette smokers received one nicotine infusion and in addition to
this, they smoked eight cigarettes of their own brand within 3
hours. Nicotine and its main metabolite cotinine were monitored in
serum and in saliva. Saliva specimens were sampled to determine
TSNA. Nicotine concentrations were about 20 times higher in saliva
from the submandibular gland and about 60 times higher in saliva
from the parotid gland as compared to serum levels. After smoking
nicotine values in pooled saliva from the parotid gland were about
10 times higher than those in serum. Cotinine values in saliva are
slightly higher than those in serum, but the differences are
marginal. TSNA could neither be detected after nicotine infusion
nor after smoking.

INTRODUCTION

Our study was carried out to investigate nicotine and cotinine
concentrations in saliva coming from the parotid gland, the
submandibular gland and in mixed saliva following intravenous

infusion of nicotine and after smoking. Additional analyses for TSNA which are similar in structure to nicotine, were carried out in mixed saliva following nicotine infusion and after cigarette smoking.

SUBJECTS AND METHODS

Our study consisted of two different experiments and was carried out according to the following protocols.

Protocol 1

Four healthy male cigarette smokers remained abstinent from smoking after admission to the laboratory on the evening before the nicotine infusion was conducted. Immediately before infusions, catheters were inserted into the ducts of the parotid and submandibular glands. An infusion pump delivered 2 µg/nicotine/min/kg body weight over a period of 10 minutes into the left antecubetal vein. This nicotine infusion was repeated twice, namely 50 and 110 min after the end of the first infusion. Saliva was continously collected in 200 µl aliquots. Specimens of mixed saliva were collected and blood samples were taken every 30 minutes and collection of both biological fluids continued for 30 minutes after the last nicotine infusion.

Protocol 2

Four healthy male cigarette smokers participated in a combined nicotine infusion/smoking experiment. The subjects received one nicotine infusion as described in protocol 1. Two hours later they smoked 4 cigarettes of their usual brand within 40 minutes. After an interval of 80 minutes the smoking session was repeated. In this experiment, only the ducts of the parotid glands were cannulated. Since saliva samples collected were to be analyzed for TSNA, larger volumes were required. The sampling period for one saliva specimen of 10 ml was 30-45 minutes. Blood and mixed saliva samples were taken prior to and following nicotine infusion and smoking sessions. Collection of blood and saliva continued for 60 minutes after the last smoking period.

RESULTS

Compared to serum, much higher levels of nicotine were noted in mixed saliva with a major contribution coming from the parotid gland compared to that of the submandibular gland. The half life of nicotine in saliva was determined to be 15-20 minutes (Table I). Following nicotine infusion, the concentration of nicotine in saliva rose to much higher levels than in serum (Table I and Table II) and rose even more significantly after the smoking periods (Table II). After smoking, nicotine concentration in saliva was 10-20 fold higher than in serum (Table II).

Significant differences between the base level concentrations of cotinine in serum and saliva were not apparent and could not be detected immediately after nicotine infusion or after smoking. Additionally, TSNA were not detected in serum and saliva following nicotine infusion nor after the smoking sessions (Table II) despite the high levels of nicotine that were observed in these fluids.

DISCUSSION

It is noteworthy that the nicotine concentration in saliva is much higher than in serum after nicotine infusion and after smoking. The increased level of nicotine in saliva is primarily due to the fact that saliva is more acidic than serum (pH 5.5 - 7.0 in saliva vs. pH 7.4 in serum (Lentner, 1985)). The higher nicotine concentration in saliva from the parotid gland can also be accounted for by a pH effect since saliva secreted from this gland

Table I: Salivary nicotine excretion after i.v. infusion of nicotine

	base	inf.1	inf.2	inf.3
Saliva from glandula parotis				
* nicotine [ng/ml]	35±18	687±252	790±251	>1000
* time		18± 3	19± 4	15± 2
* conc.f.	16±11	61±27	45±22	89±65
* half-life [min]		14± 2		19± 8
Saliva from glandula submandibularis				
* nicotine [ng/ml]	49±30	176± 86	276±143	269± 16
* time [min]		18± 5	24±14	16± 2
* conc.f.	22±18	16± 9	16±10	22±15
* half-life [min]		19± 8		13
Serum				
* nicotine [ng/ml]	2±1	11± 2	18± 6	12± 3
* cotinine [ng/ml]	203±125	204±149	217±141	227± 14

Values as mean ± SD (n=4), base: base level in mixed saliva measured before infusion 1 was started, inf.: infusion, time: number of minutes when nicotine reached its peak after the beginning of the infusion, conc.f.: concentration factor = factor by which the nicotine serum level has to be multiplied to reach the saliva level

Table II: Salivary nicotine excretion after infusion of nicotine and after smoking

	Serum	Saliva (g.parotis)
Infusion		
* nicotine [ng/ml]	13± 6	71±21
* cotinine [ng/ml]	273±114	321± 70
* TSNA [ng/ml]	nd	nd
Smoking		
* nicotine [ng/ml]	33±10	294±125
* cotinine [ng/ml]	301±112	393± 78
* TSNA [ng/ml]	nd	nd

Values as mean ± SD (n=4), TSNA: tobacco specific nitrosamines, nd: not detectable has a lower pH when compared to that of the submandibular gland (pH 5.8 vs. pH 6.5 (Lentner, 1985). The high concentration of nicotine in saliva may be explained by the nicotine pKa of 7.9. Nicotine in saliva is rapidly protonated and reabsorption of the salt is virtually impossible. From the data of Experiment 1 we have roughly estimated that up to 1 % of the nicotine infused is secreted within 2 hours into saliva from where it is swallowed, reabsorbed in the gut and metabolized by the liver.

No differences in cotinine concentrations between serum and saliva could be detected in the base levels. Although our experiments did not allow us to monitor the cotinine concentrations in serum and saliva in the course of the infusion or smoking periods, an increase in salivary cotinine similar to that found for nicotine should not be expected because cotinine is unprotonated in saliva. This is due to its pKa of 4.5. The mechanism of excretion should be similar to that reported for the renal handling process (Beckett et al., 1972).

Saliva does not contain TSNA both after nicotine infusion and after smoking inspite of its high nicotine concentration. Whether this means that the amount of TSNA taken up or synthesized is too low to be measured remains unclear. The possibility might exist that due to protein-binding of nitrosamines in serum the fraction of TSNA available for salivary excretion is reduced. This has been demonstrated for several drugs (Mühlenbruch, 1982; Lien et al., 1989). However, previous studies conducted by our group have attempted to detect TSNA in serum and urine without success.

ACKNOWLEDGEMENT: We wish to thank Dr.H.Klus of the Austria Werk, Vienna, for measuring TSNA.

REFERENCES

Beckett A.H., Gorrod J.W., Jenner P., 1972: J.Pharm.Pharmac. 24: 115-120
Lentner C., 1985: Wissenschaftliche Tabellen Geigy, Ciba-Geigy AG Basel, pp 111-120
Lien E.A., Solheim E., Lea O.A., Lundgren S., Kvinnsland S., Ueland P.M., 1989: Cancer Res., 49: 2175-2183
Mühlenbruch B., 1982: Pharmazie in unserer Zeit, 11: 41-47

Effects of Nicotine on Biological Systems
Advances in Pharmacological Sciences
© Birkhäuser Verlag Basel

PHARMACOKINETICS OF NICOTINE AFTER INTRANASAL ADMINISTRATION

P. Olsson, F. Kuylenstierna, C-J. Johansson, PO. Gunnarsson and
M. Bende

Pharmacia AB, Drug Development, Box 941, SE-251 09 Helsingborg,
ENT-dept, Central Hospital, SE-541 85 Skövde, Sweden

SUMMARY: Using a capillary gas chromatographic method, plasma
levels of nicotine was monitored after intranasal administration
of nicotine to rats and humans. In the first study nicotine was
administered intranasally to rats and a pH-dependent absorption
was observed. In the second,the bioavailability of nicotine at a
physiological pH was examined in man after application to
different anatomical regions of the nose. The absorption was rapid
and a high absolute bioavalability was found for nasal nicotine.

INTRODUCTION

Administration of nicotine in different formulations has been used
as an aid to stop smoking. Nicotine is a weak base and its
absorption can be expected to be pH dependent. As nicotine is
exposed to a large first pass hepatic metabolism and has a low
oral bioavailability, the substance is a good candidate for non-
oral routes of administration, and has been administered both
buccally and transdermally. By these routes, however, nicotine is
absorbed more slowly as compared to smoking a cigarette. Since
nicotine is a drug showing distribution kinetics with transient
high concentrations in the brain, a rapid absorption may be of
importance for achieving a central pharmacologic effect
(Benowitz, 1986).

Preliminary results on the intranasal administration of nicotine have shown a rapid, although variable, absorption (West et al., 1984). Since the high variability may be dependent on differences in intranasal deposition of the drug, one aim of the present study was to determine the influence of application sites on the absolute bioavailability of nasal nicotine.

Also, a study of the nasal absorption of nicotine in rats was initially carried out to examine the pH-dependence.

MATERIALS AND METHODS

ANIMAL EXPERIMENTS;

0.1% Nicotine solutions, buffered at pH 6 and 9, were used for the animal experiments. Male Sprague Dawley rats, weighing approximately 250 g were anesthesized with ether. 25 μl solution, containing 0.025 mg nicotine, was administered by an intranasally placed polyethylene catheter. After administration, the rat was gently rolled to disperse the solution over the nasal mucosa. Blood samples were withdrawn by cardiac puncture from three rats at each sampling time. Sampling times were 1, 3, 5, 10, 20, 30, 60 and 120 minutes after the administration.

HUMAN EXPERIMENTS;

Eight healty volunteers (2 men and 6 woman, age 38 ± 5 years) participated in the study. The subjects had a 12 hours nicotine abstinence before receiving each of the test drugs, which were administered in a randomized cross-over design. Nasally nicotine was administered in a volume of 0.1 ml at pH 7. The dose was applied by a micropipette to the nasal septum and the nasal conchae, respectively, and as a spray. As a reference, an intravenous infusion of 0.7 mg nicotine was given over 2 minutes. Plasma concentrations were followed for 6 hours post administration. For three subjects with a longer pre-study abstinence, the concentrations were followed for 24 hours after the iv and spray regimens.

ANALYTICAL METHOD

SAMPLE PREPARATION;

To 0.20 - 1.00 ml sample (adjusted to 1.00 ml with water) was 0.50 ml of 10M NaOH and 1.00 ml extraction solution (1% n-butanol in toluene) added. The mixture was carefully shaken for 10 minutes and afterward centrifuged for 10 minutes at 2500 rpm. The organic layer was transferred into an autosampler vial and sealed by a crimp cap until injected.

As internal standards N-methylanabasine and N-ethylnorcotinine were used for nicotine respectively cotinine.

The calibration curves were established by spiking human plasma with known amounts of nicotine and cotinine. The calibration curves ranged from 0.5 to 100 ng nicotine and 3.0 to 600 ng cotinine.

GAS CHROMATOGRAPHY;

3 ml of the sample extract was injected by splitless technic on a capillary gas chromatograph, equipped with a nitrogen selective detector. The chromatography was performed by means of a CP-sil 8 CB fused silica column (25m x 0.32mm) and temprature programming.

Peak height ratios between nicotine/N-methylanabasine and cotinine/N-ethylnorcotinine were calculated and the amounts of nicotine and cotinine were evaluated from calibration curves. The limits of detection were 0.2 ng and 1.3 ng for nicotine respectively cotinine.

RESULTS

ANIMAL EXPERIMENTS;

As can be seen in figure 1 there is a rapid absorption of nicotine at pH 9. The difference between the plasma concentration curves is significant during the first 5 minutes.

The Cmax of pH 9 is about 200 ng/ml compared to 50 ng/ml for pH 6. After the first 20 minutes the curves are virtually identical.

HUMAN EXPERIMENTS;

The absorption was fast for all three nasal administrations, with a mean t-max of about 10 minutes and a C-max in the range of 2.5 - 13 ng/ml. The mean bioavailabilities were in the range 56 to 71%, with no significant differences between administration sites. The coefficients of variation were about 30% for the AUCs after the nasal administrations, compared to 17% for the iv dose. After the iv dose a mean clearance of 680 ml/min and a mean volume of distribution of 2.3 l/kg were found.

In three subjects, who had a longer pre-study smoking abstinence and starting cotinine values below the limit of detection, plasma concentrations were followed for 24 hours (Fig.2). The mean ratio of the nicotine versus the cotinine AUCs for the iv and the nasal spray dose were of the same magnitude, 0.10 and 0.08 respectively, the differences being not significant.

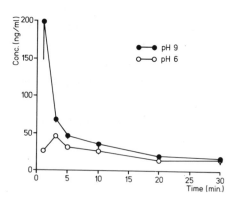

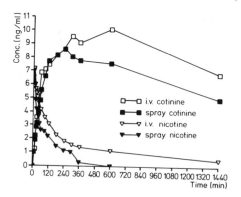

Fig. 1 Plasma nicotine concentrations after nasal administration in rats.

Fig. 2 Plasma nicotine and cotinine in humans after iv and nasal administration.

DISCUSSION

A transient effect of pH can be seen in the nasal absorption of nicotine in rats. This could be explained by the increase in the unionized moiety which penetrates the nasal mucosa better than the ionized form of nicotine.

The difference could also be explained by the irritative effects of the high pH giving rise to increased blood flow of the nasal mucosa. The very high peak plasma levels, observed when applying the alkaline nicotine solution, also should give high uptake into the brain. Further examinations of the brain levels of nicotine after intranasal administration are in progress.

In humans a rapid absorption could be seen at a physiological pH. The absorption was fast and no differences could be seen between application at different nasal sites, This indicates that the deposition of nicotine on the nasal mucosa will not greatly affect the absorption.

The bioavailability of less than 100% could be explained by some of the nicotine beeing placed in the pharynx and swallowed, thereby being exposed to a first pass metabolism. However, the ratios of nicotine/cotinine AUCs being no different in the iv and the spray dose indicates that no major part of the nicotine is swallowed as more cotinine would be formed in that case. Also, if there is any metabolism in the nasal mucosa it is propably not largely different from that in the liver.

In conclusion, by nasal administration nicotine peak plasma levels can rapidly and reliably be achieved.

REFERENCES

Benowitz, N.L., (1986) Ann. Rev. Med. 37, 21-32.
West, R.J. et al., (1984) Br.J. Addiction. 79, 443-445.

Effects of Nicotine on Biological Systems
Advances in Pharmacological Sciences
© Birkhäuser Verlag Basel

NICOTINE APPLICATION WITH AN ORAL CAPSULE

C. Conze, G. Scherer and F. Adlkofer

Analytisch-biologisches Forschungslabor Prof. Dr. F. Adlkofer, Goethestr. 20, 8000 München 2, Germany

SUMMARY: A gelantin capsule has been developed which releases 4 mg of nicotine in 0.8 ml of an alkaline (pH 10) puf-
fered solution into the mouth after breaking the capsule's
wall. In a pilot study with 5 male smokers and 5 male non-
smokers, nicotine absorption and responses after application
of one capsule were studied. The peak nicotine concentration
in plasma was reached 10 min after application of the nicotine capsule and ranged from 4 to 10 ng/ml. In most of the
subjects nicotine uptake by the capsule was accompanied by
an increase in heart rate and systolic blood pressure. Subjective response and tolerance to the nicotine capsule was
very variable from subject to subject. There was no significant difference between smokers and non-smokers in any of
the variables measured.

INTRODUCTION

Nicotine is the most pharmacologically potent substance in
tobacco smoke and is thought to control the smoking habit in
the majority of the smoking population (Warburton, 1985). In
addition, there is evidence that this alkaloid might have
beneficial effects, for example in decreasing the risks for
development of Parkinson's (Marshall & Schnieden, 1966) or
Alzheimer's (Sahakian et al., 1989) disease. For scientific
and medical reasons, it could be important to have a nicotine application which is as effective as is nicotine taken
up by inhalation.
It has been shown that nicotine is quickly absorbed by the

oral mucosa at alkaline pH (Beckett et al., 1972; Armitage & Turner, 1970; Schievelbein et al., 1973). A capsule has been developed which releases nicotine in an alkaline buffered solution into the mouth after breaking the capsule's wall. Before application of this capsule in clinical and scientific investigations, basic data on the pharmacokinetics as well as cardiovascular and other responses of nicotine administered by this route are required. For this purpose, the administration of the capsule was studied in five smokers and five non-smokers.

MATERIALS AND METHODS

Subjects and study design: Five male smokers and 5 male nonsmokers remained in the laboratory for 3 consecutive days including the nights before the first and after the last day. Smokers were abstinent from smoking for 48 h. After each meal, the subjects received 1 g of NH_4Cl in order to acidify the urine. On the first day, a placebo and 15 min later a nicotine capsule was administered to each subject. The volunteers were not aware of the type of capsule they received. The subjects were advised to break the gelatin wall of the capsule immediately after taking the capsule into the mouth, holding the contents of the capsule in the mouth for 2 min and chewing the gelatin for another 2 min. After this, 3 smokers and 3 nonsmokers were advised to swallow the whole material. The other subjects were told to spit it out. Cardiovascular and subjective responses were recorded 30 min before, during and 60 min after application of the capsules. An identical investigation was performed on the third day. However, at this time the subjects received only two placebo capsules. Blood and urine were sampled throughout the whole study at appropriate time intervals.

Nicotine capsule: The gelatin capsule contained 4 mg of nicotine in an alkaline buffered solution of pH 10. The buffer contained ethanol, polyehtylenglycol, glycerol and peppermint oil. An identical capsule without nicotine was used as placebo.

<u>Analytical methods</u>: Nicotine in plasma and urine was determined according to the method of Feyerabend and Russell (1979) by gas chromatography with nitrogen specific detection. Cotinine in plasma and urine was determined by a radioimmunoassay (Langone et al., 1973).

<u>Cardiovascular variables</u>: Heart rate and blood pressure were automatically recorded at one minute intervals during the investigational sessions on Day 1 and Day 3 (ASM 1000, Elmed, Augsburg, FRG).

<u>Subjective responses</u>: The subjects reported their feelings after administration of the capsules. The reported sensations could be categorized into the following items: "light headedness", "increasing heat", "profused perspiration", "nausea". Each item scored 1 in a summary score.

RESULTS AND DISCUSSION

The peak plasma nicotine concentration of 4 to 10 ng/ml was reached 10 min after application of the nicotine capsule (Fig. 1A). Smoking status and swallowing the capsule had no major impact on the peak height of plasma nicotine (Table I). In all but one of the subjects who swallowed the capsule, a second, smaller nicotine peak appeared 40 to 120 min after ingestion. The cotinine level in plasma increased until 2 h after application of nicotine and then decreased (Fig. 1B). Levels were higher in smokers and those who swallowed the capsule.

Nicotine excretion was nearly complete 10 h after administration of the capsule. Urinary excretion rates of nicotine were lowest in smokers who did not swallow the capsule. The total amount of cotinine excreted was higher in smokers and in those who swallowed the capsule (Table I). Assuming that 40% of the nicotine absorbed via the oral mucosa appeared in the urine as nicotine and cotinine, it can be estimated that nicotine uptake by this route amounts to about 1 mg. This figure can be derived from the data collected on nonsmokers who did not swallow the capsule (Table I).

Heart rate and blood pressure were weakly associated with

the peak nicotine concentrations in plasma. Quite unex-
pectedly, there was no higher degree of tolerance to nicotine in smokers compared to nonsmokers.

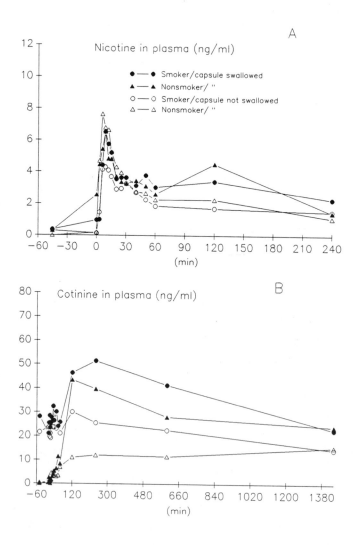

Fig. 1: Plasma profile of nicotine (A) and cotinine (B)
        after administration of a nicotine capsule at
        0 min. (The data points represent averages of the
        four subgroups).

Table I: Intra-individual data on the peak height of nicotine (nic.) in plasma, urinary excretion of
nicotine and cotinine (cot.), sujective responses,
and maximal increases in heart rate (HR) and systolic blood pressure (BP) after administration of
a nicotine capsule.

Subj. #	Smo-ker	Caps. swal-lowed	Nic. peak (ng/ml)	Amount excreted (ug/60h) Nic.	Cot.	Suject. resp. (Score)	Max. increases HR (1/min)	BP (mm Hg)
1	+	+	6.6	197	403	3	21	13
2	+	+	7.6	316	697	2	15	9
3	+	–	4.3	129	508	0	16	10
4	+	–	4.5	127	386	1	10	0
5	+	+	7.2	142	487	1	19	10
6	–	–	4.7	167	119	0	13	13
7	–	+	6.3	172	481	3	16	10
8	–	+	5.3	178	395	2	13	7
9	–	+	8.0	217	531	2	20	22
10	–	–	10.0	264	256	4	19	10

No systematic differences in pharmacokinetic parameters or
physiological effects were found between smokers and nonsmokers or those who swallowed and those who did not swallow
the capsule (Table I).
There was a high variability in the self-reported responses
of subjective effects which was weakly associated with peak
nicotine concentrations in plasma.
An important property of the capsule is that it produces a
nicotine blood profile similar to that after smoking a cigarette, although the peak nicotine levels were lower (Russell, 1978). Cardiovascular responses such as increase in
heart rate and blood pressure as well as decrease in skin
temperature (data not shown) were also comparable to those
observed after smoking (Henningfield et al., 1985). In contrast to smoking, however, capsule administration was not
accompanied by any pleasureable feelings in most subjects.
Further investigations are needed in order to confirm the

present findings. Additional points of interest would be:
eeg measurements, responses to chronic capsule administration and influence on smoking habits (nicotine compensation). The observed properties of the oral nicotine capsule
suggests that it could be applied in research and medical
studies.

REFERENCES

Armitage, A.K., and Turner, D.M. (1970) Nature 226, 1231-
    1232.
Beckett, A.H., Gorrod, J.W., and Jenner, P. (1972) J. Pharm.
    Pharmac. 24, 115-120.
Feyerabend, C., and Russell, M.A.H., (1979) J. Pharm. Phar-
    mac. 31, 73.
Henningfield, J.E., Migasato, K., and Jasinski, D.R. (1985)
    J. Pharmacol. Exp. Therap. 234, 1-12.
Langone, J., Gijika, H.B., and Van Vunakis, H. (1973) Bio-
    chem. 12, 5025-5030.
Marshall, J., and Schnieden, H. (1966) J. Neurol. Neurosurg.
    Psychiatry 29, 214-218.
Russell, M.A.H. (1978) Drug Metabolism Reviews 8(1), 29-57.
Sahakian, B.J., Jones, G.M.M., Levy, R., Gray, J.A., and
    Warburton, D.M. (1989) Br. J. Psychiatry 154, 797-800.
Schievelbein, H., Eberhardt, R., Loeschenkohl, K., Rahlfs,
    V., and Bedall, F.K. (1973) Agents and Actions 3(4), 254-
    258.
Warburton, D.M. (1985) Reviews on Environmental Health V/4,
    343-390.

Effects of Nicotine on Biological Systems
Advances in Pharmacological Sciences
© Birkhäuser Verlag Basel

CONJUGATION PATHWAYS IN NICOTINE METABOLISM

M. Curvall, E. Kazemi Vala and G. Englund

Reserca AB, P.O. Box 17007, S-104 62 Stockholm, Sweden

SUMMARY: Nicotine, cotinine and *trans*-3´-hydroxycotinine are excreted as glucuronic acid conjugates in urine of tobacco users. The average ratios between free and conjugated alkaloids in 24-hour urine samples from tobacco users are 1.0, 0.5 and 2.3 for nicotine, cotinine and *trans*-3´-hydroxycotinine, respectively. Nicotine, cotinine and *trans*-3´-hydroxycotinine and their glucuronides together with the N-oxides of nicotine and cotinine, account for more than 90 % of the ingested dose of nicotine.

INTRODUCTION

The major metabolic pathways in the phase 1 metabolism of nicotine are C-oxidation to cotinine, which in turn is extensively metabolized to *trans*-3´-hydroxycotinine, and the N-oxidation to the two diastereomers of nicotine-N´-oxide. Together with nicotine these major phase 1 metabolites account for about 65 % of the ingested dose (Benowitz, 1988).
Glucuronidation is among the most frequently used means by which humans produce polar metabolites of xenobiotics for excretion. Our recent studies have shown that nicotine, cotinine and *trans* -3´-hydroxycotinine form glucuronic acid conjugates (phase 2 metabolism), which are excreted in urine.

METHODS

Sample collections: Urine samples from 8 smokers and 7 users of Swedish moist snuff were collected during 24 hrs. The smokers consumed 10-40 middle tar cigarettes per day and the snuff-users 0.6- 40 g of moist snuff per day. Aliquots (10 ml) of the urine samples were separated and kept frozen until analyzed.

Assays of nicotine, cotinine and *trans*-3′-hydroxycotinine in urine: The concentrations of nicotine and cotinine in urine were determined using a gas chromatographic method (Curvall et al., 1982). The sensitivity of this assay is 1 ng of nicotine and cotinine per ml of urine with coefficients of variation being 10 % for nicotine and cotinine at concentrations of 5 ng/ml.
The determination of *trans*-3′-hydroxycotinine in urine could be shortly described as follows: To samples of urine (0.5 ml) were added the internal standard, 3′-hydroxy-N-propylnorcotinine (4 ug). The samples were extracted with dichloromethane (2 ml) under alkaline conditions (saturated K_2CO_3) for 30 min. The organic layer was separated and the solvent evaporated. MSTFA (N-Methyl-N-Trimethylsilyl trifluoroacetamide) was added to the residue and the mixture was heated at 70°C for 15 min. The silylethers so obtained were quantified by high resolution gas chromatography with a thermionic specific detector in the concentration range 0.1 - 40 ug/ml of urine sample and with a mass spectrometer as a detector in the range 25 - 500 ng/ml. The coefficient of variation in the determination of *trans* -3′-hydroxycotinine was less than 10 % in the concentration range 0.025 - 40 ug per ml of urine.

Assays of phase 2 metabolites of nicotine in urine: For quantification of nicotine, cotinine and *trans*-3′-hydroxy-cotinine and their glucuronides, the urine was divided into four portions. The first portion was analyzed directly for free nicotine and cotinine, the second portion for *trans*- 3′-hydroxycotinine , the third and fourth aliquots of urine were analyzed after enzymatic hydrolysis for the total amounts of nicotine, cotinine and *trans* -3′-hydroxy-cotinine, i.e. both free and conjugated alkaloids.

To urine samples (0.5 ml), which was diluted with an equal amount of acetate buffer (0.05 M, pH 4.7), was added β-glucuronidase (6000 U, *Helix Pomatia*, EC. 3.2.1.31, Sigma) in 200 ul of acetate buffer. The incubation time was 4 hrs for hydrolysis of nicotine and cotinine conjugates and 24 hrs for conjugates of *trans*-3′-hydroxycotinine (Fig. 1).

After the incubation, the reaction mixture from portion three was analyzed for the total amounts of nicotine and cotinine and portion four for *trans*-3′-hydroxycotinine. The amounts of conjugated nicotine, cotinine and *trans*-3′-hydroxycotinine were calculated from these data.

Fig. 1. Kinetic profile of the enzymatic hydrolysis of urinary glucuronides of nicotine, cotinine and *trans*-3′-hydroxycotinine.

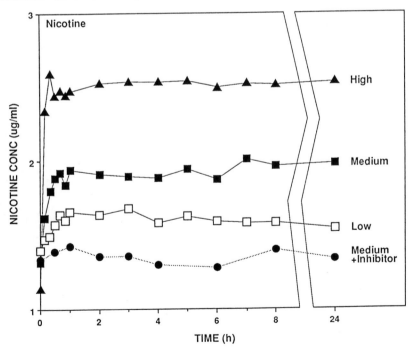

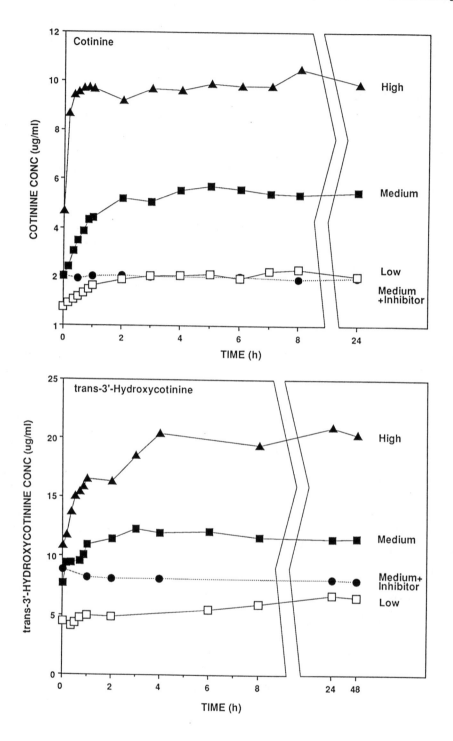

RESULTS

Precision of the assays: The precision of the different assays
for free and total nicotine, cotinine and *trans*-3′-hydroxy-
cotinine were evaluated in urine samples from a user of moist
snuff, containing 2.1, 5.3 and 10.5 ug/ml of total amount of
nicotine, cotinine and *trans*-3′-hydroxycotinine, respectively.
The coefficients of variation calculated from six determinations
are shown in the Table I. The coefficients of variation for the
instrumental analysis ranged from 1.7 to 3.2 %. the assays for
free alkaloids had coefficients ranging from 3.5 to 5 %, and
those for total amounts of alkaloids from 2.8 to 6.6 %. These
results show that these assays have good precision for urine
samples from tobacco-users and that small variations are
introduced during extraction, derivatization and enzymatic
hydrolysis.

Table I. Precision of assays, coefficients of variation(%, n=6).

COMPOUNDS	GC	ASSAY
Nicotine (free), 1.45 ug/ml	1.7	3.5
Nicotine (total), 2.12 ug/ml	2.6	5.7
Cotinine (free), 2.90 ug/ml	2.2	3.9
Cotinine (total), 5.26 ug/ml	2.8	2.8
3′-Hydroxycotinine (free), 7.9 ug/ml	3.2	5.0
3′-Hydroxycotinine (total), 10.5 ug/ml	3.2	6.6

Urinary excretion of nicotine metabolites: At steady-state the
rate of excretion reflects the generation rate. The quantitative
conversion of nicotine to its metabolites was therefore studied
in 24-hour urine collections from habitual tobacco-users by
measuring urinary excretion of total nicotine, cotinine and
trans-3′-hydroxycotinine. Eight smokers and seven snuff-users
took part in the study.

The average ratio between free and conjugated nicotine excreted in the 24-hour urine from both smokers and snuff-users was 1.0. About twice as much cotinine as nicotine is excreted in 24 hrs. The relative amount of free and conjugated cotinine was 0.5, i.e. twice as much is excreted as glucuronide conjugates as free cotinine. As trans-3′-hydroxycotinine is a rather hydrophilic compound, it is easily excreted in the urine. Accordingly, only about 30 % of the total amount is excreted as conjugate, i.e. the ratio between free and conjugated trans-3′-hydroxycotinine was found to be 2.3 (Fig. 2).

From our data on the relative amounts of free and conjugated nicotine, cotinine and trans-3′-hydroxycotinine and from the relationship between the amounts of phase 1 metabolites excreted during 24 hrs at steady-state as calculated by Benowitz (Benowitz, 1988), we propose the following quantitative disposition of nicotine in man.

Of the total dose of nicotine, 5% is excreted as nicotine-N′-oxides, 3 % as cotinine-N-oxide, 10-15 % as free and conjugated nicotine, 20-25 % as free and conjugated cotinine and 50-60 % as free and conjugated trans-3′-hydroxycotinine.

CONCLUSION: These studies show that nicotine, cotinine and trans-3′-hydroxycotinine are excreted in urine, both as free and conjugated alkaloids. The total concentrations of nicotine, cotinine and trans-3′-hydroxycotinine in the urine of tobacco users are accurately determined by quantitative analysis of urine samples subjected to enzymatic hydrolysis. The average ratios between free and conjugated alkaloids were 1.0, 0.5 and 2.3 for nicotine, cotinine and trans-3′-hydroxycotinine in the 24-hour urine of smokers and users of moist snuff. The best estimate of nicotine exposure is obtained by measuring the total amount of nicotine and its metabolites in urine of tobacco users. Nicotine, cotinine, trans-3′-hydroxycotinine and their glucuronides together with the N-oxides of nicotine and cotinine account for more than 90 % of the nicotine dose.

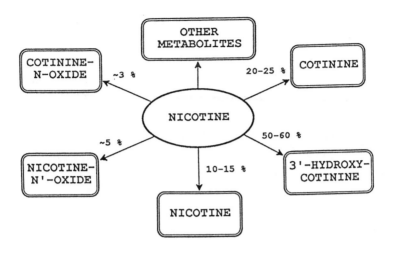

Fig. 2. Quantitative disposition of nicotine in man.

REFERENCES

Benowitz,N.L. (1988) In: The Pharmacology of Nicotine
    (M.J. Griffin and K. Thurau, Eds), IRL Press, Oxford, New
    York, pp.3-18
Curvall, M., Kazemi Vala, E. and Enzell, C.R. (1982) J. Chrom.
    Biomed. Appl. 232, 368-372

Effects of Nicotine on Biological Systems
Advances in Pharmacological Sciences
© Birkhäuser Verlag Basel

CONTRIBUTION OF TRANS 3'-HYDROXYCOTININE AND GLUCURONIDE CONJUGATES OF NICOTINE METABOLITES TO THE MEASUREMENT OF COTININE BY RIA

J. P. Richie, Jr., Ph.D., Y. Leutzinger, Ph.D., C. M. Axelrad, M.S. and N. J. Haley, Ph.D.

American Health Foundation, Valhalla, NY, USA

SUMMARY: The present study was conducted to quantitate potential cross-reactivity of the cotinine radioimmunoassay (RIA) with nicotine metabolites recently quantitated in urine. Initial experiments found that about 50% of RIA reactive material could be removed by Extralut-3 chromatography which also retained a major portion of authentic trans 3'-hydroxycotinine added to non-smokers urine. This cotinine metabolite cross-reacted with 40% efficiency in the direct RIA and was measured in the urine of smokers. When samples were treated with B-glucuronidase, no differences were observed in the RIA values or in the amounts of material retained by the column. The above results account for higher cotinine values than have been observed in urine when levels are compared to gas chromatography measurements and help account for the recovery of a higher proportion of metabolized nicotine.

INTRODUCTION

The cotinine content of various biological samples has been used as a specific and sensitive index of tobacco smoke exposure. The measurement of cotinine is commonly performed by RIA (Langone et. al., 1973; Haley et. al., 1983). However, in a recent comparison study of analytical methods for the determination of cotinine in urine (Biber et. al., 1987) and in our own observations, it has been noted that cotinine values obtained by RIA are consistently 30-40% higher than those from gas chromatography. Yet, the addition of cotinine standards to blank urine results in complete recovery by both methods.

Recent studies have identified hydroxycotinine in high concentration in smokers urine (Neurath and Pein, 1987). This metabolite had been previously over-looked since the methods for analysis often required sample processing which removed hydroxycotinine. Additionally, Curvall et. al. (1989) have found that nicotine, cotinine and hydroxycotinine can occur as glucuronic acid conjugates.

We hypothesized that the higher RIA results could arise from cross-reactivity of the cotinine antiserum with one or more of these newly identified metabolites. To test this, we conducted a series of studies to characterize our antisera and to evaluate the cross-reactivity of these nicotine metabolites in the urine of smokers with our assay.

METHODS

Urine samples were obtained from smokers and from non-smokers. The cotinine RIA method used was that of Haley et. al. (1983) and is based on rabbit antiserum raised against cotinine-BSA.

Extralut-3 columns (Merck, Germany) were utilized to extract polar metabolites from urine samples. Two ml samples were loaded onto the columns and after 10 min. they were eluted with 15 ml of methylene chloride. The extracted samples were then evaporated to dryness under nitrogen, reconstituted with 2 ml of control (non-smokers) urine and analyzed by RIA.

Smokers urine was treated with ß-glucuronidase to free conjugated metabolites. In brief, urine (2 ml) was added to 0.8 ml of 0.5 M acetate buffer, pH 5.0, containing 8 mg of ß-glucuronidase (520 U/mg). In control samples, acetate buffer without added enzyme was used. Incubation was carried out in a shaking water bath at 37°C for 12 hours.

RESULTS

Extralut-3 Column Chromatography:
In initial experiments where hydroxycotinine or cotinine was added to blank urine and passed through the columns, greater than 60% of the hydroxycotinine and less than 8% of the cotinine was retained. When smokers urine was extracted, about 50% of the RIA measurable material was retained by the column (Table I), suggesting that the RIA is sensitive to a hydrophylic metabolite in urine.

Table I.   Extraction of Smokers Urine by Extralut Columns

Subject	RIA Value (ng/ml)		% Recovery
	Before Extraction	After Extraction	
1	1150	756	65.7
2	2156	1085	50.3
3	2674	1404	52.5
4	9590	5460	56.9
5	3486	2038	58.5
6	5152	2324	45.1
7	4480	2240	50.0
8	3080	1988	64.5
		Average =	**55.4**

Hydroxycotinine Cross-Reactivity:
In order to test the cross-reactivity of the RIA with hydroxycotinine directly, 3-4000 ng/ml of authentic hydroxycotinine, synthesized at the American Health Foundation (Desai et al., 1990), was added to samples of blank urine. At all concentrations, hydroxycotinine cross-reacted with the cotinine antiserum (Figure 1). The RIA-cotinine values obtained indicated that hydroxycotinine was detected with a 40% efficiency.

CROSS—REACTIVITY OF HYDROXYCOTININE WITH COTININE ANTIBODY

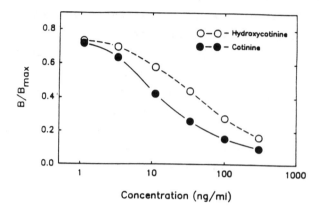

Figure 1. Cross-Reactivity of Hydroxycotinine With Cotinine Antiserum

Table II. The Effect of β-Glucuronidase Treatment of Urine Samples on the Extraction by Extralut Columns

Sample	− β-Glucuronidase		+ β-Glucuronidase	
	RIA Value of Extract (ng/ml)	Recovery (%)	RIA Value of Extract (ng/ml)	Recovery (%)
1	4795	46.3	5460	52.7
2	2324	45.1	2128	41.3
3	756	65.7	602	52.3
4	2240	50.0	1848	41.2
5	1988	64.5	2240	72.7
6	1085	50.3	1106	51.3
7	1404	52.5	1253	46.8
8	1686	41.9	2986	74.2
Mean ± SEM =		52.0 ± 3.08		54.1 ± 4.52

Glucuronide Conjugates: Unlike hydroxycotinine, authentic standards for the glucuronic acid conjugates are not yet available. Therefore, it was necessary to assess their presence indirectly by pretreating urine with ß-glucuronidase. Results revealed that there were no differences in RIA values between ß-glucuronidase treated and untreated urine. Also, no differences were found in the levels of RIA measurable material which could be extracted on Extralut columns (Table II). This suggests that either column extraction does not efficiently remove conjugates or the relative concentration of conjugates in these samples was low. It is important to note the large inter-individual variation in the levels of RIA measurable material extracted, suggesting that the conjugation of nicotine metabolites may vary greatly from subject to subject and possibly within a single subject.

DISCUSSION

Overall, the results from these studies demonstrate that the RIA is sensitive to the major nicotine metabolites found in urine, namely, cotinine and hydroxycotinine. Further, this cross-reactivity with hydroxycotinine can be responsible for the discrepancy between RIA and GC methods of cotinine analysis (Biber, 1987). Our results are consistent with the findings of Neurath and Pein (1987) that hydroxycotinine is the major metabolite found in urine.

REFERENCES

Biber, A., Scherer, G., Hoepfner, I., Adlkofer, F., Heller, W.D., Haddow, J.E., and Knight, G.J. (1987) Toxicol. Lett. 35, 45-50.

Curvall, M., Vala, E.K., Englund, G., and Enzell, C.R. (1989) 43rd Tobacco Chemists' Research Conference, Richmond, VA, USA, Oct. 2-3, 1989.

Desai, D. and Amin, S. (1990) Chem. Res. Toxicol. 3, 47-48.

Haley, N.J., Axelrad, C.M., and Tilton, K.A. (1983) Amer. J. Public Health 73, 1204-1207.

Langone, J.J., Cook, G., Bjercke, R.J. and Lifschitz, M.H. (1988) J. Immunol. Meth. 114, 73-78.

Langone, J.J., Gjika, H.B., and Van Vunakis, H. (1973) Biochem. 12, 5025-5030.

Neurath, G.B. and Pein, F.G. (1987) J. Chromatog. 415, 400-406.

Effects of Nicotine on Biological Systems
Advances in Pharmacological Sciences
© Birkhäuser Verlag Basel

EXTRAHEPATIC COMPARED WITH HEPATIC METABOLISM OF NICOTINE IN THE RAT

H. Foth, U. Walther, H. Looschen, H. Neurath and G.F. Kahl

Department of Pharmacology and Toxicology, University of Göttingen, Robert-Koch-Strasse 40, D-3400 Göttingen, F.R.G.

SUMMARY: Primary oxidation of nicotine exhibits similar rates in isolated rat lung and liver. Secondary metabolism, however, is restricted to the liver as demonstrated by the elimination parameters of ^{14}C-cotinine in isolated lung and liver. Phenobarbital (PB) induction markedly enhances the elimination of cotinine only in liver and not in lung.
In conscious rats extrahepatic sites significantly contribute to the elimination of ^{14}C-nicotine. Hepatic first-pass extraction of nicotine and the rate of cotinine turnover are highly induced by PB. This effect is paralleled by a shift in the ratio of norcotinine to cotinine in the urine.

Nicotine is readily absorbed by lung epithelium to a substantial amount. The rapid decline of systemic concentrations is attributed mainly to metabolic clearance within the liver (Schievelbein, 1982). The ability of the lung to cope with nicotine, although quantitatively low, has been demonstrated in isolated perfused lungs of rabbit and dog (McGovren et al., 1976; Turner et al., 1975). Comparing isolated perfused organs of rats, we have obtained evidence that the lung contributes substantially to the elimination of nicotine (Foth et al., 1988). Phenobarbital induction enhances nicotine oxidation in rat liver 8-fold while 5,6-benzoflavone has no effect.

The purpose of this study was to compare the metabolism of

nicotine in perfused liver and lung. The contribution of the liver to the total body clearance of nicotine was estimated by a pharmacokinetic approach in conscious non-induced and PB-induced rats.

MATERIALS AND METHODS

Male Wistar rats were pretreated with phenobarbital (PB, 80 mg/kg, daily ip on 3 consecutive days). Silicone tubings were implanted into the superior vena cava for serial blood sampling in conscious rats. [2-^{14}C-pyrrolidine]-nicotine was obtained from Dupont, NEN division, Dreieich. ^{14}C-cotinine was prepared by oxidation of ^{14}C-nicotine with $K_3Fe(CN)_6$ and purified to 99 %. Rat liver and lung perfusions were performed in situ in a recirculating manner as described previously (Rüdell et al., 1987; Foth et al., 1988). Conscious rats received nicotine either iv or po by a rubber stomach tube. Serial blood samples of 150 µl each were drawn and the loss of volume was compensated for. Prior to analysis the samples of perfusate and plasma were alkalized and extracted with chloroform. Metabolites were separated by thin-layer chromatography (TLC) either on Al_2O_3-coated plastic sheets (Polygram^R Alox N/UV$_{254}$) in chloroform/methanol/ 25 % ammonia (95:5:0.2 v/v/v) or on silicagel F_{254}-coated glass sheets in methanol/acetone/toluene/triethylamine (4:40:50:3 v/v/v/v). HPLC was performed on a nucleosil 120 C_{18}, 5 µm column using H_2O/methanol/0.1 M acetate buffer/acetonitrile (65:29:4:2 v/v/v/v), adjusted to pH 7.4 with triethylamine, as the mobile phase. The flow rate was set to 0.7 ml/min (0-7 min) and 1.2 ml/min (7-30 min).

RESULTS

Comparing isolated perfused rat lung and liver of rats, we have shown a much higher relative availability of the pul-

monary tissue (5-fold per g of wet weight) to convert nicotine
(Foth et al., 1988). The time course of the main metabolite of
nicotine in the perfusate, cotinine, is almost identical in
non-induced rat lung and liver (Fig. 1). Pulmonary elimination
of cotinine is very low which is demonstrated by the kinetics
of cotinine as the metabolite of nicotine as well as of ^{14}C-
cotinine as the substrate. The elimination of cotinine,
reaching values of clearance of 0.16 (saline) and 0.13 (PB),
is not affected in lung after PB induction. Cotinine kinetics
in non-induced rat liver is 4-fold higher (0.6 ml/min) than in
lung. After PB-induction the hepatic clearance of cotinine is
induced 7-fold (4.6 ml/min).

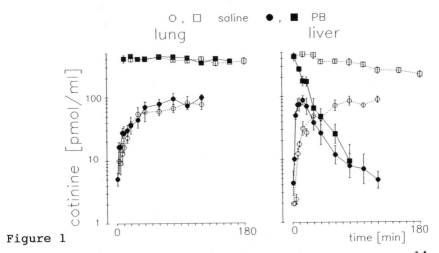

Figure 1

Elimination of cotinine in rat lung and liver after ^{14}C-
cotinine (0.5 μM, □ , ■) or ^{14}C-nicotine (0.5 μM, O , ●) as the
substrate. Geometric means ± SE (n = 3-4)

In the perfusate of non-induced (PB-induced) rat lung ^{14}C-
nicotine-derived radioactivity consists of 4% (2%) nicotine,
50% (51%) cotinine, 22% (24%) nicotine-N'-oxide, 5% (5%)
nornicotine, 8% (3%) cotinine-N-oxide, 4% (6%) norcotinine at
120 min. In non-induced (PB-induced) rat liver the metabolites
of nicotine are composed of 9% (0%) nicotine, 25% (1%) cotinine, 44% (9%) nicotine-N'-oxide, 2% (4%) nornicotine, 10%
(60%) cotinine-N-oxide and 5% (7%) norcotinine.

In conscious control rats the oral availability of unchanged nicotine was 89 % indicating low hepatic extraction of about 10 %. The first-pass extraction of oral nicotine is increased to almost 99 % after PB treatment. The total body clearance of iv nicotine, however, is only twofold induced by PB. The AUC values of cotinine demonstrate a 4- (po) to 7-fold (iv) increase of cotinine elimination (Table I).

71 - 85 % of nicotine is excreted into the urine mainly as polar metabolites. The fraction of non-polar metabolites, although low in quantity, also indicates the PB-induced increase of cotinine turnover: the ratio of norcotinine to cotinine correlated with the PB-induced state.

Table I

Increase of hepatic nicotine elimination after phenobarbital induction in the conscious rat

	iv		po	
	saline	PB	saline	PB
Dose, nmol	427 ± 33	413 ± 41	404 ± 5	378 ± 20
AUC, (nmol/ml) x min				
nicotine	24 ± 4	11 ± 3 **	21 ± 3	0.16 ± 0.02 **
cotinine	211 ± 17	40 ± 16 **	136 ± 10	34 ± 6 **
total ^{14}C	409 ± 29	197 ± 46 **	65 ± 20	184 ± 13 **
urinary excretion, % of dose/48 h				
polar	57 ± 6	56 ± 2	57 ± 8	74 ± 4
non-polar	22 ± 3	17 ± 2	14 ± 1	11 ± 1
norcot/cot	0.19 ± 0.04	0.97 ± 0.34 **	0.33 ± 0.09	4.9 ± 1.3 **

Arithmetic means ± SE, n= 3-4, ** $p < 0.005$
AUC = area under the concentration - time curve

DISCUSSION

Contrary to the common expectation that nicotine undergoes extensive metabolism by the liver, our data, showing an oral availability of nicotine in non-induced rats of 90 %, indicate a hepatic extraction ratio of only 10 % in the rat liver in vivo. Nicotine elimination in isolated rat lungs approaches the values of isolated non-induced liver (Foth et al., 1988). In view of the unique position of the lung in the circulation it must be assumed that a substantial part of systemic nicotine clearance and metabolism is mediated by the lung.

The hepatic oxidation of nicotine is markedly increased. PB-induction (Rüdell et al., 1987) and, as shown in this study, secondary metabolic pathways via turnover of cotinine are also severalfold increased in liver. A PB-related increase of nicotine clearance, though much smaller, is also observed lung. Cotinine elimination in induced and uninduced lung remains marginal. The marked increase of hepatic nicotine metabolism is masked in vivo, when extrahepatic organs are also exposed to nicotine iv. These discrepancies of PB-related response might be interpreted as the result of organ-specific differences in the rate of constitutive expression and inducibility of cytochrome(s) P-450 in lung and liver (Omiecinski, 1986).

CONCLUSION: Due to its unique position in the circulation, the removal of nicotine by lung should have a large impact compared to the elimination by the liver.

ACKNOWLEDGEMENTS: The Forschungsrat Rauchen und Gesundheit is thanked for financial support.

REFERENCES

Schievelbein, H., (1982) Pharmacol. Ther. 18, 233-248.
McGovren, J.P., Lubaway, W.C., and Kostenbauder, H.B. (1976) J. Pharmacol. exp. Ther. 190, 198-207.
Turner, D.M., Armitage, A.K., Briant,R.H., and Dollery, C.T. (1975) Xenobiotica 5, 539-551.
Foth, H., Rüdell, U., Ritter, G., and Kahl, G.F. (1988) Klin. Wochenschr. 66, 98-104
Rüdell, U., Foth, H., and Kahl, G.F. (1987) Biochem. biophys. Res. Commun. 148, 192-198
Omiecinski, C.J. (1986) Nucleic Acids Res. 14, 1525-1539

RELEVANCE OF NICOTINE-DERIVED N-NITROSAMINES IN TOBACCO CARCINO-
GENESIS

D. Hoffmann, A. Rivenson, F.-L. Chung, and S.S. Hecht, American
Health Foundation, Naylor Dana Institute for Disease Prevention,
1 Dana Road, Valhalla, New York, 10595 / U.S.A.

SUMMARY: Tobacco-specific N-nitrosamines (TSNA) are formed from
the Nicotiana alkaloids by N-nitrosation. Among identified TSNA
are powerful animal carcinogens, namely N'-nitrosonornicotine
(NNN), 4-(methylnitrosamino)-1-(3-pyridyl)-1-butanone (NNK) and
4-(methylnitrosamino)-1-(3-pyridyl)-1-butanol (NNAL). In animals
and in human tissues these nitrosamines are metabolically
activated by α-hydroxylation to highly reactive species which
form adducts with protein and DNA. These three TSNA contribute
significantly to the increased risk of snuff dippers for oral
cancer and to the risk of smokers for cancer of the lung and oral
cavity and likely also for cancer of the esophagus and pancreas.
Chemopreventive approaches towards reduction of risk of TSNA-
related cancer are discussed, as is the need for developing
methods for the reduction of TSNA in tobacco and tobacco smoke.
Future research should concentrate on biomarkers for the exposure
of tobacco users to TSNA, the DNA binding of TSNA in humans, and
on the question of endogenous formation of TSNA.

INTRODUCTION

At the 1987 Symposium on "The Pharmacology of Nicotine" we
presented an overview on the chemistry, biochemistry and
bioassays of the nicotine-derived N-nitrosamines (Hoffmann et
al., 1988). Since then our knowledge about the tobacco-specific
N-nitrosamines (TSNA) has made remarkable progress, testifying to
the biological significance attributed to TSNA. New facts have
emerged regarding the formation, biochemistry and carcinogenicity

of TSNA; biomarkers have been developed for the uptake of TSNA by
tobacco chewers and smokers, transplacental effects of TSNA have
been studied, and chemopreventive agents have been examined for
for their potential to reduce the genotoxicity of nicotine-derived nitrosamines. These research programs focus on the key question, the relevance of tobacco-specific N-nitrosamines to human
cancer.

TSNA: Carcinogenicity. During tobacco processing the Nicotiana
alkaloids undergo N-nitrosation to form TSNA. Seven TSNA have
been identified in tobacco, five of these are formed from nicotine (Fig. 1). Three of the nicotine-derived TSNA,

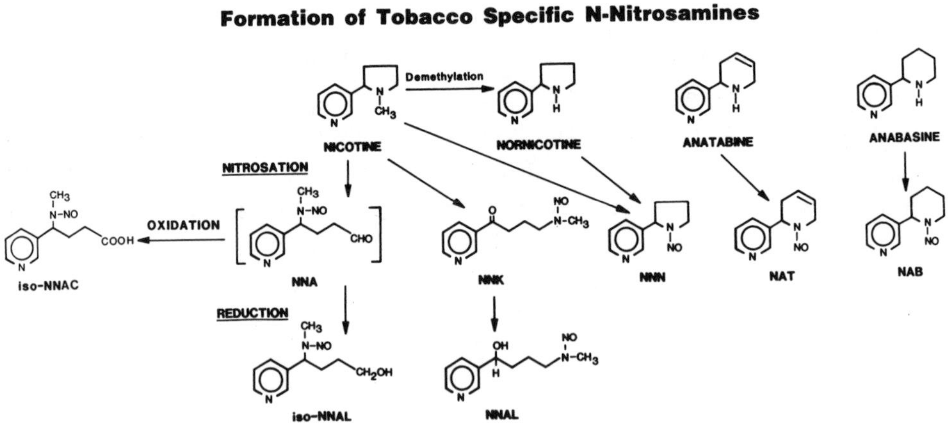

Formation of Tobacco Specific N-Nitrosamines

Figure 1.

namely N'-nitrosonornicotine (NNN), 4-(methylnitrosamino)-1-(3-
pyridyl)-1-butanone (NNK) and 4-(methylnitrosamino)-1-(3-pyri-
dyl)-1-butanol (NNAL) are powerful carcinogens in mice, rats and
hamsters; they induce benign and malignant tumors of the lung,
upper aerodigestive tract, exocrine pancreas and/or liver. N'-
Nitrosoanabasine (NAB) is a weak carcinogen in the rat, N'-ni-

trosoanatabine (NAT) and 4-(methylnitrosamino)-4-(3-pyridyl)-butyric acid (iso-NN-acid) are not carcinogenic and the genotoxic 4-(methylnitrosamino)-4-(3-pyridyl)-1-butanol (iso-NNAL) has not been bioassayed for carcinogenicity (Brunnemann et al., 1987; Hecht and Hoffmann, 1989; Rivenson et al., 1989).

Recent bioassay results have demonstrated that TSNA are also transplacental carcinogens in mice and hamsters. NNK given i.p. (100 mg/kg) during days 14-18 of gestation to three strains of mice induced lung adenoma and/or liver tumors in the progeny (Anderson et al., 1989). NNK given by s.c. injection (50-300 mg/kg) during days 13-15 of gestation to Syrian golden hamsters resulted in tumors at various sites in up to 76% of the offspring within one year after treatment. These tumors were found pre-dominantly in the respiratory tract (Correa et al., 1990). Al-though the NNK doses given during the gestation phase were high, these data support, nevertheless, the suggestion that smoking during pregnancy increases the risk for cancer in the offspring (Stjernfeldt et al., 1986).

TSNA: Formation and Analysis. The TSNA in processed tobacco are primarily formed after harvesting during curing, fermentation and aging. The major precursors for the TSNA are nicotine and nitrate. During processing, nitrate is partially reduced to the nitrosating agent nitrite. The higher the nitrate and/or nico-tine concentrations in tobacco, the higher is the TSNA yield in the finished product. The significantly higher concentrations of the TSNA in snuff (NNN 2-154 µg/g; NNK 0.1-14 µg/g) compared to those in chewing tobacco (NNN 0.3-19 µg/g; NNK 0.03-0.38 µg/g) and in cigarette tobacco (NNN 0.3-19 µg/g; NNK 0.1-1.5 µg/g) are not only due to the high nitrate content of the snuff tobacco but also to factors associated with the fermentation and aging pro-cesses (Hecht and Hoffmann, 1989; Djordjevic et al., 1989; Tricker and Preussmann, 1989).

In the case of cigarette smoke, the TSNA yields are also deter-mined by the nitrate and nicotine content of the tobacco. The mainstream smoke (MS) of nonfilter cigarettes made exclusively from burley or black tobaccos, which are rich in nitrate, is significantly higher in TSNA (NNN 85-590 ng/cigarette; NNK 70-430 ng/cigarette) than the MS of nonfilter cigarettes made exclusively from oriental and bright tobacco (NNN 3-77 ng/cig; NNK 0-91 ng/cig). Blended U.S. cigarettes (NNN 120-950 ng/cig ; NNK 80-770 ng/cig) and West German cigarettes (NNN 5-625 ng/cig; NNK 4-432 ng/cig) have medium to high TSNA levels (Hecht and Hoffmann, 1989; Spiegelhalder et al; 1989). When cigarettes are smoked under standard laboratory conditions, the release of TSNA into sidestream smoke (SS) exceeds the levels generated in MS. The release into SS is at least 3 times higher than that into MS (Adams et al., 1987).

Model studies with ^{14}C-labeled TSNA have determined that 40-46% of NNN and 26-37% of NNK in MS originate by transfer from the tobacco, whereas the remainder is pyrosynthesized during smoking (Hoffmann et al., 1980; Adams et al., 1983). Recently this concept has been challenged by Fischer et al. (1990). When these investigators added nitrate, nicotine, or ^{14}C-nicotine to the tobacco column and analyzed the MS of the cigarettes, the results of their TSNA analysis suggested that NNN and NNK in MS derive almost exclusively from the tobacco from where they transfer into the smokestream. Since the reduction of TSNA in tobacco smoke is most desirable, the origin of the TSNA requires re-evaluation.

Cancer of the Oral Cavity. It is known from epidemiological studies that chewing of tobacco, and especially dipping of snuff increases the risk for cancer of the oral cavity (Winn et al., 1981; IARC, 1985; U.S. Dept. Health Human Services, 1986; WHO, 1988). Long-term daily applications of snuff into a surgically created lip canal have induced oral cancer in rats (Hirsch and Johansson, 1983; Hecht et al., 1986; Johansson et al., 1989). Among the carcinogenic N-nitrosamines in oral snuff, NNN (2-154 µg/g) and NNK (0.1-14 µg/g) occur in the highest concentrations (Hecht and Hoffmann, 1989; Tricker and Preussmann, 1989). Other known carcinogens in snuff are volatile nitrosamines (N-nitrosodimethylamine $\leq$0.22 µg/g), traces of benzo(a)pyrene ($\leq$0.09 µg/g), formaldehyde, acetaldehyde and crotonaldehyde, as well as polonium-210 (^{210}Po, <0.64 pCi/g; Hoffmann et al., 1987). NNN and NNK are the only tobacco carcinogens known to induce oral tumors in laboratory animals. They are present in the saliva of snuff dippers (Hecht and Hoffmann, 1989) and are metabolically activated by cultured human buccal mucosa (Castonguay et al., 1983).

The leading U.S. commercial snuff contains NNN at levels of 64 µg/g while NNK amounts to 3.1 µg/g. Since the "average snuff dipper" consumes about 10 g of snuff daily, it is estimated that the lifetime exposure of a snuff dipper corresponds to 9,340 mg NNN (0.70 mmol/kg) and 450 mg NNK (0.03 mmol/kg). These levels are comparable to the dose of 1.6 mmol/kg of a mixture of NNN and NNK which induced tumors in the mouth of rats by oral swabbing (Hecht et al., 1986). In agreement with such considerations are recent studies which have demonstrated the occurrence of NNN- and NNK-derived hemoglobin adducts in the blood of snuff dippers (Carmella et al., 1990). Taken together, the available evidence strongly supports the role of TSNA as causative agents for oral cancer in snuff dippers.

Smoking of cigarettes, cigars and pipes is strongly associated with cancer of the oral cavity, especially in conjunction with chronic alcohol use (IARC, 1986; U.S. Dept. Health Human Services, 1989). A smoker of 40 cigarettes per day is subject to lifetime exposure of about 285 mg NNN (0.02 mmol/kg) and 180 mg NNK (0.01 mmol/kg), a lifetime smoker of 5 cigars/day to 228 mg NNN (0.018 mmol/kg) and 135 mg NNK (0.009 mmol/kg). Although the relative amounts of carcinogenic TSNA in tobacco smoke appear low relative to the doses used to induce oral cancer in rats (1.6 mmol/kg), consideration should be given to their combined effects with cocarcinogens in tobacco smoke such as catechol (up to 300 µg/cigarette; Mirvish et al., 1985; Yamaguchi et al., 1989).

Lung cancer. A fair number of chemical, biochemical and bioassay data support the concept that the tobacco-specific N-nitrosamines contribute significantly to the causal association of tobacco smoke with cancer of the lung. Smokers of cigarettes made exclusively from black tobacco, such as those commonly smoked in France, North Africa and Cuba, are at greater risk for lung cancer than are the smokers of cigarettes made from bright

(Virginia) and oriental tobaccos and from tobacco blends (Joly et al., 1983; Benhamou et al., 1985). The TSNA are present in the highest concentrations in the smoke from black cigarettes compared to the other types of cigarettes (Hecht and Hoffmann, 1989; Tricker and Preussmann, 1989). The lowest total doses of NNK that have been shown to induce tumors in the respiratory tract in animals are 0.5 mg (0.13 mmol/kg) in mice, 3.0 mg (0.03 mmol/kg) in rats and 1 mg (0.03 mmol/kg) in hamsters (Hecht and Hoffmann, 1990; Belinsky et al., 1990). As discussed earlier, a heavy cigarette smoker is exposed to a cumulative dose of about 0.01 mmol/kg, which is comparable to about one third of the total dose which induces tumors of the respiratory tract in hamsters, and one third of the total dose required to elicit squamous cell carcinoma and lung ademona in rats (Rivenson et al., 1988; Hecht and Hoffmann, 1990). This estimation does not take into consideration that there is likely endogenous formation of NNK upon smoke inhalation and retention.

As in the case of cultured human buccal mucosa, human lung tissue can also metabolically convert NNK to active species (Castonguay et al., 1983) which form DNA-adducts and activate the K-ras oncogene in A/J mouse lung. Activation occurs by a point mutation on codon 12 (Belinsky et al. 1988). Activated K-ras oncogenes with mutations on codon 12 have indeed been detected in adenocarcinomas of the lung in 13 of 45 cigarette smokers (Rodenhuis et al., 1988).

These data strongly support the concept that the nicotine-derived nitrosamine, NNK, contributes to lung cancer in smokers. The relative importance of NNK as a lung carcinogen must be weighed against that of other lung carcinogens in cigarette smoke, such as polynuclear aromatic hydrocarbons (PAH), especially benzo(a)pyrene, and ^{210}Po. It is also likely that the carcinogenicity of NNK, certain PAH and ^{210}Po is greatly enhanced by certain other smoke constituents such as volatile aldehydes, carcinogenic PAH, volatile phenols and catechols (Hoffmann and Hecht, 1990).

Esophageal and Pancreas Cancer. Table I is a listing of the

Table I.   Probable Causative Agents in Tobacco Carcinogenesis

Site	TSNA	Other Tumor Initiators and Carcinogens	Tumor Enhancing Factors
Oral Cavity Snuff-dippers	NNN, NNK	^{210}Po (?)	HSV-1 and HSV-2, Nutrition, Irritation, Nicotine (?)
Smokers	NNN, NNK	^{210}Po (?)	HSV-1 and HSV-2, Nutrition, Alcohol
Lungs Smokers	NNK, NNAL	PAH, ^{210}Po (minor) Volatile Aldehydes 1,3-Butadiene (?) Cr, Cd, Ni (?)	Tumor promoters and co-carcinogens, Aldehydes
Esophagus Smokers	NNN, NAB	Volatile N-nitrosamines (?)	Catechols, Alcohol
Pancreas Smokers	NNK, NNAL		Nutrition, Nicotine (?)
Bladder		4-Aminobiphenyl 2-Naphthylamine 2-Toluidine Other Aromatic Amines (?)	Infection (?)

likely contributions of TSNA and other tobacco components and factors to the increased risks of tobacco consumers for cancer at various sites. This tabulation includes a listing of TSNA that are likely contributors to cancer of the esophagus and pancreas.

NNN induces tumors of the esophagus in rats. Esophageal tumorigenicity of NNN and other unknown carcinogens is likely enhanced by the presence of high amounts of catechol in tobacco smoke (<300 µg/cigarette; Mirvish et al., 1985; Yamaguchi et al. 1989). Other nitrosamines in tobacco smoke, known to induce tumors in the esophagus of rats, are N'-nitrosoanabasine (0.1-4.6 µg/cigarette), N-nitrosodiethylamine ($\leq$25 ng/cigarette) and N-nitrosoethylmethylamine ($\leq$13 ng/cigarette; Hoffmann and Hecht, 1990). Although epidemiological studies have clearly established a synergism for the carcinogenic activity of smoking and alcohol consumption in humans (U.S. Dept. Health Human Services, 1989), bioassays in hamsters and rats have so far failed to provide evidence for an increase in the carcinogenicity of NNN by ethanol (McCoy et al., 1981; Castonguay et al., 1984; Griciute et al., 1986).

NNK and its enzymatic reduction product, NNAL, are the only tobacco components known to cause benign and malignant tumors of the pancreas in laboratory animals (Rivenson et al., 1988; Pour and Rivenson, 1989). The tumors induced by these TSNA occur mostly in the exocrine pancreas, the same location in which the tumors of the pancreas of cigarette smokers are seen (U.S. Health Dept. Health Human Services, 1989). The lowest dose of NNK that induces pancreas tumors in rats is 0.17 mmol/kg. The lifetime (40-year) exposure of a cigarette smoker is 0.01 mmol/kg. While the cumulative exposure of the smoker appears to be low by comparison, one must recognize that other smoke components as well as nutritional factors may enhance the carcinogenicity of NNN and NNK.

Reduction of the Carcinogenic Effect of TSNA. The likely association of exposure to TSNA with an increased risk of tobacco users for cancer of the lung, oral cavity, esophagus and pancreas has led to several approaches towards a reduction of these carcinogens in chewing tobacco, snuff and in tobacco smoke.

Modified preparation and a different composition of tobacco products have a potential for reduction of N-nitrosamines. In fact, some success has already been attained in this area. Levels of nitrosomorpholine (NMOR), a powerful animal carcinogen, have drastically decreased in U.S. snuff since the manufacturers have avoided contamination with morpholine, the precursor for NMOR (Brunnemann et al., Hoffmann et al., 1987). Similar reduction has been observed for N-nitrosodiethanolamine in all types of U.S. tobacco products by the elimination of maleic hydrazide-diethylamine as a sucker-growth inhibitor in the cultivation of tobacco (Brunnemann et al., 1981; Brunnemann and Hoffmann, 1990). Other approaches towards the reduction of N-nitrosamines are avoidance of the nitrate-rich burley ribs (Neurath and Ehmke, 1964; Brunnemann et al., 1983), and the bacterial degradation of nitrate in burley ribs before incorporating the latter into tobacco products. The drastic reduction of nicotine in tobacco by supercritical extraction with carbon dioxide appears to be a promising approach towards the reduction of TSNA (Grubbs and Havel, 1987). These are but a few examples towards the reduction of TSNA. A number of other methods should be explored and developed, so as to yield products with reduced carcinogenic potential.

Inhibition of nitrosamine carcinogenesis has also been explored with chemopreventive agents. TSNA, like most nitrosamines, are procarcinogens. They are metabolically activated by α-hydroxylation to alkylating agents which then bind to protein and to cellular DNA. Chemopreventive agents may either inhibit the metabolic activation and/or the DNA binding, or they may accelerate the repair of lesions induced by N-nitrosamines. In rats, the in vivo α-hydroxylation of N-nitrosopyrrolidine is significantly inhibited by isothiocyanates, including the naturally occurring phenylethyl isothiocyanate (Chung et al., 1984). Isothiocyanates inhibit the metabolic activation, DNA-binding, and tumorigenicity of NNK in mice and rats (Morse et al., 1989a, 1989b; Morse et al., 1990a). The induction of lung tumors in mice by NNK is increasingly inhibited with the increase of the chain length from benzyl- to 4-phenylbutyl isothiocyanate (Morse et al., 1989c). The lung tumor formation in mice by NNK has also been significantly inhibited by indole-3-carbinol (Morse et al.,1990b). These findings on the inhibition of the tumorigenicity of NNK in mice and rats by chemopreventive agents require additional biochemical studies and bioassays. Nevertheless, these investigations are most encouraging and are not only of academic interest.

OUTLOOK

In this overview we have summarized data in support of the concept that the increased risk for cancer of the oral cavity in snuff dippers and for cancer of the lung, upper aerodigestive tract and pancreas in smokers is most likely associated with exposure to TSNA. To strengthen the concept that the TSNA play a major role in tobacco carcinogenesis, it is important that future studies deal with application of biomarkers for the uptake of TSNA by tobacco consumers, the binding in vivo of TSNA to human DNA, and the dependence of the DNA-binding on other factors. The question of endogenous formation of TSNA in tobacco chewers and smokers is a key question that is currently under intensive study. We hope that the encouraging experiments for chemoprevention of the carcinogenic effect of TSNA will be continued and intensified. The exposure to TSNA will certainly continue for many years to come; thus, product modification and chemopreventive methods are strongly indicated.

ACKNOWLEDGEMENTS

We appreciate the many contributions of our colleagues at the American Health Foundation to the chemistry, biochemistry and biological activity of the tobacco-specific N-nitrosamines. We thank Bertha Stadler and Ilse Hoffmann for editorial assistance. Our studies are supported by Grants No. CA-29580, CA-46535, CA-44161, and CA-44377 from the U.S. National Cancer Institute.

REFERENCES

Adams, J.D., Lee, S.J., Vinchkoski, N., Castonguay, A. and Hoffmann D. (1987) Cancer Letters 17, 330-346.

Adams, J.D., O'Mara-Adams, K.J., and Hoffmann, D. (1987) Carcinogenesis 8, 729-731.

Anderson, L.M., Hecht, S.S., Dixon, D.E., Dove, L.F., Kovatch, R., Amin, S., Hoffmann, D., and Rice, J.M. (1989) Cancer Res. 49, 3770-3775.

Benhamou, S., Benhamou, E. Tirmarche, M., and Flamant, R. (1985) J. Natl. Cancer Inst. 74, 1169-1175.

Belinsky, S.A., Devereux, T.R., Stoner, G.D., and Anderson, M.W. (1988) Cancer Res. 49, 5305-5311.

Belinsky, S.A., Foley, J.F., White, C.M., Anderson, M.W., and Maronpot, P.R. (1990) Cancer Res. 50, 3772-3780.

Brunnemann, K.D., Genoble, L., and Hoffmann, D. (1987) Carcinogenesis 8, 465-469.

Brunnemann, K.D. and Hoffmann, D. (1981) Carcinogenesis 2, 1123-1127.

Brunnemann, K.D. and Hoffmann, D. (1990) Submitted.

Brunnemann, K.D., Masaryk, J., and Hoffmann, D. (1983) J. Agr. Food Chem. 31, 1221-1224.

Brunnemann, K.D., Scott, J.C., and Hoffmann, D. (1983) Carcinogenesis 3, 693-696.

Carmella, S.G., Kagan, S.S., Kagan, M., Foiles, P.G., Palladino, G., Quart, A.M., Quart, E., and Hecht, S.S. (1990) Cancer Res. In press.

Castonguay, A., Stoner, G.D., Schut, H.A.J., and Hecht, S.S. (1983) Proc. Natl. Acad. Sci. U.S.A. 80, 6694-6697.

Castonguay, A., Rivenson, A., Trushin, N., Reinhardt, J., Stathopoulis, S., Weiss, C., Reiss, B., and Hecht, S.S. (1984) Cancer Res. 44, 2285-2290.

Chung, F.L., Juchatz, A., Vitarius, J., and Hecht, S.S. (1984) Cancer Res. 44, 2924-2928.

Correa, E., Joshi, P.A., Castonguay, A., and Schüller, M. (1990) Cancer Res. 50, 3435-3438.

Djordjevic, M.V., Gay, S.L., Bush, L.P., and Chaplin, J.F. (1989) J. Agr. Food Chem. 37, 752-756.

Fischer, S., Spiegelhalder, B., Eisenbarth, J., and Preussmann, R. (1990) Carcinogenesis 11, 723-730.

Griciute, L., Castegnaro, M., Bereziat, J.C., and Cabral, J.R. (1986) Cancer Letters 31, 267-275.

Grubbs, H.J. and Havel, T.M. (1987) European Pat. Office Publ. No. 0280817:2.

Hecht, S.S. and Hoffmann, D. (1989) Cancer Surveys 8, 275-294.

Hecht, S.S. and Hoffmann, D. (1990) In: Relevance to Human Cancer of Nitroso Compounds, Tobacco and Mycotoxins (I.K. O'Neill, J.S. Chen, and H. Bartsch, eds.) IARC Sci. Publ. 105, in press.

Hecht, S.S., Rivenson, A., Braley, J., DiBello, J., Adams, J.D., and Hoffmann, D. (1986) Cancer Res. 46, 4162-4166.

Hirsch, J.M. and Johansson, S.L. (1983) J. Oral Pathol. 12, 187-198.

Hoffmann, D., Adams, J.D., LaVoie, E.J., and Hecht, S.S. (1988) In: The Pharmacology of Nicotine (M.J. Rand and K. Thurau, eds.), IRL Press/Oxford-Washington, D.C., pp. 43-60.

Hoffmann, D., Adams, J.D., Lisk, D., Fisenne, I., and Brunnemann, K.D. (1987) J. Natl. Cancer Inst. 79, 1281-1286.

Hoffmann, D., Adams, J.D., Piade, J.J., and Hecht, S.S. (1980) Intern. Agency Res. Cancer Sci. Publ. 31, 507-516.

Hoffmann, D. and Hecht, S.S. (1990) In: Handbook of Experimental Pharmacology (C.S. Cooper and P.L. Grover, eds.) Springer-Verlag, Berlin, Vol. 94/I, 63-102.

International Agency for Research on Cancer (1985) IARC Monogr. 37, 291 p.

International Agency for Research on Cancer (1986) IARC Monogr. 38, 421 p.

Johansson, S.L., Hirsch, J.M., Larsson, P.A., Saidi, J., and Oesterdahl, B.G. (1989) Cancer Res. 49, 3063-3069.

Joly, O.G., Lubin, J.H., and Caraballoso, M. (1983) J. Natl. Cancer Inst. 70, 1033-1039.

McCoy, G.D., Hecht, S.S., Katayama, S., and Wynder, E.L. (1981) Cancer Res. 41, 2849-2854.

Mirvish, S.S., Salmasi, S., Lawson, T.A., Pour, P., and Sutherland, D. (1985) J. Natl. Cancer Inst. 74, 1283-1290.

Morse, M.A., Amin, S.G., Hecht, S.S., and Chung, F.L. (1989b) Cancer Res. 49, 2894-2857.

Morse, M.A., Eklind, K.I., Amin, S.G., Hecht, S.S., and Chung, F.L. (1989c) Carcinogenesis 10, 1757-1759.

Morse, M.A., LaGreca, S.D., Amin, S.G., and Chung, F.L. (1990b) Cancer Res. 50, 2613-2617.

Morse, M.A., Reinhardt, J.C., Amin, S.G., Hecht, S.S., Stoner, G.D., and Chung, F.L. (1990a) Cancer Letters 49, 225-230.

Morse, M.A., Wang, C-X., Stoner, G.D., Mandal, S., Conran, P.B., Amin, S.G., Hecht, S.S., and Chung, F.-L. (1989a) Cancer Res. 49, 549-553.

Neurath, G. and Ehmke, H. (1964) Beitr. Tabakforsch. 2, 333-344.

Pour, P.M. and Rivenson, A. (1989) Am. J. Pathol. 134, 627-631.

Rivenson, A., Djordjevic, M.V., Amin, S., and Hoffmann, D. (1989) Cancer Letters 47, 111-114.

Rivenson, A., Hoffmann, D., Prokopczyk, B., Amin, S., and Hecht, S.S. (1988) Cancer Res. 48, 6912-6917.

Rodenhuis, S., Sletos, R.J.C., Boot, A.J.M., Evers, S.G., Moor, W.J., Wagenaar, S.S.C., VanBodegan, P.C.H., and Bos, J.L. (1988) Cancer Res. 48, 5738-5741.

Spiegelhalder, B., Fischer, S., and Preussmann, R. (1989) In: Tobacco and Cancer (A.P. Maskens, R. Molinard, R. Preussmann, and J.W. Wilner, eds.) Excerpta Medica/Amsterdam-New York-Oxford, pp. 27-34.

Stjernfeldt, M., Berglund, K., Lindsten, J., and Ludvigsson, J. (1986) Lancet I, 1350-1352.

Tricker, A.R. and Preussmann, R. (1989) In: Tobacco and Cancer (A.P. Maskens, R. Molinard, R. Preussmann, and J.W. Wilner, eds.) Excerpta Medica/Amsterdam-New York-Oxford, pp. 35-47.

U.S. Department of Health and Human Services (1986) NIH Publ.
   No. 86-2874, 195 p.
U.S. Department of Health and Human Services (1989) DHHS Publ.
   No. (CDC) 89-8411, 703 p.
Winn, D.M., Blot, W.J., Shy, C.M., Pickle, L.W., Toledo, A., and
   Fraumeni, J.R., Jr. (1981) New Engl. J. Med. 394, 745-749.
World Health Organization (1988) Tech. Rpt. Ser. 773, 81 p.
Yamaguchi, S., Hirose, M., Fukushima, S., Hasegawa, R., and Ito,
   N. (1989) Cancer Res. 49, 6015-6018.

Effects of Nicotine on Biological Systems
Advances in Pharmacological Sciences
© Birkhäuser Verlag Basel

NO PYROSYNTHESIS OF N'-NITROSONORNICOTINE (NNN) AND 4-(METHYLNITROSAMINO)-1-(3-PYRIDYL)-1-BUTANONE (NNK) FROM NICOTINE

Sophia Fischer, Bertold Spiegelhalder and Rudolf Preussmann

Institute of Toxicology and Chemotherapy, German Cancer Research Center
Im Neuenheimer Feld 280, 6900 Heidelberg, FRG

SUMMARY: Nicotine, a tertiary amine, cannot easily be nitrosated. A synthesis of NNN and NNK from nicotine during the smoking procedure (pyrosynthesis) does not occur for standard cigarette types. The NNN and NNK burden of smokers is determined by the amount of the preformed nitrosamines NNN and NNK in tobacco and their transfer into the mainstream smoke. Thus the TSNA intake of smokers can be reduced by manufacturing cigarettes with low levels of preformed TSNA, i.e. only tobaccos with low nitrate levels should be used and the portion of the nitrate rich stems and Burley tobaccos in blends should be reduced.

INTRODUCTION

N'-Nitrosonornicotine (NNN) and 4-(methylnitrosamino)-1-(3-pyridyl)-1-butanone (NNK) are powerful and organspecific carcinogens, which have been found in tobacco and tobacco smoke in relatively high concentrations (Hoffmann & Hecht, 1985; Fischer et al., 1989a; Fischer et al., 1989b; Fischer et al., 1990b). NNN is either formed by nitrosation of the major tobacco alkaloid nicotine under loss of a methylgroup or by nitrosation of the minor alkaloid nornicotine. NNK can only derive from nicotine by oxidative N-nitrosation (Figure 1). In order to reduce smokers burden with NNN and NNK it is essential to know origin and pathways of their formation. During curing and fermentation of the tobacco, the tobacco alkaloids can react with nitrite, which is formed by microbial reduction of nitrate, to yield TSNA (Chamberlain et al., 1986). Nitrate also gives rise to nitrogen oxides in smoke which are supposed to react with the tobacco alkaloids during smoking (pyrosynthesis) (Adams et al., 1983; Hoffmann et al., 1977; Adams et al., 1984) Thus the presence of NNN and NNK in mainstream smoke was explained by their direct transfer from tobacco into the mainstream smoke and by their formation during smoking. The purpose of this study was to investigate the contribution of the pyrosynthesis to the TSNA burden of smokers.

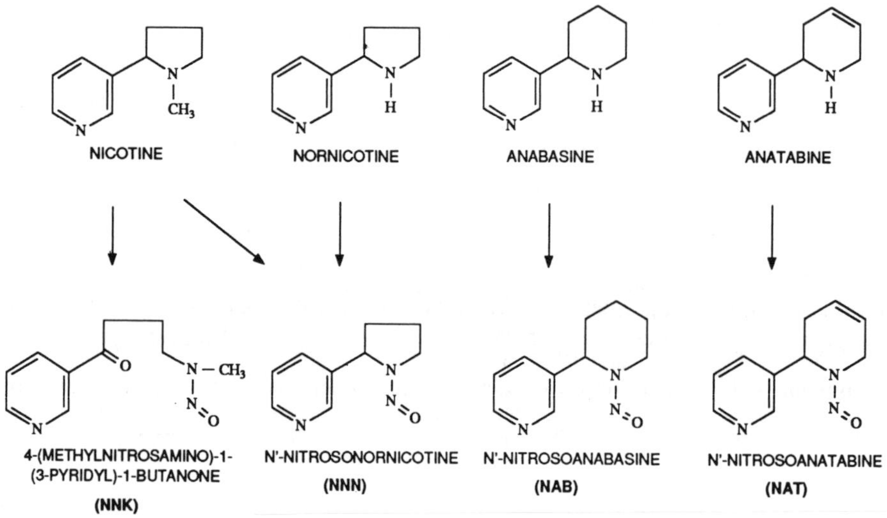

Figure 1: Tobacco alkaloids and the corresponding tobacco-specific nitrosamines (TSNA)

RESULTS

Four different cigarette brands (blend filter, Oriental nonfilter, dark nonfilter, Virginia nonfilter) with different original nitrate and nicotine levels were spiked with the TSNA precursor nitrate prior to smoking. The nitrate spiking level was between

4 and 20 mg/cigarette, whereas most commercially available cigarettes show original nitrate levels below 10 mg/cigarette. However, no significant increase in the mainstream smoke concentration of NNN and NNK could be observed (Fischer et al., 1990a).

Spiking of a blend filter cigarette with the NNN and NNK precursor nicotine at a concentration of 10 mg/cigarette prior to smoking, which doubles the original nicotine level, did not result in an increased mainstream smoke concentration of NNN and NNK (Fischer et al., 1990a).

After spiking cigarettes with radioactive labelled nicotine prior to smoking ([pyrrolidine-2-^{14}C]-nicotine at a total activity of 1.4×10^8 dpm and 2.6×10^8 dpm) no significant increase in radioactivity could be determined for peaks with the retention time of NNN and NNK (a pyrosynthesis rate of 0.0001 % would have been detectable) (Fischer et al., 1990a).

Furthermore German nonfilter cigarettes were analyzed for TSNA in tobacco and in mainstream smoke as well as for nitrate in tobacco. The mainstream/tobacco-ratios were calculated and related to the nitrate levels in tobacco and the smoke nicotine yields, which is graphically presented in Figures 2 and 3 (Fischer et al., 1990a). The analyzed cigarettes had different sizes and ventilation ratios. In the case of ventilation the mainstream smoke is diluted with air according to the ventilation ratio. Therefore the mainstream smoke/tobacco-ratios for NNN and NNK were corrected for this comparison.

As can be seen in Figure 2 and 3 the mainstream smoke/tobacco-ratios are constant and do not depend on the level of the NNN and NNK precursors nitrate and nicotine except for NNK in dark tobacco type cigarettes.

The nitrosation potential of freshly generated mainstream smoke was investigated in different solvents using tobacco alkaloids as precursor amines. Addition of the secondary amine nornicotine to the trapping fluids in amounts of 1 mg/cigarette resulted in an increase of NNN depending on the trapping fluid (Table I). The highest nitrosation of nornicotine with freshly generated mainstream smoke was observed in cyclohexane. Whereas after addition of the tertiary amine nicotine to cyclohexane as trapping fluid in amounts of 1 and 10 mg per cigarette smoked (i.e. 15 and 150 mg nicotine were added to the trapping fluid, when 15 standard cigarettes were smoked on a smoking machine), no additional NNN and NNK was observed (Fischer & Spiegelhalder, 1989c).

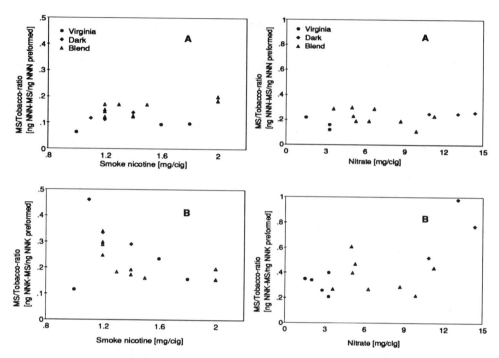

Figure 2: Dependence of the mainstream smoke/tobacco-ratios (MS/Tobacco-ratios) for NNN and NNK on the nicotine delivery for West German nonfilter cigarettes. The ratios have been corrected for the different cigarette lengths (A = NNN, B = NNK).

Figure 3: Dependence of the mainstream smoke/tobacco-ratios (MS/Tobacco-ratios) for NNN and NNK on the nitrate level of the tobacco. The ratios have been corrected for the different cigarette lengths and the different ventilation ratios (A = NNN, B = NNK).

Table I: NNN concentration after addition of 15 mg nornicotine to different solvents in the traps when 15 standard cigarettes were smoked on a smoking machine

Solvent	NNN (ng/cigarette)
Citrate-phosphate buffer + ascorbic acid without addition of nornicotine	99
Cyclohexane	259
2 M NaOH	142
Citrate-phosphate buffer + ascorbic acid	103
H_2O	130

DISCUSSION AND CONCLUSION

The results of this study clearly demonstrate that the mainstream smoke concentrations of NNN and NNK are independent on the level of nitrate and nicotine. Provided, pyrosynthesis would occur, chemical kinetics would require a direct correlation between precursors and the corresponding nitrosamines. This, however, could not be observed in any experiment presented in this study. The presence of NNN and NNK in mainstream smoke can only be explained by their direct transfer from the tobacco into the mainstream smoke. The transfer rate is dependent on the presence of a filter tip, the inhalation volume and the ventilation ratio, since in the case of ventilation the mainstream smoke is diluted with air (Fischer et al., 1990a; Fischer et al., 1989c). Thus the NNN and NNK burden of smokers is determined by the amount of preformed TSNA in tobacco, which can be reduced by manufacturing cigarettes with low levels of preformed TSNA in tobacco, i.e. only tobaccos with low nitrate levels should be used and the portion of the nitrate rich stems and Burley tobaccos in blends should be reduced.

REFERENCES

Adams, J.D., Lee, S.L., Vinchkoski, N., Castonguay, A., Hoffmann, D. (1983) Cancer Lett 17, 339-346.

Adams, J.D., Lee, S.J., Hoffmann, D. (1984) Carcinogenesis 5, 221-223.

Chamberlain, W.J., Bates, J.C., Chortyk, O.T., Stephenson, M.G. (1986) Tobacco Sci. 81, 38-39.

Fischer, S., Spiegelhalder, B., and Preussmann, R. (1989a) Carcinogenesis 10, 169-173.

Fischer, S., Spiegelhalder, B., and Preussmann, R. (1989b) Carcinogenesis 10, 1511-1517.

Fischer, S., Spiegelhalder, B., and Preussmann, R. (1989c) Carcinogenesis 10, 1059-1066.

Fischer, S. and Spiegelhalder, B. (1989) Beitr. Tabakforsch. 14, 145-153.

Fischer, S., Spiegelhalder, B., Eisenbarth, J., and Preussmann; R. (1990a) Carcinogenesis, 11, 723-730.

Fischer, S., Spiegelhalder, B., and Preussmann, R. (1990b) Arch. Geschwulstforsch., 60, 169-177.

Hoffmann, D., Dong, M., and Hecht, S.S. (1977) J. Natl. Canc. Inst. 58, 1841-1844.

Hoffmann, D. and Hecht, S.S. (1985) Cancer Res. 45, 935-944.

Effects of Nicotine on Biological Systems
Advances in Pharmacological Sciences
© Birkhäuser Verlag Basel

EXPOSURE TO NICOTINE-DERIVED N-NITROSAMINES FROM SMOKELESS
TOBACCO AND EVIDENCE AGAINST THEIR ENDOGENOUS FORMATION

A.R. Tricker and R. Preussmann

Institute of Toxicology and Chemotherapy, German Cancer
Research Center, Im Neuenheimer Feld 280, D-6900 Heidelberg

SUMMARY: The exposure to carcinogenic nicotine-derived
N-nitroso compounds from smokeless tobacco products is
confined to preformed N-nitroso compounds (exogenous exposure)
and not from the in vivo nitrosation of nicotine per se
(endogenous exposure).

The nitrosation of the tertiary amine nicotine (Fig. 1)
results in low yields of both N-nitrosonor-nicotine (NNN) and
4-(N-nitrosomethylamino)-1-(3-pyridyl)-1- butanone (NNK), two
potent organ-specific genotoxic carcinogens in experimental
animals for the nasal cavity and lung, respectively (Hecht &
Hoffmann, 1988). Under redox conditions,
4-(N-nitrosomethylamino)-4-(3-pyridyl)butyric acid (iso-NNAC),
4-(N-nitrosomethylamino)-1-(3-pyridyl)-1-butanol (NNAL) and
4-(N-nitrosomethylamino)-4-(3-pyridyl)-1-butanol (iso-NNAL)
are also formed (Brunnemann et al., 1987; Djordjevic et al.,
1989). Unlike nicotine, secondary amines undergo rapid
nitrosation giving high yields of N-nitroso compounds.
Nornicotine is readily nitrosated to yield only NNN. The
endogenous formation of N-nitroso compounds from secondary
amines has been reported in the oral cavity, gastrointestinal
and urinary tracts. The endogenous nitrosation of nicotine has
often been speculated. The presence of the nicotine-derived
N-nitroso compounds NNN, NNK and iso-NNAL in a variety of
smokeless tobacco products is reviewed and the exogenous
exposure to these compounds estimated. Experimental evidence
against the potential endogenous nitrosation of nicotine is
presented.

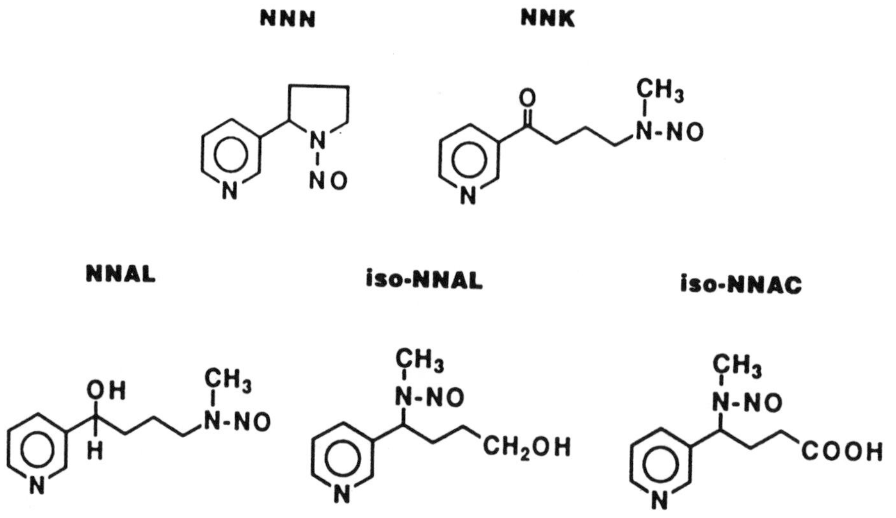

NNN **NNK**

NNAL **iso-NNAL** **iso-NNAC**

Figure 1. Nitrosation products of nicotine

MATERIALS AND METHODS

The potential formation and analysis of nicotine-derived
N-nitrosamines in artificial saliva and simulated gastric
juice was determined in vitro using our previously reported
methods (Tricker et al.,1988). Nicotine (99.5 % purity, Carl
Roth GmbH, Karlsruhe, FRG) was analysed and found not to
contain N-nitroso impurities. Fresh pooled human saliva (31
ml, pH 7.6, 6.4 ppm NO_2^-, 39.5 ppm NO_3^- and gastric juice (25
ml, pH 6.2, 0.6 ppm NO_2^-, 7.4 ppm NO_3^-) obtained by endoscopy
from volunteers with no current tobacco habits was maintained
at 37 ± 2°C. Aliquots (2 ml) of both saliva and gastric juice
were spiked with nicotine (1.0 mg) and incubated at 37°C for
1 hr. The pH of the reaction mixture was determined using a
glass electrode and corrected to pH 2.5, 4.0 and 7.0 (gastric
juice) and pH 2.5 (saliva), nitrite (1.0 mg) was added and the
reaction mixture incubated for a further hour prior to
analysis for nicotine-derived N-nitrosamines. Experiments were

repeated using heat denatured (110°C for 30 min) saliva and
gastric juice with suitable reagent blank controls.

RESULTS

The concentrations of preformed nicotine-derived nitrosamines
NNK, iso-NNAL and NNN (derived from both nitrosation of
nicotine and nornicotine) in a range of different smokeless
tobacco products are presented in Table I. The exogenous
exposure to these compounds occurring during smokeless tobacco
use for some tobacco-using populations was estimated by
multiplying the concentrations of individual compounds by the
average amount of tobacco consumed by the corresponding
population assuming that 100 % extraction into saliva occurs

Table I. N-Nitroso compounds in smokeless tobaccos

Tobacco type [number of samples] (country of origin)	Concentration range (µg/g)					
	NNK		iso-NNAL		NNN[1]	
	Mean	(Range)	Mean	(Range)	Mean	(Range)
Nasal snuff [5] (UK)	1.86	(0.97–4.32)	0.66	(ND[2]–1.59)	7.69	(3.00–15.7)
Nasal snuff [5] (FRG)	3.01	(0.58–6.43)	0.06	(ND–0.20)	7.98	(2.83–18.8)
Oral snuff [6] (UK)	2.80	(0.40–8.25)	0.06	(ND–0.15)	3.96	(1.09–7.63)
Oral snuff [6] (Sweden)	0.79	(0.40–1.04)	0.03	(ND–0.08)	3.36	(2.10–4.80)
Oral snuff [6] (India)	0.03	(0.13–0.60)	0.02	(ND–0.06)	0.86	(0.47–1.74)
Masheri [3] (India)	0.42	(0.11–0.96)	0.09	(0.07–0.12)	1.71	(0.26–3.75)
Kiwam [11] (India)	0.56	(0.10–1.03)	0.69	(0.16–1.86)	4.30	(2.50–8.95)
Zarda [11] (India)	4.03	(0.22–24.1)	1.42	(0.12–8.10)	13.4	(0.40–79.0)
Pan masala with zarda [7] (India)[3]	ND		ND		0.21	(0.09–0.33)
Naswar [3] (Pakistan)	0.61	(0.26–0.92)	0.02	(ND–0.07)	0.95	(0.38–2.07)

[1] Also derived from the nitrosation of nornicotine
[2] ND, not detected. Limit of detection 0.005 µg/g tobacco.
[3] Mixture of Areca (betel) nut and zarda tobacco

(Table II). Österdahl and Slorach (1988) have reported that
this type of calculation may underestimate the actual exposure
situation since endogenous nitrosation of secondary amino
groups may occur in the oral cavity.

Table II. Exposure to N-nitroso compounds from smokeless tobacco use.

Tobacco using population	Usage (g/day)	Daily nitrosamine exposure (µg/day)		
		NNK	iso-NNAL	NNN
Nasal snuff (UK)	2.0	3.7	1.32	15.4
Nasal snuff (FRG)	2.0	6.0	0.12	16.0
Oral snuff (UK)	4.5	5.6	0.12	7.9
Moist snuff (Sweden)	15.9	12.6	0.48	53.4
Zarda (India)	5.0	20.2	7.10	67.0

Experiments involving nitrosation in artificial saliva and
gastric juice showed that nicotine was not nitrosated.
Nitrosation of tobacco extracts (Fig. 2) yielded the

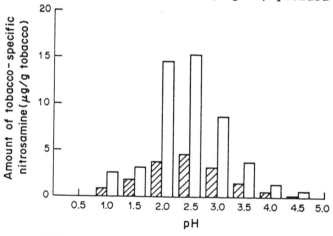

Figure 2. Effect of pH on the nitrosation of tobacco alkaloids
under simulated gastric conditions. Nitrosation resulted in
the formation of NAB/NAT ▨ , and NNN ☐ . Nitrosation of
nicotine to yield NNK and iso-NNAL was not observed.

tobacco-specific nitrosamines N-nitrosoanabasine (NAB) and N-nitrosoanatabine (NAT) as well as NNN. The latter is most probably formed from the nitrosation of nornicotine.

In vitro incubation of nicotine in saliva and gastric juice (prior to and after heat denaturation) followed by nitrosation over a range of pH values did not result in nitrosamine formation.

CONCLUSIONS

Nicotine per se is not nitrosated in vivo in both saliva or gastric juice. Furthermore, non-specific demethylases which may be present in saliva and gastric juice do not demethylate nicotine to produce nornicotine, an easily nitrosatable precursor to NNN. Exposure to the nicotine-derived N-nitroso compounds NNK and iso-NNAL (but not NNN) is limited to exogenous exposure from preformed N-nitroso compounds present in tobacco and is not increased by endogenous nitrosation of nicotine.

REFERENCES

Brunnemann, K.D., Genoble, L. and Hoffmann, D. (1987)
    Carcinogenesis, 8, 465-469.
Djordjevic, M.V., Brunnemann, K.D. and Hoffmann, D. (1989)
    Carcinogenesis, 10, 1725-1731.
Hecht, S.S. and Hoffmann, D. (1988) Carcinogenesis, 9,
    875-884.
Österdahl, B.-G. and Slorach, S. (1988) Food Add. Contam., 5,
    581-586.
Tricker, A.R., Haubner, R., Spiegelhalder, B. and Preussmann,
    R. (1988) Fd. Chem. Toxicol., 26, 861-865.

TOXICOKINETICS OF NICOTINE-DERIVED NITROSAMINES IN THE RAT

B.L. Pool-Zobel, U. Liegibel, R.G. Klein, P. Schmezer

Institute for Toxicology and Chemotherapy, German Cancer
Research Center, Im Neuenheimer Feld 280, 6900 Heidelberg, F.R.G.

SUMMARY: It was the aim of this study to elucidate in which
manner active metabolites of tobacco specific nitrosamines are
distributed systemically. The experimental approach was to
assess genotoxic effects by 4-(N-methyl-N-nitrosamino)-1-(3-
pyridyl)-1-butanone (NNK) and N-nitrosonornicotine (NNN). For
this, induced DNA single strand breaks (SSB) were determined by
two different methods in isolated cells of different organs
following single p.o. application and 1 h exposure to the test
chemicals. So far, with the alkaline elution technique it was
found that NNK induces SSB beginning at 25 mg/kg in the liver.
NNN was not genotoxic in the liver up to 100 mg /kg, and the
compound was only moderately positive in doses up to 500 mg/kg.
No genotoxicity was observed in testes, or thymus by either
compound. First results with a single cell analysis of DNA
damage by in situ nick translation indicate both the compounds
to be genotoxic in esophageal cells.

INTRODUCTION :

The two tobacco specific nitrosamines NNK (4-(N-methyl-N-
nitrosamino)-1-(3-pyridyl)-1-butanone) and NNN (nitrosonor-
nicotine) are potent carcinogens which may induce tumors locally
or systemically (rev. Hecht & Hoffman, 1989; Hoffman and Hecht,
1990). At the present state of knowledge, it is desirable to
study the toxicokinetics of these carcinogens in animal models
and to compare the data to human dosimetry studies. These
analyses should aid in determining in which manner the active
metabolites of NNN and NNK contribute to the human systemic
carcinogenic properties of tobacco related products. So far,
systemic genotoxicity has mainly been studied by detecting

adducts which result from methylation of macromolecules and a good correlation between afflicted tumor organs and distribution of methylated DNA adducts has been found for NNK (Hecht et al., 1986; rev. Hoffman and Hecht, 1990). Methods to detect adducts which result from the 4-(3-pyridyl)-4-oxobutyl diazohydroxide-intermediates, however, are not yet available. This study was directed at determining whether an alternative experimental approach, which was to assess toxic and genotoxic effects by NNK and NNN in different organs of the rat, could be used for analysing the toxicokinetics of the nicotine derived nitrosamines. For this, induced DNA single strand breaks (SSB) were first determined in hepatocytes *in vitro* (Sina et al., 1983). Then SSB were assessed in isolated cells of target and non targed following single *p.o.* application and 1 h exposure to the test chemicals (Pool et al., 1990).

MATERIALS AND METHODS:

Isolation of primary cells: Primary rat hepatocytes were isolated from male Sprague Dawley rats with a two step *in situ* perfusion method (Berry and Friend, 1969). The liberated cells were centrifuged at 30 x g for purification. For the isolation of the other cells, the organs were removed from the animal, freed from the surrounding tissue and minced. The oesophagus was filled with collagenase solution and incubated to ease the liberation of the epithelium prior to mincing. The obtained suspensions were filtered and centrifuged at 800 r.p.m. for 5 minutes at 4^oC. All pellets containing the isolated cells were resuspended in a Ca^{++}, Mg^{++}-free buffer and adjusted to 2 x 10^6 cells per ml.

Exposure to N-nitroso-compounds: The hepatocytes are treated *in vitro* by incubating a 1 ml suspension culture containing 5- 20 μl of the test compound solution in DMSO for 1 h in a shaking water bath at 37° C. After the incubation the cells are centrifuged to pellets and resuspended in FCS free buffer for

determination of viability by trypan blue exclusion and DNA SSB
as described below.

For the *ex vivo* analysis 1-2 animals are treated with the
compound in physiol. NaCl or corn oil by oral gavage. After 1 h
exposure, cells of different organs are isolated and viability
and genotoxicity is determined with the isolated cells.

Determination of DNA SSB: SSB were determined with the method of
alkaline filter elution (*AE*) of Kohn et al. (1981).   As has
been specified previously, 0.75 ml aliquotes of the 1ml
suspension cultures are lysed on polycarbonate filters.   The DNA
is eluted at alkaline pH and the DNA in the resulting fractions
is determined fluorimetrically (Pool et al., 1990;   Schmezer et
al., 1990).   The % DNA retained on filters after the elution
(control-test) is calculated for evaluation of the results.

With the second technique, *in situ* nick translation (*NT*) may be
utilized to visualize DNA damage (`nicks`) microscopically (Anai
et al., 1988).   For this, the cells are fixed on slides, treated
with lysolecithine and alkaline buffer and then incubated for 10
minutes at 21°C with E. coli DNA polymerase I, ATP, CTP, GTP,
and [^{3}H]TTP. Incorporation of nucleoside triphosphates into DNA
is detected by autoradiography.

RESULTS AND DISCUSSION:

a. *in vitro* effects in primary rat hepatocytes: These studies
were performed to elucidate to which extent the chosen end
points of DNA damage could be induced by NNK and NNN.   Table I
shows the representative results of an individual reproduced
experiment, in which both detection methods were performed with
the same cell batches.

The results of the *AE* method show clearly that NNK and NNN are
both inducers of DNA SSB in hepatocytes, whereby the potency of
NNK is higher than of NNN. The effects of NNK are comparable to
those of N-nitrosodimethylamine (NDMA) which induces SSB at

B. L. Pool-Zobel et al.

similar potency in the concentration range of 6.25 to 50 μmoles per ml (Schmezer et al., 1990). Whereas NNN generates only 4-(3-pyridyl)-4-oxobutyl diazohydroxide, NNK additionally yields methyldiazohydroxide (rev,. Hecht and Hoffman, 1989).

Table I: Assessment of genotoxic effects by NNK und NNN in primary rat hepatocytes *in vitro*, Comparison of induced DNA SSB as detected by *AE* (C-T) and by *in situ NT* (T/C).

μmoles/ml	% V abs/rel	% DNA retained	(C-T)	Silver grains/ 100 Cells	(T/C)
0	60/100	73±1		5.9±1	
NNK					
6.25	56/93	42±3	31	5.1±1	0.9
12.5	59/98	22±5	51	8.2±2	1.4
18.75	47/78	14±2	59		
25	39/65	13±2	60	8.0±3	1.4
NNN					
6.25	58/91	69±7	4	8.4±2	1.4
12.5	47/78	56±8	17	7.1±2	1.2
18.75	43/72	52±9	21		
25	40/67	49±10	24	6.2±1	1.1
PC	51/85	5±1	68	100	17

%V, % of viable cells, determined by trypan blue exclusion; abs/rel, absolute % of viable cells/relative, % viable cells based on 100 % viable in control; *Mean and standard deviation of 3 determinations / concentration; (C-T), % DNA retained on filter in control groups minus % DNA retained on filters in treated groups; (C-T)> 20 is evaluated as positive; (T/C), Number of silver grains per 100 cells test/control. A positive effect is present at >2. The values were obtained from 3 different slides, 100 cells each. PC, positive controls were 2.5 μmoles N-Nitrosomethylbenzylamin in the *AE* and 20μg DNAase in the *in situ* Nicktranslation methods.

Therefore, when assuming that all three compounds are metabolized at roughly similar rates by the hepatocytes, the higher responses in DNA SSB may be due to the methylating properties of NNK and NDMA. End point specificity may also be the cause of the non- responsiveness of NNN and NNK when using *in situ* nick translation. This method seems relatively insensitive to the genotoxic activities of these nitrosamines. However the technique is still very new, and more studies will have to be performed to elucidate whether an enhanced susceptibility for the genotoxic action of these nitrosamines can be achieved.

b. Genotoxic effects of NNK and NNN *in vivo*: Genotoxic effects induced by the compounds after their oral application to the rats are also readily detectable in the liver cells with the *AE* method. Table II shows the results obtained so far.

Table II: Toxicokinetics of NNN and NNK 1h after oral application of the compounds to Sprague-Dawley rats. Detection of induced DNA SSB by the *AE* and in situ *NT* methods are compared (preliminary results with single and mean values of 3-6 determinations from 1-3 animals per dose).

% LD$_{50}$	mg /kg	μmoles /kg	Leber AE/NT	Testes AE/NT	Thymus AE/NT	Esophagus AE/NT
NNK						
*	25	121	++/±			
*	35	169	++/±			
*	50	242	++/			
*	100	483	++/	-/	-/	
*	250	1208	++/-			/+
*	500	2415	+/±	-/	-/	/±
NNN						
2.5	25	141	-/+			
3.5	35	198	-/±			
5	50	282	±/±	-/-	-/-	
10	100	565	+/±	-/	-/	/±
25	250	1412	+/			
50	500	2825	+/+	-/	-/	/+

AE, alkaline filter elution, (C-T) <10 (-); 10-19 (±); 20-39 (+); 40-59 (++); ≥60 (+++); NT, *in situ* nick translation, (T/C) <1.19 (-); 1.2-1.9 (±) 2.0-4.9 (+); 5.0-20 (++); >20-99 (+++); *Oral LD$_{50}$ is not available, doses were chosen analogous to those of NNN-

In vivo NNK is again more potent than NNN, when regarding the results obtained with *AE*. In this case, the effects may be correlated to carcinogenic potencies and target organ susceptibilities, where NNK but not NNN has been shown to induce tumors of the liver. Also no genotoxicity is observed in the non-target organs of carcinogenesis, thymus and testes by either NNN or NNK. Similar results have been found previously for the strong liver carcinogen NDMA which under certain conditions may induce tumors of lung and kidney. *Ex vivo*, NDMA is a potent inducer of DNA SSB in liver (>0.1 mg/kg), and 20 fold higer doses (>2 mg/kg) are needed to induce DNA damage in lung, kidney (Pool et al., 1990). No genotoxicity by NDMA was observed in testes or thymus. NNK induces SSB at 25 mg/kg in the liver and lower doses are being assessed. 25-50 mg/kg NNN were not genotoxic in the liver. This dose range is only 2.5-5% of its reported LD_{50}, but is equitoxic to the extremely genotoxic dose range of 1-2 mg/kg NDMA (oral LD_{50} in rats is 40 mg/kg). Accordingly NNK, but even more so NNN are less potent genotoxins *in vivo* than NDMA.

An important result of Table II is that both compounds were genotoxic in the esophagus using the new *in situ NT* technique. This is significant for the general development of methods in genetic toxicolgy, since techniques for detecting DNA changes in esophageal cells are extremely rare. Efficient methods to monitor DNA damage in single cells of remote organs (e.g. esophagus-, urinary bladder- and nasal cavity epithelia) are highly needed (Pool et al., 1990), and the elegant method of *in situ NT* is a promising approach. The second significance of the findings is the correlatability to target organs in carcinogensesis, since NNN is carcinogenic in the esophagus, although NNK is apparently not. However, when comparing the results of these single doses and one hour exposure it must be born in mind that most carcinogenicity experiments are performed with repetetive dosing and life long exposures. Therefore, it is not expected that tumors must occur in all organs in which DNA damage is detectable. However those organs in which DNA damage is detectable must be considered to be more susceptible for tumorigenesis than those organs in which DNA damage is not detectable. We are presently studying the extent of genotoxicity in other target organs of these nitrosamines (pancreas, nasal cavity and lung) to ensure that susceptible organs are properly identifiable. Accordingly, we should have a rapid, efficient and animal saving approach to study numerous practical problems of environmental carcinogenesis (e.g. dosimetry after different exposure routes and concentrations, influence of diet, combination effects with other smoke components, possibilities of chemoprevention etc.). This type of approach in toxicokinetics may aid not only in elucidating the role of individual compounds of the complex carcinogenic mixtures tobacco and tobacco smoke, but may also serve as a useful model system for studying specific mechanisms.

Acknowledgements: We are very grateful to Ms. Michaela Fröstel, Mr. Reinhard Gliniorz and Ms. Roswitha Zimmermann for their excellent technical assistance. We thank the Research Council on Smoking and Health for financing these studies.

REFERENCES

Anai, H. Meahara, Y., Suginachi, K. (1988) Cancer Letters 40, 33-38.
Berry , M.N., Friend, D.S. (1969) J. Cell biology 43, 506-520
Hecht, S.S., Hoffman, D. (1989) Cancer Surveys 8, 273-295.
Hecht, S.S., Trushin, N., Castonguay, A., Rivenson, A. (1986) Cancer Research 46, 498-502.
Hoffman, D., Hecht, S. (1990) In: Chemical Carcinogenesis and Mutagenesis I, (C. S. Cooper and P.L. Grover, eds.) Springer, Berlin, Heidelberg, New York, London, Paris, Tokyo, Hong Kong, pp. 63-102.
Kohn, K.W., Ewing, R.A:G:, Erickson, L.C. Zwelling, L.A., (1981) In: Friedberg, E.C. and Hanawalt, P.C. (eds) DNA repair, Lab.- Manual, Marcel Decker Inc., New York pp 379-401.
Pool, B.L., Brendler,S.Y., Liegibel, U.M., Tompa,A., Schmezer, P. (1990) Environm. Mol. Mutagen. 15, 24-35 .
Schmezer, P., R. Preussmann, D. Schmähl, Pool, B.L. (1990) Mutation Research, in press.
Sina, J.F., Bean, C.L., Dysart, G.R., Taylor, V.I., Bradley M.O. (1983) Mutation Research 113, 357-391.

Effects of Nicotine on Biological Systems
Advances in Pharmacological Sciences
© Birkhäuser Verlag Basel

INTESTINAL FIRST PASS METABOLISM OF NNK IN THE MOUSE

J. Schulze, G. Hunder and E. Richter

Walther Straub-Institut für Pharmakologie und Toxikologie, Ludwig-
Maximilians-Universität München, Nussbaumstr. 26, D-8000 München 2
Federal Republic of Germany

SUMMARY: The tobacco specific nitrosamine 4-(N-methyl-N-nitrosamino)-1-(3-pyridyl)-1-butanone, NNK, undergoes a high first-pass metabolism in isolated small intestinal segments of mice which was
further stimulated significantly by starvation and/or acetone pretreatment. Three major metabolites were tentatively identified by
cochromatography with authentical reference compounds as 4-(methylnitrosamino)-1-(3-pyridyl)-1-butanol, 4-oxo-4-(3-pyridyl)-1-
butanol and 4-oxo-4-(3-pyridyl)butyric acid.

NNK, a tobacco specific nitrosamine, has been shown to be a potent
carcinogen inducing tumors of the lung in mice (Hecht & Hoffmann,
1988). The metabolism of NNK is known to occur by carbonyl reduction, N-oxidation, or by α-hydroxylation either at the methyl or
N-methylene carbon (Hecht et al., 1980). In Clara cells from rat
lung the in vitro formation of 4-(methylnitrosamino)-1-(3-pyri-
dyl)-1butanol, NNAL, has been shown to be one order of magnitude
higher than the formation of NNK-N-oxide and products of α-hydroxylation, 4-oxo-4-(3-pyridyl)-1-butanol, keto alcohol, and 4-oxo-
4-(3pyridyl)butyric acid, keto acid (Belinsky et al., 1989).

In a series of studies we have previously shown a high capacity
of the small intestine in metabolizing dialkyl nitrosamines (Rich-

ter et al., 1986; Schulze & Richter, 1989; Feng & Richter, 1989).
Although the small intestine is the main site of nitrosamine uptake in smokeless tobacco users and an important one even in smokers, nothing is yet known about intestinal first pass metabolism
of NNK.

MATERIALS AND METHODS

Animals: Male NMRI mice, 25 - 30 g b.w. (Interfauna, Tuttlingen,
FRG) were either fed ad libitum or starved for 24 h. An additional
group of mice was starved for 48 h and given acetone, 5 ml/kg,
24 h before the experiment.

Chemicals: 4-(N-methyl-N-nitrosamino)-1-(3-pyridyl)-1-[carbonyl-
^{14}C]butanone, [^{14}C]NNK, spec.act. 29 mCi/mmol was from Chemsyn
Sci.Lab. (Lenexa, KS, USA). Possible NNK metabolites were a gift
from D. Hoffmann (American Health Foundation, Valhalla, NY, USA).
Biochemical reagents were supplied by Sigma (Taufkirchen, FRG).
All other chemicals used were from Merck (Darmstadt, FRG).

Intestinal perfusion: Jejunal segments of about 10 cm length were
reperfused for 2 h in an all glass perfusator (Richter & Strugala,
1985) with a phosphate buffer containing 15 mM glucose and 1 μM
[^{14}C]NNK. Transfer of glucose and water was determined to control
the viability of the segments. Samples of absorbate and perfusate
were taken for determination of glucose and total [^{14}C] and depro-
teinized for HPLC analysis by addition of TCA.

Analytical procedure: [^{14}C]NNK and its metabolites were separated
in samples of absorbate and perfusate by reversed phase HPLC
(LiChroCART 250-4, RP18, 5μ, Merck) with gradient from 10 to 80%
methanol in buffer pH 7. Detection was by radioactivity monitoring
(Ramona, Raytest, Straubenhardt, FRG). Retention times of reference compounds were determined by UV-spectrophotometry at 230 nm and
270 nm (UVD 160-2, Gynkotek, Germering, FRG).

RESULTS

Isolated perfused jejunal segments of mice concentrated total ^{14}C in absorbate about 1.5 fold. This small but significant accumulation of radioactivity was further increased by starvation with or without acetone pretreatment (Table 1).

Table 1: Transfer of water, glucose and ^{14}C in jejunal segments of mice perfused for 2 h with 1 μM [14C]NNK. Values represent means ± S.E. of 8 experiments.

Treatment	Water absorption (ml/cm)	S/M[a] Glucose	Total ^{14}C
Fed	1.74 ± 0.27	9.1 ± 0.3	1.41 ± 0.05
Starved	1.80 ± 0.11	10.9 ± 0.6[b]	1.81 ± 0.08[b]
Starved + acetone	2.25 ± 0.33	11.4 ± 0.7[b]	1.73 ± 0.13[b]

[a]Serosal to mucosal concentration ratio of glucose and total radioactivity, respectively
[b]Significantly different from fed animals, $p < 0.05$

After 2 h perfusion more than 95% of the ^{14}C in perfusates was unmetabolized NNK. In contrast, a high percentage of metabolites of about 60% was observed in absorbates from jejunal segments of mice fed ad libitum (Figure 1). There were three major metabolites separated by HPLC which coeluted with **keto acid**, **keto alcohol** and **NNAL** (Table 2). Further analysis of pooled samples of absorbate by a more slowly increasing gradient revealed additional small metabolite peaks which were tentatively identified as 4-hydroxy-4-(3-pyridyl)butyric acid, **hydroxy acid**, **NNAL-N-oxide** and 4-hydroxy-4-(3-pyridyl)-1-butanol, **diol**, and/or **NNK-N-oxide**.

In mice starved for 24 h the jejunal metabolism of NNK increased to 75%. This was mainly due to an increase in **keto acid** (Table 2). Starvation for 48 h with additional acetone pretreatment did not result in a further increase of **keto acid** formation. However, the relation of the two major metabolites besides the keto acid was changed in favor of the **keto alcohol** (Table 2).

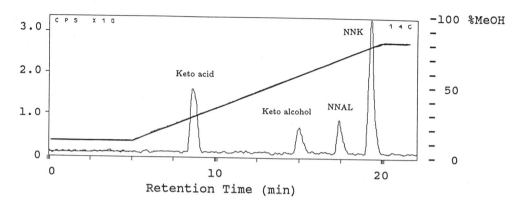

Figure 1: HPLC radiochromatogram of absorbate from a mouse jejunal
segment perfused for 2 h with 1 μM [14C]NNK.

Table 2: Metabolism of NNK in mouse jejunal segments perfused for
2 h with 1 μM [14C]NNK.
Values represent means ± S.E. of 8 segments.

Treatment	Percentage of metabolites in absorbate			
	keto acid	keto alcohol	NNAL	Total[a]
Fed	28.2±2.8	12.7±1.5	12.2±1.1	58.8±2.7
Starved	42.0±1.9[b]	16.9±1.1[bc]	8.8±1.0[bc]	74.8±2.2[b]
Starved + acetone	45.1±3.4[b]	25.1±1.9[b]	4.0±0.4[b]	76.1±5.2[b]

[a]Including minor metabolites
[b]Significantly different from fed animals, p < 0.05
[c]Significantly different from acetone treated animals, p < 0.05

DISCUSSION

Jejunal segments of mice are very effective in metabolizing NNK.
Two of the major metabolites identified are consistent with a high
capacity of small intestine in α-hydroxylation of NNK at the N-methyl carbon leading to **keto alcohol** and/or the N-methylene carbon
leading to **keto acid** (Hecht et al., 1980). Because of the possible
further metabolism of **keto alcohol** to **keto acid** the relative importance of these two possible pathways cannot be answered from
this experiment. Previous studies with N-nitrosodibutylamine have

shown a high capacity of small intestine not only in ω-hydroxyla-
tion but also in further metabolism of the hydroxide to the carbo-
xylic acid (Richter et al., 1988). Therefore the conversion of keto alcohol to keto acid by small intestine cannot be excluded.

The increase of α-hydroxylation in fasted mice may result from
the induction of a specific isozyme of cytochrome P450 (P450IIE1)
which has been implicated in the α-hydroxylation of N-nitrosodimethylamine, whereas its role in NNK metabolism is not yet clear
(Yang et al., 1990). To our knowledge neither the presence nor the
inducibility of P450IIE1 has been investigated in small intestine.
The further increase iń keto alcohol formation after prolonged
starvation with additional acetone pretreatment could be taken as
a proof of the hypothesis that cytochrome P450IIE1 might be involved in α-hydroxylation of NNK at the N-methyl carbon.

NNAL, the major metabolite formed in liver (Yang et al., 1990)
and in target cells of rat lung (Belinsky et al., 1989), was only
a minor metabolite in absorbates from mouse intestine. Its formation was reduced after fasting and/or acetone pretreatment.

CONCLUSION: The capacity of small intestine in α-hydroxylation of
NNK can be considered as a detoxification pathway because the reduced uptake of the carcinogenic nitrosamines NNK and NNAL.

Acknowledgements: This study was supported by a grant from the
Forschungsrat Rauchen und Gesundheit. We are grateful to D. Hoffmann for the gift of reference compounds of NNK metabolites.

REFERENCES

Belinsky, S.A., White, C.M., Trushin, N., and Hecht, S.S. (1989)
    Carcinogenesis 10, 2269-2274
Feng, X., and Richter, E. (1989) Arch.Toxicol.Suppl. 13, 227-229.
Hecht, S.S., and Hoffmann, D. (1988) Carcinogenesis 9, 875-884.
Hecht, S.S., Young, R., and Chen, C.B. (1980) Cancer Res. 40,
    4144-4150.
Richter, E., Richter-Cooberg, U., Feng, X., Schulze, J., and
    Wiessler, M. (1986) Carcinogenesis 7, 1207-1213.
Richter, E., and Strugala, G.J. (1985) J.Pharm.Meth. 14, 297-304.
Richter, E., Zwickenpflug, W., and Wiessler, M. (1988) Carcinoge-
    nesis 9, 499-506.
Schulze, J., and Richter, E. (1989) Progr.Pharmacol.Clin.Pharma-
    col. 7, 148-154.
Yang, C.S., Smith, T., Ishizaki, H., Yoo, J.S.H., and Hong, J-Y.
    (1990) In: N-Oxidation of Drugs (P. Hlavica, L.A. Damani and
    J.W. Gorrod, Eds), Chapman & Hall Ltd., London, in press

III.

General Pharmacology of Nicotine

EFFECTS OF NICOTINE ON INSULIN: ACTIONS AND IMPLICATIONS

Neil E. Grunberg and Margarita Raygada

Medical Psychology Department
Uniformed Services University of the Health Sciences
4301 Jones Bridge Road
Bethesda, Maryland 20814-4799

SUMMARY: In 1988, we reported that nicotine decreases plasma levels of insulin in a dose-response fashion. More specifically, two weeks of continuous administration by osmotic minipump of 6 mg and 12 mg nicotine/kg body weight/day to rats resulted in significant decreases in plasma insulin. Recently, we found that plasma insulin increases after cessation of this type of nicotine administration, slightly overshooting control values for a few days. We have continued to explore the effects of nicotine on insulin. This paper reviews our findings regarding nicotine's effects on insulin and presents data indicating that nicotine administration somewhat decreases pancreatic levels of insulin and slightly increases levels of insulin in the hypothalamus. The implications of these findings for nicotine's effects on body weight, specific food preferences, energy expenditure, and nicotine reinforcement are discussed.

INTRODUCTION

It is now well documented that cigarette smoking and body weight are inversely related. Smokers weigh less than comparably aged, same sex nonsmokers, and most smokers who quit smoking gain weight

(Grunberg, 1986; 1990; USDHHS, 1988, 1990). The relationship between smoking and body weight has practical and basic science implications. Clinically, many smokers, particularly women, report that a major reason that they smoke is to control body weight and that concern about body weight gain is a leading reason to continue to smoke (Klesges & Klesges, 1988). If the mechanisms that control the smoking-body weight relationship could be elucidated, then ways to avoid unwanted weight gain after cessation of smoking might be developed. In turn, smoking cessation might be increased and relapse partially prevented. In addition to these pragmatic concerns, it is important to determine the mechanisms underlying the smoking-body weight relationship to better understand the reinforcing actions of smoking.

This paper reviews a series of studies designed to determine the role of insulin in the smoking-body weight relationship. In addition, the implications of these findings are discussed.

BACKGROUND

In a series of human and animal experiments, we determined that: (a) the relationship between cigarette smoking and body weight is based on actions of nicotine (Grunberg, 1986, 1990); (b) nicotine acts on energy intake by decreasing consumption of sweet-tasting, high carbohydrate foods (Grunberg, 1982; Grunberg, Bowen & Morse, 1984); (c) nicotine additionally increases energy expenditure (Grunberg et al., 1984; Grunberg, 1986); (d) cessation of nicotine results in a reversal of the energy intake and expenditure effects of nicotine (Grunberg, 1986, 1990). As part of this line of research, we have found that nicotine alters insulin levels in the body in ways that are relevant to energy expenditure, energy intake, and body weight. This paper reviews our series of experiments about nicotine and insulin that follow directly from our presentation at the 1987 International Symposium on Nicotine (Grunberg, 1988).

NICOTINE AND INSULIN

Establishing a role for insulin. Studying nicotine and body weight, we have found Alzet miniosmotic pumps to be a useful and reliable way to administer nicotine to rats and produce results consistent with studies of human smokers (see Grunberg, 1982, 1988, for a detailed discussion of this paradigm). Basically, we subcutaneously implant miniosmotic pumps filled with saline or various dosages of nicotine in rats and measure behavioral and biological variables before, during, and after drug administration. We typically use pumps that administer nicotine or saline for two weeks. This technique allows chronic nicotine administration in animals without the stress of repeated injections.

In 1988, we reported two experiments designed to study the effects of nicotine on plasma insulin using this animal model (Grunberg et al., 1988). In the first experiment, subjects were 108 Sprague-Dawley rats that weighed about 400 g at the beginning of the study. Animals were individually housed in a controlled environment of 22° C, 50% relative humidity, and a 12 hour light/dark cycle. Standard rat chow and tap water were continuously available.

After a gentling period, animals were anesthetized by IP injections of sodium pentobarbital and an Alzet miniosmotic pump (Model 2002, Alza Corp., Palo Alto, CA) was implanted SC in each rat between the shoulders. There were 36 animals in each of three groups: 12 mg nicotine/kg/day, 6 mg nicotine/kg/day, or saline. In addition to pump implantation, catheters were implanted in the femoral artery of half of the animals in each treatment group. Daily blood samples (3 ml) were drawn from these subjects during the first week of drug administration. After one week, catheters were implanted in the subjects that had not yet had blood samples drawn and daily blood samples were collected from these subjects during the second week. Samples were centrifuged to separate plasma that was stored at -70° C until later assay. Plasma samples were assayed by radioimmunoassay for insulin, by glucose oxidase reaction procedure for glucose, and by a radioenzymatic procedure for epinephrine and norepinephrine.

During the first week of drug infusion, all groups had similar insulin levels. In contrast, during the second week of drug infusion, there was a significant inverse dose-effect relationship between nicotine and insulin levels. In this chronic nicotine administration paradigm, there was no difference among groups in plasma glucose values after one and two weeks. Epinephrine and norepinephrine were significantly elevated during week 1, but were similar to saline controls during the second week. To confirm the insulin and glucose findings, we repeated the study during the second week in 36 additional animals, 12 in each of the three treatment groups. Again, we. found a significant effect of nicotine to reduce plasma insulin but not to alter plasma glucose in this paradigm at the end of two weeks.

We interpreted these results to indicate that the effects of chronic nicotine on body weight may act partially through effects of nicotine on insulin. Insulin, a small protein (molecular weight 5808 for human insulin), is composed of two amino acid chains connected by disulfide linkages. For insulin to exert its function, it must attach to cell membranes. In the periphery, insulin enhances rate of glucose metabolism, decreases blood glucose concentration, and increases glycogen stores in the tissues. The effects of insulin on fat storage and utilization are especially relevant to the nicotine-body weight relationship. Under the influence of insulin, and when an excess of glucose is available simultaneously, a great amount of the glucose that enters the body is converted in the liver into fat. In addition, insulin strongly enhances the transport of glucose into the fat cells. The presence of excess glucose inside the fat cells promotes fat storage. Therefore, one of the most rapid and most potent effects of insulin is to promote fat storage in adipose tissue. In the absence of insulin, fat is not stored in fat cells and it is released in the form of free fatty acids. The decreased plasma insulin in response to nicotine in a paradigm that produces decreased body weight is consistent with the interpretation that nicotine acts to decrease insulin which, in turn, decreases body weight via energy expenditure.

In addition to the effects of insulin on energy expenditure, the effect of nicotine to decrease insulin may also help to explain the effects of nicotine to decrease consumption of sweet-tasting foods. This interpretation is consistent with reports that hyperinsulinemia results in increased ratings of pleasantness of sweet solutions (Rodin, Wack, Ferranninia, & Defronzo, 1985). In addition, it has been demonstrated that after ingestion of glucose, humans report that formerly pleasant sweet solutions become unpleasant (Cabanac, 1971). In contrast, Jacobs (1958) and Briese and Quijada (1979) reported that insulin injections to rats resulted in increased preference for sweet foods. Effects of nicotine on sweet preferences, therefore, may be mediated through actions of insulin, or actions of insulin on glucose.

Nicotine cessation and insulin. The preceding studies of nicotine and insulin indicated that plasma insulin decreases with nicotine administration, but provided no data regarding effects of nicotine cessation. Therefore, we designed an experiment to directly examine the effects of cessation of chronic nicotine administration (Grunberg, Raygada, Popp, Nespor, Sibolboro, & Winders, 1988).

Subjects were 144 Sprague-Dawley rats weighing 300-350 g at the beginning of the study. Subjects were individually housed under conditions identical to our previous studies of nicotine and insulin with bland food and water continuously available. Alzet miniosmotic pumps (Model 2002) were implanted SC to deliver saline, 6 mg nicotine/kg/day, or 12 mg nicotine/kg/day. Two weeks after pump implantation, subjects were anesthetized with sodium pentobarbital, minipumps were removed, and catheters were implanted in the femoral artery of 72 of the animals. Daily blood samples were collected from these animals for approximately 5 days. Samples were stored at -70° C until later assay. Seven days after cessation of drug administration, all 72 animals were anesthetized by IP injection of sodium pentobarbital and blood samples were taken from the abdominal aorta. The other 72 animals were divided into two groups of 36 rats (12 from each treatment group) that were anesthetized as above, and blood samples were drawn from the abdominal aorta either one day or seven days after cessation of nicotine.

In the catheterized animals, plasma insulin values for the three treatment groups were similar on the first two days after cessation. In contrast, beginning on the third day after cessation, plasma insulin levels were higher in the groups that had received nicotine compared to saline controls. The mean insulin values for the 12 mg nicotine group were 135%, 133%, and 167% of controls on cessation days 3, 4, and 5, respectively. The mean insulin values for the 6 mg nicotine group were 120%, 140%, and 174% of saline controls on cessation days 3, 4, and 5, respectively. On day 7 after cessation of nicotine or saline, blood samples were gathered by aortic stick from the catheterized animals and were assayed for insulin. Based on this analysis, the group that had received 12 mg of nicotine was 114% of controls, but the 6 mg group was 89% of controls. The 72 animals that were not catheterized had blood drawn by aortic stick on day 1 or 7 after cessation of nicotine or saline. Based on these subjects, plasma insulin were elevated (approximately 108% of controls) at days 1 and 7 for the 12 mg and 6 mg nicotine cessation groups.

Glucose was suppressed 13% (i.e., 87% of controls) in the catheterized 12 mg nicotine cessation animals on the first day after cessation of drug administration. These differences disappeared by day 3 after cessation. Based on the samples drawn by aortic stick, glucose in nicotine cessation animals was 93% of controls on day 1 after cessation and was attenuated to 96% of controls by day 7.

The catheter preparation allowed blood draws with minimal stress to the animals. Therefore, we believe that the values obtained from this preparation, rather than by aortic stick, are more indicative of the in situ biochemical status. Considering these studies together: (1) nicotine clearly decreases plasma insulin, and (2) nicotine cessation results in an increase in plasma insulin levels that somewhat overshoots controls; (3) the nicotine suppression of insulin develops over time and continues.

How does nicotine alter insulin? Although results of the preceding studies of nicotine and insulin are consistent with the effects of nicotine on body weight and specific food preferences, these studies do not establish how nicotine alters insulin levels in the periphery. Nicotine may affect production, secretion, metabolism, elimination, or binding of insulin. Because insulin is manufactured in the islets of Langerhans in the pancreas, the levels of insulin in the pancreas are an indicant of this production. Therefore, we performed a study to measure insulin levels in the pancreas as well as in the plasma (Raygada, Nespor and Grunberg, 1990). To further our investigation of whether the effects of nicotine on plasma insulin underlie the effects of nicotine on body weight and food consumption, we included a low nicotine dose treatment group that should not alter body weight and food consumption. If insulin values were reduced at this lower dosage, then our argument regarding insulin's role would be weakened. If the lower dosage group showed no changes in any of these variables, then our interpretation of previous results would hold.

Subjects were 60 male Sprague-Dawley rats weighing 300-350 grams at the beginning of the study. Animals were maintained under identical housing conditions as in the previous studies except that the chow was sweetened ground rat chow (60% rat show, 40% sugar). Alzet pumps were implanted to deliver 12 mg, 6 mg, 1.5 mg nicotine/kg/day, or saline to groups of 12 animals. Body weight and eating behavior were measured daily for 6 days before drug administration and for 14 days during drug administration. Then, animals were sacrificed, trunk blood was collected, and the pancreas was removed from each animal. Samples were later assayed for insulin.

As in previous studies (cf. Grunberg, 1986, 1990), animals receiving 6 mg or 12 mg nicotine/kg/day had significantly less weight gain and ate significantly less sweet food than did control animals. The animals receiving 1.5 mg nicotine/kg/day were similar to saline controls on body weight and food consumption. With regard to plasma insulin 2 weeks into drug administration, the controls and 1.5 mg nicotine group were similar, and the 6 mg and 12 mg groups were significantly decreased to 77% and 72% of controls, respectively. These results are consistent with the interpretation that decreased plasma insulin is associated with effects of nicotine on body weight and food consumption. When nicotine dosage (1.5 mg/kg/day) was administered that does not alter body weight or food intake, there was no change in plasma insulin.

The assay of pancreatic insulin revealed lower values for the 6 mg and 12 mg nicotine groups (74% and 88% of controls, respectively). This finding suggests that nicotine decreases plasma insulin partially by decreasing production of insulin in the pancreas. Whereas the plasma insulin decrease was dose-dependent, the pancreatic decrease was not. Therefore, the effect of nicotine to decrease plasma insulin is only partially mediated by a decreased production of insulin in the pancreas. There must be an additional "downstream" mechanism to decrease plasma insulin.

Effects of nicotine on insulin in the brain. Based on the available data, nicotine decreases circulating insulin which, in turn, should increase lipolysis, decrease lipogenesis, and decrease consumption of sweet foods. After cessation of nicotine, this process reverses and somewhat overshoots baselines. Nicotine's effects on insulin are mediated partially via pancreatic production of insulin. Exactly how this phenomenon works and what additional mechanisms contribute to the nicotine-plasma insulin relationship are not known.

The effects of nicotine mediated by insulin may operate elsewhere besides or in addition to the periphery. Insulin is present in the brain (Baskin et al., 1987), and it has been suggested that insulin in the brain decreases food intake and is related to body weight (Woods & Porte, 1983). Therefore, we designed a study to examine the effects of nicotine administration and cessation on insulin levels in the hypothalamus (Raygada, Nespor, & Grunberg, 1989), the area of the brain most relevant to body weight control.

Subjects were 113 male Sprague-Dawley rats weighing 300-350 mg at the beginning of the study. Housing conditions were identical to previous studies. Standard rat chow and water were continuously available. Alzet miniosmotic pumps were implanted SC to deliver saline, 6 mg, or 12 mg nicotine/kg/day. Half of the animals were sacrificed after two weeks of drug administration. Half of the animals were sacrificed seven days after a two-week drug administration period. Brains were removed and stored at -70° C for later assay. The hypothalamus was removed from each brain, insulin was extracted from the hypothalamus with 0.2 M HCl/75% ethanol, homogenates were neutralized and reconstituted in assay buffer, and insulin was measured by RIA.

The assays revealed a positive dose-effect relationship between nicotine and hypothalamic insulin during drug administration. The 6 mg group had hypothalamic insulin levels that were 105% of controls; the 12 mg group had levels that were 109% of controls. After cessation of nicotine, there was a decrease in hypothalamic insulin to 97% and 95% of saline controls for the 6 mg and 12 mg nicotine groups, respectively. Although the magnitude of these effects may appear small, they are consistent with concentrations of insulin that produce changes in eating and body weight when infused into the lateral cerebral ventricles of baboons (Porte & Woods, 1981).

CONCLUSIONS AND IMPLICATIONS

Nicotine administration decreases plasma insulin levels, and increases hypothalamic insulin levels. These actions of nicotine reverse during nicotine cessation. It is noteworthy that our findings, using a chronic nicotine administration paradigm in rats, are consistent with work of Tjalve and Popov (1973) using rabbit pancreas pieces and with a report of Florey, Milner, and Miall

(1977) comparing plasma insulin in human smokers versus nonsmokers. These changes in peripheral and central insulin also are consistent with the changes in body weight and food consumption that occur during and after nicotine administration, but do not establish a causal relationship. The fact that a low dose of nicotine did not alter insulin values, body weight, or food intake supports the interpretation that insulin mediates the relationship between nicotine and body weight. To unequivocally establish a causal role for insulin in the nicotine-body weight relationship, however, the effects of nicotine on body weight and food consumption must be examined while separately manipulating insulin.

With regard to the mechanism by which nicotine affects insulin, nicotine decreased insulin in the pancreas; however, this decrease was not dose-dependent. Therefore, reduction in pancreatic insulin production probably is only one part of the mechanism underlying effects of nicotine on insulin. It would be useful to examine effects of nicotine on insulin metabolism, distribution, binding, and elimination to fully determine how nicotine alters plasma insulin levels.

Nicotine is known to affect many biochemicals in the body, including acetylcholine, catecholamines, indoleamines, and endogenous opioids (USDHHS, 1988). Many of these actions have been suggested to underlie nicotine reinforcement (Clarke et al., 1988; DiChiara & Imperato, 1988; Imperato, Mulas, & DiChiara, 1986; USDHHS, 1988). The effects of nicotine on insulin also should be examined from this perspective. Insulin may act as a reinforcer for nicotine by mediating effects on eating and body weight, or it may alter energy availability in specific brain regions and thereby affect reward systems, or it may modulate and interact with other reinforcing neurotransmitter systems. These possibilities deserve empirical and theoretical attention. Moreover, it would be valuable to examine effects of nicotine on brain levels of other biochemicals, including cholecystokinin, bombesin, and neuropeptide Y, that are known to affect body weight and food intake and that may mediate reinforcing effects of nicotine. Based on the current findings, it also is worth exploring whether use of nicotine-containing products contributes to the risk of developing diabetes. This possible risk is especially relevant in light of the increased use of nicotine replacement therapy to help people abstain from tobacco products.

Acknowledgement: This work was supported by the Uniformed Services University of the Health Sciences Protocol No. CO7223. The opinions or assertions contained herein are the private ones of the authors and are not to be construed as official or reflecting the views of the Department of Defense or the Uniformed Services University of the Health Sciences.

REFERENCES:

Baskin, D. G., Figlewicz, D. P., Woods, S. C., Porte, D., Jr., & Dorsa, D. M. (1978). Insulin in the brain. Ann. Rev. Physiol., 49:335-347.

Briese, E., & Quijada, M. (1979). Positive alliesthesia after insulin, Experientia, 35:1058-1059.

Cabanac, N. (1971). Physiological role of pleasure, Science 173:1058-1059.

Clarke, P., Fu, D., Jakubovic, A., & Fibiger, H. (1988). Evidence that mesolimbic dopaminergic activation underlies the locomotor stimulant action of nicotine in rats. Journal of Pharmacology and Experimental Therapeutics, 246:701-708.

DiChiara, G., & Imperato, A. (1988). Drugs abuse by humans preferentially increase synaptic dopamine concentrations in the mesolimbic system of freely moving rats. Proceedings of the National Academy of Sciences, 85:5274-5278.

Florey, C., Milner, R. D. G., & Miall, W. E. (1977). Serum insulin and blood sugar levels in a rural population of Jamaican adults. Journal of Chronic Disease, 30:49-60.

Grunberg, N. E. (1982). The effects of nicotine and cigarette smoking on food consumption and taste preferences. Addictive Behaviors, 7, 317-331.

Grunberg, N. E. (1986). Behavioral and biological factors in the relationship between tobacco use and body weight. In E. S. Katkin and S. B. Manuck (Eds.). Advances in Behavioral Medicine, Vol. 2 (97-129). Greenwich, CT: JAI Press.

Grunberg, N. E. (1988). Nicotine and body weight: Behavioral and biological mechanisms. In M. J. Rand & K. Thurau (Eds.), The pharmacology of nicotine: ISCU Symposium Series (pp. 97-110). Washington, DC: IRC Press.

Grunberg, N. E. (1990). The inverse relationship between tobacco use and body weight. In L. T. Kozlowski et al. (Eds.), Research Advances in Alcohol and Drug Problems, Vol 10 (pp. 273-315). New York: Plenum Press.

Grunberg, N. E., Bowen, D. J., & Morse, D. E. (1984). Effects of nicotine on body weight and food consumption in rats. Psychopharmacology, 83, 93-98.

Grunberg, N. E., Popp, K. A., Bowen, D. J., Nespor, S. M., Winders, S. E., & Eury, S. E. (1988). Effects of chronic nicotine administration on insulin, glucose, epinephrine, and norepinephrine. Life Sciences, 42, 161-170.

Grunberg, N. E., Raygada, M., Popp, K. A., Nespor, S. M., Sibolboro, & E. C., & Winders, S. E. (1988, August). Nicotine cessation affects plasma insulin and glucose levels in rats. Poster presented at the meeting of the American Psychological Association, Atlanta, GA.

Guyton, A. C. (1976). Textbook of Medical Physiology. Philadelphia, Saunders, Co.3

Imperato, A., Mulas, A., & DiChiara, G. (1986). Nicotine preferentially stimulates dopamine release in the limbic system or freely moving rats. European Journal of Pharmacology, 132:337-338.

Jacobs, H. L. (1958). Studies on sugar preference: I. The preference for glucose solutions and its modification by injections of insulin. Journal of Comparative Physiological Psychology, 51:304-310.

Klesges, R. C., & Klesges, L. M. (1988). Cigarette smoking as a dieting strategy in a university population. International Journal of Eating Disorders, 7: 413-419.

Porte, D., Jr., & Woods, S. C. (1981). Regulation of food intake and body weight by insulin. Diabetologia, 20:274-280.

Raygada, M., Nespor, S. M., & Grunberg, N. E. (1989). Nicotine alters hypothalamic insulin in rats. Poster presented at the meeting of the American Psychological Association, New Orleans, LA.

Raygada, M., Nespor, S. M., & Grunberg, N. E. (1990). Nicotine alters pancreatic insulin in rats. Poster presented at the meeting of the American Psychological Association, Boston, MA.

Rodin, J., Wack, J., Ferrannini, & Defronzo, R. A. (1985). Metabolism, 34:826-831.

Tjalve, H., & Popov, D. (1973). Effect of nicotine and nicotine metabolites on insulin secretion from rabbit pancreas pieces. Endocrinology, 92:1343-1348.

USDHHS (1988). The Health Consequences of Smoking: Nicotine Addiction. A Report of the Surgeon General, Department of Health and human services. Public Health Service, Office on Smoking and Health, DHHS Pub 1. No. (CDC) 88-8406.

USDHHS (1990). The Health Benefits of Smoking Cessation: A Report of the Surgeon General, Department of Health and Human Services, Public Health Service Office on Smoking and Health, DHHS.

Woods, S. C., & Porte, D., Jr. (1983). Advances in Metabolic disorders, 10:457-468.

Effects of Nicotine on Biological Systems
Advances in Pharmacological Sciences
© Birkhäuser Verlag Basel

SMOKING AND ADIPOSE TISSUE LIPOPROTEIN LIPASE

T. CHAJEK-SHAUL, E.M. BERRY, E. ZIV, G. FRIEDMAN, O. STEIN, G. SCHERER* & Y. STEIN

Lipid Research Laboratory, Department of Medicine B, Hadassah University Hospital, Jerusalem, Israel and *Clinical Chemistry Institute, German Heart Centre, Munich, FRG.

SUMMARY: Adipose tissue lipoprotein lipase activity was measured in eight non-smokers and 17 smokers on fasting and four hours after oral glucose load. Fasting LPL activity was higher in the smokers when corrected for cell weight. Four hours post oral glucose load, a two fold increase in enzyme activity in the non-smoking subjects was observed while in the smokers, a 30% reduction occurred. The degree of reduction of adipose tissue LPL activity negatively correlated with the insulin release in response to the oral glucose. It is suggested that the reduction in adipose tissue LPL activity after oral glucose accounts for the reduced body weight observed in smokers. This reduction reflects a relative insulin resistance of smokers' adipose tissue. The later may be developed due to the known increase in catecholamine in smokers.

INTRODUCTION   Smokers weigh less than non-smokers and gain weight upon cessation of smoking. De novo synthesis of fatty acid by human adipose tissue is low, and as a result, the adipose tissue uptake of fatty acid generated from plasma triglyceride by lipoprotein lipase is an important step in maintenance and development of adipose tissue.

METHODS   Healthy young male subjects were recruited for this study.   All subjects signed an informed consent form for the protocol which was approved by the ethical committee for Medical Research of Hadassah University Hospital, Jerusalem, Israel.   There were eight non—smokers and 17 smokers.   Seven of the smokers quit smoking and were examined five to nine weeks after they stopped smoking.   Fasting blood samples and adipose tissue from the buttock area were taken and the subjects received an oral glucose load.   Blood samples were taken at 15—60 minute intervals for four hours.   At this point, a second adipose tissue aspiration was performed.   Analysis of serum glucose insulin cotinine as well as adipose tissue LPL activity and cell weight were performed.

RESULTS   Fasting adipose tissue LPL activity was higher in the smokers as compared to the non—smokers only when the LPL values were corrected for cell weight ($p<0.001$).   This correction is necessary, smokers adipocytes are smaller than those of non—smokers.   However, the effect of oral glucose load varied considerably between smokers and non—smokers, while in non—smokers and almost two—fold increase in adipose tissue LPL activity was observed four hours post oral glucose load 50.5 to $91.6mU/10^6$ cells ($p<0.02$).   A 30% decrease in adipose tissue LPL was observed in 17 smokers 63.8 to 45.8 $mU/10^6$ cells ($p<0.002$).

The effect of oral glucose load on plasma glucose and insulin was similar in the smoker and the non—smoker subject groups.   However, the degree of decrease in adipose tissue LPL activity observed in smokers after oral glucose load was negatively correlated with insulin release ($p<0.005$); i.e., the greater the insulin release, the smaller the decrease in LPL activity following oral glucose load.

For seven smokers who stopped smoking for five to nine weeks, the adipose tissue LPL activity did not vary significantly before and after oral glucose load, in contrast to the decrease in activity observed in these subjects four hours after oral glucose load during the smoking period.

The ex-smokers gained 2-6.5kg of body weight on cessation of smoking. This increase positively correlates with fasting adipose tissue LPL activity determined during the smoking period.

DISCUSSION Uptake of plasma triglyceride fatty acids is important for storage of lipid energy in adipose tissue and occurs mainly after feeding. Therefore, the decrease in adipose tissue LPL activity after oral glucose load in smokers may account for the weight difference between smokers and non-smokers. The negative correlation between the release of plasma insulin and the degree of decrease in adipose tissue LPL activity following glucose load suggests that a resistance to insulin occurs in smokers' adipose tissue. This may be related to the increase in catecholaminos in smokers.

Acknowledgements: This study was supported in part by a grant from the Forschungsrat Rauchen und Gesundheit, Mbh.

EFFECTS OF NICOTINE ON PULMONARY DEFENCE SYSTEMS

Hans-L. Hahn

Medizinische Poliklinik der Universität, Klinikstraße 8, 8700 WÜRZBURG

SUMMARY: To protect itself against inhaled noxious agents the respiratory tract employs a vital defence system containing several organelles: the mucociliary system, sensory neuro-receptors controlling the pattern of breathing including cough, bronchial smooth muscle, and mucosal blood vessels. Nicotine has effects on all of these systems. Acute effects of nicotine given locally or systemically appear to facilitate clearance. Effects of nicotine applied chronically have only partly been studied but those known appear to impede mucociliary clearance.

INTRODUCTION

As a rule, nicotine enters our body through the airways. Those same airways represent the body's primary line of defence against inhaled noxious gases, vapours, aerosols and other materials. The question arises how this system functions after contact with nicotine which occurs in two ways: initially from the luminal (airway) side and, after systemic absorption, distribution in the body and dilution, from the submucosal or tissue side. Thus we have to distinguish local from systemic effects of nicotine. Since all nicotine contained in the body at one time must have entered through the narrow tubing system of trachea and bronchi, local concentrations in these airways must be considerably higher (though shorter lasting) than systemic concentrations of nicotine measured after dilution of the same amount of nicotine in body fluids although these lower concentrations of nicotine will last longer. Thus, two types of action of nicotine on pulmonary defence systems have to be considered: local and systemic effects.

Table 1 lists some of the more important components of the pulmonary defence system. The epithelium is the layer nearest to the lumen. Here the first contact with particles, bacteria and noxious agents is made. The epithelium contains the cilia which propel the mucus layer cranially. They can be defective in structure leading to inefficient movement or, even if they beat efficiently, their beat frequency can be altered. The cilia beat in the periciliary fluid which needs to have the right depth and the right rheologic properties. Above this sol layer there is a gel layer which contains trapped particles. Cilia

Table 1: COMPONENTS OF THE PULMO-
NARY DEFENCE SYSTEM:

EPITHELIUM:
CILIA: Structure, beat frequency
PERICIL.FLUID: Depth, rheol. prop.
GOBLET CELLS: Mucous glycopro-
teins, mucus layer under resting
conditions

SUBMUCOSAL GLANDS:
MUCUS SECR.: Fast response, cough
MUCUS LAYER: Depth, rheologic
properties, immunologic properties

CONTROL OF RESPIRATION:
COUGH: Clearance in upper airways
PATTERN OF BREATHING: Rapid shallow breathing limits aerosol penetration.

AIRWAY MECHANICS:
BRONCHOCONSTRICTION: Limits depth
of aerosol penetration. Stabilizes
trachea during cough. Reflex bronchoconstriction.

beating in the periciliary (sol) layer move the gel layer towards the pharynx where it is swallowed. The gel layer is maintained mainly by surface epithelial cells, e.g., goblet cells under resting conditions and normally contains more proteoglycans than mucous glycoproteins (1). Only in chronic inflammatory bronchial disease are mucous glycoproteins found in larger quantities (2,5).

While surface epithelial cells provide the bulk of the mucus layer under resting conditions submucosal glands can produce substantial amounts of mucus in a very short time and they seem to reserve their function for emergency situations. Their function is coupled closely to cough. Cough cannot move particles sticking to a dry mucosa. A minimum amount of mucus is necessary which can be loosened by the shearing forces of the cough manoeuvre which produces high linear velocities of air. This layer can be produced quickly by submucosal glands. Functional coupling is accompanied by an anatomic linkage. In cat, rat and mouse the distribution of intraepithelial nerves (thought to mediate coughing) is the same as that of submucosal mucus glands. Both are restricted to the larger cartilagenous airways in cat and rat (6, 9, 20, 21). The function of glands is not limited to supporting cough. For ciliary transport, mucus needs to have certain rheologic properties in order to be moved cranially by the cilia (24, 25). In addition, mucus has bactericidal properties. Thus, it contains lysozyme which is produced by the serous type of submucosal glands.

Important is the control of respiration which depends on sensory receptors in the upper and lower airways and on centres in the CNS, e.g., rhythm generators. Two instances are of particular relevance: one is the generation of cough or of the expiration reflex (rapid expiratory effort not preceded by a deep inspiration, often elicited from laryngeal sensory receptors (11)) and the other is the rapid shallow breathing pattern evoked by irritants upon contact with sensory receptors. It may serve to limit the penetration of materials into the periphery of the lungs. If the stimulus is very strong, the pattern is not just rapid, shallow breathing but it is long lasting apnoea, followed by rapid shallow breathing. Apnoea obviously limits the penetration of particles into the bronchial tree and lungs. Both patterns: rapid shallow breathing and apnoea, followed by rapid shallow breathing, are typical responses seen after stimulation of nonmyelinated C-fibre endings (4), which also elicit gland secretion (7), but it is likely that sensory irritant receptors are also involved.

Finally, there is bronchoconstriction which is evoked by stimulation of the same receptors, i.e., nonmyelinated C-fibre endings and myelinated irritant receptors (10, 27). Bronchocon-striction may limit the depth of aerosol penetration and by tightening the posterior membrane of the trachea opposes complete occlusion of the trachea which can be produced by invagination of the posterior membrane (18, 19). Another advantage of reflex bronchoconstriction is speed of response: reflex bronchoconstriction is a fast response just as cough. It could thus serve as a stabilizing force for the trachea increasing cough efficiency (18).

METHODS:

1. Ciliary Beat Frequency: In order to study the effects of nicotine on the frequency of ciliary beating we peeled strips of epithelium off the surface of ferret tracheae. This tissue contains hardly any goblet cells

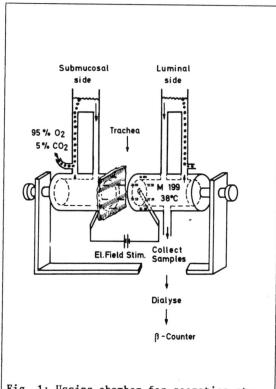

95 % O_2
5 % CO_2

Submucosal side

Luminal side

Trachea

M 199
38°C

El.Field Stim.   Collect Samples

Dialyse

β-Counter

Fig. 1: Ussing chamber for secretion studies *in vitro*

and, since glands are left behind in the submucosa epithelial strips from the ferret trachea are suitable for studying direct effects of nicotine on the cilia without interference from mucus secreting structures. Epithelial strips were put in a culture chamber, placed on a microscope stage and the beating cilia rhythmically interrupted or attenuated the light beam of the microscope. These fluctuations in light intensity were magnified by a photomultiplier tube and subjected to frequency spectrum analysis (fast Fourier transformation). Recordings were taken for 10 min. During the initial 5 min the chamber was perfused with fresh medium M-199 containing nicotine. Although the perfusion itself changed ciliary beating (progressive decrease in beat frequency, cf. Fig. 1), we recorded the ciliary beat frequency from the beginning because of possible tachyphylaxis and because of one report in the literature that effects of nicotine on cilia were stimulatory, but transient (17).

2. Mucus Secretion: To study mucus secretion in vitro we mounted segments of ferret trachea between the two halves of Ussing chambers perfused with medium M-199 aerated with 95% O_2 and 5% CO_2 (Fig. 1). A radioactive precursor of sulfated mucins ($Na_2^{35}SO_4$) was applied to the submucosal side, labeled material was collected from the luminal side at 15 min intervals, was dialyzed against nonradioactive Na_2SO_4 (molecular weight cutoff, 12 000 Dalton) and labeled macromolecules were counted in

a β-counter before and after application of nicotine to the luminal side of the trachea (10^{-6} – 10^{-4} M).

To study mucus secretion in vivo an isolated perfused segment of trachea extending from the larynx to one centimeter above the tracheal bifurcation was created in an open chest ferret ventilated artificially through the tracheal stump (Fig. 2). Radioactive precursor was

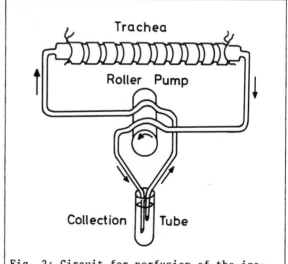

Fig. 2: Circuit for perfusion of the isolated segment of trachea *in vivo*.

applied locally to the segment and was also given intravenously into the systemic circulation. After 60 min, the radioactive perfusate was drained, the segment washed with fresh medium M-199, and 5 min-collections were started and analyzed for radioactive macromolecules secreted into the airway lumen. This was done before and after systemic application of nicotine 10^{-7} to 10^{-5} M· kg^{-1}.

3. Bronchoconstriction: To study bronchoconstriction induced by locally applied nicotine in vitro we suspended strips of tracheal smooth muscle or tracheal rings in organ baths filled with medium M-199 aerated with 95% O_2 and 5% CO_2. Smooth muscle tension was recorded with force transducers (FTO3, Grass) and the initial tension was set at a level producing maximal contraction to the subsequent addition of acetylcholine (10^{-5} M). Nicotine-induced muscle tension was expressed either in grams or as a percentage of the maximum tension achieved by ACh.

Bronchoconstriction induced by systemic nicotine in vivo was measured in the model described (Fig. 2) as the increase in the amplitude of the oscillations in transpulmonary pressure induced by the respirator.

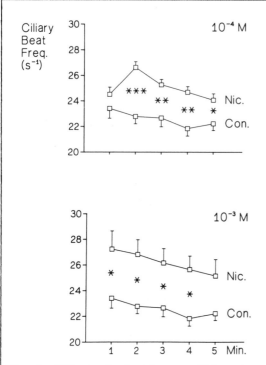

Fig. 3: Effect of nicotine on ciliary beat frequency *in vitro*.

4. Chronic effects of nicotine on gland structure and function: In 28 ferrets we created a chronic tracheostomy and, starting one week later, exposed 14 animals (1254±156 g, x±SD) to an aerosol of 5.4×10^{-2} M nicotine bitartrate generated from an ultrasonic nebulizer (DeVilbiss # 65) while 14 control animals (1282±98 g) inhaled the solvent only (0.9% NaCl). Inhalation was 10 min 10 times a day for 35 days. 24 h after the last exposure animals were killed and the structure and function of submucosal glands were compared between groups.

RESULTS:

1. Cilia: Fig. 3 shows that ciliary beat frequency was higher during perfusion with nicotine in medium M-199 than during perfusion with medium M-199. By the time the perfusion was stopped (beyond 5 min, not shown) the stimulatory effect of nicotine was no longer significant, confirming that the effect was transient.

2. Mucus Secretion: Fig. 4 shows the effect of systemic nicotine on mucus secretion measured in vivo. Nicotine increased mucus secretion in a dose dependent manner. Effective concentrations ranged from 5×10^{-7} to a peak effect at 10^{-5} M where the increase was approximately 300%. The dose response curve to nicotine applied locally in vitro was similar in shape but was shifted 1½ orders of magnitude to the right, secretory

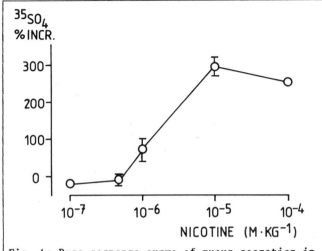

Fig. 4: Dose response curve of mucus secretion in response to nicotine applied systemically.

effects starting at 10^{-5} M and being maximal at 3×10^{-4} M (not shown).

3. Bronchoconstriction: In vitro, nicotine applied locally caused a short lasting bronchoconstriction of muscle strips and tracheal rings. The range of effective concentrations was the same as was observed for mucus secretion, i.e., above 10^{-5} M. At its peak, nicotine-induced bronchoconstriction was approximately 30% of the maximum contraction induced by ACh. Systemic nicotine applied in vivo was effective at far lower concentrations than local concentrations applied in vitro, the dose response curve being shifted to approximately 10-fold lower concentrations. Fig. 5 shows the effect in vivo of 10^{-6} $M \cdot kg^{-1}$ on mucus secretion (top) and on transpulmonary pressure amplitude (bottom) and also shows that both effects are specific effects of nicotine because they are blocked by the specific nicotine antagonist hexamethonium (right half of both panels).

4. Chronic effects of nicotine on gland structure and function: Nicotine applied chronically had different effects on male and female animals. An increase in gland size occurred only in female, but not in male animals whose submucosal glands were much larger to start with (Fig. 6, left vs. right half). However, when the mucous and the serous types of glands were assessed separately both sexes showed a similar shift in gland structure: serous glands decreased, mucous glands increased, therefore the serous: mucous ratio decreased considerably (Fig. 7). Autoradiographic studies (not shown) showed that the turnover of radioactive material was much faster in glands from nicotine treated than from control animals.

DISCUSSION

We have shown in these studies that the acute effects of nicotine on components of the pulmonary defence system, i.e., increase in ciliary beat frequency, increase in mucus secretion and moderate bronchoconstriction, would all be expected to improve defence functions of the lungs. We have not done studies on the control of respiration ourselves but the literature shows unequivocally that effects of nicotine on the control of respiration are such that defence mechanisms would be improved rather than compromised, the typical responses being cough or apnoea, followed by rapid shallow breathing. One of the

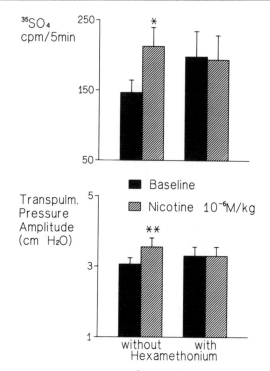

Fig. 5: Effect of 10^{-6} M nicotine on mucus secretion (top, left half of panel) and on bronchomotor tone measured as the transpulmonary pressure amplitude during artificial ventilation (bottom, left half of panel). The right half of each panel shows complete blockade of effects by hexamethonium.

most prominent actions of nicotine inhaled as an aerosol by man is cough (3). Experimentally, only 1-2 breaths of cigarette smoke inhalation were enough to cause long lasting apnoea. This was followed by rapid shallow breathing and these responses were mediated by nonmyelinated C-fibre endings (13), although nicotine has been shown to stimulate other types of receptor as well (22, 26). These changes in the control of respiration were "imitated" by inhalation of nicotine, but not of other components of cigarette smoke (14, 15, 16), which confirmed that it was the nicotine in cigarette smoke that elicited these responses.

Were the concentrations of nicotine applied in vitro or in vivo relevant to real life situations? We have previously presented

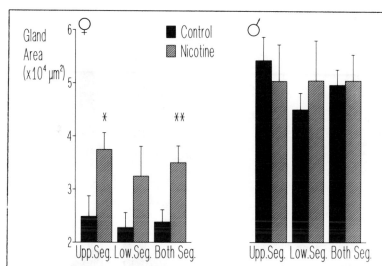

Fig. 6: **Effect of chronic application of nicotine on gland volume measured as gland area in histologic sections. Significant increase in gland area only in female (left), not in male animals (right).**

calculations based on plasma levels of nicotine that show concentrations used in vitro and in vivo were probably realistic although no measurements are available (8, 12). In addition, systemic concentrations measured in plasma are not representative of tissue concentrations because nicotine concentrations are higher in most organs than in plasma. Organs with the highest content of nicotine are adrenal glands, uterus, liver and salivary glands where nicotine concentrations are approximately 10 times higher than in plasma. In the lungs, concentrations are still five times higher than in plasma (23). Calculations based on plasma levels will therefore underestimate tissue levels of nicotine and will tend to underestimate local tracheal concentrations of nicotine derived from tissue levels. Concentrations used were thus relevant.

While acute effects of nicotine are likely to improve mucociliary clearance long term effects of nicotine on the pulmonary defence system are largely unknown. We have performed a first long term study to determine effects of chronic application of nicotine on submucosal glands. It appears that, regardless of a possible increase in gland size the turnover of mucus is increased and that it is the product of mucous acini, which increases most. The effect would be twofold: There would be an increase in the total load of mucus the system has to cope with and it would be the more viscous variety of mucus that predominates. The net effect on clearance is difficult to predict but clearance may well be decreased in the long term.

ACKNOWLEDGEMENTS: We wish to thank the Research Council on Smoking and Health for supporting the studies described in this paper.

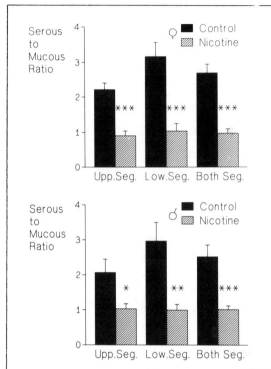

Fig. 7: Significant decrease in the ratio of serous to mucous glands after chronic inhalation of nicotine. This shift in gland composition was present in male (bottom) as in female (top) animals.

REFERENCES

1. BHASKAR K.R., D.D. O'SULLIVAN, H. OPASKAR-HINCMAN, L.M. REID, S.J. COLES. Density gradient analysis of secretions produced in vitro by human and canine airway mucosa: identification of lipids and proteo-glycans in such secre-tions. Exp. Lung Res. 10: 401-422, 1986.

2. BHASKAR K.R., D.D. O'SULLIVAN, J. SELTZER, T.H. ROSSING, J.M. DRAZEN, L.M. REID. Density gradient study of bronchial mucus aspirates from healthy volunteers (smokers and nonsmokers) and from patients with tracheostomy. Exp. Lung Res. 9: 289-308, 1985.

3. BURCH S.G., M.L. ERBLAND, L.P. GANN, F.CH. HILLER. Plasma nicotine levels after inhalation of aerosolized nicotine. Am. Rev. Respir. Dis. 140: 955-957, 1989.

4. COLERIDGE J.C.G., H.M. COLERIDGE. Afferent vagal C fibre innervation of the lungs and airways and its functional significance. Rev. Physiol. Biochem. Pharmacol. 99: 1-110, 1984.

5. COLES S.J., L.R. LEVINE, L. REID. Hypersecretion of mucus glycoproteins in rat airways induced by tobacco smoke. Am. J. Pathol. 94: 459-471, 1979.

6. DAS R.M., P. JEFFERY, J.G. WIDDICOMBE. The epithelial inner-vation of the lower respiratory tract of the cat. J. Anat. 126: 123-131, 1978.

7. DAVIS B., A.M. ROBERTS, H.M. COLERIDGE, J.C.G. COLERIDGE. Reflex tracheal gland secretion evoked by stimulation of bronchial C-fibers in dogs. J. Appl. Physiol. 53: 985-991, 1982.

8. HÜMMER B., I. PURNAMA, H.-L. HAHN. Stimulation of submucosal glands by nicotine applied locally to the airway mucosa. Klin. Wochenschr. 66(Suppl.XI): 161-169, 1988.
9. JEFFERY P., L. REID. Intra-epithelial nerves in normal rat airways: a quantitative electron microscopiy study. J. Anat. 114: 35-45, 1973.
10. KARLSSON J.-A., G. SANT'AMBROGIO, J. WIDDICOMBE. Afferent neural pathways in cough and reflex bronchoconstriction. J. Appl. Physiol. 65: 1007-1023, 1988.
11. KORPAS J., TOMORI Z. Cough and other respiratory reflexes Progr. Respir. Res., Basel: Karger, 1979, vol. 12, p. 1-335.
12. LANG M., B. HÜMMER, H.-L. HAHN. Effect of systemic nicotine on mucus secretion from tracheal submucosal glands and on cardiovascular, pulmonary, and hematologic variables. Klin. Wochenschr. 66(Suppl.XI): 170-179, 1988.
13. LEE L.-Y., E.R. BECK, R.F. MORTON, Y.R. FRAZER, D.T. KOU. Role of bronchopulmonary C-fiber afferents in the apneic response to cigarette smoke. J. Appl. Physiol 63: 1366-1373, 1987.
14. LEE L.Y., R.F. MORTON, D.T. FRAZIER. Influence of nicotine in cigarette smoke on acute ventilatory responses in awake dogs. J. Appl. Physiol. 59: 229-236, 1985.
15. LEE L.-Y., R.F. MORTON, A.H. HORD, D.T. FRAZIER. Reflex control of breathing following inhalation of cigarette smoke in conscious dogs. J. Appl. Physiol. 54: 562-570, 1983.
16. LEE L.-Y., R.F. MORTON. Hexamethonium aerosol prevents cigarette smoke-induced pulmonary reflexes in dogs. Respir. Physiol. 66: 303-314, 1986.
17. LINDBERG S., J.C. HYBBINETTE, U. MERCKE. Effects of nicotine bitartrate on mucociliary activity. Eur. J. Respir. Dis. 66: 40-46, 1985.
18. OLSEN C.R., M.A. DEKOCK, H.J.H. COLEBATCH. Stability of airways during reflex bronchoconstriction. J. Appl.Physiol. 23: 23-26, 1967.
19. OLSEN C.R., A.E. STEVENS, N.B. PRIDE, N.C. STAUB. Structural basis for decreased compressibility of constricted tracheae and bronchi . J. Appl. Physiol. 23: 35-39, 1967.
20. PACK R.J. AL-UGAILY L.H., G. MORRIS, J.G. WIDDICOMBE. The distribution and structure of cells in the tracheal epithelium of the mouse. Cell Tissue Res. 208: 65-84, 1980.
21. PACK R.J., H. AL-UGAILY, G. MORRIS. The cells of the tracheobronchial epithelium of the mouse. A quantitative light and electron microscope study. J. Anat. 132: 71-84, 1981.
22. PAINTAL A.S. Impulses in vagal afferent fibres from specific pulmonary deflation receptors. The response of these receptors to phenyl diguanide, potato starch, 5-hydroxytryptamine and nicotine, and their role in respiratory and cardiovascular reflexes. Quart. J. Exptl. Physiol. 40: 89-111, 1955.
23. ROWELL P.P., H.H. HURST, C. MARLOWE, B.D. BENNETT. Oral administration of nicotine: its uptake and distribution after chronic administration to mice. J. Pharmacol. Meth. 9: 249-261, 1983.
24. SADE J., N. ELIEZER, A. SILBERBERG, A.C. NEVO. The role of mucus in transport by cilia. Am. Rev. Respir. Dis. 102: 48-52, 1970.
25. SPUNGIN B., A. SILBERBERG. Stimulation of mucus secretion, ciliary activity, and transport in frog palate epithelium. Am. J. Physiol.(Cell Physiol. 16) 247: C299-C308, 1984.

26. TAYLOR R.F., L.Y. LEE, L.A. JEWELL, D.T. FRAZIER. Effect of nicotine aerosol on slowly adapting receptors in the airways of the dog. J. Neurosci. Res. 15: 583-593, 1986.
27. WIDDICOMBE J.G. Nervous receptors in the tracheobronchial tree: airway smooth muscle reflexes. in: Airway smooth muscle in health and disease, edited by R.F. Coburn. New York: Plenum Publishing Corporation, 1989, p. 35-53.

EFFECT OF NICOTINE ON THE ELASTINOLYTIC ACTIVITY OF HUMAN LUNG
TUMOR CELLS AND HUMAN ALVEOLAR MACROPHAGES

G. Trefz, G. Klitsche, V. Schulz, and W. Ebert

Thoraxklinik Heidelberg-Rohrbach, Amalienstraße 5, D-6900
Heidelberg, FRG

SUMMARY: The elastinolytic activity of two human lung tumor cell
lines as well as human alveolar macrophages isolated from
bronchoalveolar lavage fluid of patients with different lung
diseases was calculated by digestion of ^{3}H-labeled elastin. In
the case of the tumor cell lines differences in the elastinolytic activity as well as in the ability of the cells to be
stimulated by nicotine were detected. From the different lung
diseases macrophages of sarcoidosis and emphysema patients
showed the highest elastinolytic activities.

Proteolytic enzymes are thought to play a role in tumor cell
invasion (Tryggvason et al., 1987) by cleaving extracellular ma-
trix and basement membrane components. Many tumor cell lines as
well as malignant tissues are reported to express proteinases.
In this context human breast carcinoma cells are reported to se-
cret elastases (Kao & Stern, 1986). We investigated the elastinolytic activity expressed by two human non-small lung cancer
cell lines.

Elastases from human alveolar macrophages (AMs) (Sibille &
Reynolds, 1990) as well as PMN-granulocytes (Baugh & Travis,
1976) are thought to play a role in other lung diseases especially in emphysema (Taylor & Mittman, 1986). The activity of PMN-
elastase in the lung is regulated by two inhibitors, α_1-pro-
teinase inhibitor (Carrell, 1986) and mucus proteinase inhibitor
(Boudier & Bieth, 1989). An imbalance between these inhibitors
and the elastases is thought to be a main reason for the pathogenesis of emphysema. However, we found, that individuals develop emphysema although having high PMN-elastase inhibitory activity in their lungs (Trefz et al., 1990a) Therefore, the

elastases from AMs, which are not inhibited by α_1-proteinase inhibitor and mucus proteinase inhibitor might be of importance.
Because of this we started to isolate AMs from bronchoalveolar
lavage fluid (BALF) of patients with emphysema and other lung
diseases to calculate their elastinolytic activity.

MATERIAL AND METHODS

Cells: HS 24 was established from a human squamous cell carcinoma of the lung, SB 3 from a metastasis of a primary adenocarcinoma of the lung into the adrenal gland (Erdel et al., 1990).
AMs were isolated fom BALF and purified and cultured as described (Albin et al., 1987). BALF was performed as recommended by
the european task group (Klech & Pohl, 1989).
[^{3}H]Elastin Assay: [^{3}H]Elastin was prepared as described (Banda
et al., 1981). For the determination of the elastinolytic activities of the cells, 60 µl of a [^{3}H]Elastin suspension (16
mg/ml) were coated on the wells of microtiter plates and air-
tried. 0.5×10^6 cells suspended in medium (RPMI 1640 for tumor
cells , DMEM for AMs) were seeded on the elastin and the elastinolytic activity was calculated after 72 h/37 $^\circ$C from the released radioactivity. In control experiments [^{3}H]Elastin was
treated with medium alone.
Nicotine stimulation:
In stimulation experiments nicotine base (50, 100 and 200 ng/ml)
was added directly to the medium.

RESULTS

Elastinolytic activity of human lung tumor cells:
The results of the determination of the elastinolytic activities
of HS 24 and SB 3 as well as their ability to be stimulated by
nicotine are shown in figures 1 and 2. HS 24 cells express a
higher elastinolytic activity compared to SB 3 cells. The elastinolytic activity of the HS 24 cells could be stimulated by
nicotine (Fig.1). The maximum of stimulation was achieved with a

concentration of 100 ng nicotine /ml. In contrast to HS 24, the elastinolytic activity of SB 3 could not be stimulated by nicotine (Fig.2).

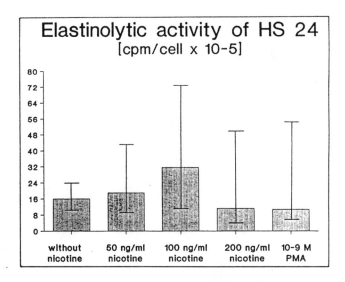

Fig.1: Elastinolytic activities of HS 24 cells

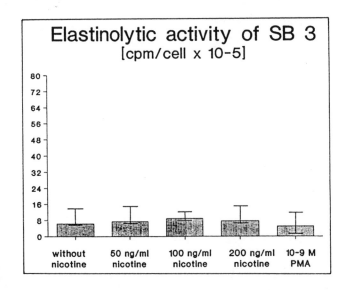

Fig.2: Elastinolytic activities of SB 3 cells

Elastinolytic activity of AMs:

AMs from patients suffering from lung tumors expressed the lowest elastinolytic activity, followed by fibrosis, alveolitis,
and emphysema (Fig.3). The highest elastinolytic activities of
AMs were found in sarcoidosis. The effect of nicotine on the
elastinolytic activity of macrophages investigated so far was
found to be ambiguous. In the case of smokers (n=11) the elastinolytic activity seemed not to be influenced or even decreased
after addition of nicotine (100 ng/ml). In the case of nonsmokers (n=6) more an increase of the elastinolytic activity after
nicotine addition was observed (data not shown).

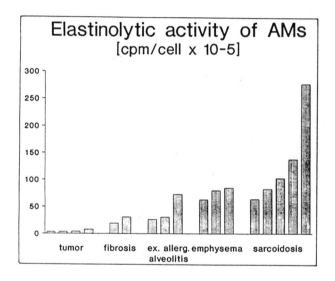

Fig.3: Elastinolytic activities of AMs from patients suffering
from different lung diseases (n = 17).

DISCUSSION

Elastinolytic activity of human lung tumor cells:

HS 24 cells, which were established from a primary squamous cell
carcinoma of the lung showed a higher elastinolytic activity
than SB 3 cells, which were established from a metastasis deri-

ved from a primary lung adenocarcinoma. Therefore elastinolytic enzymes might play more a role in tumor growth at primary sites by digestion of the connective tissue of the lung than in tumor invasion where other proteinases like cathepsins might be more important (Trefz et al. 1990b). Stimulation experiments showed that primary lung tumor cells (HS 24) seemed to be more effected by nicotine than cells which already had metastasized (SB 3). Although one can speculate that a receptor is involved in the action of nicotine on the elastinolytic activity of the lung tumor cells, the underlaying mechanism is yet unknown.

Elastinolytic activity of AMs:

Lowest elastinolytic activities of AMs were found in BALF of tumor patients, indicating that the elastinolytic capacity of AMs might play a minor role in this disorder. Higher activities were found in alveolitis and emphysema. This might reflect the presence of activated AMs. In emphysema, elastinolytic enzymes of AMs must be considered to be involved in the formation of emphysematous lesions. Highest activities were found in sarcoidosis. Because the highest elastinolytic activities were found in stage three of sarcoidosis, which is known show fibrotic disorders of lung tissue, it is possible that the elastinolytic enzymes are involved in the generation of peptids, which can stimulate proliferation of fibroblasts. The effect of nicotine on the elastinolytic activity of AMs seemed to depend on individual variations as well as smoking behaviour.

Acknowledgements: This work was supported by a grant from the German Research Council on Smoking and Health.

REFERENCES

Albin, R.J., Senior, R.M., Welgus, H.G., Connolly, N.L., Campbell E.J. (1987) Am. Rev. Respir. Dis. 135, 1281-1285.
Banda, M.J., Dovey, H.F., Werb, Z. (1981) In: Methods for studying mononuclear phagocytes, pp. 603-618. Academic Press, New York.
Baugh, R.J., and Travis, J. (1976) Biochemistry 15, 836-841.
Boudier, C., and Bieth, J.G. (1989) Biochim. Biophys. Acta 995, 36-41.

Carrell, R.W. (1986) J. Clin Invest. 78: 1427–1431.

Erdel, M., Peter, W., Spiess, E., Trefz, G., Ebert, W. (1990) Cancer Genet. Cytogenet. 48: in press.

Kao, R.T., and Stern, R. (1986) Cancer Res. 46, 1355–1358.

Klech, H., Pohl, W. (eds.) (1989) Eur. Respir. J. 2, 561–585.

Mason, R.W., Johnson, D.A., Barrett, A.J., Chapman, H.E. (1986) Biochem. J. 233, 925–927. Tryggvason, K., Höyhtyä, M., Salo, T. (1987) Biochim. Biophys. Acta 907, 191–217.

Sibille, Y., and Reynolds, H.Y. (1990) Am. Rev. Respir. Dis. 141, 471–501.

Taylor, J.C., Mittman, C. (eds.) (1986) Pulmonary emphysema and proteolysis. Academic Press, London.

Trefz, G., Schliesser, J., Heck, B., Geiger, R., Schulz, V., and Ebert, W. (1990a) Eur. Respir. J.: submitted for publication.

Trefz, G., Erdel, M., Spiess, E., and Ebert, W. (1990b) Biol. Chem. Hoppe Seyler 371, in press.

A NEW ROLE FOR NICOTINE: SELECTIVE INHIBITION OF THROMBOXANE
FORMATION BY DIRECT INTERACTION WITH THROMBOXANE SYNTHASE

Matthias Goerig[1,2], Volker Ullrich[3], Gotthard Schettler[4],
Christina Foltis[1], and Andreas Habenicht[4]

1   [4]th Medical Clinic, University of Erlangen-Nürnberg,
    Kontumazgarten 14-18, D-8500 Nürnberg 80, F. R. G.
2   formerly Medical Clinic, University of Heidelberg
3   Department of Biology, University of Konstanz
4   Medical Clinic, University of Heidelberg

SUMMARY: Animal cells can convert arachidonic acid into
prostaglandins, thromboxane, and leukotrienes. These locally
produced mediators of inflammatory and hypersensitivity
reactions have been implicated in several clinically important
disease processes. Thromboxane, the major oxygenated arachidonic
acid metabolite of macrophages, is the most potent
vasoconstricting and proaggregatory molecule known and has also
been related to host defense mechanisms. We have studied effects
of nicotine on thromboxane formation using cultured
macrophage-like cells, microsomal assays, and purified
thromboxane synthase. In intact macrophage-like cells nicotine,
cotinine, and methylnicotine, at submicromolar concentrations,
inhibited the rate of conversion of endoperoxide prostaglandin
H_2 into thromboxane but not into prostacyclin. This indicated
that nicotine selectively inhibited thromboxane synthase at
concentrations that are readily observed in the circulation of
smokers. Microsomal assays revealed that nicotine decreased the
maximal velocity (V_{max}) of thromboxane synthase without affecting
the apparent affinity of the enzyme for its substrate (K_m). No
effect of nicotine on kinetic parameters of prostacyclin
synthase could be observed. To elucidate the molecular mechanism
of nicotine's effect we obtained difference spectra using
purified thromboxane synthase. The difference spectra revealed
that nicotine directly interacted with the enzyme presumably by
binding of the nitrogen of the nicotine ring structure to the
iron of the cytochrome p-450 component of thromboxane synthase.

The biological effects of nicotine are believed to be mediated
by its activity in the central nervous system and its ability to

induce catecholamine release from adrenal chromaffin cells (Bassenge et al., 1988; Benowitz, 1986; Benowitz, 1988). Several observations that have been made in smokers or experimental animals, however, indicate that nicotine also exerts effects in vivo that appear to be independent from its classical mechanisms of action. For example, smokers suffer from an increased susceptibility to acute and chronic infections (Haynes et al., 1966; Parnell et al., 1966) and exhibit severe alterations of morphologic and metabolic parameters of white blood cells (Balter et al., 1989; Harris et al., 1970; Sasagawa et al., 1985) while chronic nicotine application using nicotine containing chewing gum does not lead to an elevation of the concentration of catecholamines in the circulation (Benowitz, 1988). In addition, smokers or experimental animals that receive nicotine for extended periods of time by the use of implanted minipumps show an increased bleeding time, impaired platelet aggregability towards ADP, and reduced thromboxane formation from exogenous arachidonic acid (Becker et al., 1988; Brinson, 1974; Mansouri et al., 1982; Marasini et al., 1986, Siess et al., 1982). The underlying molecular mechanisms of the catecholamine-independent effects of nicotine remain largely to be defined.

The focus of the present report has been the effects of nicotine on eicosanoid metabolism and in particular on thromboxane formation.

Most eicosanoids (prostanoids, thromboxane and leukotrienes) are biologically active substances that are derived from the metabolism of arachidonic acid, a long-chain polyunsaturated fatty acid with 20 carbon atoms (see Goerig et al., 1985 for review). These molecules have been shown to be potent mediators of inflammatory and immunological reactions (see Samuelsson et al., 1987 for review). Thromboxane is mainly produced by two cell types: platelets and monocytes / macrophages. While both cell types form thromboxane the biological role of platelet-derived and macrophage-derived thromboxane might very well differ significantly. For platelet-derived thromboxane it is generally believed that its formation should be reduced in order to block the exaggerated platelet activation that occurs in patients afflicted with cardiovascular disease and in particular during plasminogen activator therapy of acute myocardial infarction (Fitzgerald et al., 1989). In sharp contrast to the situation in platelets, there is evidence that macrophage-derived thromboxane has a protective role for the host by its ability to localize bacterial infections (Tripp et al., 1987).

Eicosanoids are formed in response to a large variety of agonists that interact with the cell surface to trigger the release of arachidonic acid from cellular phospholipid stores. Nonesterified arachidonic acid is subsequently metabolized by one of two major pathways of the cascade, the prostaglandin (PG)H synthase (or cyclooxygenase) and the 5-lipoxygenase pathway. In the first of the two pathways, the released arachidonic acid is initially converted by the PGH-Synthase into the shortlived endoperoxide intermediate PGH_2. Subsequently PGH_2 is converted by several 'branchpoint enzymes' into the biologically active end products. In particular, thromboxane synthase converts PGH_2 into thromboxane, PGH_2/PGE_2-isomerase converts PGH_2 into PGE_2, and prostacyclin synthase converts PGH_2 into prostacyclin. In the other major pathway, arachidonic acid is converted by the enzymes of the 5-lipoxygenase pathway to leukotriene B_4 or the peptidoleukotrienes C_4, D_4, and E_4.

DIFFERENTIAL EFFECTS OF NICOTINE ON ARACHIDONIC ACID METABOLISM IN CULTURED WHITE BLOOD CELLS:

In the present investigation we used a model culture system of differentiating macrophages, the promyelocytic human leukemia cell line HL-60 (Harris and Ralph, 1985), to study the effects of nicotine and its main metabolite cotinine on eicosanoid metabolism. These cells acquire several morphological and functional characteristics of macrophages such as phagocytosis of bacteria and adherence to plastic surfaces upon incubation with activators of protein kinase C (12-0-tetradecanoylphorbol-13-acetate, TPA) (Harris and Ralph, 1985 for review). In published studies we have already shown that several enzymes of the arachidonic acid cascade, in particular PGH synthase and 5-lipoxygenase, are coordinately induced early after incubation of the undifferentiated precursor cells with activators of protein kinase C (see Goerig et al., 1987; Goerig et al., 1988). We therefore considered this culture system to be suitable for the study of the effects of nicotine and cotinine on selective enzymes of arachidonic acid metabolism (see Goerig and Habenicht, 1988).

Differentiating macrophages, that had acquired the ability to form high amounts of thromboxane from exogenous arachidonic acid (Goerig et al., 1987; Goerig et al. 1988), were incubated with increasing concentrations of nicotine or cotinine (0.1 - 1000 nM) in the presence of an optimal dose of nonesterified arachidonic acid. Subsequently the concentrations of several eicosanoids including thromboxane were determined. We found that 1) nicotine at concentrations that are readily observed in the circulation of smokers (10-500 nM) (Gritz et al., 1981; Scherer et al., 1988) led to an up to 90% inhibition of thromboxane formation; 2) no effects of nicotine were observed on the rate of formation of prostacyclin, PGE_2, or leukotrienes B_4 and C_4.; 3) cotinine and methylnicotine but not nornicotine had similar effects when compared with nicotine. These results indicated that nicotine or cotinine could not have affected either PGH synthase or prostacyclin synthase activities because PGE_2 and prostacyclin synthesis rates remained unaffected by nicotine or cotinine. Rather it seemed that nicotine and cotinine might have selectively affected the 'branchpoint enzyme' thromboxane synthase. We therefore incubated differentiating macrophages with the endoperoxide intermediate PGH_2 and determined the concentrations of thromboxane and prostacyclin as convenient and specific intact cell assays for thromboxane and prostacyclin possibility that nicotine directly binds to this enzyme thereby inhibiting its catalytic activity. Alternatively, nicotine might have affected one or more cofactors that are required for thromboxane synthase activity. To address the former possibility directly we used spectroscopic analyses of highly purified thromboxane synthase preparations derived from human blood platelets or platelet microsomes and studied potential nicotine/thromboxane synthase interactions. Difference spectra of the cytochrome p-450 type enzyme, thromboxane synthase (Haurand and Ullrich, 1985), were obtained in the absence and presence of increasing concentrations of nicotine. We found that nicotine directly affected light absorbance without interfering with the Soret constants (see Haurand and Ullrich, 1985, for more detailed information). These results clearly established that nicotine directly interacts with the cytochrome p-450 component of thromboxane synthase. It is noteworthy in this connection that Barbieri et al. (Barbieri et al., 1986) have previously shown that nicotine binds to another cytochrome p-450 enzyme, the aromatase (estrogen synthase). These authors suggested that nicotine's binding to this enzyme is responsible for the decrease in the concentration of estrogen that had been observed in pregnant women who smoke (Barbieri et al., 1986). Likewise it is conceivable to assume, based on the difference spectra obtained in our present experiments using platelet-derived thromboxane synthase, that nicotine binds to the cytochrome p-450 type enzyme thromboxane synthase in vivo.

CONCLUSIONS: We have identified a new pharmacological mechanism of action of nicotine. It will now be important to demonstrate that the effects in cultured cells and purified enzyme preparations occur in vivo and thus have physiological relevance. Finally, it remains to be demonstrated that the selective inhibition of thromboxane synthase by nicotine and possibly cotinine (spectroscopic analyses with cotinine have not been performed until now) is causally related to the increased bleeding time and the impaired host defense both of which have been observed in smokers.

REFERENCES

Balter, M. S., Toews, G. B., and Peters-Golden, M. (1989) J. Lab. Clin. Med. 114, 662-673

Barbieri, R. L., Gochberg, J., and Ryan, K. J. (1986) J. Clin. Invest. 77, 1727-1733

Bassenge, E., Holtz, J., and Strohschein, H. (1988) Klin. Wochenschr. 66 (Suppl. XI), 12-21

Becker, B. F., Terres, W., Kratzer, M., and Gerlach, E. (1988) Klin. Wochenschr. 66 (Suppl. XI), 28-36

Benowitz, N. L. (1986) Ann. Rev. Med. 37, 21-32

Benowitz, N. L. (1988) N. Engl. J. Med. 319, 1318-1330

Brinson, K. (1974) Atherosclerosis 20, 137-140

Fitzgerald, D. J., and FitzGerald, G. A. (1989) Proc. Natl. Acad. Sci. USA 86, 7585-7589

Goerig, M., Habenicht, A., Schettler, G. (1985) Klin. Wochenschrift 63, 293-311

Goerig, M., Habenicht, A. J. R., Heitz, R., Zeh, W., Katus, H., Kommerell, B., Ziegler, R., and Glomset, J. A. (1987) J. Clin. Invest. 79, 903-911

Goerig, M., Habenicht, A. J. R., Zeh, W., Salbach, P., Kommerell, B., Rothe, D. E. R., Nastainczyk, W., Glomset, J. A. (1988) J. Biol. Chem. 263, 19384-19391

Goerig, M., Habenicht, A. (1988) Klin. Wochenschrift 66 (Suppl. XI), 117-119

Gritz, E. R., Baer-Weiss, V., Benowitz, N. L., van Vunakis, H., and Jarvik, M. E. (1981) Clin. Pharm. Ther. 30, 201-209

Harris, J. O., Swanson, E. W., and Johnson III, J. E. (1970) J. Clin. Invest. 49, 2086-2096

Harris, P., Ralph, P. (1985) J. Leukocyte Biol. 37, 407-422

Haurand, M., and Ullrich, V. (1985) J. Biol. Chem. 260, 15059-15067

Haynes, W. F. Jr., Kristulovic, V. J., and Bell A. L. L. Jr. (1966) Am. Rev. Respir. Dis. 93, 730-735

Mansouri, A., and Perry, C. A. (1982) Thromb. Haemostas. 48, 286-288

Marasini, B., Biondi, M. L., Barbesti, S., Zatta, G., and Agostoni, A. (1986) Thromb. Res. 44, 85-94

Parnell, J. L., Anderson, D. O., and Kinnis, C. (1966) N. Engl. J. Med. 274, 979-984

Samuelsson, B., Dahlen, S.-E., Lindgren J. A., Rouzer, C. A., Serhan, C. N. (1987) Science 237, 1171-1175

Sasagawa, S., Suzuki, K., Sakatani, T., and Fujikura, T. (1985) J. Leukocyte Biol. 37, 493-502

Scherer, G., Jarczyk, L., Heller, W., Biber, A., Neurath, G. B., and Adlkofer, F. (1988) Klin. Wochenschr. 66 (Suppl. XI), 5-12

Siess, W., Lorenz, R., Roth, P., and Weber, P. C. (1982) Circulation 66, 44-48

Tripp, C. S., Needleman, P., and Unanue, E. R. (1987) J. Clin. Invest. 79, 399-403

EFFECTS OF NICOTINE AND COTININE ON PROSTACYCLIN BIOSYNTHESIS

R. Chahine, R. Kaiser and A. Pham Huu Chanh

Research Center, Hopital du Sacré-Coeur, 5400 boul Gouin O, Montréal, H4J1C5 (Qc) Canada and Institut de Physiologie, 2 rue F. Magendie, 31400 Toulouse, France

SUMMARY : In the present investigation, nicotine dose-dependently inhibited the biosynthesis of prostacyclin from arachidonic acid in horse aorta microsomes, while cotinine caused stimulation. These two compounds have no action on thromboxane A_2 in horse platelet microsomes. Furthermore, cotinine preincubated before nicotine with horse aorta microsomes can prevent the inhibitory effect of nicotine on prostacyclin biosynthesis in vitro. Our results suggest that the biotransformation of nicotine to cotinine may be a beneficial mechanism.

Extensive evidence indicates that smokers have an increased risk to develop cardiovascular disease (Schievelbein & Eberhardt, 1972). A variety of explanations have been proposed. However, the increased aggregability of their platelets in vitro, may be a possible explanation (Levine et al., 1973). Since endothelium and platelets transform arachidonic acid (AA) to several prostaglandins (PG), special attention has been attributed to prostacyclin (PGI_2) and thromboxane (TXA_2) which have an importance in homeostasis (Moncada & Vane, 1979). Nicotine, a major constituent of cigarette smoke, has been shown to decrease the release of PGI_2 from heart and aorta (Alster et al., 1986) contributing thereby to acute cardiovascular effects of tobacco. It has been also reported that cotinine arises as a result of nicotine metabolism in a variety of mammalian species and during the fermentation of tobacco as well (Gorrod & Jenner, 1975). Since the half life of cotinine in blood is longer than that of nicotine, and consequently cotinine is found in the blood and urine in larger amounts and for longer periods than nicotine (Benowitz et al., 1983), it was of interest to compare the effects of nicotine and cotinine on the PG system in vitro.

MATERIALS AND METHODS

AA (32.8 µM) was incubated with increasing concentrations of horse platelet microsomes (HPM), source of TXB_2 synthetase and horse aorta microsomes (HAM) source of PGI_2 synthetase as descreibed elsewhere (Pham Huu Chanh et al., 1987). Incubation was carried out at 37°C for 2 min. the reaction was stopped by addition of 0.2 M citric acid (pH 3.5). Various concentrations of nicotine and/or cotinine were preincubated with enzyme sources for 10 min at 37°C. At the end of incubation, TXB_2 and 6-keto PG $F_{1\alpha}$ stable metabolites of TXA_2 and PGI_2 respectively were extracted by diethylether and determined by RIA.

RESULTS

Incubation of AA (32.8 µM) with HPM, at concentrations ranging from 25 to 640 µg in protein equivalent, induced the biosynthesis of increased amounts of TXB_2. When preincubated 10 min with HPM, nicotine and cotinine (0.01 to 1 mM) did not change significantly the amounts of TXB_2 formed. When AA (32.8 µM) was incubated with increasing doses of HAM, 6-keto PG $F_{1\alpha}$ increased with HAM concentrations. A 10 min preincubation of nicotine prior to AA addition increased the 6-keto PG $F_{1\alpha}$ formation (Fig 1 left), while cotinine under the same

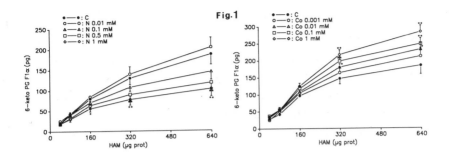

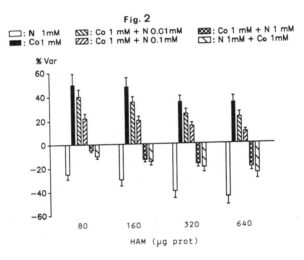

Fig 1. Effects of nicotine and cotinine on PGI_2 biosynthesis in vitro: Quantities of 6-keto PG $F_{1\alpha}$ formed in pg versus increased horse aorta microsomes (HAM) concentrations in µg of protein content. Left panel: in the absence (control) and presence of nicotine (N). right panel: in the absence (control) and presence of cotinine (Co). All values are represented as means ± ESM, n = 7, * p < 0.05, ** p <0.01

Fig 2. Interference effects of nicotine and cotinine on PGI_2 biosynthesis: % of variation of 6-keto PG $F_{1\alpha}$ formation in comparision with control values, versus increased HAM concentrations.
Cotinine 1 mM was preincubated 5 min with HAM before nicotine (0.01, 0.1, 1) mM. In the last experiment nicotine (1 mM) was preincubated 5 min before cotinine (1 mM). n = 7.

experimental conditions decreased the formation of 6-keto PG $F_{1\alpha}$ (Fig 1 right). When cotinine was preincubated with HAM prior to the addition of nicotine, the inhibitory effect of nicotine was abolished with regard to the doses used. However, the addition of cotinine after nicotine incubation was not able to reverse the inhibitory effect of the latter (Fig 2).

DISCUSSION

Our study provides a new approach in the comprehension of nicotine metabolism. In fact administration of nicotine induces often unpredictable changes in the body due to its complex biphasic action. On the other hand, cotinine the major metabolite by-product of nicotine, has been shown to induce a vascular muscular relaxation, hypotension and to reverse the pressor action of nicotine (Borzelleca, et al., 1962 ; Chahine, 1989). Therefore, a comparative action of these two compounds was interesting at organ and cellular level.

In our experimental conditions, neither nicotine nor cotinine had any effect on TXA2 biosynthesis from AA in horse platelet microsomes. On the other hand both these compounds had an effect on PGI2 biosynthesis generated by the incubation of AA with horse aorta microsomes: Nicotine inhibits and cotinine stimulates PGI_2 synthesis. Our results on nicotine corroborated the findings of Wennmalm (1980) in rabbit heart and Stoel et al. (1979) in human umbilical artery. Using various concentrations of AA, nicotine and cotinine effects on PGI_2 synthetase were non-competitive , which does not exclude an action on other enzymes of the AA cascade present in the milieu. However, the most important finding is that cotinine incubated before nicotine with HAM seems to reverse the inhibitory effects of nicotine on prostacyclin synthetase. It is important to note that cotinine possesses a preventive action against the inhibitory effect of nicotine but not a curative effect. In fact, added after the incubation of nicotine with HAM, cotinine is not able to remove the inhibitory effect of nicotine on PGI_2 biosynthesis <u>in vitro</u>.

<u>CONCLUSION</u>: The conversation of nicotine to cotinine seems to minimize the harmful effects of the parent compound. These results suggest that the metabolic pathway from nicotine to cotinine may be a beneficial mechanism.

<u>Acknowledgements</u>: This work is dedicated to the memory of Dr Pham Huu Chanh.

<u>REFERENCES</u>

Alster, P., Brandit, R., Koul, B.L., Nowak, J., Sonnenfeld,.T. (1986)
    Gen. Pharmacol. 17, 441-444.
Benowitz, N. (1983)
    Clin. Pharmacol.Ther. 309, 139-142.
Borzelleca, J., Bowman, E., and McKennis, H., Jr. (1962)
    J. Pharmacol. Exp. Ther. 137, 313-318.
Chahine, R. (1989) In : Recherches sur les effets de la métabolisation de la nicotine. Thesis (Doctorat Etat), Toulouse, p.150-200.
Gorrod, J.W., and Jenner, P. (1975)
    Essays Toxicol. 6, 35-37.
Levine, P.H. (1973)
    Circulation 48, 619-623.
Moncada, S., and Vane, J.R. (1979)
    Pharmacol. Rev. 30, 293-331.
Pham Huu Chanh, Chahine, R., Pham Huu Chanh, A., Dossou-Gbete, V., and Navarro-Delmasure, C. (1987)
    Prostaglandins Leukotrienes and Med. 28, 243-254.
Schievelbein, H., and Eberhardt, R. (1972)
    J.Nat.Cancer.Inst. 48, 1785-1794.
Stoel, I., Giessen, N., Zwolsman, E., Verheugt, F., Ten Hoor, F., Quadt, F., and Hugenholtz, P. (1982)
    Br.Heart J. 48, 493-496.
Wennmalm, A. (1980)
    Br.J.Pharmacol. 69, 545-549.

Effects of Nicotine on Biological Systems
Advances in Pharmacological Sciences
© Birkhäuser Verlag Basel

EFFECTS OF NICOTINE ON THE GENERATION OF LIPOXYGENASE PRODUCTS

M. Raulf & W. König

Lehrstuhl für Med. Mikrobiol. & Immunol., AG
Infektabwehrmechanismen, RUHR-Universität Bochum, 4630 Bochum,
FRG;

Summary: The effects of nicotine on human polymorphonuclear
granulocytes (PMN) and platelets were studied with regard to
the generation of lipoxygenase-derived arachidonic acid
metabolites. The treatment of PMN with nicotine-tartrate leads
to a concentration-dependent decrease of leukotriene B4
generation induced by Ca-ionophore A23187. In contrast, with
opsonized zymosan as leukotriene inducing stimulus the same
concentration of nicotine-tratrate significantly increased
leukotriene B4 generation. Only nicotine-tratrate
concentrations higher than $1x10^{-5}$ M showed these modulatory
effects. High concentrations of nicotine-tratrate increased the
metabolism of LTB4 to its omega-oxidated products; the
intermediate products of omega-oxidation, the 20-OH-LTB4, was
significantly enhanced, whereas the amount of 20-COOH-LTB4, was
decreased. The results obtained by incubation of human
platelets with nicotine-tratrate demonstrate no effect in
platelet aggregation, the metabolism of platelet activating
factor (PAF) and release of serotonin. Only high concentrations
of nicotine tratrate decreased the generation of 12-HETE.

INTRODUCTION:

Environmental agents and drugs lead to a number of alterations
in both humoral and the cellular components of the immune
system (Phillips et al., 1985). They may induce an activation
of immunological and non-immunological mechanisms of host
defense which includes a variety of cell types and mediators.
The function of polymorphonuclear granulocytes (PMN) which
constitute the first line of defense against bacteria is
affected by different drugs, e.g. extracts of smoke (Corberand
et al., 1980). In addition to chemotaxis, phagocytosis and the
release of granular enzymes, an important property of PMNs is
the ability to generate and release potent mediators of
inflammation (König et al., 1990). Stimulation of the
granulocytes activates a complex signaltransduction cascade,
which results e.g. in the generation of leukotrienes. LTB4, the
main metabolite of human PMNs, stimulates the chemotaxis,

chemokinesis and degranulation in vitro and is catabolized rapidly via omega-oxidation into the less active products, the 20-hydroxy- and 20-carboxy-LTB4 (Brom et al., 1987). Platelets play a vital role in haemostasis and thrombosis. Stimulation of platelets leads to a change in their shape, in the release of their granules (e.g. serotonin) and aggregation. They also produce and metabolize bioactive mediators such as 12-hydroxyeicosatetraenoic acid (12-HETE). The purpose of the study was to investigate the effects of nicotine on leukotriene generation and metabolism of human neutrophils. Additional experiments were carried out to determine the in-vitro effects of nicotine on the cellular responses of platelets.

MATERIALS AND METHODS:

Preparation of the cells: Human polymorphonuclear leukocytes (PMN) were isolated from 200 ml heparinized (15/ml) blood using Ficoll-metrizoate gradient centrifugation followed by dextran sedimentation (Böyum, 1968). Platelets were isolated from platelet rich plasma (PRP) (Brom et al., 1989).

Analysis of leukotrienes and mono-HETEs: Cells (1×10^7 PMNs or 1×10^8 platelets) were incubated in the presence of calcium (1 mM) and magnesium (0.5 mM) with the indicated stimuli over different times at 37^0C. The reaction was terminated by the addition of methanol/acetonitril (50:50, v/v) and centrifugation at 1900xg for 15 min. After lyophilization, the remainder was analysed by RP-HPLC (Knöller et al., 1988).

RESULTS:

Effects of nicotine-tartrate on leukotriene B4 generation and metabolism: Human PMNs were preincubated for 5 min with different concentrations of nicotine-tartrate or the same volume of phosphate buffered saline (PBS) as control, followed by incubation of the Ca-ionophore A23187 (5 uM) or opsonized zymosan (2mg) for additional 15 min. Nicotine-tartrate alone, in the absence of any additional stimuli was not sufficient to induce leukotriene generation. Our data demonstrate that the treatment of PMN with nicotine-tartrate leads to a concentration-dependent decrease of leukotriene B4 generation induced by the Ca-ionophore. In contrast, with opsonized zymosan as leukotriene inducing stimulus the same concentration of nicotine-tartrate significantly increased the leukotriene B4 release. The effects were obtained with high concentrations of nicotine-tartrate (150-15 uM) under non-cytolytic conditions.

Experiments were then carried out to analyze the effects of nicotine tartrate pretreatment of PMNs on the metabolism of exogenously added LTB4. These experiments allow to analyze whether an exogenously added compound, such as nicotine-tartrate may affect the chemotactic activity of LTB4. For this purpose, PMNs were incubated with different concentrations of nicotine-tartrate or PBS (as control), followed by the addition of LTB4. The metabolism of LTB4 to 20-OH-LTB4 and 20-COOH-LTB4 was calculated by RP-HPLC. Our data demonstrate (Fig. 1) that only high concentrations of nicotine-tartrate increased the metabolism of LTB4 to its omega-oxidated products, the intermediate product of omega-oxidation, the 20-OH-LTB4, was significantly enhanced, whereas the amount of 20-COOH-LTB4, was decreased.

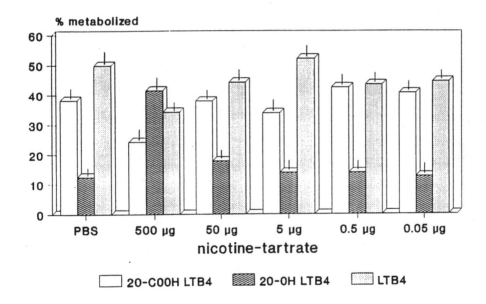

Fig. 1: Effect of nicotine-tartrate on the metabolism of exogenously added LTB4. Each value represents the mean of four independent experiments (mean +/- SEM, n=4).

Effects of nicotine-tartrate on the cellular responses of
platelets: Incubation of platelets from nonsmokers with
nicotine-tartrate did not affect platelet aggregation induced
by thrombin or the metabolism of PAF. Nicotine-tartrate alone,
was not able to induce aggregation or 12-HETE-generation. Only
high concentrations of nicotine-tartrate (150-15 uM) decreased
the generation of 12-HETE induced by thrombin. In contrast,
similar concentrations of cotinine did not affect the thrombin-
induced 12-HETE formation.

DISCUSSION:

Recent studies have shown that cigarette smoke inhibits the
cyclooxygenase-related metabolism and decreases the Ca-
ionophore-induced LTB4 production of alveolar macrophages
obtained from bronchial lavages of smokers (Laviolette et al.,
1981; 1985; Tardif et al., 1990). The components of cigarette
smoke responsible for the modulation of the cyclooxgenase- and
lipoxygenase-pathway are not yet identified. Our data show that
treatment of PMNs from nonsmokers with nicotine-tartrate
modulates the LTB4-generation in a stimulus-dependent manner:
LTB4-generation was inhibited after stimulation with Ca-
ionophore, whereas with opsonized zymosan it was enhanced.
Comparable results as observed for in-vitro treatment of PMNs
with nicotine-tartrate were obtained with PMNs of smokers (data
not shown). In this regard PMNs of smokers generate reduced
amounts of LTB4 (combined amounts of LTB4, 20-OH-LTB4 and 20-
COOH-LTB4) after stimulation with the Ca-ionophore and
increased values when stimulated with opsonized zymosan.
     Nicotine-tartrate modulates the metabolization of LTB4 to
its omega-oxidated products; it influences the ability of the
PMNs to decrease the chemotactic potential. These results
suggest that 1.) nicotine-tartrate may stimulate the P-450-
cytochrome-enzyme, which is the LTB4-omega-hydroxylase or 2.)
nicotine-tartrate may inhibit the enzymatic activity of the 20-
OH-dehydrogenase. One also may suggest, a downregulation of the
LTB4-binding sites.
     The results observed with platelets demonstrate that only
high concentrations of nicotine-tartrate modulate the thrombin-
induced 12-HETE-generation.
     Our data suggest, that only high concentrations of
nicotine-tartrate modify the mediator generation depending on

the subsequent stimulus. Further studies have to elucidate which concentrations of nicotine-tartrate interact with the target cells and subsequently modulate the cellular response.

REFERENCES:

Böyum, A. (1968) Scand. J. Lab. Invest. 97 (Suppl) 77-89
Brom, C., Köller, M., Brom, J., & König, W. (1989) Immunology
    68, 240-246
Brom, J., König, W., Stüning, M., Raulf, M. & Köller, M. (1987)
    Scand. J. Immunol. 25, 283-294
Corberand, J., Laharrague, P., Nguyen, F., Dutan, G.,
    Fontanilles, A.M., Gleizes, B. & Gyard, E. (1980) Infect.
    Immun. 30, 649-655
Knöller, J., Schönfeld, W., Köller, M., Hensler, T. & König, W.
    (1988) J. Chromatography 427, 199-208
König, W., Schönfeld, W., Raulf, M., Köller, M., Knöller, J.,
    Scheffer, J. & Brom, J. (1990) Eicosanoids 3, 1-22
Laviolette, M., Chang, J. & Newcombe, D.S. (1981) Am. Rev.
    Respir. Dis. 124, 397-401
Laviolette, M., Coulombe, R., Picard, S., Braquet, P. &
    Borgeat, P. (1985) J. Clin. Invest. 77, 54-60
Phillips, B., Marshall, M.E., Brown, S. & Thompson, J.S. (1985)
    Cancer 56, 2789-2792
Tardif, B., Borgeat, P. & Laviolette, M. (1990) Am. J. Resp.
    Cell Mol. Biol. 2, 155-161

EFFECTS OF NICOTINE ON URETERIC MOTILITY

M. İ. CİNGİ, K. EROL, R. S. ALPAN, M. ÖZDEMİR

Department of Pharmacology, Medical Faculty, University of Anadolu, Meşelik 26480 Eskişehir, Turkey

SUMMARY: The effects of nicotine on the ureteric motility of the sheep were studied in vitro. Indomethacin inhibited and finally stopped rhythmic motility. ACh increased dose-dependently both frequency and amplitude of contractions. Nicotine caused stimulatory effect on frequency and amplitude of contractions. Additionaly it initiated the ureteral peristaltism inhibited by indomethacin and dexamethasone.

Nicotine is an important ingredient of tobacco smoke, representing over 95% of its alkaloidal content. It causes the release of catecholamines in a number of isolated organs. This action results in a sympathomimetic response (Goodman Gillman 1985).

It was shown that arachidonic acid metabolites are involved in the physiological regulation of micturition in vivo (Maggi et al 1984). Therefore, it seems reasonable to assume that prostaglandins also play a role in ureteric motility. Spontaneous rhythmic contractile activity of ureter was blocked by non-steroidal antiinflammatory drugs (Thulesius and Angelo-Khattar 1985, Angelo-Khattar et al 1985).

The aim of this study was to determine the direct effect of nicotine and the interaction of indomethacin and nicotine on ureteric motility.

MATERIALS AND METHODS

Sheep ureters were obtained immediately after slaughter from the local abattoire and transported in an ice-cold modified De-Jalon solution to the laboratory within 30 minutes. The proximal part of the ureter, 2-5 cm from the renal pelvis, was selected and properly dissected, freed of adherent fat and cut spirally. Ureteral strips were suspended in 10 ml organ baths filled with modified De-Jalon solution maintained at 37 $^{\circ}$C and gassed with 95% O_2 and 5% CO_2. The preparations were connected to an isotonic transducer (TD 112 S) (load 1 g) or an isometric transducer (TB-651 T) (load 2 g). The responses were recorded using a recorder (Nihon Kohden). The bathing medium was modified De-Jalon solution containing (mM): NaCl:154.1, KCl:5.6, $CaCl_2$:1.8, $NaHCO_3$:5.9, glucose:5.5. Paired t test is used for statistical analysis. Data are expressed as mean ± SEM.

<u>Drugs:</u> Indomethacin (Sigma), Acetylcholine (Sigma), Nicotine (Sigma), Dexamethasone (Dekort amp.), Hexamethonium (Sigma).

<u>RESULTS:</u>

After a short latency period, isolated ureteral strip preperations showed spontaneous rhytmic motility. Nicotine caused an insignificant increase of the frequency and a dose dependent increase of the amplitude on the ureteric rhythmic contractions (Table I, Fig. 1). Acetylcholine increased the ureteral tonus and peristaltism (Fig. 2). The spontaneous rhytmic contractile activity of sheep ureter was dose-dependently blocked by indomethacin (Fig. 3) and dexamethasone (Fig. 4). Nicotine stimulated the rhythmic contractions of ureter inhibited by indomethacin and dexamethasone (Fig. 5), but not acetylcholine. The amplitude of ureteral contraction was decreased, but not completely inhibited by hexamethonium (Fig. 6).

Table I. Effect of nicotine on the amplitude of ureteric motility in vitro. (n=9)

	Ni 10 -6 M	Ni 10-5 M	Ni 10-4 M	Ni 10-3 M
Before nicotine	13.37 ± 1.82	12.27 ± 2.02	9.68 ± 0.89	10.66 ± 1.89
After nicotine	14.50 ± 1.89*	13.31 ± 2.23**	11.18 ± 1.02***	15.61 ± 2.22***
% difference	8.3	8.5	15.49	46.43

    * $p < 0.01$ ** $p < 0.05$ *** $p < 0.001$

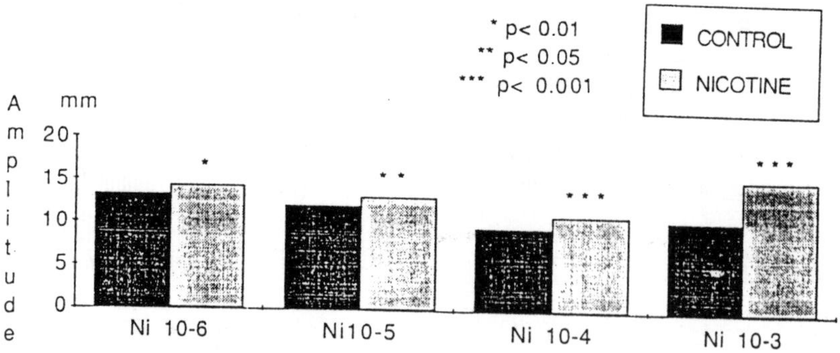

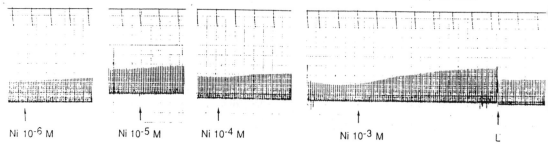

Fig. 1a. Effect of nicotine on the amplitude of ureteric motility in vitro.

Fig. 1b. The effect of nicotine on ureteric rhythmic motility (10^{-6}-10^{-3} M)

Fig. 2: The effect of Acetylcholine on ureteric rhythmic motility.

Fig. 3: The effect of indomethacin on ureteric

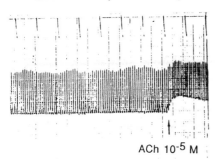

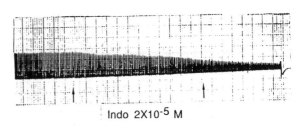

ACh 10^{-5} M

Indo 2X10^{-5} M

Fig. 4: The effect of dexamethasone on ureteric rhythmic motility.

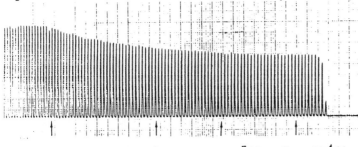

Dexa 10^{-7} M   Dexa 10^{-6} M   Dexa 10^{-5} M   Dexa 10^{-4} M

Fig. 5: Nicotine stimulated the rhythmic contractions inhibited by dexamethasone and indomethacin.

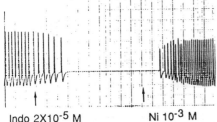

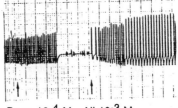

Indo 2X10^{-5} M   Ni 10^{-3} M

Dexa 10^{-4} M   Ni 10^{-3} M

Fig. 6: The effect of nicotine on the response of hexamethonium.

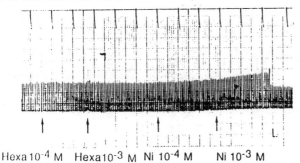

Hexa 10^{-4} M   Hexa 10^{-3} M   Ni 10^{-4} M   Ni 10^{-3} M

DISCUSSION:

As in the previous studies (Thulesius and Angelo-Khattar 1985, Angelo-Khattar et al 1985) we were able to show that indomethacin inhibits ureteral motility in vitro. This was taken as indirect evidence in support of a motility-promoting role of prostaglandins.

It may seem surprising to find PGI_2 and TXA_2 synthesis occuring in uretheral tissue, since these arachidonic acid derivatives are mainly associated with endothelial cells or platelets, respectively (Higgs and Moncoda1983). Extravascular production of PGI_2 and TXA_2 has however, also been reported to occur in a variety of tissues such as the rat uterus (Fenwick et al 1977), guinea-pig trachea (Burka 1981), bovine gastric mucosa (Ali and Mc Donald, 1980) and sheep lungs (Mc Donald 1983).

Our results on inhibition of ureteric motility by indomethacin were similar to those of certain investigators (Angelo-Khattar et al 1985, Thulesius et al, 1987).

CONCLUSION: Nicotine potentiated the normal ureteric peristaltism and initiated the ureteric motility inhibited by indomethacin and dexamethasone. These results may suggest that there is an important role of prostaglandins on ureteric motility and nicotine stimulates prostaglandin production and/or ureteric parasympathetic ganglion.

REFERENCES:

Ali, M., and Mc Donald, J.W.D. (1980). Synthesis of thromboxaneB_2 and 6-ketoprostaglandinF$_1\alpha$ by bovine gastric mucosal and muscle microsomes. Prostaglandins 20, 245-254.

Angelo-Khattar, M., Thulesius, O., Nilson, T., Cherian, T., and Joseph, L., Motility of the human ureter, with special reference to the effect of indomethacin. (1985). Scand. J. Urol. Nephrol. 19, 261-265.

Burka, J. F., Ali, M., Mc Donald, W. D., and Paterson, A. M. (1981). Immunological and non immunological synthesis and release of prostaglandins and thromboxanes from isolated guinea pig-trachea. Prostaglandins. 22, 683-691.

Fenwick, L., Jones, R. L., Naylor, B., Poyser, N. L., and Wilson, N. H. (1977).Production of prostaglandins by the pseudopregnant rat uterus in vitro and the effect of tamoxifen with the identification of 6-keto prostaglandin F$_1\alpha$ as a major product. Br. J. Pharmacol 59, 191-199.

Goodman Gilman, A., Goodman, L. S., Rall, T. W., Murad, P. (1985). The pharmacological Basis of Therapeutics 7th ed., Macmillan Publishing company, New York. pp. 217-218.

Higgs, E. A., and Moncado, S. (1983). Prostacyclin-physiology and clinical uses. Gen. Pharmacol. 14, 7-11.

Maggi, C. A., Evangelista, S., Grimaldi, G., Santicioli, P., Gioliotti, A., Meli, A. (1984). Evidence for the involvement of arachidonic acid metabolites in spontaneous and drug-induced contractions of rat urinary bladder. J. Pharmacol. Exp. Ther. 230, 500-513.

Mc Donald, J.W.D., Ali, M., Morgan, E.,Townsend, E.R., and Dropper, J.D. (1983). Thromboxane synthesis by sources other than platelets in association with complement-induced pulmonary leucostasis and pulmonary hypertension in sheep. Circ. Res. 52, 1-6.

Thulesius, M., Angelo-Khattar, M., and Ali, M. (1987). The effect of prostaglandin synthesis inhibition on motility of the sheep ureter. Acta Physiol. Scand. 131, 51-54.

Thulesius, M., and Angelo-Khattar, M. (1985). The effect of indomethacin on the motility of isolated sheep ureters. Acta Pharmacol. Toxicol. 56, 298-301.

Effects of Nicotine on Biological Systems
Advances in Pharmacological Sciences
© Birkhäuser Verlag Basel

NICOTINE INDUCED ALTERATION IN PHOSPHATASE ACTIVITY IN PLASMA, LIVER AND KIDNEY OF RATS.

S. Dasgupta, C. Mukherjee and S. Ghosh

Department of Physiology, University of Calcutta,
92, A.P.C.Road, Calcutta-700 009, India.

Acid and alkaline phosphatase activities have been estimated in plasma, liver and kidney of male wistar rats treated with nicotine, to assess the functional significance of toxic reactions of nicotine. Rats received nicotine by subcutaneous injection at a dose of 400 μg/kg^{-1} b.wt. for 7 consecutive days. Enzyme activities were estimated by the standard method of Bessey, data being statistically analysed using student's 't' test. There is an observed increase in plasma ACP and decrease in hepatorenal enzyme activity. ALP increased in plasma and kidney while liver enzyme showed a decrease, all results being significant in experimental group when compared to normal control. These point to a probable mechanism of action of nicotine, involving the metabolic alteration of hepatorenal system.

INTRODUCTION

Nicotine is of considerable medical significance because of its toxicity and tendency to produce coronary heart disease and lung Cancer (Downey et al, 1977). Induction of enzymes in liver by tobacco smoke and alteration of resting metabolic rate by nicotine (Perkins et al 1989) have been mentioned. Liver and kidney accumulate nicotine and are prone to damage by this toxic alkaloid.

Acid and alkaline phosphatases (ACP and ALP) are two enzymes having significant applicability is different metabolic reactions of the organism ACP, a lysosomal enzyme, controls a reconversion of phosphorylated glucose to glucose and inorganic phosphate (Lloyd, 1978). ALP, derived in serum, mainly from liver cells (Righetti et al, 1971) may also be found in the kidneys where it acts as an esterase (Lloyd, 1978).

The present study has therefore been designed to assess the effects of nicotine on hepatorenal phosphatase activity which might throw some light on the metabolic effect of nicotine.

MATERIALS AND METHODS

Normal, adult, male Wistar rats, 170-200 gms b.wt. were maintained at normally illuminated laboratory under controlled temperature. They were provided with laboratory stock diet and water ad libitum for a week. The acclimatized animals were divided into two groups of 10 animals each fed normal diet. One group served as control while the other as experimental. For 30 days they were maintained on normal diet after which the experimental group received nicotine, dissolved in physiological, saline at a dose of 400 μg kg^{-1} b.wt. (Rosecrans, 1970). Subcutaneously for seven consecutive days. The animals of the control group served as pairfed control and were injected with the vehicle for the same priod. After completion of treatment, all   animals   were sacrificed by decapitation between 8.00 and 11.00 hours to minimize any diurnal variation.

Blood was collected in heparinized tubes and plasma was separated by immediate centrifugation. Liver and Kidney were dissected out and adhering blood and tissue fluid were blotted dry. Each tissue was weighed and homogenized in 0.25M sucrose in a glass homogenizer maintained at 0°-4°C. Activities of ACP and ALP were estimated by the standard method (Bessey, et al, 1946) using p-nitrophenyl phosphate as the substrate. Data were statistically analyzed using students' 't' test.

RESULTS

Table-I shows that a significant increase in ACP occurs in plasma decrease in liver and kidney in experimental animals when compared to control ones.

Table-I : Effect of nicotine (400 μg kgm^{-1} b.wt.,S.C.) on ACP activity in rats (Means ± SEM; n = 10)

	NORMAL CONTROL	EXPERIMENTAL (NICOTINE TREATED)	P Value
PLASMA (mmol/h/100 ml)	0.95 ± 0.08	1.21 ± 0.09	< 0.05
LIVER (mmol/h/g of wet tissue)	2.88 ± 0.14	1.59 ± 0.20	< 0.001
KIDNEY (mmol/h/g of wet tissue)	3.13 ± 0.27	1.83 ± 0.14	< 0.001

Table-II shows a significant increse in alkaline phosphatase activity in plasma and kidney while liver enzyme shows a significant decrease in activity in the experimental group as compared with untreated control.

Table-II : Effect of nicotine (400 μg kgm^{-1} b.wt; S.C.) on ALP activity in rats (Means ± SEM; n = 10)

	NORMAL CONTROL	EXPERIMENTAL (Nicotine Treated)	P Value
PLASMA (mmol/h/100 ml)	9.05 ± 0.93	13.105 ± 0.89	< 0.01
LIVER (mmol/h/g of wet tissue)	6.48 ± 0.42	4.08 ± 0.35	< 0.001
KIDNEY (mmol/h/g of wet tissue)	194.52 ± 8.49	276 ± 9.27	< 0.001

DISCUSSION

Serum is usually very low in ACP activity. It is a lysosomal marker enzyme, localized in the lysosomal membranes. Nicotine increases ACP activity in the serum, which is probably derived from kidney as well as liver. This might very well suggest that nicotine has a damaging effect upon the metabolic organs. Damage to liver and kidney may take place which causes release of ACP from these tissues to serum.

    Liver is a very important site of metabolic activity as well as the site of synthesis of phosphatases (Abderhalden, 1961). It is widely distributed in the plasma membranes and act to regulate the dimensions of the membrane, thus involved indirectly in the transport processes of the liver (Fishman et al, 1973). Short term nicotine treatment causes a significant elevation of serum ALp with a commitant significant decrease in liver ALP. This suggests that nicotine may play some role in altering the rate of movement of substances across hepatic cells. It might also have some effect on the synthesizing machinery of liver cells for ALP. Hepatic damage is well indicated because liver ALP decreases upon nicotine treatment, which suggests a loss of ALP from the liver to the plasma.

ALP is localized in the microvilli of the proximal tubular cells, forming an integral part of the membrane. Nicotine causes an increase in kidney ALP indicating an accumulation of the enzyme in renal tissue. Any alteration in membrane structure may lead to this accumulation. Nicotine may thus exert its effects upon renal tissue.

CONCLUSION

It may, therefore, be surmised that nicotine appears to affect the metabolically active tissues as liver and kidney of rats. Probable mechanism of action may inivolve changes in enzyme synthesis, change in the metabolic status of liver and alteration of structural integrity of both hepatic and renal tissue.

REFERENCES

Abderhalden,R. (1961) Clinical Enzymology, p.58, D. Van Nostrand Company, Inc., Princeton, New Jersey.

Bessey,O.A., Lowry, O.H. and Brock, M.J. (1946). J. Biol. Chem.,164, 321-329.

Downey,M.F.,Bashour,C.A.,Boutros,I.S.,Bashour,F.A. and Parker, P.E.ïï (1977) J. Pharmacol. Exp. Ther., 202 : 55-68.

Fishman,W.H. and Lin,C.W.(1973) Membrane Phosphohydrolases, Metabolic Conjugation, Metabolic hydrolysis p.387, N.Y., Acad.Press.

Lloyd,L.E., McDonald,B.E., Crampton,E.w.(1978) In : 'Fundamentals of Nutrition' 2nd ed., pp.237-241, W.H.Freeman & Company, San Francisco.

Perkins,K.A., Epstein,L.H.,Stiller,R.L.,Marks B.L., Jacob,R.G.,(1989) Am. J. Clin. Nutr. 50(3) : 545-550.

Righetti,A.B. and Kaplan,M.K.(1971). Biochim Biophys Acta, 230, 504-509.

Rosecrans,J.A.(1970) Fred. Proc. 29(2) : 748.

CARDIOVASCULAR EFFECTS OF REPEATED NICOTINE ADMINISTRATIONS: CONDITIONS WHERE EFFECTS ARE MAINTAINED OR WHERE SELF-BLOCKADE TAKES PLACE.

S.L. Cruz, L.A. Salazar and J.E. Villarreal.

Sección de Terapéutica Experimental, CINVESTAV, IPN. Apartado Postal 22026, 14000 México, D.F., México.

SUMMARY: The pharmacologic conditions were examined where repeated doses of nicotine produce either sustained stimulant effects or self-blockade on the cardiovascular system. Repeated nicotine showed quantitatively maintained effects in unanesthetized spinal rats, at a dose range from 0.025 to 0.4 mg/kg i.v. Sodium pentobarbital (60 mg/kg i.p.) produced a tenfold decrease in sensitivity to single initial doses of nicotine and thus brought about a requirement of larger doses for cardiovascular stimulation. Nicotine self-blockade was observed only with very large doses under special conditions of pentobarbital anesthesia.

INTRODUCTION

Cigarette smoking is a major risk factor for the production of cardiovascular disease. The tobacco alkaloid nicotine has undesirable effects which predispose to ischemic heart conditions: stimulation of sympathetic ganglia and adrenal glands with increases in heart rate and in blood pressure (See Holbrook, 1987). Clinical studies have reported that repeated cigarette smoking produces sustained increases in heart rate (Benowitz,

1986). However, classical pharmacology describes the autonomic actions of nicotine as an initial stimulation of nicotinic synapses followed by a more sustained depression (Langley, 1889; Dale, 1914; Acheson, 1954; Domino, 1967). In fact, there is a general notion in the current pharmacological literature that the stimulant synaptic actions of nicotine show self-blockade (see for instance Goldstein et al., 1974).

Our group has previously reported that in unanesthetized spinal rats, the cardiovascular stimulatory actions of nicotine show no self-blockade when nicotine is given in repeated administrations (Cruz et al., 1988). In experimental work from another group, tachyphylaxis was noted as absent with regard to the effects of repeated low doses of nicotine on heart rate and blood pressure, in rats anesthetized with low doses of sodium pentobarbital (30 mg/kg i.p.) (Kachur et al. 1986).

The purpose of the experiments presented here was to solve the apparent discrepancy between the strong concepts of classical pharmacological literature about nicotine self-blockade, with the results of clinical studies on maintained effects of repeated cigarette smoking and with our own results of sustained sympathetic cardiovascular stimulation with repeated nicotine administrations.

MATERIALS AND METHODS

Male Wistar rats weighing between 200-300 g were used. The studies were performed under the following four types of experimental conditions: a) Unanesthetized rats with spinal section between the first cervical vertebra and the foramen magnum; b) Spinal rats anesthetized with sodium pentobarbital (60 mg/kg i.p.); c) Intact rats anesthetized with the same dose of pentobarbital; and d) Intact CNS rats, vagotomized, anesthetized with the same dose of pentobarbital.

For the preparation of spinal rats, the brain was destroyed under ether anesthesia by inserting a short steel tube through the foramen magnum. Recovery was allowed for one hour.

In all the groups, an endotracheal cannula was connected to a Palmer pump maintaining a respiratory volume of 2cc/100g and a frequency of 54 rpm. Blood pressure recording was made via a cannula (PE50 tubing) in the right carotid artery attached to a pressure transducer GOULD Statham P23 1D and a Grass polygraph. Heart rate was derived from the blood pressure recording. Drugs were administered via a cannula (PE10 tubing) in the right jugular vein. Each injection was flushed with 0.2 ml of physiological saline.

RESULTS

Repeated doses of nicotine produced different effects when administered to rats under different experimental conditions. Unanesthetized animals responded to nicotine in the dose range from 0.025 to 0.4 mg/kg i.v. Repeated injections were studied in pairs, or in consecutive sequences, at different intervals with the same doses or with progressively increasing doses. The effects of a test dose of nicotine were compared in animals with and without nicotine pretreatment administrations. Under none of the conditions listed above was the response to the second dose less than the response to the first dose of nicotine.

In anesthetized rats (pentobarbital 60 mg/kg i.p.), a tenfold decrease in sensitivity to nicotine was found for the different experimental preparations studied.

Figure 1 shows sample records of blood pressure responses to pairs of nicotine administrations. In the unanesthetized spinal rat, 0.2 mg/kg of nicotine produced an important increase in blood pressure (group mean $\pm$ S.E.M. was 54 $\pm$ 5.6 mmHg). The same dose given 5 minutes later produced a further increase in blood pressure which reached a peak averaging 116 $\pm$ 6 mmHg above the pre-nicotine baseline. Figure 1 also shows that under pentobarbital, a tenfold higher dose of initial nicotine (2.0 mg/kg) produces a similar increase in pressure (57 $\pm$ 7.7 mmHg).

EFFECTS OF REPEATED DOSES
OF NICOTINE

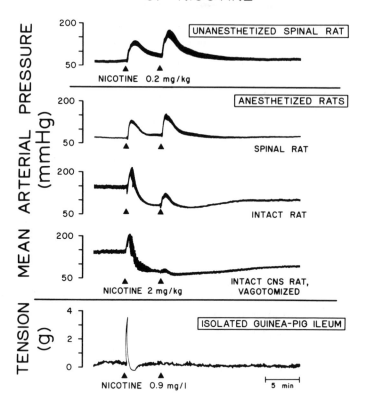

Figure 1. Effects of pairs of nicotine administrations under different experimental conditions on the blood pressure of rats and on the longitudinal muscle tension of isolated guinea-pig ileum (see text).

Self-blockade occurred with the high dose, but then it was only clearly observed in anesthetized rats with intact CNS and it was, in fact, best seen after vagotomy. For reference, figure 1 also shows the self-blockade of nicotine stimulant neurogenic actions in the isolated guinea-pig ileum. Here, the dose of nicotine was that giving responses approximately 90 % of maximum.

DISCUSSION

In the unanesthetized state, the cardiovascular system is very sensitive to nicotine's sympathetic stimulant effects. The doses of nicotine which produce stimulant effects are in a range relevant to nicotine concentrations obtained with heavy cigarette smoking (Benowitz, 1986). These doses of nicotine do not produce self-blockade but in fact produce sustained cardiovascular stimulant effects on blood pressure and heart rate.

With regard to cardiovascular stimulation, the classical self-blockade by nicotine can be reached only with very high doses of nicotine and under special conditions of general pentobarbital anesthesia.

REFERENCES

Acheson, G.H. (1954) In: Pharmacology in Medicine (V.A. Drill, Ed.), McGraw-Hill, New York, pp 29/1-29/11.
Benowitz, N.L. (1986) Ann. Rev. Med. 37, 21-32.
Cruz, S.L., Salazar, L.A., and Villarreal, J.E. (1988) Memorias del XII Congreso Nacional de Farmacología, México, p 152.
Dale H.H. (1914) J. Pharmacol. Exp. Ther. 6, 147-190.
Domino, E.F. (1967) Ann. N.Y. Acad. Sci. 142, 216-244.
Goldstein, A., Aronow, L. and Kalman, S.M. (1974) Principles of Drug Action, John Wiley & Sons, Inc., New York.
Holbrook, J.H. (1987) In: Harrison's Principles of Internal Medicine, 11th ed. McGraw-Hill, New York, pp 855-858.
Kachur, J.F., May, E.L., Awaya, H., Egle, J.L., Aceto, M.D. and Martin, B.R. (1986) Life Sci. 38, 323-330.
Langley, J.N. and Dickinson, W.L. (1889) Proc. Roy. Soc. B. 46, 423-431.

Effects of Nicotine on Biological Systems
Advances in Pharmacological Sciences
© Birkhäuser Verlag Basel

NICOTINE-INDUCED RELEASE OF NORADRENALINE AND NEUROPEPTIDE Y IN THE
HEART: DEPENDENCE ON CALCIUM, POTASSIUM, AND pH

G. Richardt, M. Haass, Th. Brenn, E. Schömig*, and A. Schömig

Department of Cardiology, University of Heidelberg, Germany
Department of Pharmacology, University of Würzburg, Germany*

SUMMARY: Dependence of nicotine-induced noradrenaline and
neuropeptide Y release on extracellular calcium, potassium, and pH
has been investigated in isolated perfused guinea pig hearts.
Noradrenaline and neuropeptide Y were determined by high pressure
liquid chromatography and by a specific radioimmunoassay,
respectively. Nicotine (1-100 μmol/l) induced a concentration
dependent release of both sympathetic transmitters. The release was
absolutely dependent on the extracellular presence of calcium.
Reduction of extracellular potassium resulted in a markedly
increased overflow of noradrenaline and neuropeptide Y. A rise of
potassium above physiological levels caused a suppression of the
release. Likewise, a rise of extracellular proton concentrations
caused a reduction of noradrenaline and neuropeptide Y release.

Cardiovascular effects of cigarette smoking, such as tachycardia
and peripheral vasoconstriction, are related to a nicotine-induced
release of noradrenaline from sympathetic nerve terminals.
Recently, it has been reported that nicotinic noradrenaline release
in guinea pig heart is closely associated with the liberation of
neuropeptide Y (Richardt et al., 1988), a sympathetic peptide
transmitter with potent constrictive properties in peripheral and
coronary vessels. There is now good evidence that nicotine causes
a co-release of both noradrenaline and neuropeptide Y by exocytosis
which is comparable to that achieved by electrical nerve
stimulation (Haass et al., 1989).

The nicotine-induced noradrenaline and neuropeptide Y release
may be significantly affected by the ionic environment of the
cardiac sympathetic nerve terminals, which is influenced by
myocardial ischemia, heart failure, and treatment with various

drugs. In order to define the effects of extracellular ionic
changes on nicotine-induced noradrenaline and neuropeptide Y
release in the heart, the dependence on extracellular calcium,
potassium, and pH was investigated in isolated perfused guinea pig
hearts.

MATERIALS AND METHODS

Guinea pig hearts were
cannulated for isolated
coronary perfusion
according to Langendorff
(1895). Hearts were
perfused at a constant
flow of 5 ml/g/min. The
initial perfusion medium
was a modified Krebs-
Henseleit solution
(composition in mmol/l:
NaCl 125, NaHCO$_3$ 16.9,
Na$_2$HPO$_4$ 0.2, KCL 4.0, CaCl$_2$
1.85, MgCl$_2$ 1.0, glucose
11, EDTA 0.027). The
buffer was gassed with 95%
O$_2$ and 5% CO$_2$ and pH was
initially adjusted to 7.4.
After a perfusion of at
least 20 min, the changes
of the ionic composition
were made and pH-levels

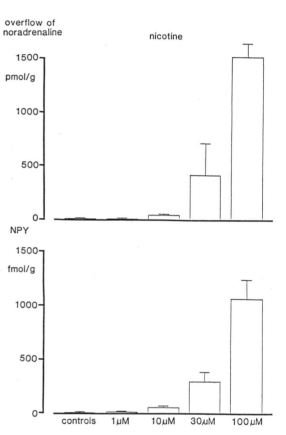

Figure 1. Concentration dependent effect
of nicotine on noradrenaline and neuro-
peptideY release in guinea pig heart

were adjusted by changing of the content of CO$_2$ and bicarbonate.
Sodium was reduced in the experiments with high potassium in order
to preserve isoosmotic conditions. After 10 min of perfusion with
changed ionic composition of the buffer, the hearts were subjected
to nicotine tartrate throughout the rest of the experiments.

Samples for determination of endogenous noradrenaline and neuropeptide Y were taken cumulatively for 2 min periods from the effluent of the hearts. Nicotine-induced release of noradrenaline and neuropeptide Y is given as cumulative overlow in a sampling period of 10 min, a period sufficient to cover the release of both transmitters, since the release is determined by rapid tachyphylaxis to nicotine (Richardt et al. 1988). Noradrenaline was measured by high pressure liquid chromatography and electrochemical detection according to Schömig et al. 1987. Endogenous neuropeptide Y was measured by specific radioimmunoassay (Haass et al. 1989). In text and figures, results are expressed as means ±SEM (n=5-8, each concentration).

RESULTS

The concentration dependence of nicotine-induced noradrenaline and neuropeptide Y revealed a steep increase of the effect between 10 and 100 μmol/l as shown in Figure 1. Further rise of the nicotine concentration did not cause additional release (Richardt et al. 1988).

The nicotine-induced release of both noradrenaline and neuropeptide Y was dependent on extracellular calcium (Fig. 2). Omission of calcium from the perfuion buffer suppressed the release of noradrenaline (21±6 pmol/g) and neuropeptide Y (46±11 fmol/g) to nearly basal levels, while 0.2 and 0.6 mmol/l of calcium resulted in a concentration dependent increase of transmitter liberation. Increase of extracellular potassium above physiological levels (10, 25, 50 mmol/l) reduced nicotine-induced noradrenaline and neuropeptide Y release in parallel.

Reduction of potassium below 4 mmol/l resulted in a marked increase of the nicotinic noradrenaline and neuropeptide Y release (Fig 3). At 10, 25, and 50 mmol/l of potassium, a small concentration dependent release of noradrenaline and neuropeptide Y was observed following the change of the potassium concentration alone (50 mmol/l potassium: noradrenaline 35±10 pmol/g/min,

overflow of
noradrenaline   nicotine 100µM

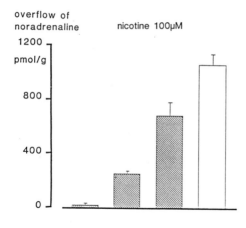

NPY

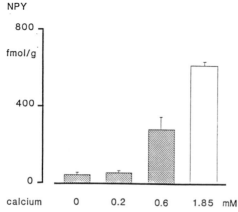

Figure 2. Effect of calcium on nicotine-
induced release of noradrenaline and
neuropeptide Y in guinea pig heart.

neuropeptide   Y: 13±3
fmol/g/min).

The investigation of the nicotinic release on pH dependence demonstrated a significant reduction of both transmitters with increasing extracellular proton concentrations (Fig. 4). At a pH of 6.0, the overflow of both substances was nearly abolished. Reduction of pH in the perfusion medium alone, did not affect the overflow of noradrenaline and neuropeptide Y.

DISCUSSION

Nicotine-induced cardiovascular actions are thought to be mediated by a liberation of noradrenaline from sympathetic nerves in the heart and the blood vessels. However, knowledge of the events leading to this neuronal release of noradrenaline and of sympathetic cotransmitters is still incomplete (for review: Löffelholz 1979).

The present findings show a nicotine-induced release of both noradrenaline and neuropeptide Y in guinea pig heart. The release of both substances was closely correlated at different concentrations of nicotine. Moreover, the overflow of both sympathetic transmitters was affected by ionic changes in parallel which supports the concept of a common release of both substances. Calcium dependence of the release provides strong evidence for an exocytotic release mechanism. Exocytosis would result in a proportional liberation of co-stored transmitters from the synaptic vesicles. In fact, co-storage of catecholamines and neuropeptide Y in sympathetic nerve terminals has been reproted (Lundberg et al, 1983; Everitt et al. 1984).

With the exception of calcium, there is little information about the influences of ionic changes on nicotine induced release of noradrenaline and, particularly, of neuropeptide Y. These changes may become

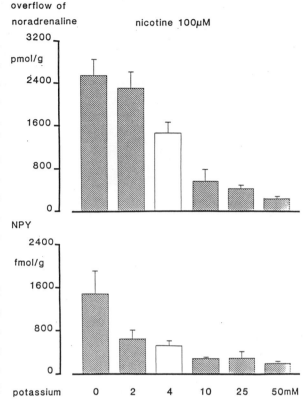

Figure 3. Effect of potassium on nicotine-induced release of noradrenaline and neuropeptide Y in guinea pig heart

relevant in patients with myocardial ischemia and heart failure or those receiving drugs which alter potassium or proton homeoestasis of the organism. The present results demonstrate a marked influence of changes in extracellular potassium on nicotine-induced noradrenaline and neuropeptide Y release. The activity of Na^+, K^+ -ATPase is known to be linked with calcium availability of the neuron such that Na^+, K^+ -ATPase activity leads to a reduction of intracellular calcium (Blaustein and Wiesmann, 1970). Thus, the latter finding may be due to activation of Na^+, K^+ -ATPase by increasing potassium concentrations.

Another interesting feature of nicotine-induced noradrenaline and neuropeptide Y release is dependence on pH in the perfusion medium. Our results show, that the release is significantly reduced by acidosis. In contrast, earlier studies about exocytotic catecholamine release in bovine adrenal medullary cells (Knight and Baker, 1982) and about stimulation-induced noradrenaline release in the rat heart (Dart et al., 1989) found that exocytosis was not altered by comparable acidosis. Therefore, pH dependence may be influenced by the inducing factors of exocytosis or may differ in various species.

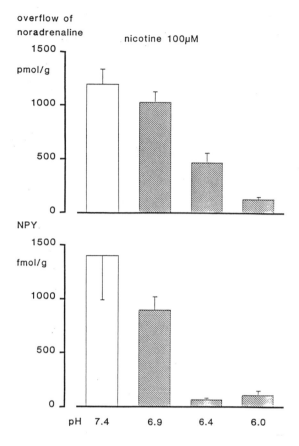

Figure 4. Effect of acidosis on nicotine-induced relese of noradrenaline and neuro-peptide Y in guinea pig heart.

CONCLUSION

Nicotine induces a concentration dependent co-release of noradrenaline and neuropeptide Y in the guinea pig heart. Calcium-dependence of the release strongly suggests an exocytotic release mechanism. The nicotine-induced release was markedly affected by changes of potassium, and pH. The suppression of nicotine-induced noradrenaline and neuropeptide Y release by low pH and high potassium may be of relevance in myocardial ischemia.

REFERENCES

Blaustein M.P., Wiesmann W.P. (1970) Proc. Natl. Acad. Sci. 66, 664-671
Dart A.M. (1989) In: Adrenergic system and ventricular arrhythmias in myocardial infarction (J. Brachmann, A. Schömig, Eds) Springer Verlag, pp 34-43
Everitt B.J., Höfkelt T., Terenius L., Tatemoto K., Mutt V., Goldstein M. (1984) Neurosci. 11, 443-462
Haass M., Cheng B., Richardt G., Lang R.E., Schömig A. (1989) Naunyn-Schmiedeberg's Arch. Pharmacol. 339, 71-78
Knight D.E., Baker P.F. (1982) J. Membrane Biol. 68, 107-140
Langendorff O (1895) Arch. Ges. Physiol. 61, 291-332
Löffelholz K. (1979) In: The release of catecholamine from adrenergic neurones (D.M. Paton, Ed) Pergamon Press, pp 275-301
Lundberg J.M., Terenius L., Höfkelt T., Goldstein M. (1983) Neurosci. Lett. 42: 167-172
Richardt G., Haass M., Neeb S., Hock M., Lang R.E., Schömig A. (1988) Klin. Wochenschr. 66, (Suppl XI) 21-27
Schömig A., Fischer S., Kurz Th., Richardt G., Schömig E. (1987) Circ. Res. 60, 194-205

Acknowledgement : The study has been supported by a grant from the 'Forschungsrat Rauchen und Gesundheit'.

EVIDENCE FOR NORADRENALINE-MEDIATED CHANGES IN PLATELET AND HEART FUNCTION FOLLOWING CHRONIC TREATMENT OF RATS AND GUINEA PIGS WITH NICOTINE

B.F. Becker, P. Dominiak and E. Gerlach, Physiologisches Institut der Universität München, Pettenkoferstr. 12, 8000 München 2, FRG

SUMMARY: Chronic application of nicotine to rats and guinea pigs for 8 weeks induces species dependent changes in blood platelet and cardiac function, which can be ascribed to elevated levels of noradrenaline in plasma and tissue. Platelet aggregability in the rat is attenuated by a ß-adrenoceptor mediated action on adenylyl cyclase, there being no effect on the adrenoceptor-devoid platelets of the guinea pig. The ß-adrenoceptor density of rat myocardium is down-regulated in 1 week upon nicotine application, affording protection of the heart from any untoward metabolic stress due to elevation of catecholamine release. In contrast, ß-receptor down-regulation occurs more slowly in the guinea pig myocardium (4 weeks). A transitory depression of function can be observed in working heart preparations isolated after 2 weeks of pretreatment. This is associated with lower levels of ATP and NAD$^+$ in the ventricular tissue. The metabolic signs of stress disappear again with progression of down-regulation.

INTRODUCTION

Chronic treatment of rats and guinea pigs with nicotine was found to lower platelet responsiveness towards the aggregatory stimulus adenosine diphosphate (ADP) in the rat, but not in the guinea pig (Becker et al., 1988b). Furthermore, simultaneous treatment of rats with the ß-blocker propranolol prevented the platelet stabilizing action of nicotine, while the nicotine effect was not mimicked by adrenaline (A). These observations may be explained on the basis of enhanced levels of noradrenaline (NA) in blood and tissues of animals pretreated with nicotine (Dominiak et al., 1984). NA is known to stimulate platelet adenylyl cyclase via ß-adrenoceptors, the subsequent rise in cAMP content suppressing platelet aggregation. Adrenaline, in contrast, inhibits adenylyl cyclase via platelet α-adrenoceptors (Kerry & Scrutton, 1985). Since the guinea pig platelets are devoid of adrenoceptors, no catecholamine-mediated effects are to be expected. A different species specific effect prevailed when hearts isolated from nicotine pretreated animals performed pressure-volume work. Whereas myocardial function remained unchanged in rats, guinea pig hearts revealed a transitory depression of function after about 2 weeks of nicotine application (Becker & Gerlach, 1988).

To establish whether these actions of nicotine were due to a selective chronic elevation of the noradrenergic tone, platelet function, platelet cAMP and plasma thromboxane (TxB_2) levels were assessed in rats chronically treated with NA or nicotine. Furthermore, the density of ß-adrenoceptors was determined in hearts of both species during chronic nicotine application, as were the myocardial tissue levels of high energy phosphates and of oxidized NAD^+. The latter served to ascertain the degree of metabolic stress (Becker et al., 1988a).

MATERIALS AND METHODS

Rats (Sprague-Dawley) and guinea pigs (Pirbright-White), all males, were obtained at 8-10 weeks of age. Noradrenaline (0.05 and 0.5mg NA·HCl/kg/d), adrenaline (0.5mg/kg/d), or nicotine (10mg/kg/d) were administered continuously, for up to 8 weeks, from osmotic minipumps (Alzet 2002) implanted subcutaneously every 2 weeks. The filling solutions all contained 0.1% ascorbic acid.

Platelet aggregation rates in response to ADP were established in rat platelet rich plasma (Becker et al., 1988b). TxB_2 was assessed in platelet poor plasma containing $2 \times 10^{-5}M$ indomethacin with the RIA kit TRK.780 from Amersham.

Isolation and perfusion of hearts were performed according to Becker & Gerlach, 1984. At a constant left ventricular filling pressure of 8 mmHg, rat and guinea pig hearts were required to work against a mean aortic pressure of 68 and 98 mmHg, resp.. The ß-adrenoceptor binding sites were determined with ^{3}H-dihydroalprenolol (DHA) in pooled ventricular tissue of 6 - 8 hearts each (see Dominiak et al., 1984). Adenine nucleotides (cAMP, ATP, ADP, AMP), NAD^{+} and creatine phosphate were measured, using HPLC, in perchloric acid extracts of sedimented platelets or lyophylized ventricular tissue (Becker et al., 1988a).

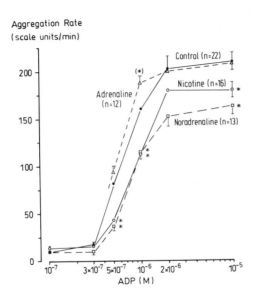

Figure 1: ADP-induced aggregation of platelets from control rats and rats pretreated for 8 weeks with adrenaline (0.5mg/kg/d), nicotine (10mg/kg/d) or noradrenaline (0.05 or 0.5mg/kg/d). Mean values ± SEM, * p<0.01, (*) p<0.05 vs control by t-test

RESULTS

Chronic application of NA reduced both the sensitivity of rat platelets towards ADP and the maximal rate of aggregation attainable (Fig.1). There was no significant difference in effect between the two NA doses investigated. A similar attenuation of

function was brought about by nicotine, whereas adrenaline activated platelets. Accordingly, the concentration of ADP required
to elicit half-maximal aggregation was significantly increased by
nicotine and NA (0.80 and 0.74μM, resp.) as compared to control
(0.62μM) and A (0.56μM).

The basal cAMP content of platelets from rats pretreated with
adrenaline was significantly lower than in controls (Fig.2).
Since neither nicotine nor NA measurably enhanced cAMP levels,
only an α-, but no ß-adrenoceptor mediated effect is clearly
discernible.

In support of enhanced platelet stability also in vivo, TxB_2
levels tended to be lower in plasma of rats pretreated with
nicotine or NA for 8 weeks, though the decreases were not
statistically significant (control: 0.34±0.05 (n=11); nicotine:
0.22±0.07 (n=4); NA: 0.24±0.04 (n=12); ng/ml ±SEM).

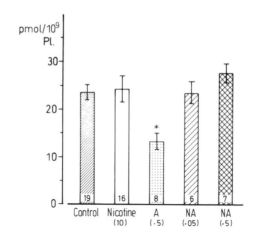

<u>Figure 2</u>: Platelet cAMP contents after 8 weeks of pretreatment of
rats with nicotine, adrenaline (A) and noradrenaline (NA). The
mg-doses/kg/d are given in brackets. Mean values ± SEM, *p<0.05
vs control by t-test

The density of ß-adrenoceptors in hearts of both species was
markedly affected by nicotine, but with different time-courses
(Fig.3). Down-regulation in rat hearts had occurred already after
1 week, but there was no further change, while the density of

control hearts had decreased to the same extent by 8 weeks. In the guinea pig heart, down-regulation developed more slowly, and was still pronounced after 8 weeks.

The stroke volume ejected by hearts isolated from pretreated animals under standardized conditions of perfusion was unchanged by nicotine in the case of the rat (Fig.4). Hearts from guinea pigs, however, showed a transitory decline in function after about 2 weeks of nicotine application. Similarly, tissue levels of adenine nucleotides determined in guinea pig hearts immediately after isolation were altered in comparison to control levels after 2 weeks, but had fully recovered by six weeks. Significant changes were found for ATP (control: 4.83±0.29, n=9; 2 weeks: 4.59±0.19**, n=7; 6 weeks: 4.97±0.24, n=8, μmol/g ±SD, **p<0.01 vs control and 6 weeks nicotine), as well as for NAD$^+$ (0.62±0.06; 0.53±0.03**; 0.66±0.05, sequence and units as above).

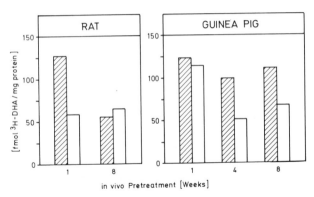

Figure 3: Density of ß-adrenoceptors in myodardium of animals pretreated with nicotine (hatched columns = control hearts)

DISCUSSION

Taken together with our previous findings (Becker et al., 1988b), the fact that chronic application of noradrenaline could elicit the same platelet stabilizing effect as nicotine strongly suggests a mediatory role of this catecholamine in the chronic actions of nicotine. Indeed, application of nicotine has been reported to dose-dependently elevate tissue and plasma levels of

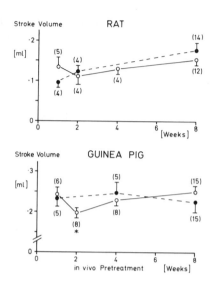

<u>Figure</u> 4: Function of hearts isolated from pretreated animals
(filled circles = controls, open circles = nicotine). Mean values
± SEM, *p < 0.05 vs control by analysis of variance

NA (Dominiak et al, 1984; Grunberg et al., 1988). Though there
seems to be some adaptation in the course of several weeks, this
was far more pronounced for adrenaline, only NA remaining elevated after 2 weeks (Grunberg et al., 1988).

Platelet activation by A obviously derives from inhibition of
adenylyl cyclase, as evidenced by the fall in cAMP. The failure
to detect clearly elevated cAMP in quiescent platelets under the
chronic influence of NA and nicotine does not argue against an
action via ß-adrenoceptors. Due to the auto-catalytic nature of
platelet aggregation (release of ADP, formation of TxA_2, etc.),
even a slight preactivation of adenylyl cyclase can increase the
threshold for aggregation (Kerry & Scrutton, 1985). An additional
effect on platelet aggregability could stem from a direct
inhibition of thromboxane synthetase by nicotine (Goerig &
Habenicht, 1988). However, nicotine did not reduce plasma TxB_2
levels to an extent exceeding that of NA.

In keeping with a selectively enhanced ß-adrenergic stimulation, the density of ß-adrenoceptors in the hearts of both investigated species declined in the course of nicotine application. The very rapid drop (to 50%) observed for rats already in the first week leads one to expect a high degree of cardioprotection against sympathetic stimulation of metabolism, as encountered, e.g., during the isolation of hearts (Becker et al., 1988a). Accordingly, the data on heart function show, if anything, a slightly better performance of rat hearts after 1 week. The fall in ß-adrenoceptor density also of the control hearts over 8 weeks is a physiological phenomenon in growing rats (Baker & Potter, 1980). Due to the slower down-regulation in the guinea pig, it may be assumed that full (or even exacerbated) sympathetic stimulation of the heart is still possible after 1-2 weeks of nicotine application. Cardioprotection would become manifest after about 4 weeks, when ß-adrenoceptor density decreases below controls. The changes in heart function and in energy rich ATP and oxidized NAD^+ in the myocardial tissue detected in the course of nicotine pretreatment fully comply with this interpretation.

CONCLUSION: Chronic application of nicotine to rats and guinea pigs enhances noradrenergic tone. The effects, however, vary with the animal species, the organ or cell type, and the duration of exposure.

Acknowledgements: We thank the German Research Council on Smoking and Health for financial support of this study.

REFERENCES

Baker, S.P., and Potter, L.T. (1980) Br. J. Pharmacol. 68, 65-70
Becker, B.F., and Gerlach, E. (1984) Klin. Wochenschr. 62 (Suppl II), 58-66
Becker B.F., and Gerlach, E. (1988) In: The Pharmacology of Nicotine (M.J. Rand and K. Thurau, Eds), IRL Press, Oxford, pp. 140-141
Becker, B.F., Heier, M., and Gerlach, E. (1988a) In: New Concepts in Viral Heart Disease (H.-P. Schultheiß, Ed), Springer-Verlag, Berlin, pp. 465-474

Becker, B.F., Terres, W., Katzer, M., and Gerlach, E. (1988b)
    Klin. Wochenschr. 66 (Suppl XI), 28-36
Dominiak, P., Kees, F., and Grobecker, H. (1984) Klin.Wochenschr.
    62 (Suppl II), 76-80
Goerig, M., and Habenicht, A.J.R. (1988) Klin. Wochenschr. 66
    (Suppl XI), 117-119
Grunberg, N.E., Popp, K.A., Bowen, D.J., Nespor, S.M., Winders,
    S.E., and Eury, S.E. (1988) Life Sciences 42, 161-170
Kerry, R., and Scrutton, M.C. (1985) In: The Platelet: Physio-
    logy and Pharmacology (G.L. Longenecker, Ed), Academic Press,
    Orlando, pp. 113-157

Effects of Nicotine on Biological Systems
Advances in Pharmacological Sciences
© Birkhäuser Verlag Basel

DOPPLER FLOW MEASUREMENTS AND SMOKING IN PREGNANCY

Schmidt W, Rühle W, Ertan K, I.Stohrer
Department OB/GYN, University of Homburg/Saar, West-Germany

The acute effect of smoking during pregnancy was investigated using pulsed doppler ultrasound. Doppler-flow-parameters (A/B-ratio, Pulsatility Index) were monitored in 3 minute-intervals following smoking of one cigarette of the commonly used brand. In uncomplicated pregnancies no significant changes could be observed.

The chronic effect of smoking during pregnancy was investigated in several studies. A significant increased rate of dystrophic newborns and preterm delivery was found. Only little is yet known about the pathophysiological mechanism of this effect in pregnancy.

Few previous studies were performed to evaluate the acute effect of smoking one cigarette on fetal circulation. Measurements were made at only two different times (before and 10 or 30 minutes after smoking of one cigarette) and in only one or two fetal vessels. The results were different (1,2,3).

The aim of this study was the serial monitoring of fetal **and** maternal vessels after smoking of one cigarette. 35 pregnant smokers were evaluated by doppler ultrasound to assess the possible effect of smoking one cigarette of the commonly-used brand.

A duplex doppler scanner (ADR 5000, ATL, D-Solingen) was used to evaluate three vessels (fetal aorta, umbilical arteries, uteroplacental vessels). The vessels were monitored consecutively to get a result from each vessel within 3 minute intervals. The registration was evaluated using the A/B-ratio and the pulsatility index (fig.1).

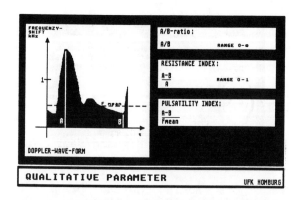

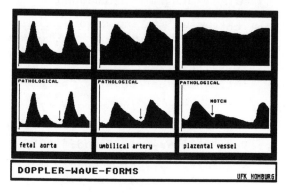

Fig.1: Definition of Doppler-Flow Parameters and description of normal and pathological signals in fetal and maternal vessels:

After smoking of one cigarette, maternal pulse and blood pressure were registered in three minute intervals too. Additionally, continuous CTG monitoring was included (fig.2).

The examinations took between 27-52 minutes (mean: 37). All doppler-flow and laboratory parameters were evaluated before smoking (including nicotine and cotinine levels). Following that, the patient was smoking a cigarette in approximately 3-5 minutes. Immediately afterwards continuous registration was started. 30 minutes later a second blood sample for laboratory analysis was taken.

Fig.2: Sites of measurements:

Mean nicotine-/cotinine serum levels before smoking were 5.7/100.5 ng/ml, 30 minutes later 9.6/111.3 ng/ml. In pregnancies without previous pathology (N=27) cigarette smoking had no discernible influence on doppler parameters in fetal vessels.

With reservations this is also true for complicated pregnancies (N=3). The A/B-Ratio in these cases decreased markedly in the first 20 minutes of monitoring. This effect is possibly a consequence of the small number of patients, because the pulsatility index remained unchanged over the 30 minutes (Fig.3).

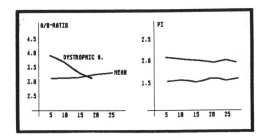

Fig.3: Results in the umbilical artery:

It was an interesting finding that blood flow in maternal
uteroplacental vessels remained constant during the monitoring
time and that there was no difference between normal and
complicated pregnancies (fig.4).

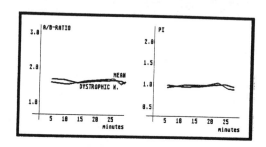

Fig.4: Results in maternal uteroplacental vessels:

Fetal heart rate in cases with uncomplicated pregnancy was
found to be unchanged during the monitoring period (30 minutes). In cases with dystrophic fetuses/newborns there seemed to
be a slight increase during the monitoring time (fig.5). Mater-

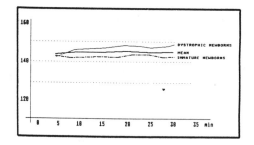

Fig.5: Results: Fetal heart rate

nal systolic blood pressure was constant. In normal pregnancies
and in cases with dystrophic fetuses maternal heart rate fell
significantly (fig.6)

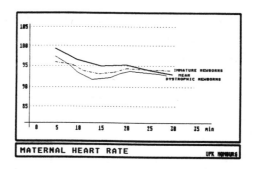

Fig.6: Results: Maternal heart rate

We conclude that there are no acute changes in doppler flow parameters to be observed after cigarette smoking in un-complicated pregnancies. There were no differences in doppler-flow-parameters compared to the control group (N=10). Perhaps the increase in smoke-related substances in the fetal organism after smoking of one cigarette was too small to induce a measurable effect. The decline of maternal heart rate seems to be a nicotine-related effect.

Acknowledgement: This study was supported by the Research Council on Smoking and Health, Bonn, West-Germany.

References:

Lehtovira P, Forss M (1978) The acute effect of smoking on intervillous blood flow of the placenta. Br J Obstet Gynaecol 85:729-731
Jouppila P, Kirkinen P, Eik-Nes S (1983) Acute effect of materanl smoking on the human fetal blood flow. Br J Obstet Gynaecol 90:7-10
Sindberg-Eriksen P, Marsal K (1987) Circulatory changes in the fetal aorta after maternal smoking. Br J Obstet Gynaecol 94:301-305

Effects of Nicotine on Biological Systems
Advances in Pharmacological Sciences
© Birkhäuser Verlag Basel

EFFECT OF NICOTINE ON RED CELL DEFORMABILITIY IN DIABETES MELLITUS

K.Salbaş, V.Ulusoy, T.Akyol

Department of Cardiology and Cardiology Research Center, Medical School of Ankara University, Ankara-Turkey

SUMMARY: Red blood cell (RBC) deformability was assessed by using microfiltration technique and expressed in terms of filtration time of 0.5 ml of 20% erythrocyte suspension. The mean filtration time of diabetic group was 35.5% longer than that of control group. Nicotine (10^{-5}-10^{-2}M) produced biphasic alterations in filtration times of both control and diabetic erythrocytes. An initial increase was followed by a slight but significant decrease. The extents of these alterations were concentration-dependent and greater in diabetic group.

INTRODUCTION:RBC deformability is one of the major factors determining blood flow through capillaries. RBC deformability decreases under several physiological and pathophysiological conditions such as aging, sicle cell anemia, occlusive arterial and coronary artery disease. It has been reported that erythrocyte deformability reduced in diabetes mellitus (Schmid-Schönbein and Volger,1976; Mc Millan et al.,1978; Juhan et al.,1978; Juhan et al.,1982).

It has also been shown that smoking caused a decrease in RBC deformability (Galea and Davidson,1978; Belch et al.,1984; Ernst and Matrai,1987).

Our aim in this study was to investigate in-vitro effects of nicotine on RBC deformability and to test whether the adverse actions of diabetes mellitus and nicotine on RBC deformability additive or not.

MATERIALS AND METHODS:Heparinized blood samples (20 ml) obtained
from 15 healthy volunteers and 28 non-insulin dependent diabetic
patients (with mean$\pm$SD duration of 6$\pm$2 years of diabetes
mellitus). Some characteristics of controls and diabetic patients
are shown in table I.

Table I:Some characteristics of controls and diabetic
        patients(mean$\pm$SD)

Group	Age(year)	Sex	Fasting blood glucose (mg/dl)
Control (N=15)	46$\pm$9	8 men 7 women	85$\pm$7
Diabetic (N=28)	48$\pm$12	14 men 14 women	265$\pm$46

The method used to evaluate RBC deformability is based on that
described by Ekeström et al.,(1983) and Yamaguchi et al.,(1984).
Briefly, washed RBC suspended in phosphate buffered glucose
solution,pH 7.4 containing 0.5 g albumin per 100 ml to yield a
haematocrit of 20%. The 20% erythrocyte suspension was filled
in a 1 ml syringe (with no plunger) which was connected to the
inlet of a Nuclepore filter holder via a three-way stopcock and
allowed to pass through the filter by gravity. Deformability was
determined by measuring the filtration time of 0.5 ml of 20% RBC
suspension through a Nuclepore polycarbonate filter with a pore
diameter 5 μm. First, filtration times of control and diabetic
RBC were determined. Then nicotine was added to both control and
diabetic RBC suspensions in a cumulative manner to yield
concentrations of 10^{-5}, 10^{-4}, 10^{-3} and 10^{-2}M. Measuremets were
made after 0.5,1,...5 and 10min after addition of nicotine.

Results are presented as mean values $\pm$S.E.M. Statistical
significance of the group differences was evaluated by using
paired and unparied Student's t-test and $p<0.05$ was assumed to be
significant.

RESULTS:The mean filtration time of 0.5 ml of 20% RBC suspension was 35.5% longer in diabetic group than that in control group. The measurements made after 0.5,1,...5 and 10min after additions of 10^{-5}, 10^{-4}, 10^{-3} and 10^{-2}M nicotine to RBC suspension revealed that, nicotine altered the deformability of erythrocytes in a time-dependent manner both in control and diabetic groups. The alteration was biphasic: An initial decrease in deformability (which was measured as an increase in filtration time) was followed by a slight but significant increase (this increase was not significant at 10^{-5} and 10^{-4}M in control group). The extents of both phases of the effect were concentration-dependent. At 10^{-5}M, the mean$\pm$SD of filtration time was 6.8$\pm$1.4 min in control and 10$\pm$1.7 min in diabetic groups. At 10^{-2}M these values were 7.5$\pm$1.4 min and 10.8$\pm$1.4 min in control and diabetic groups respectively. The results are summarized in table II.

Table II:Changes in filtration times (% of pre-nicotine value) of 0.5 ml of 20% RBC suspensions in the presence of nicotine (10^{-5}-10^{-2}M) in control and diabetic groups (means$\pm$S.E.M.)

Time (min)	Group	Changes in filtration time (% pre-nicotine) at nicotine concentrations of:			
		10^{-5}M	10^{-4}M	10^{-3}M	10^{-2}M
0.5	C	3.5+0.4	4.1+0.5	6.5+1.2	6.6+1.0*
	DM	4.2+0.8	4.7+1.2	6.9+1.1*	8.5+1.8
1	C	4.8+0.7	8.0+1.9	8.0+1.5	9.6+1.9
	DM	5.9+1.4	9.2+2.1*	10.0+2.3	14.2+2.4
2	C	9.6+2.0	11.2+2.0	12.9+2.1	16.1+3.4
	DM	12.0+4.4	13.4+3.9	14.5+3.0	20.2+4.5
3	C	10.6+1.8*	16.1+3.8	16.8+3.6	20.9+5.2
	DM	14.3+5.1	22.0+4.8	23.1+4.6	24.1+4.0
4	C	6.2+0.6*	8.8+1.0*	17.6+2.9	21.0+3.5
	DM	10.4+2.4	15.0+3.4	24.2+4.3*	29.0+4.4
5	C	N.S	N.S	-8.2+1.8	-9.6+2.0
	DM	-9.2+1.8	-10.5+2.1	-15.4+2.4	-18.2+3.8
10	C	N.S	N.S	-8.5+1.8	-10.1+2.1
	DM	-9.0+2.0	-10.6+1.8	-15.6+2.6*	-18.0+4.1

C:Control, DM:Diabetes mellitus, NS:Non significant, *:p<0.001, for the values without asterix p<0.05.

As it is seen in table II, the effect of nicotine on diabetic group is greater.

DISCUSSION:Our results showed that RBC deformability reduced in diabetes mellitus. This is in agreement with previous reports (Schmid-Schönbein and Volger,1976; Mc Millan et al.,1978; Juhan et al.,1978; Juhan et al.,1982). The exact mechanism responsible for the reduction of RBC deformability in diabetes mellitus is not clear. High blood glucose concentration, hyperosmolarity (Schmid-Schöhbein and Volger,1976) and direct effect of insulin (Juhan et al.,1981; Juhan et al.,1982; Bryszewka and Leyko,1983) were thought to be involved. Whatever the exact mechanism is, it seems plausible that the membrane alterations in diabetes mellitus (Vali et al.,1988) make the RBC less flexible.

Smoking is another factor that was shown to disturb blood rheology (Galea and Davidson,1978; Belch et al.,1984; Ernst and Matrai,1987). It was reported that, in peripheral venous blood, the maximum nicotine concentrations during smoking were in the range of 10^{-7}M (Spohr et al.,1979). We studied in-vitro effects of much higher concentrations (10^{-5}-10^{-2}M) of nicotine. Within the concentration range that we studied, nicotine caused biphasic changes in both healthy and diabetic RBC deformability.In a few minutes after addition of nicotine a decrease in deformability was observed. This decrease was concentration-dependent and greater in diabetic group. The mean+SD of filtration times of control and diabetic groups were 6.2+1.2 and 8.4+1.5 min respectively. That means, in diabetes mellitus RBC deformability decreased by 35.5%. 4 min after addition of 10^{-2}M nicotine, the mean filtration times were found to be 7.5+1.3 and 10.8+1.4 min in control and diabetic groups respectively. As it will be seen in table II, the nicotine-induced increases in filtration times (%pre-nicotine value) are 21+3.5 in control and 29+4.4 in diabetic groups. That is, the effect of nicotine on RBC deformability is 38% greater in diabetic group than that in control group. These results lead us to conclude that in the first 4 minutes after addition of nicotine, the adverse actions of diabetes mellitus and nicotine on RBC deformability add up. The nicotine-induced increase in filtration time was transient and followed by a slight but significant decrease (table II). The initial phase of the effect was not surprising, since both diabetes mellitus and nicotine individually reduce erythrocyte deformability. But why does nicotine then cause an increase in deformability in a concentration-dependent manner and more effectively in diabetic group? Does affinity of nicotine binding sites on erythrocyte membrane to nicotine vary with time and increase in diabetes mellitus? Or/and does the number of nicotine binding sites increase in erythrocytes of diabetic patients? It is not possible to answer to these questions by using the results of our study, because in our study we aimed to see the effects of nicotine on RBC deformability, not to analyse its mechanism of action. Therefore the subject merits further investigations.

CONCLUSION: Diabetes mellitus reduced the deformability of RBC. Nicotine had a biphasic effect (a transient decrease which was followed by a slight increase) on the deformabilities of both healthy and diabetic erythrocytes but the diabetic erythrocytes were found to be more sensitive to nicotine.

REFERENCES

Belch,J.J.F., Mc Ardle,B.M., Burns,P., Lowe,G.D.O., Forbes,C.D. (1984) Thromb. Haemostas. 51:6-8.
Bryszewska,M., Leyko,W. (1983) Diabetologia 24:311-313.
Ekeström,S., Koul,B.L., Sonnenfeld,T. (1983) Scand.J.Thor.Cardiovasc.Surg.17:41-44.
Ernst,E., Matrai,A. (1987) Atherosclerosis 64:75-77.
Galea,G., Davidson,R.J.L. (1985) J.Clin.Path. 38:978-982.
Juhan,I., Bayle,J., Vague,P., Juhan,C. (1978) Nouv.Pr.Med. 7:759.
Juhan,I., Vague.P., Buonocore,M., Moulin,J.P., Calas,M.F., Vialettes,B., Verdot,J.J. (1981) Scand.J.Clin.Lab.Invest. 41(Suppl 156): 159-164.
Juhan,I., Buonocore,M., Jouve,R., Vague,P., Moulin,J.P., Vialettes,B. (1982) Lancet 1: 535-537.
Mc Millan,D.E., Utterback,N.G., La Puma,J. (1978) Diabetes 27:895-901.
Schmid-Schönbein,H., Volger,E. (1976) Diabetes 25(Suppl2):897-902.
Spohr,U., Hofmann,K., Steck,W., Herdenberg,J., Walter,E., Hengen,N., Augustin,J., Mörl,H., Koch,A., Horsch,A., Weber,E. (1979) Atherosclerosis 33:271-283.
Vali,R.K.,Jaffe,S.,Kumar,D.,Kalra,V.K.(1988) Diabetes 37:104-111
Yamaguchi,H., Allers,M., Roberts,D.(1984)Scand.J.Thor.Cardiovasc. Surg. 18:119-122.

Effects of Nicotine on Biological Systems
Advances in Pharmacological Sciences
© Birkhäuser Verlag Basel

PRENATAL NICOTINE EXPOSURE AND SEXUAL BRAIN DIFFERENTIATION

W.Lichtensteiger, N. von Ziegler and M. Schlumpf

Institute of Pharmacology, University of Zürich, Zürich, Switzerland

SUMMARY: Drugs may interfere with the organizing action of circulating hormones such as gonadal and adrenal steroids on the developing brain and by this mechanism induce long-lasting alterations in brain function. In the rat, prenatal exposure to nicotine on gestational days 12-19 affects sexually dimorphic behavior and causes sex-dependent neurochemical changes. Prenatal nicotine suppresses a characteristic peak of plasma testosterone in the male rat fetus and results in depressed, female-like activity of brain aromatase which converts testosterone to estradiol, at a particular, early postnatal (PN) stage (PN 6).

INTRODUCTION

The early organizing action of hormones on the developing brain represents a potentially sensitive process which may be disturbed by drugs. In this way, even transient drug exposure may cause long-lasting changes in brain function. Alterations of sexual brain differentiation and of adrenal function have been observed in rats, mice and hamsters (Lichtensteiger and Schlumpf, 1985a, for review).

PRENATAL EFFECTS OF NICOTINE ON SEXUALLY DIMORPHIC BEHAVIOR AND ON CENTRAL TRANSMITTER SYSTEMS

Since nicotine interacts in many ways with adult neuroendocrine systems, it seemed conceivable that this drug might affect brain-endocrine relationships during ontogeny. In a first step, we studied sexually dimorphic behavior in adult offspring after prenatal nicotine exposure. In most mammals, females show a higher preference for sweet nutrients than do males. In order to avoid additional effects of calory intake, this sexual dimorphism is tested by giving the animals a choice between normal and saccharin-containing drinking water.

We administered nicotine to time-pregnant rats by means of an osmotic minipump (Alzet 2001 with nicotine tartrate or tartaric acid in controls) delivering nicotine for one week from gestational day (GD) 12. The dose used in the initial investigation, 0.25 mg nicotine base/kg x hr, has been used in several rat studies and, from our earlier electrophysiological data, was expected to yield a moderate activation of neuronal systems such as the nigrostriatal dopamine system. This dose did not affect the outcome of pregnancy (duration, weight gain of dams, litter size, sex ratio; Lichtensteiger and Schlumpf, 1985b). Body weight of preweaning nicotine-exposed offspring was reduced up to 8% in one study (Lichtensteiger and Schlumpf, 1985b) and normal in a second study (Ribary and Lichtensteiger, 1989). In order to allow for comparisons across different behavioral, neurochemical and endocrine parameters, the same dose was administered also in subsequent investigations.

When exposed to nicotine between GD 12 and 19, adult male offspring exhibited the high saccharin preference of their female littermates, i.e., sexual dimorphism was abolished (Lichtensteiger and Schlumpf, 1985b). Sexual activity of males is also reduced (Ribary, 1985; Segarra and Strand, 1989).

An influence of sex was noted in the effect of nicotine on developing central catecholamine systems (Ribary and Lichtensteiger, 1989). The prenatal development of nicotinic cholinergic receptor sites coincides with the ontogeny of norepinephrine and dopamine nerve cell groups; a clear-cut co-localization of nicotinic sites and dopamine nerve cells can be demonstrated in the fetal mesencephalon (Lichtensteiger et al., 1988). Nicotine exposure between GD 12 and 19 initiates a complex, sex-dependent pattern of changes in the metabolism of central dopamine and norepinephrine systems, with certain alterations persisting until adulthood (Ribary and Lichtensteiger, 1989). Males exhibit an imbalance of dopamine and norepinephrine (NE) turnover and increased activity at PN 15 (Schlumpf et al., 1988); a relation is suggested by models of developmental alterations of locomotor activity (Raskin et al., 1984). In adulthood, a sex difference is found in the NE system, with reduced NE turnover (Ribary and Lichtensteiger, 1989) and increased beta-adrenergic receptor density restricted to prenatally nicotine-exposed males (Peters, 1984; nicotine in drinking water). The suppression of normal sexual dimorphisms in behavior and the induction of sex diffe- rences in drug response may be based on different processes (see Lichtensteiger et al., 1988), but they both point to an inter- action of nicotine with sexual brain differentiation.

A SEARCH FOR MECHANISMS UNDERLYING THE ACTION OF NICOTINE ON SEXUAL BRAIN DIFFERENTIATION: TESTOSTERONE AND BRAIN AROMATASE

Male sexual brain differentiation depends upon the presence of testosterone which is increased in the midfetal human fetus and in the late fetal and early postnatal rat (refs. in Lichten- steiger and Schlumpf, 1985a). In the rat, a peak of plasma testo- sterone at GD 18 appears to be of particular importance (Ward and Weisz, 1980). Prenatal exposure to nicotine according to the treatment schedule that abolishes sexual dimorphism of saccharin preference, suppresses the GD 18 testosterone peak in male rat fetuses (Lichtensteiger and Schlumpf, 1985b), suggesting that a disturbed regulation of the fetal male gonadal axis is involved in the alteration of the differentiation process. A similar asso- ciation of disturbances in sexually dimorphic behavior and reduc- tion of fetal testosterone levels has been observed after prena- tal exposure to other centrally active drugs (Lichtensteiger and Schlumpf, 1985a).

An important part of sexual brain differentiation is based on the action of estradiol formed in some brain areas from testoste- rone by aromatase (Reddy et al., 1974; Selmanoff et al., 1977; Gorski, 1984; Hutchison and Steimer, 1984; Schleicher et al., 1986). It seemed conceivable that drug exposure might interfere with steroid action during the perinatal period by an effect on aromatase. We determined the activity of this enzyme in rat fetu- ses and pups by measuring the conversion of 1ß-3H-androstenedione to estrone (modification of Steimer and Hutchison, 1989) in a brain piece comprising preoptic, hypothalamic and amygdaloid areas. In untreated animals, the development of aromatase activi- ty between GD 18 and PN 15 was very similar in both sexes, except for a significant decrease of activity in female brain at PN 6, i.e., in the later part of the critical period of sexual brain

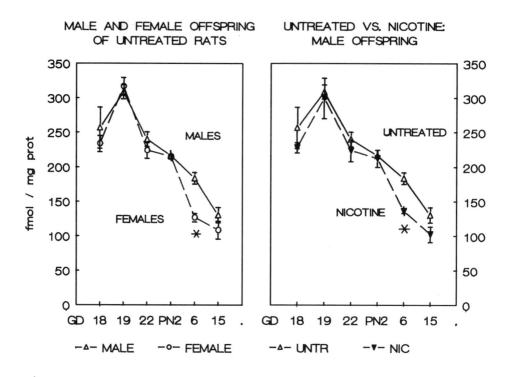

Fig. 1. Brain aromatase activity in offspring of untreated LE rats (1a) and in male offspring of untreated or nicotine-treated dams (1b; additional controls cf. text). Mean values of 4-8 litters + S.E.M. (2-3 pups/litter), * = p < 0.01.

differentiation (Fig. 1a). After nicotine treatment for 1 week from GD 12 or for 2 weeks from GD 8, male offspring showed a selective reduction of aromatase activity at PN 6 to the normal female level (Fig. 1b). No effect was seen in female offspring. Enzyme activity in the 3 control groups, i.e., offspring from 1) untreated dams, 2) dams sham-operated at GD 12 or 8, 3) dams implanted with tartaric acid-containing Alzet pumps, was the same (von Ziegler and Lichtensteiger, 1990 and in preparation).

These data indicate that in the rat, prenatal nicotine exposure interferes in a selective way with brain aromatase activity which plays a key role in sexual brain differentiation. The delayed effect, a reduction of enzyme activity in male offspring at PN 6, was consistently found after both treatment schedules; it would reduce the capacity of aromatization of testosterone.

Since the tissue examined in this study contained three brain areas, each developmental stage represents the sum of the developmental curves of enzyme activity in these three areas (which differ from each other, von Ziegler and Lichtensteiger, in preparation). Hence, a number of processes can be imagined that might cause the low enzyme activity seen in normal females and prenatally nicotine-exposed males at PN 6, e.g., a reduced enzyme activity per cell, a reduced number of cells expressing the enzyme in one or several cell populations, or a difference in the developmental time-course of cell populations containing the enzyme. On a more general level, it is evident from the various sets of data that nicotine exposure during fetal life can interfere with the early organization of neuroendocrine processes.

ACKNOWLEDGEMENTS
The investigations were supported by several Swiss National Science Foundation grants (aromatase project: no. 3.048-0.87).

REFERENCES
Gorski, R.A. (1984) Progress in Brain Research 61: 129-146, Elsevier.
Hutchison, J.B., and Steimer, T. (1984) Progress in Brain Research 61: 23-51, Elsevier.
Lichtensteiger, W., and Schlumpf, M. (1985a) Progress in Neuroendocrinology 1: 153-166, VNU Science Press.
Lichtensteiger, W., and Schlumpf, M. (1985b) Pharmacol. Biochem. Behav. 23: 439-444.
Lichtensteiger, W., Ribary, U., Schlumpf, M., Odermatt, B., and Widmer, H.R. (1988) Progress in Brain Research 73: 137-157, Elsevier.
Peters, D.A.V. (1984) Res.Commun.Chem.Pathol.Pharm. 46: 307-317.
Raskin, L.A., Shaywitz, B.A., Shaywitz, S.E., Cohen, J.D., and Anderson, G.M. (1984) In: Early Brain Damage, Vol. 1 (C.R. Almly and S. Finger, Eds.), Academic Press, Orlando, pp. 111-125.
Reddy, V.V., Naftolin, F., and Ryan, K.J. (1974) Endocrinology 94: 117-121.
Ribary, U., and Lichtensteiger, W. (1989) J. Pharmacol. Exp. Ther. 248: 786-792.
Ribary, U. (1985) Thesis No. 7939, Federal Institute of Technology, Zürich.
Schleicher, G., Stumpf, W.E., Morin, J.K., and Drews, U. (1986) Brain Res. 397: 290-296.
Schlumpf, M., Gähwiler, M., Ribary, U., and Lichtensteiger, W. (1988) Pharmacol. Biochem. Behav. 30: 199-203.
Segarra, A.C., and Strand, F.L. (1989) Brain Res. 480: 151-159.
Selmanoff, M.K., Bradkin, L.D., Weiner, R.I., and Siiteri, P.K. (1977) Endocrinology 101: 841-848.
Steimer, T., and Hutchison, J.B. (1989) Brain Res. 480: 335-339.
von Ziegler, N.I., and Lichtensteiger, W. (1990) Neuroendocrinology 52, S1: 140.
Ward, I.L., and Weisz, J. (1980) Science 207: 328-329.

DESIPRAMINE INHIBITS THE NICOTINE-INDUCED EXOCYTOSIS IN RAT PHAEOCHROMOCYTOMA CELLS (PC12)

S. Baunach[1], H. Russ[1], M. Haass[3], G. Richardt[3], H.G. Kress[2], A. Schömig[3], and E. Schömig[1]

[1] Dept. of Pharmacology, University of Würzburg, Versbacherstr. 9, 8700 Würzburg, Germany
[2] Dept. of Anaesthesiology, University of Würzburg, Germany
[3] Dept. of Cardiology, University of Heidelberg, Germany

SUMMARY: In clonal PC12 cells (differentiated with nerve growth factor), nicotine (100 µmol/l) induces exocytotic release of ^{3}H-noradrenaline, since release (1) depends absolutely on the presence of extracellular calcium and (2) is associated with release of neuropeptide Y. The tricyclic antidepressant desipramine potently inhibits nicotine-induced noradrenaline release. However, desipramine is a much less potent inhibitor when release is elicited by the calcium ionophore A23187 or by potassium-induced depolarization. Desipramine most probably blocks the nicotinic acetylcholine receptor or the receptor-related cation channel.

INTRODUCTION

Already in 1970, Su and Bevan reported the tricyclic antidepressant desipramine to inhibit the nicotine-induced release of noradrenaline from adrenergic neurons. However, the mechanism has not yet been clarified. Therefore, we investigated the effect of desipramine on the exocytotic release of noradrenaline. As experimental model the clonal rat phaeochromocytoma cell line PC12 was chosen. PC12 cells express many properties characteristic of adrenergic neurones, including the ability to release noradrenaline after the stimulation of nicotinic acetylcholine receptors (Greene and Rein, 1977).

METHODS

The PC12 cells were grown in surface culture on plastic dishes coated with collagen (Schwarz et al., 1989). Prior to the experiments, the cells were differentiatd by 200 µg/l nerve growth factor (NGF) for 3 days (Greene and Rein, 1977). The cells were loaded with ^{3}H-noradrenaline (10 nmol/l) for 60 min. After a wash-out period of 30 min the release of ^{3}H-noradrenaline into the incubation medium was stimulated by various methods (see below). The fraction of ^{3}H-noradrenaline which appeared in the incubation buffer (125 mmol/l NaCl, 4.8 mmol/l KCl, 1.2 mmol/l $MgSO_4$, 1.2 mmol/l $CaCl_2$, 1.2 mmol/l KH_2PO_4, 25 mmol/l HEPES·NaOH, pH 7.4) during 1 min of stimulation was measured. Exocytosis was stimulated by the addition of either nicotine (100 µmol/l), the calcium ionophore A23187 (10 µmol/l, or potassium chloride (56 mmol/l). In some experiments, the release of neuropeptide Y (NPY) – an indicator for exocytosis – was measured by radioimmunoassay (Haass et al., 1989). Various compounds were tested for their ability to inhibit exocytosis. These compounds were present during five minutes of preincubation and during one minute of stimulation. Fractional rate of loss (FRL) is defined as the rate of release/ intracellular content of tritium or NPY at the beginning of the stimulation period. Intracellular calcium was measured fluorometrically by the fura-2 method (Williams et al., 1985, Kress et al., 1987).

RESULTS AND DISCUSSION

PC12 cells were preloaded for 60 min with 10 nmol/l ^{3}H-noradrenaline. After a wash-out of 30 min, release of tritium was stimulated for 1 min by 100 µmol/l nicotine. Column chromatography revealed that the fraction of ^{3}H-noradrenaline amounted to more than 90% of the released tritium. Nicotine induced exocytotic release of ^{3}H-noradrenaline, since release depended absolutely on the presence of extracellular calcium (Fig. 1) and

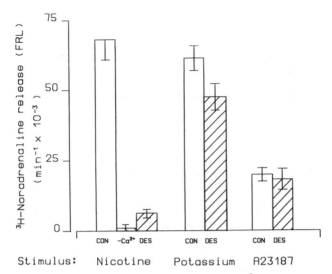

Fig. 1: Exocytotic release of ^{3}H-noradrenaline was induced by 100 μmol/l nicotine, by 56 mmol/l potassium or by 10 μmol/l of the calcium ionophore A23187. The open columns represent the controls (CON) and the hatched columns the release either in the absence of extracellular calcium ($-Ca^{2+}$) or in the presence of 1 μmol/l desipramine (DES). Shown are means ± SEM (n=4-6).

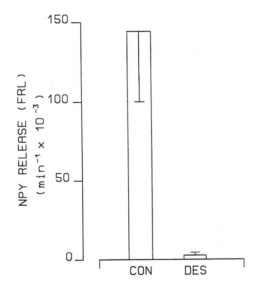

Fig. 2: Inhibition of nicotine-induced (100 μmol/l; CON) release of NPY by desipramine (1 μmol/l; DES). Shown are means ± SEM (n=4).

went hand in hand with the release of NPY (Fig. 2). The effect
of nicotine was due to stimulation of nicotinic acetylcholine
receptors, since ^{3}H-noradrenaline release was inhibited by 30
µmol/l hexamethonium (23 ± 4% of control; n=11) but not by 0.3
µmol/l atropine (87 ± 12% of control; n=4).

Interestingly enough, the tricyclic antidepressant desipramine potently inhibited nicotine-induced release of [3]H-
noradrenaline (Fig. 1, 3). The IC_{50} of this interaction was 61
nmol/l (95% confidence limits: 35, 107; n=6). The Hill
coefficient was about unity. Desipramine inhibited not only the
release of ^{3}H-noradrenaline but also the release of NPY (Fig.
2). Figure 4 shows a hypothetical model of the sequence of
events which eventually lead to the exocytosis of the storage
vesicles from PC12 cells. There is experimental evidence that
desipramine inhibits nicotine-induced exocytosis at an early
step. (1) Desipramine was a potent inhibitor (see above) when
release was induced by nicotine but a much less potent inhibitor
when release was induced more directly either by depolarization-
induced (56 mmol/l potassium) opening of calcium channels or by
the calcium ionophore A23187 (10 µmol/l), the IC_{50}'s of desipra-

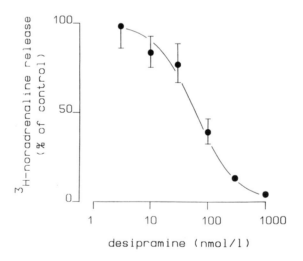

Fig. 3: Inhibition of nicotine-induced (100 µmol/l) [3]H-
noradrenaline release from PC12 cells by desipramine. The data
are given relative to controls in the absence of desipramine.
Shown are ± SEM (n=6).

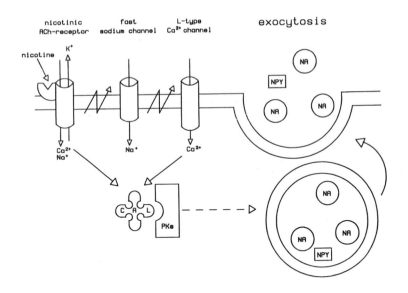

Fig. 4: Proposed mechanism of nicotine-induced noradrenaline release from PC12 cells by exocytosis (modified from Marley, 1988). Binding of nicotine to the nicotinic acetylcholine receptor opens the associated non-selective cation channel. The ion flux causes depolarization which activates voltage-sensitive ion channels, namely the fast sodium channels and the L-type calcium channels. The rise of intracellular calcium activates the exocytotic machinery. The activation of calmodulin (CAL)-dependent protein kinases (PKs) is most probably involved in that step eventually leading to exocytosis of storage vesicles which contain noradrenaline (NA) and the co-transmitter neuropeptide Y (NPY).

mine being above 3 μmol/l (Fig. 1). (2) 1 μmol/l desipramine reduced the nicotine-induced rise of intracellular calcium by 89 ± 4% (n=8) but failed to reduce (21 ± 11%, n=8) the rise of intracellular calcium caused by the addition of 56 mmol/l potassium.

The results suggest that desipramine inhibits nicotine-induced exocytosis at an early step in the sequence of events which eventually lead to exocytosis. Most probably desipramine blocks the nicotine acetylcholine receptor or the cation channel which is closely associated with that type of receptor.

Acknowledgement: We thank Michaela Hoffmann and Elsbeth Fekete for skilful technical assistance.
This study was supported by the Research Council on Smoking and Health and by the Deutsche Forschungsgemeinschaft (Scho 373; SFB 176).

REFERENCES

Greene, L.A., Rein, G. (1977) Brain Res 129: 247-263
Haass, M., Cheng, B., Richardt, G., Lang, R.E., Schömig, A. (1989) Naunyn-Schmiedeberg's Arch Pharmacol 339: 71-78
Kress, H.G., Eckhardt-Wallasch, H., Tas, P.W.L., Koschel, K. (1987) FEBS 221: 28-32
Marley, P.D. (1988) Trends Pharmacol Sci 9: 102-107
Schwarz, M.A., Brown, P.J., Eveleth, D.D., Bradshaw, R.A. (1989) J Cell Physiol 138: 121-128
Su, C., Bevan, J.A. (1970) J Pharmacol Ther 175: 533-540
Thoa, N.B., Costa, J.L., Moss, J., Kopin, I.J. (1974) Life Sci 14: 1705-1719
Williams, D.A., Fogarty, K.E., Tsien, R.Y., Fay, F.S. (1985) Nature 318: 558-561

IV.
Mechanisms of Nicotine Actions in the CNS

GLUCOSE METABOLISM: AN INDEX OF NICOTINE ACTION IN THE BRAIN

Edythe D. London

Neuropharmacology Laboratory, Neuroscience Branch, Addiction Research Center, National Institute on Drug Abuse, P.O. Box 5180, Baltimore, MD 21224, U.S.A.

SUMMARY: Several experimental approaches have been used to delineate the initial sites and the distributions of action of nicotine in the brain. Radioligand binding studies have revealed a heterogeneous distribution of high affinity, specific binding sites for nicotine in brain. Other studies have employed metabolic mapping with 2–deoxy–D–[1–^{14}C]glucose (DG) to measure of the regional cerebral metabolic rate(s) for glucose (rCMRglc). In rats, l–nicotine stimulates rCMRglc in brain areas which contain specific binding sites for [^{3}H]nicotine, indicating that the sites are true receptors, linked to functional activity. Affected areas include limbic structures, components of the visual system, brain stem nuclei important in cardiovascular reflexes, and areas involved in motor function. In general, rats show tolerance to the effects of nicotine on rCMRglc. The distribution of *in vivo* effects of nicotine on rCMRglc implicates various brain regions in the behavioral and physiological effects of the drug.

INTRODUCTION

Despite an epidemic in substance abuse, nicotine use in the U.S. far exceeds consumption of any other abused substance, other than ethanol. In a recent survey conducted by the U.S. National Institute on Drug Abuse on 8,814 households (National Household Survey of Drug Abuse, 1990), 67 million individuals reported recent use of nicotine, as compared with 2.9 million reporting recent use of cocaine, or 12 million indicating that they had used marijuana within the month before the survey. Just as striking is a report that 346,000 deaths in the U.S. were attributable to the use of tobacco (Henningfield et al., 1990). This estimate far exceeds the number of deaths attributable to the use of illegal substances, such as cocaine, or the combination of ethanol with other drugs. These statistics reinforce a commitment to research on factors leading to the compulsive use of nicotine, and on therapeutic approaches for smoking cessation.

Nicotine produces various behavioral effects, including enhanced arousal, which may contribute to reinforcement produced by this drug (Larson and Silvette, 1971; Aceto and Martin, 1982; Henningfield, 1984). It is a tenable hypothesis that reinforcement, which supports the compulsive use of nicotine, depends, to some extent, upon actions of nicotine in the brain. This rationale has inspired studies to delineate the local sites of action as well as the distribution of action of nicotine in the brain. Autoradiographic ligand binding studies have been performed to reveal the distribution of specific binding sites for nicotine in the brain. In addition, 2-deoxy-D-[1-^{14}C]glucose (DG) has been used to map and quantitate cerebral metabolic responses to acute and chronic nicotine in rats. The purpose of this chapter is to review information from these studies and to provide a view of possible future directions for this research.

Studies of Nicotinic Cholinergic Receptors in Brain: Several radioligands have been used to study nicotinic receptors in the brain. In early studies, [^{125}I]α-bungarotoxin was the most widely used ligand for putative nicotinic receptors (Schmidt et al., 1980). Subsequent studies have employed [^{3}H]acetylcholine or [^{3}H]nicotine (Romano and Goldstein, 1980; Abood et al., 1980; Schwartz et al., 1982; Marks et al., 1986; Martino-Barrows and Kellar, 1987), and, most recently, [^{3}H]methylcarbamylcholine (Abood and Grassi, 1986; Boksa and Quirion, 1987; Yamada et al., 1987; Takayama et al., 1989). The conclusion from most of these studies is that [^{3}H]acetylcholine, [^{3}H]nicotine, and [^{3}H]methylcarbamylcholine label high affinity nicotinic cholinergic receptors, but that [^{125}I]α-bungarotoxin labels other molecules in brain (for review, see Clarke, 1987).

In *vitro* autoradiography has been used to study the distribution of nicotinic receptors in brain (Clarke et al., 1984, 1985; London et al., 1985$_b$). High densities of specific sites are seen in the interpeduncular nucleus, medial habenula, thalamic nuclei, components of the visual system, and the cerebral cortex. Marked similarities in the binding patterns of [^{3}H]nicotine and those of [^{3}H]acetylcholine support the view that nicotine binds to cerebral nicotinic receptors (Clarke et al., 1984).

Recent efforts to image nicotinic receptors have involved *in vivo* procedures. Studies performed in mice have demonstrated that radiolabeled nicotine can be used as an *in vivo* ligand (Broussolle et al., 1989). [^{3}H]Nicotine enters the brain almost instantaneously, and brain radioactivity declines rapidly. As in *in vitro* studies, *in vivo* specific binding of [^{3}H]nicotine is inhibited by nicotinic agonists much more effectively than by antagonists. However, despite the affinity of nicotinic receptors for agonists, the use of agonists as inhibitors of radioligand binding (to estimate nonspecific binding) could be problematic in noninvasive human studies involving positron emission tomography (PET), as pharmacological doses would be required to inhibit specific binding completely. Nonetheless, PET studies using [^{11}C]nicotine have been performed in the monkey and human brain (Nybäck et al., 1989). Full quantitation of receptor densities and affinities using [^{11}C]nicotine have not yet been reported. However, these parameters ultimately might be quantitated using modeling techniques that have been applied in the study of D_2 dopamine receptors using [^{11}C]N-methylspiperone (Wong et al., 1986).

DG Studies of Nicotine Action: Studies with DG provide a method to map functional drug responses in the brain of an intact animal or human subject (McCulloch, 1982; London et al., 1986; London, 1989). DG is used as a tracer to determine rCMRglc in discrete anatomical structures (Sokoloff et al., 1977). Unlike receptor studies, which reveal initial loci of drug action, metabolic mapping can visualize responses propagated by afferents to sites remote from primary interactions. Glucose is the major substrate for oxidative metabolism of the adult brain; therefore, rCMRglc provides an index of local brain function (Sokoloff et al., 1977). A close relation between functional activity and rCMRglc has been shown in many physiological and experimental paradigms.

Following the intravenous injection of DG, rCMRglc in discrete areas can be determined from the time course of the radiotracer in the arterial plasma, the plasma glucose concentration, and brain radioactivity, using an operational equation (Sokoloff et al., 1977). In animals, the radiotracer, 2-deoxy-D-[1-^{14}C]glucose (DG), is used with quantitative autoradiography of brain sections. Adaptation of the method for human studies employs PET and 2-deoxy-2-[^{18}F]fluoro-D-glucose (Phelps et al., 1979, Reivich et al., 1979).

Acute nicotine stimulates rCMRglc in rats (Grünwald et al., 1987; London et al., 1988$_a$, 1988$_b$; McNamara et al., 1990). The effects occur mainly in brain areas with specific binding sites for [^{3}H]nicotine. Subcutaneous (s.c.) nicotine increases rCMRglc by 100% or more in the medial habenula, fasciculus retroflexus, superior colliculus, and median eminence (Fig. 1). Substantial increases also occur in the cerebellum, interpeduncular nucleus, and some thalamic nuclei, with less prominent effects in the reticular nucleus of the medulla, paramedian lobule, subicular areas, various brain stem nuclei, and the midbrain. Correlation of nicotine-induced stimulation of rCMRglc with the presence of specific [^{3}H]nicotine binding sites implies that the sites are true receptors, coupled to energy metabolism.

There are some discrepancies between nicotinic receptor densities and rCMRglc effects (London et al., 1988$_a$). Some regions lack [^{3}H]nicotine binding sites, but show nicotine-induced rCMRglc increases (e.g., red nucleus, median eminence). Such stimulation may reflect propagation of nicotine effects from receptor-rich areas. Other areas, such as the caudate-putamen and neocortical areas, show binding (London et

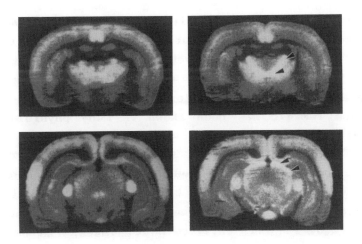

Figure 1. Effects of nicotine (0.3 mg/kg s.c., 2 min before DG) on rCMRglc (London et al., 1988$_a$). Pseudogray glucograms in the top panel illustrate the effect of nicotine (right) compared to saline (left) in thalamic nuclei. Arrows indicate the anteroventral (upper) and interanteromedial (lower) nuclei. Nicotine effects (right) in the visual system are shown in the bottom panel. Arrows indicate the superficial gray layer of the superior colliculus (top) and the nucleus of the optic tract (below). The lightest shades of gray indicate the highest rCMRglc, and the darkest shades signify the lowest rates.

al., 1985$_b$) but no rCMRglc effects of nicotine. The lack of effect may reflect dilution of nicotine action on responsive cells in regions with heterogeneous cell populations.

The visual system exhibits marked effects of nicotine on rCMRglc, suggesting a role for nicotinic cholinergic transmission in visual processing and visual–motor function. The visual cortex, nucleus of the optic tract, lateral geniculate bodies, and superior colliculus show significant densities of nicotinic binding sites (Oswald and Freeman, 1981; London et al., 1985$_b$), as does the inner plexiform layer of the retina (Schwartz and Bok, 1979).

Another study addressed whether nicotine effects in the visual system reflected direct interactions with nicotinic receptors in brain. Unilateral enucleation abolished nicotine-induced stimulation throughout visual pathways in the contralateral side of the brain, suggesting that effects in the retina are necessary for stimulation of rCMRglc by nicotine in visual pathways. Similarly, enhancement of rCMRglc in the superior colliculus by the acetylcholinesterase inhibitor diisopropyfluorophosphate blocked by intraocular injection of kainic acid, which depleted acetylcholinesterase in the retina (Gomez-Ramos et al., 1982). These findings suggest that retinal input is required for cholinomimetic-induced stimulation of rCMRglc in visual pathways.

Mecamylamine, a ganglionic blocking drug which enters the brain readily, antagonizes nicotine's effects on behavior and on rCMRglc, but hexamethonium, a quaternary ganglionic blocking drug, has no effect (London et al., 1985_a, London et al., 1988_a, 1988_b). Therefore, these effects of nicotine are attributable to specific central actions of the drug.

More recent studies have focused on chronic effects of nicotine (Grünwald et al., 1988; London et al., in press). In one of these studies (London et al., in press), l-nicotine was injected s.c. twice daily for 10 days and once in the morning on the eleventh day. On the following afternoon, rats received either nicotine challenge (0.3 mg/kg) or saline s.c. 2 min before DG. As in other studies (London et al., 1985_a, 1988_a, 1988_b, Grünwald et al., 1987), drug-naive rats showed increases in rCMRglc due to the nicotine challenge. Animals subjected to chronic nicotine showed reduced responses in all regions in which stimulation occurred. Some regions, including components of visual pathways (Fig. 2), the cerebellum, and vestibular nuclei, showed complete tolerance. As the chronic nicotine treatment regimen used had previously increased the densities of nicotinic receptors in several brain regions (Schwartz and Kellar, 1985), cerebral metabolic responses to nicotine appear to depend on factors other than nicotinic receptor densities.

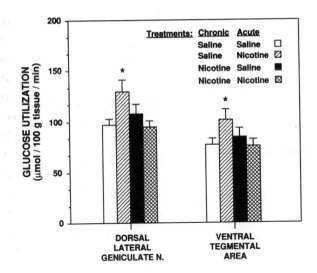

Figure 2. Chronic nicotine produces tolerance to nicotine challenge. Rats received nicotine bitartrate (1 mg/kg s.c.) or saline twice daily for 10 days and on the morning of the eleventh day. Two min before the DG injection, rats received saline or a nicotine challenge (0.3 mg/kg s.c.). Data are graphed as means ± S.E.M. Tolerance to nicotine challenge was observed in several regions including those for which the data are graphed. * indicates that mean rCMRglc for the group receiving daily injections of saline and a nicotine challenge differed significantly from other means (p < 0.05, significant interaction of chronic and acute treatments by 2-way analysis of covariance, and Duncan's new multiple range test).

Another study, that assessed the effect of nicotine by chronic infusion (Grünwald et al., 1988), found that some brain regions that had shown enhanced rCMRglc in response to acute nicotine also showed stimulation with the chronic infusion. In contrast to findings with repeated nicotine injections (London et al., in press), infusion produced stimulation rather than tolerance in the lateral geniculate body. However, stimulation in the lateral geniculate body and other areas may have resulted from high levels of nicotine present at the time of DG experiment rather than from a response to chronic treatment. In this regard, rats that received chronic infusion in the Grünwald et al. (1988) study exhibited plasma concentrations of nicotine that were similar to levels achieved with acute challenge in the study by London et al. (in press), but substantially higher than those in rats that received chronic nicotine by intermittent injections (London et al., in press). Nonetheless, stimulation in the lateral geniculate body suggests that complete tolerance was not produced in this region by chronic infusion. The findings suggest that treatment regimen is an important determinant of chronic nicotine effects.

A promising future avenue for research on nicotine involves human studies using PET scanning. Although PET does not offer the fine anatomical resolution achievable with autoradiography, it allows noninvasive assay of biochemical processes in the brain and concurrent measurements of drug effects on mood. PET studies have shown that morphine or cocaine, at euphorigenic doses, reduce cortical rCMRglc, suggesting that reduced cortical activity is a common feature of drug–induced euphoria (London et al., 1990_a, 1990_b). According to this hypothesis, euphorigenic doses of nicotine should reduce cortical rCMRglc in humans.

CONCLUSIONS: Nicotinic receptors in the rat brain show high densities in components of the limbic and visual systems, thalamic nuclei, and the cerebral cortex. They can be visualized and quantitated with radiolabeled nicotine *in vivo*, suggesting that quantitative studies ultimately might be performed noninvasively in humans using PET.

Metabolic mapping has delineated the distribution of the *in vivo* cerebral response to nicotine. In rats, nicotine stimulates rCMRglc in regions with high densities of [^{3}H]nicotine receptors, demonstrating that these receptors are coupled to functional activity. Chronic nicotine produces tolerance. A potential research frontier is the extension of this work to include human studies. PET offers a promise of relating effects of nicotine on mood to discrete functional changes in the human brain.

Acknowledgements

Supported in part by a grant from the Council for Tobacco Research U.S.A., Inc. Special thanks to Drs. Alane Kimes and Michael Morgan for their input regarding this manuscript.

REFERENCES

Abood, L.G. and Grassi, S. (1986) Biochem. Pharmacol. 35, 4199–4202.

Abood, L.G., Reynolds, D.T., and Bidlack, J.M. (1980) Life Sci. 27, 1307–1314.

Aceto, M.D. and Martin B.M. (1982) Med. Res. Rev. 2, 43–62.

Boksa, P. and Quirion, R. (1987) Eur. J. Pharmacol. 139, 323–333.

Broussolle, E.P., Wong, D.F., Fanelli, R.J., and London, E.D. (1989) Life. Sci. 44, 1123–1132.

Clarke, P.B.S. (1987) Trends Pharmacol. Sci. 8, 32–35.

Clarke, P.B.S., Pert, C.B., and Pert, A. (1984) Brain Res. 323, 390–395.

Clarke, P.B.S., Schwartz, R.D., Paul, S.M., Pert, C.B., and Pert, A. (1985) J. Neurosci. 5, 1307–1315.

Gomez-Ramos, P., Nelson, S., Walter, D., Cross, R., and Samson, F.E. (1982) J. Neurosci. Res. 7, 297–303.

Grünwald, F., Schröck, H., and Kuschinsky, W. (1987) Brain Res. 400, 232–238.

Grünwald, F., Schröck, H., Theilen, H., Biber, A., and Kuschinsky, W. (1988) Brain Res. 456, 350–356.

Henningfield, J.E. (1984) In: Advances in Behavioral Pharmacology (T. Thompson et al., Eds.), Academic Press, New York, pp. 31–210.

Henningfield, J.E., Clayton, R., and Pollin, W. (1990) Brit. J. Addiction 85, 279–292.

Larson, P.S. and Silvette, H. (1971) Tobacco: Experimental and Clinical Studies, Suppl. II, Williams & Wilkins, Baltimore.

London, E.D. (1989) J. Neuropsychiatry 1, S30–S36.

London, E.D., Connolly, R.J., Szikszay, M., and Wamsley, J.K. (1985a) Eur. J. Pharmacol. 110, 391–392.

London, E.D., Waller, S.B., and Wamsley, J.K. (1985b) Neurosci. Lett. 53, 179–184.

London, E.D., Szikszay, M., and Dam, M. (1986) In: Problems of Drug Dependence (L.S. Harris, Ed.) Washington, DC, pp. 26–35.

London, E.D., Connolly, R.J., Szikszay, M., Wamsley, J.K., and Dam, M. (1988a) J. Neurosci. 8, 3920–3928.

London, E.D., Dam, M., and Fanelli, R.J. (1988b) Brain Res. 20, 381–385.

London, E.D., Broussolle, E.P.M., Links, J.M., Wong. D.F., Cascella, N.G., Dannals, R.F., Sano, M., Herning, R., Snyder, F.R., Rippetoe, L.R., Toung, T.J.K., Jaffe, J.H., and Wagner, H.N. Jr (1990a) Arch. Gen. Psychiat. 47, 73–81.

London, E.D., Cascella, N.G., Wong, D.F., Phillips, R.L., Dannals, R.F., Links, J.M., Herning, R., Grayson, R., Jaffe, J.H., and Wagner, H.N. Jr. (1990$_b$) Arch. Gen. Psychiat. <u>47</u>, 567-574.

London, E.D., Fanelli, R.J., Kimes, A.S., and Moses, R.L. (in press) Brain Res.

Marks, M.J., Stitzel, J.A., Romm, E., Wehner, J.M., and Collins, A.C. (1986) Mol. Pharmacol. <u>30</u>, 427-436.

Martino-Barrows, A.M. and Kellar, K.J. (1987) Mol. Pharmacol. <u>31</u>, 169-174.

McCullough, J. (1982) In: Handbook of Psychopharmacology. (L.L. Iversen et al., Eds.), Plenum, New York, pp. 321-410.

McNamara, D., Larson, D.M., Rapoport, S.I., and Soncrant T.T. (1990) J. Cereb. Blood Flow Metab. <u>10</u>, 48-56.

National Household Survey on Drug Abuse (1990) DHHS Publication No. 90-1681, Alcohol, Drug Abuse and Mental Health Adm.

Nybäck, H., Nordberg, A., Långström, B., Halldin, C., Hartvig, P., Ahlin, A., Swan, C.-G., and Sedvall, G. (1989) In: Progress in Brain Research (A. Nordberg et al., Eds.) Elsevier Science Publishers, pp. 313-319.

Oswald, R.E. and Freeman, J.A., (1981) Neuroscience <u>6</u>, 1-14.

Phelps, M.E., Huang, S.C., Hoffman, E.J., Selin, C., Sokoloff, L., and Kuhl, D.E. (1979) Ann. Neurol. <u>6</u>, 371-388.

Reivich, M., Kuhl, D., Wolf, A., Greenberg, J., Phelps, M., Ido, T., Casella, V., Fowler, J., Hoffman, E., Alavi, A., Som, P., and Sokoloff, L. (1979) Circ. Res. <u>44</u>, 127-137.

Romano, C. and Goldstein, A. (1980) Science <u>210</u>, 647-650.

Schmidt, J, Hunt, S., and Polz-Tejera, G. (1980) In: Neurotransmitters, Receptors and Drug Action (W.B. Essman, Ed.), Spectrum, New York, pp. 1-45.

Schwartz, I.R. and Bok, D. (1979) J. Neurocytol. <u>8</u>, 53-66.

Schwartz, R.D. and Kellar, K.J. (1983) Science <u>220</u>: 214-216.

Schwartz, R.D. and Kellar, K.J. (1985) J. Neurochem. <u>45</u>, 427-433.

Sokoloff, L., Reivich, M., Kennedy, C., Des Rosiers, M.H., Patlak, C.S., Pettigrew, K.D., Sakurada, O., and Shinohara, M. (1977) J. Neurochem. <u>28</u>, 897-916.

Takayama, H., Majewska, M.D., and London, E.D. (1989) J. Pharmacol. Exp. Ther. <u>253</u>, 1083-1089.

Wong, D.F., Gjedde, A., Wagner, H.N. Jr., Dannals, R.F., Douglass, K.H., Links, J.M., and Kuhar, M.J. (1986) J. Cereb. Blood Flow Metab. <u>6</u>, 147-153.

Yamada, S., Gehlert, D.R., Hawkins, K.N., Nakayama, K., Roeske, W.R., and Yamamura, H.I. (1987) Life Sci. <u>41</u>, 2851-2861.

EFFECTS OF A CHRONIC NICOTINE INFUSION ON THE LOCAL CEREBRAL
GLUCOSE UTILIZATION

W. Kuschinsky, F. Grünwald and H. Schröck

Department of Physiology, University of Heidelberg, Im Neuenheimer Feld 326, D-6900 Heidelberg, Fed. Rep. Germany

SUMMARY: Local cerebral glucose utilization was measured as an
indicator of local brain function in 45 brain regions in conscious rats using the quantitative 2-deoxy-(^{14}C)-glucose method. L-nicotine was infused by osmotic minipumps for 14 days
resulting in a plasma nicotine concentration of 77 ng/ml plasma. Local cerebral glucose utilization was significantly increased in 6 brain structures. 4 of these structures had already revealed increased glucose utilization in previous experiments perfomed during acute nicotine infusion; the 2 other
structures were activated during chronic infusion only. It is
concluded that chronic nicotine infusion has distinct effects
on the functional activity of several brain structures which
are partly congruent with those affected during acute nicotine
infusion and partly divergent from them.

In previous studies (Grünwald et al, 1987) we have measured
local cerebral glucose utilization (LCGU) in conscious rats
during an acute infusion of L-nicotine. LCGU as measured by the
quantitative 2-deoxy-(^{14}C)-glucose method of Sokoloff et al
(1977) is taken as an indicator of brain function. During acute
nicotine infusion, we have observed an increased LCGU in several structures of the limbic system (anteroventral and antero-
medial nucleus of thalamus, interpeduncular nucleus, mammilary

body, cingulate cortex), the visual system (optic chiasm, superior colliculus, lateral geniculate body) and the substantia nigra, compact part. These effects indicate a functional activation of these structures by an acute administration of nicotine. Since chronic effects may be different, the present study had the aim to investigate LCGU in conscious rats during chronic infusion of nicotine.

MATERIALS AND METHODS

The experiments were performed on rats. LCGU was measured using the quantitative 2-deoxy-(^{14}C)-glucose method of Sokoloff et al (1977) as described previously (Grünwald et al, 1987). During halothane/N_2O anesthesia, osmotic minipumps (Alzet 2002), filled with L-nicotine, which had been diluted in saline, were implanted subcutaneously. Control animals received minipumps filled with saline.

On the 14th day after implantation, catheters were implanted into a femoral artery and vein during anesthesia. The rats were then placed into loose-fitting plaster casts and partly, although comfortably, immobilized. After recovering from anesthesia, a bolus of 2-deoxy-(^{14}C)-glucose was injected via the femoral vein and timed arterial blood samples were taken. After 45 minutes the brain was taken out, frozen and processed for quantitative autoradiography. Local tissue concentrations of ^{14}C were determined from autoradiographs by densitometric analysis using precalibrated standards.

RESULTS

In 39 of the 45 brain structures investigated, LCGU was unchanged. In the remaining 6 brain structures, LCGU was significantly increased (p < 0.05, t-test) compared to the saline infusion control group. No significant decrease was observed in any

structure. The structures affected by chronic nicotine treatment can be grouped into different systems. Concerning the
sensorimotor and extrapyramidal systems, the only structure
which was significantly affected was the globus pallidus. Three
structures of the visual system were activated significantly:
the optic chiasm, lateral geniculate body and superior colliculus. Finally, remarkable and significant increases in LCGU were
also observed in the septal nucleus and interpeduncular nucleus. These structures are part of the limbic system.

Plasma concentrations of nicotine were 77 ± 17 and of cotinine were 540 ± 137 ng/ml plasma (means ± SEM). Blood pressure,
heart rate, acid base status, hematocrit, body temperature and
plasma glucose concentration were unchanged during the period
of LCGU measurement as compared to control rats.

DISCUSSION

The present study shows effects of chronic nicotine treatment
on part of the visual system (superior colliculus, lateral
geniculate body, optic chiasm), the limbic system (interpeduncular nucleus, septal nucleus) and the globus pallidus. These
results are in partial agreement with previous studies of our
group performed during acute nicotine infusion (Grünwald et al,
1987). Congruent with the present results, an increased LCGU
has been shown in the superior colliculus, lateral geniculate
body, optic chiasm and interpeduncular nucleus. The functional
activity in these brain structures is apparently increased
during acute as well as chronic infusion of nicotine.

A dissociation between acute and chronic effects was found
in other structures. In the septal nucleus and the globus
pallidus, LCGU was found to be increased during chronic nicotine infusion only. This dissociation of effects could be explained by an up-regulation of cerebral nicotinic receptors by
chronic nicotine administration as shown in in vitro binding
studies (Ksir et al, 1987; Schwartz and Kellar, 1985; Hulihan-

Giblin et al, 1990), however, up-regulation is not confined to these structures but is rather a general phenomenon. Up-regulation is likely to have occured in other structures as well, although this did not result in higher values of LCGU; several of the structures which were found to be activated during acute nicotine infusion (compact part of substantia nigra, cingulate cortex, anteroventral and anteromedial nucleus of the thalamus) were not affected by the chronic nicotine treatment. We conclude from this that changes observed in receptor binding studies may not necessarily show up as changes in functional activity.

Acknowledgement: Supported by the Forschungsrat Rauchen und Gesundheit

REFERENCES

Grünwald, F., Schröck, H., and Kuschinsky, W. (1987) Brain Res 400, 232-238
Hulihan-Giblin, B.A., Lumpkin, M.D., and Kellar, K.J. (1990) J. Pharmacol. Exp. Therap. 252, 21-25
Ksir, C., Hakan, R., and Kellar, K.J. (1987) Psychopharmacology 92, 25-29
Schwartz, R.D., and Kellar, K.J. (1985) J. Neurochem. 45, 427-433
Sokoloff, L., Reivich, M., Kennedy, C., Des Rosiers, M.H., Patlak, C.S., Pettigrew, K.D., Sakurada, O., and Shinohara, M. (1977) J. Neurochem. 28, 897-916

ION CHANNEL BLOCKERS AND NICOTINIC AGONISTS AND ANTAGONISTS TO
PROBE BRAIN NICOTINIC RECEPTORS

LG Abood, S Banerjee, and JS Punzi

Dept of Pharmacology, University of Rochester Medical Center,
Rochester, NY 14642

Summary: In an effort to investigate the nature of brain
nicotinic receptors receptor binding studies have been
undertaken using nicotinic radioligands. With ^{3}H-nicotine or the
^{3}H-azetidine analogue of nicotine the Scatchard plot yielded a
high and lower affinity site; whereas, with ^{3}H-
methylcarbamylcholine (^{3}H-MCC) only the lower affinity site was
observed. Based on the observation that the addition of a methyl
group on carbamylcholine yielded a potent and specific nicotinic
agonist, a number of nicotinic agonists and antagonists have
been synthesized and their structure-activity relationships
described. Studies are described using ^{3}H-mecamylamine as a
ligand for investigating the ion channel associated with brain
nicotinic receptors. A good correlation was observed between the
K_i values of a variety of mecamylamine and pempidinge analogues
and their ability to antagonize the psychotropic effects of
nicotine. Since nicotine competed for ^{3}H-mecamylamine binding,
it was inferred that nicotine may be interacting with the ion
channel as well as with the recognition site.

Various radioligands for the studying brain nicotinic recognition sites: During the past decade a number of radioligands have

been developed for investigating the characteristics of nic-

otine's action in brain and interaction with nicotinic choliner-

gic receptors. Included among such radioli-gands are ^{3}H-nicotine

(Romano and Goldstein, 1980; Abood et al, 1981), (S, R)-^{3}H-1-

methyl-2-(3-pyridyl)-azetidine (Abood et al, 1987), and ^{3}H-

methylcarbamylcholine (Abood and Grassi, 1986); Bokas et al,

1987); Bokas et al, 1989). Scatchard analysis of ^{3}H-pyridylaze-

tidine (^{3}H-PAz) binding to rat brain membranes yielded curvilinear curves, resolvable into two distinct binding sites with K_d values of 7 x 10^{-11} and 2 x 10^{-9} M, and B_{max} values of 0.3 x 10^{-14} and 2.5 x 10^{-14} mol/mg protein respectively. Comparable results were achieved with ^{3}H-nicotine; however, since the azetidine analogue had a higher affinity than nicotine, it proved to be a more convenient radioligand for Scatchard anaylsis. On the other hand with ^{3}H-methylcarbamylcholine (^{3}H-MCC) as the radioligand, a linear Scatchard plot was obtained with only the lower affinity K_d of 1 x 10^{-9} M and a B_{max} of 2 x 10^{-14} mol/mg.

A re-examination of ^{3}H-PAz Scatchard plots using a 100-fold greater concentration of unlabeled (S)-nicotine or MCC (previous published data (3) utilized a 1000-fold greater concentration of unlabeled ligand) than ^{3}H-MCC yielded single Scatchard curves with K_d values of about 1 x 10^{-10} and B_{max} values of 2 x 10^{-14} mol/mg (Figure 1).

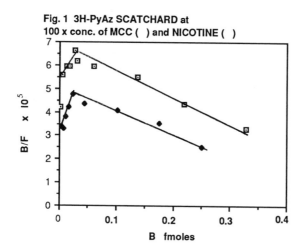

Fig. 1 3H-PyAz SCATCHARD at
100 x conc. of MCC ( ) and NICOTINE ( )

At very low concentrations of either nicotine or MCC positive cooperativity was observed i.e., the binding of ^{3}H-pyridylazetidine was enhanced. Such positive cooperativity at very low concentrations of ligand has been reported for both (R)- and (S)-nicotine as well as other nicotinic agents (Sloan et al, 1987).

When the Scatchard analyses were performed with a 10^{-5} M concentration of either unlabeled (S)-nicotine or MCC, curvilinear Scatchards were obtained for ^{3}H-pyridylazetidine either unlabeled ligands (Figure 2).

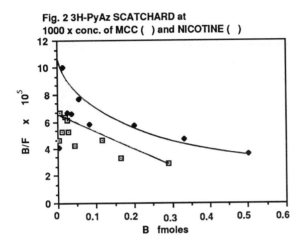

Fig. 2 3H-PyAz SCATCHARD at 1000 x conc. of MCC ( ) and NICOTINE ( )

It can be concluded from the Scatchard analyses with the various nicotinic radioligands that nicotine and its analogues appear to be recognizing more that one binding site; whereas, MCC, similar to acetylcholine (Schwartz and Kellar, 1985) appears to bind to only one site.

Structure-activity relationships of analogues of MCC: It has been noted that the addition of a methyl substituent on the carbamyl N of carbamylcholine increased nicotinic potency a 100-fold greater while the compound, MCC, was devoid of carbamylcholine's muscarinic cholinergic properties (Abood et al, 1988; Bokas et al, 1989). The selective nicotinic cholinergic activity of MCC has been determined by receptor binding studies utilizing prototypic radioligands, such as 3H-nicotine, ^{3}H-MCC, and ^{3}H-3-quinuclidinylbenzilate (Banerjee and Abood, 1989), and by employing a wide variety of behavioral and pharmacologic tests characteristic of nicotinic action and demonstrating their reversal by hexamethonium and mecamylamine (Bokas et al, 1989).

Additional analogues of MCC have recently been synthesized to determine the effect of varying chain length of the alkyl substituent R on the binding affinity and pharmacologic potency of MCC analogs:

$$RNCOOCH_2CH_2N(CH_3)_n$$

where R = CH_3, $(CH_3)_2$, C_2H_5, C_3H_7, or C_4H_9 and n = 2 or 3.

Receptor binding studies were performed using ^{3}H-nicotine and ^{3}H-QNB along with their ability to produce prostration in rats following their administration into the lateral ventricles, a behavioral response which is characteristic of nicotine-like activity. An excellent correlation was observed between the psychotropic potency of the agents and decreasing chain length of the alkyl substituent on carbamyl N. The most potent of the series were the methyl and dimethyl derivatives and the least potent was the butyl derivative. As the chain length is increased beyond C-4, the compounds exhibit nicotinic antagonism, as determined by their ability to block the nicotine-induced prostration in rats (2). All of the derivatives had a very low affinity for the muscarinic cholinergic site as determined by ^{3}H-QNB binding. The quaternary derivatives had nicotinic binding affinities and psychotropic potencies which were about 2 orders of magnitude greater than the tertiary derivatives.

Competitive antagonists to nicotine: In the course of investigating the structure-activity requirements for the nicotinic characteristics of carbamate esters related to MCC , it was noted that replacement of the alkyl group on the carbamyl N by phenyl or cycloalkyl resulted in nicotinic antagonists (Abood et al, 1988). A variety of carbamate and thiocarbamate esters were synthesized having the general formula:

$$RNCOOCH_2CH_2N(CH_3)_n \quad \text{or} \quad RNCOSCH_2CH_2N(CH_3)_n$$

where R = phenyl, cyclobutyl, cyclopentyl, or cyclohexyl and n = 2 or 3.

In addition various other esters and thioesters related to acetylcholine were synthesized having the general structure:

$$RCOOCH_2CH_2N(CH_3)_n \quad \text{or} \quad RCOSCH_2CH_2N(CH3)_n$$

where R = phenyl, cycloalkyl, naphthyl, or pyridyl and n = 2 or 3.

Also synthesized were a number of heterocyclic amino esters of benzoic acid and methylcarbamate, such as 3-quinuclidinylbenzoate and 3-quinuclinylmethyl carbamate; both being effective antagonists. The quaternary derivatives in both series were over 10-100-fold more active both in receptor binding studies and as antagonists to the psychotropic effects of nicotine; however, in order to antagonize the psychotropic effects (prostration and seizures) of nicotine they had to be administered intraventricularly. Among the most potent antagonists were the benzoic and cycloalkylcarboxylic acid esters of choline and dimethylaminoethanol and the phenyl- and cycloalkyl carbamyl esters of choline and dimethylaminoethanol.

Studies on the ^{3}H-Mecamylamine (^{3}H-MEC) binding site in rat brain: Mecamylamine, which had been originally developed as a ganglionic blocking agent (Stone et al, 1956), effectively antagonizes the central and peripheral actions of nicotine by interfering with its action at nicotinic cholinergic receptors. Since mecamylamine exhibits a low affinity for the nicotinic recognition site, it has been suggested that it may be exerting its action at the ion channel associated with the receptor (Lingle, 1983; Eldefrawi et al, 1982). Electrophysiologic studies have demonstrated that mecamylamine blocks the acetycholine-induced currents in crustacean muscle in a concentration and voltage-dependent manner, and recovery from the blockade required the presence of the agonist (Veranda et al, 1985). Evidence for the action of mecamylamine at ion channels is also derived from receptor binding studies on membranes from Torpedo electroplax utilizing 3Hperhydrohistionicotoxin as a radioligand (Eldefrawi et al, 1982).

We have identified a saturable, specific mecamylamine binding site which is distinct from the nicotinic recognition site and appears to be associated with an ionic channel (Banerjee et al, 1990). Scatchard analysis reveals a curvilinear plot with K_d values of 96 nM and 1 μM and B_{max} of 7 and 30 pmoles/mg (Figure 3).

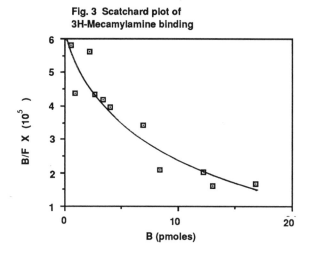

Fig. 3 Scatchard plot of 3H-Mecamylamine binding

B/F X (10^5)

B (pmoles)

^{3}H-MEC binding was destroyed completely by heating at 100 degrees by trypsin; therefore, the site appears to be proteinaceous. It is also inhibited by low concentrations of inorganic cations , particularly Ni, Ca, and other divalent cations.

A variety of mecamylamine and pempidine derivatives as well as other agents known to block ion channels were evaluated for their affinities to the 3H-mecamylamine binding site and for their ability to the prevent the nicotine-induced prostration in rats. Various doses of the test agent were administered i.p. to rats 5 minutes prior to the administration of nicotine into the lateral ventricles of rats via chronically implanted cannulae. A good correlation was observed between the binding affinities of various agents and their ED_{50} values in preventing the nicotine-induced prostration in rats (Figure 4).

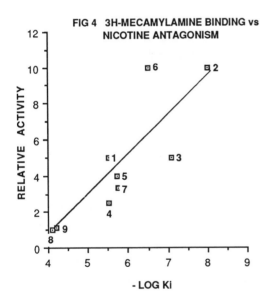

MECAMYLAMINE ANALOGS: R = H (**1**); R = CH$_3$ (**2**); R = (CH$_3$)$_2$ (**3**); R = benzyl (**4**)

PEMPIDINE ANALOGS: R = none, R$_2$ = H (**5**); R$_1$ = none, R$_2$ = CH3 (**6**) R$_1$ = NH$_2$, R$_2$ = H (**7**) ; Cyclohexylamine (**8**); exo-aminonorbornane(**9**). Numbers refer to figure 4.π

Within the mecamylamine series the most effective agent was 3-dimethylaminoisocamphene (#3) which had a K$_i$ of 7 x 10^{-8} M and

ED_{50} of 15 mg/kg in blocking the nicotine-induced prostration. Mecamylamine (#2) was slightly less effective, while normecamylamine (#1) had a K_i value 30-fold greater than mecamylamine. The N-benzyl derivative of mecamylamine (#4) exhibited 1/25 the affinity of mecamylamine.

Within the pempidine series. although the K_i of pempidine (#6) was about 5-fold greater than that for mecamylmine, it was slightly more potent in preventing prostration. Norpempidine and tetramethyl-4-aminopiperidine (#5) had about 1/5 the affinity of pempidine. Other agents such as cyclohexylamine (#8) and exo-aminobornane (#9) were considerably less active. Other compound exhibiting affinity in the submicromolar range included verapamil, a Ca channel blocker, and phencyclidine, a K channel blocker.

An interesting finding was that nicotine and its analogues also exhbited a high affinity for the ^{3}H-MEC binding site. This finding is consistent with the observation that at lower concentrations nicotine acts as an agonist by enhancing 3H-perhydrohistionicotoxin binding; while at higher concentrations nicotine acts as antagonist to inhibit binding (Eldefrawi, 1982) Since mecamylamine has a very low affinity for the nicotinic recognition site, one can infer that nicotine may be acting at ionic channels in addition to its action at the recognition site.

Although ^{3}H-mecamylamine appears to be useful for investigating the ion channel associated with nicotinic receptors, it is not an entirely suitable ligand, since complete binding saturation is not obtainable. This shortcoming might possibly be due to the fact that the radioligand is chemically unstable. We are presently developing more suitable radioligand for investigating this site.

We have synthesized an affinity ligand for the binding site: 4-bromoacetamido-2,2,6,6-tetramethylpiperidine:

$$\text{NHCOCH}_2\text{Br}$$

(chemical structure: piperidine ring with NHCOCH$_2$Br at top, H$_3$C and H$_3$C groups on left carbon, CH$_3$ and CH$_3$ groups on right carbon, NH at bottom)

We have shown that the compound exhibits a high affinity for the ^{3}H-MEC site and that it irreversibly inhibits the site at 10 uM. Various other ligands of the pempidine series are presently being investigated as more suitable ligands for investigation the ion channel associated with the brain nicotinic receptors.

REFERENCES

Romano, C. and Goldstein, A.   (1980) Science 210, 647-650.
Abood, L.G., Reynold, D.T., Booth, H., and Bidlack, J.M.   (1981)
      Neurosci.Biobehav. Rev. 5, 479-486.
Abood, L.G., and Grassi, S.   (1986) Biochem. Pharmacol. 35, 4199-
      4202.
Abood, L.G., Lu, X, and Banerjee, S.   (1987) Biochem. Pharmacol.
      36, 2337-2341.
Abood, L.G., Salles, K.S., and Maiti, A.   (1988) Pharmacol.
      Biochem. Behav.30, 403-408.
Banerjee, S and Abood L.G. (1989) Biochem. Pharmacol. 38: 2933-
      2935.
Banerjee, S., Punzi, J.S., Kreilick, K., and Abood, L.G.   Biochem.
      Pharmacol.in press, 1990.
Bokas, P., and Quiron, R.   (1987) Eur. J. Pharmacol. 139, 323-
      333.
Bokas, P., Maryka, Q., Mitchell, J.B., Collier, B., O'Neill, W.,
      and Quiron, R.   (1989)   Eur. J. Pharmacol. 173, 93-198.

Eldefrawi, A.T., Miller, E.R., and Eldefrawi, M.E.   (1982)   .
      Biochem. Pharmacol. 31, 1819-1822.
Schwartz, R.D. and Kellar, K.J. (1985) J. Neurochem. 45, 427-
      433..
Lingle, C. (1983) J. Physiol. 339, 395-417.
Sloan, J.W., Martin, W.R., and Smith, W.T. (1987) In: Tobacco
      Smoking and Nicotine: A Neurobioligical Approach (W.R.
      Martin, G.R. Van Loon, E.T.Iwamoto and D.L. Davis, Eds),
      Plenum Press, New York, pp. 451-465.

Stone, C.A., Torchiana, M.L., Navarrro, A., and Beyer, K.H..
    (1956) J.Pharmacol. Exp. Ther. 117, 169-183.
Veranda, W.A., Arcava, Y., Sherby, S.M., Van Meter, W.G.,
    Eldefrawi, M.E.,and Albuquerque. (1980) Mol.Pharmacol.28,
    128-137.

INCREASED SENSITIVITY OF BRAIN MUSCARINIC RECEPTOR TO ITS AGONISTS AS A RESULT OF REPEATED NICOTINE PRETREATMENT IN RATS, MICE AND RABBITS

Hai Wang and Chuan-gui Liu

Institute of Pharmacology and Toxicology, Academy of Military Medical Sciences, P.O.Box 130, Beijing 100850, China

SUMMARY: In gallamine-immobilized and artificially ventilated rats, the dose-reponse curves of M-agonists and AntiChE rather than GABA or glycine receptor antagonists inducing EEG seizures shifted leftward after the brain nAChR was desensitized by acute or chronic nicotine pretreatment.This phenomenon could be prevented by N-antagonist mecamylamine.Similar results were obtained in acute nicotine tolerant mice and rabbits.In addition, the effect of M-agonist arecoline for producing brain mAChR down-regulation was potentiated in acute nicotine tolerant rats. Apparent dissociation constant values of brain mAChR high-affinity site to its agonists arecoline and oxotremorine were decreased in the presence of nicotine. It is concluded that brain nAChR desensitization as a result of repeated nicotine pretreatment selectively increases the sensitivity of brain mAChR high affinity binding sites to its agonists.

INTRODUCTION

It has been demonstrated by our laboratory that in rats anticholinesterase(AntiChE) induced EEG seizures are characterized by initial tonic seizures of short period and subsequent long-lasting clonic ones. The brain nAChR and mAChR were suggested to be responsible for the tonic and clonic seizure discharges respectively. Brain nAChR can be desensitized by AntiChE or repeated injections of nicotine. After pretreatment of nicotine, AntiChE could only induce clonic EEG seizures as a result of muscarinic effects of endogenous ACh on mAChR. The central cholinolytics with both M- and N-antagonistic effects rather than M-antagonists or N-antagonists are more potent against AntiChE-induced EEG seizures (Zhang & Liu,1987, 1988).It is interesting to know the roles of desensitized brain nAChR, and the regulatory relationship between brain nAChR and mAChR in generating the EEG seizures by AntiChE. In this paper, we present some recent results of our laboratory about the effect of nAChR desensitization on the sensitivity of brain mAChR to its agonists.

MATERIALS AND METHODS

Animals:Wistar rats (180–220g), Swiss mice (18–22g) and New Zealand rabbits(1.5–3.0kg) of either sex were provided by Experimental Animal Center of our Academy.

EEG recording:Rats or rabbits were paralyzed by gallamine (60mg / kg ip) or d–tubocurarine (0.5 mg / kg iv)and artificially ventilated with room air under positive pressure. Surgical proceduces were performed under local aneasthesia (procaine 2%). EEG electrods were fixed on the skull correspoonding to frontal, parietal and occipital cortex and placed stereotaxically in hippocampus bilaterally. A reference electrod was placed on the skull over frontal sinus. Animals were allowed to recover for 30 min. Mice were mechanically immobilized. Two electrods were chronically fixed as in rats but only corresponding to motor cortex. EEG recording began 3 days after surgery.

Chronic tolerance tests:A multifactorial test battery was used to assess the responses of the rats to nicotine during the chronic nicotine treatment (2.0 mg / kg sc bid 10 days).

1.Seizure susceptibility: Each rat was placed in a separate observation box and its behavior was observed for 30 min after nicotine. Seizure intensity was assessed on a scale from 1 to 7: 1:gustatory movement and scratching; 2:tremor and hindlimb extension; 3:head bobbing and backward walking; 4:rearing and clonus; 5: rearing, clonus and falling; 6:whole body jerks; 7:tonic–clonic seizures.

2.Rotarod test: Rotation speed was 15 rpm. Rats were trained to walk on rotarod for 180 sec or more. 3 min after nicotine, the animal was placed on the rotating rod until it fell from the device, in which case the time was recorded.

3.Body temperature: Body temperature was measured with a rectal thermometer 15 min after nicotine.

Biochemical measurement:Rats were decapitated and the cerebral cortex was homogenized in 10 volumes (w / v) of ice–cold 0.32M sucrose. The homogenate was centrifuged at 1000g for 10 min at 4°C, and the supernatant fluid was recentrifuged at 20000g for 30 min at 4°C. The pellet (P_2) was suspended in Na / K phosphate buffer (50 mM, pH 7.4). The P_2 fractions of brain membranes (0.1 mg protein) were incubated for 30 min at 35°C in a final 1.0 ml of buffer containing [^{3}H]QNB (0.02–2.0 nM). Specific binding was calculated as the total binding minus that occurring in the presence of 10 μM QNB. The M–agonist binding to mAChR was determined from the competition by arecoline or oxotremorine of specific [^{3}H]QNB binding.Tubes containing brain membranes (0.1 mg protein), [^{3}H]QNB (0.2 nM),arecoline or oxotremorine (0.1 nM–3.0 mM), were incubated for 30 min at 35°C. Upon completion of the incubation, samples were filtered through glass–fiber filters. The radioactivity was determined on Beckman 1215 Liquid Scintillation Spectrometer at an efficiency of 36%.

RESULTS

Ⅰ.Effects of acute and chronic nicotine pretreatment on EEG seizure sensitivity to M−agonists in rats, mice and rabbits.

In gallamine immobilized and artificially ventilated rats, nicotine 1.0 mg / kg iv produced EEG tonic seizure discharges lasted about 30 sec (n = 20). Nicotine 0.5−0.75 mg / kg iv produced EEG desynchronization (n = 100), but 10 min after that nicotine 2.0−5.0 mg / kg iv produced only transient EEG desynchronization (n = 100).10 min later additional larger doses of nicotine 5.0−10.0 mg / kg iv produced neither EEG desynchronization nor seizures (n = 10). This indicated that acute tolerance to nicotine resulted after two doses of nicotine (0.75 and 2.0 mg / kg iv at 10 min interval).

After iv M−agonists arecoline 25−150 mg / kg or pilocarpine 100−380 mg / kg, the number of rats with EEG clonic seizures increased dose−dependently. Arecoline 150 mg / kg or pilocarpine 380 mg / kg iv produced EEG clonic seizures in all tested rats (n = 15 or 10). After two doses of nicotine

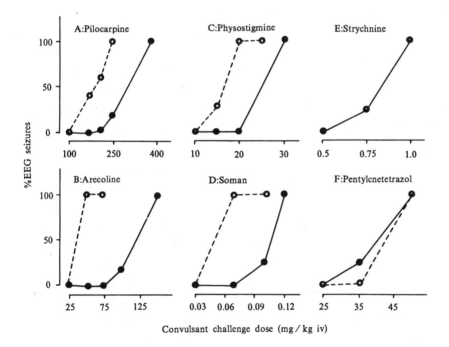

Fig.1. Dose−response curves of convulsant−induced EEG seizures in naive(—●—) and nicotine pretreated (···○···)rats. The convulsants were given 10 min after nicotine (0.75 and 2.0 mg / kg iv at 10 min interval)

pretreatment (0.75 and 2.0 mg / kg iv at 10 min interval), the dose—response curves of arecoline and pilocarpine for producing EEG clonic seizures shifted leftward (Fig.1—A,B). Now arecoline 50 mg / kg (n = 20) or pilocarpine 250 mg / kg iv (n = 5) could produce EEG seizures in all tested rats. This indicated that EEG seizure sensitivity to M—agonists increased in acute nicotine tolerant rats. This phenomenon was also observed in AntiChE—induced EEG seizures. The threshold doses of physostigmine and soman for producing EEG clonic seizures in all tested rats decreased from 30 (n = 5) to 20 mg / kg iv (n = 5) and from 0.12 (n = 5) to 0.07 mg / kg iv (n = 7) after two doses of nicotine pretreatment (Fig.1—C,D). But this was not the case in noncholinergic convulsants.

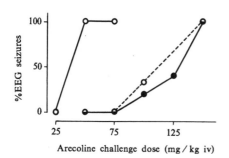

Fig.2. Dose—response curves of arecoline induced EEG seizures in naive (—●—), nicotine pretreated (— ○ — , as in Fig.1) and mecamylamine—nicotine pretreated (··· ○ ··· , mecamylamine 0.5 mg / kg iv 5 min prior to nicotine) rats

After the same precedures for producing brain nAChR desensitization by repeated pretreatment of nicotine, the dose response curves for producing EEG clonic or tonic seizure remained unchanged for both GABA receptor antagonist pentylenetetrazol (EEG seizure threshold dose 50 mg / kg iv, n = 10), and glycine receptor antagonist strychnine (EEG seizure threshold dose 1.0 mg / kg iv, n = 5) (Fig.1—E,F).

According to the receptor desensitization model, the initial exposure to nicotine produced nAChR activation and subsequent desensitization. If that is the case, after mecamylamine (0.5 mg / kg iv) had prevented the activation of brain nAChR, the desensitization of brain nAChR did not occur. In this case, the leftward shift of dose—response curve of arecoline for producing EEG clonic seizures was abolished. As in naive rats, only iv arecoline 150 mg / kg could produce EEG seizures in all tested rats (n = 10) (Fig.2).

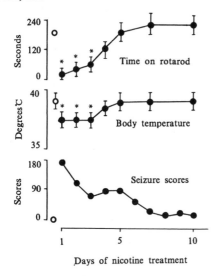

Fig.3. Time course of chronic nicotine tolerance development. Nicotine(●) or saline (○) was given to rats at 2.0 mg / kg or 1.0 ml / kg sc bid. Results represent mean ± SD of 10 rats. Significant difference from control *p < 0.05

It is demonstrated that chronic tolerance to nicotine could also be obtained in rats (Schwartz & Keller,1983). In this study nicotine 2.0 mg / kg

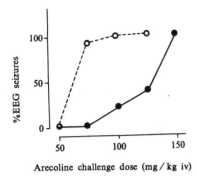

Fig.4. Dose—response curves of arecoline induced EEG seizures in saline—pretreated (—●—)
and chronic nicotine pretreated rats (···○···).
Nicotine was given at dose of 2.0 mg / kg sc bid
for 10 days. Arecoline was given 48 hours after
the last dose of nicotine

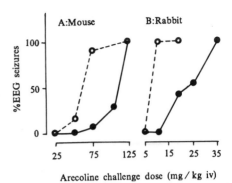

Fig.5. Dose—response curves of arecoline induced EEG seizures in naive (—●—) and nicotine pretreated (···○···) animals. Nicotine 0.5
mg / kg iv or 0.1 mg / kg × 6 at 5 min interval
was given to mice or rabbits 5 min before
arecoline

sc bid was given to rats for 10 days. As shown in Fig.3, nicotine induced high seizure scores,decrease of
body temperature and decrease of time on rotarod gradually returned to control level after 5—day nicotine treatment. After 10—day nicotine treatment, nicotine 2.0—5.0 mg / kg iv could not induce EEG seizures (n = 5).Chronic tolerance to nicotine resulted. In this case, the dose—response curves of arecoline
for producing EEG seizures shifted leftward. Now iv arecoline 75 mg / kg could produce EEG seizures
in 6 out of 7 tested rats, in stead of 150 mg / kg arecoline in control animals(n = 5), treated with saline
1.0 ml / kg sc bid 10 days (Fig.4). It is suggested that rat brain nAChR desensitization, as a result of
nAChR excitation by acute or chronic nicotine pretreatment, can selectively increase the sensitivity of
brain mAChR to its agonists for generating EEG seizures.

It is desirable to know whether this phenomenon also occurs in other species of experimental
animals. In mice, acute tolerance to nicotine could be developed after one subconvulsive dose of nicotine 0.5 mg / kg iv, and then convulsive dose of nicotine 1.0 mg / kg iv could not further induce the
EEG seizures(n = 30). In these acute nicotine tolerant mice, the dose—response curve of arecoline for
producing EEG seizures shifted leftward. Now arecoline 75 mg / kg iv produced EEG seizures in 13
out of 15 tolerant mice instead of 125 mg / kg iv in all tested naive mice (n = 10) (Fig.5—A). In rabbits,
acute tolerance to nicotine could also be developed but only after 6 injections of nicotine (0.1mg / kg ×
6 iv at 5 min interval), and then convulsive dose of nicotine (1.0—1.5mg / kg iv)could not further induce
EEG seizures(n = 5). In these tolerant rabbits, the dose—response curve of arecoline for producing EEG
seizures again shifted leftward(Fig.5—B). Now arecoline 12 mg / kg iv could produce EEG seizures in
all tolerant rabbits (n = 5) instead of 35 mg / kg iv in naive ones (n = 5). Thus, increased sensitivity of

brain mAChR to its agonists could also be observed in acute nicotine tolerant mice and rabbits.

II. Effects of acute nicotine pretreatment on cerebral mAChR down−regulation induced by its agonist in rats.

It has been reported that excessive excitation of mAChR to its agonists can produce the loss of mAChR from synaptic membranes known as receptor down−regulation, and the down−regulation of mAChR is linked to the pharmacological efficacy of its agonists exposed to the cholinergic neurons (Freedman et al., 1988). As shown in Tab. I , arecoline 200 mg / kg ip produced the loss of [^{3}H]QNB binding sites of cerebral cortex in naive rats, in which B_{max} values decreased from 713.3 ± 79.4(n = 4) to 572.0 ± 73.0 pmol / mg pr.(n = 4) (p < 0.05). This down−regulation of brain mAChR could be prevented by pretreatment of atropine 5.0 mg / kg ip(n = 3). Arecoline 50 mg / kg ip had no effects on B_{max} and K_d values (n = 4). After pretreatment of nicotine (0.5, 1.0 × 2, 2.0 × 2 mg / kg ip at 10 min interval), 50 mg / kg arecoline now could decrease the B_{max} values to 456.0 ± 90.1 pmol / mg pr. (n = 7) (p < 0.01), and increase the K_d values from 0.14 ± 0.02 (n = 4) to 0.18 ± 0.01 nM (n = 7) (p < 0.05). The facilitatory effect of acute nicotine pretreatment on down−regulation of mAChR induced by its agonist arecoline suggested that the pharmacological efficacy of arecoline increased after nicotine pretreatment. These results were consistent with those of EEG experiments.

Table I . [^{3}H]QNB binding to membrane fractions (P$_2$) derived from cerebral cortex of rats treated with nicotine or / and arecoline[#]

Drugs (mg / kg ip)			n	B_{max}(pmol / mg pr.)	K_d (nM)
Atropine	Nicotine	Arecoline			
—	—	—	4	713.3 ± 79.4	0.14 ± 0.02
—	—	200	4	572.0 ± 73.0*	0.18 ± 0.05
5.0	—	200	3	783.7 ± 61.8	0.11 ± 0.03
—	—	50	4	726.3 ± 177.9	0.16 ± 0.04
—	0.5, 1.0 × 2, 2.0 × 2	—	4	676.2 ± 102.0	0.16 ± 0.03
—	0.5, 1.0 × 2, 2.0 × 2	50	7	456.0 ± 90.1**	0.18 ± 0.01*

[#][^{3}H]QNB binding experiments were carried out 12 hours after nicotine or / and arecoline treatment. Results (mean ± SD) were calculated by linear regression analysis of Scatchard plots of binding data. Significant difference from naive animals *p < 0.05; **p < 0.01

III. Effects of nicotine on the binding profiles of rat brain mAChR to its agonists.

M−agonists arecoline and oxotremorine were used to displace the binding of [^{3}H]QNB to mAChR of rat cerebral cortex. Binding parameters derived from computer analysis significantly (p < 0.01) fitted a two−site model. Hill coefficients were 0.41 ± 0.04(n = 5) or 0.54 ± 0.11(n = 3) for arecoline or oxotremorine respectively. As shown in Tab. II , in the presence of nicotine 100μM, the apparent dissociation constant values of mAChR high−affinity sites (K_1) to arecoline decreased from 0.35 ± 0.11

μM (n = 5) to 0.03 ± 0.01 μM (n = 7), (p < 0.01). And in the presence of nicotine 1.0μM, K_1 values of mAChR high–affinity sites to oxotremorine decreased from 7.00 ± 3.95 (n = 3) to 0.48 ± 0.44 nM (n = 4) (p < 0.05). But the apparent dissociation constants of mAChR low–affinity binding sites (K_2) were unchanged. In addition, the amount of specific [^{3}H]QNB binding sites was not influenced by the presence of nicotine (0.1nM–1.0mM) (data not showed). The percentage of high–affinity binding sites in the total binding sites (R_1) was also unchanged. This indicated that in vitro experiments, affinity of mAChR high–affinity binding sites to M–agonists could be increased in the presence of nicotine.

Table II. Parameters for the inhibition of specific [^{3}H]QNB binding to membrane fractions (P_2) of rat cerebral cortex by arecoline and oxotremorine

	n	n_H	IC_{50} (μM)	K_1	K_2 (μM)	R_1 (%)
arecoline:						
Control	5	0.41 ± 0.04	30.07 ± 10.52	0.35 ± 0.11 (μM)	146.13 ± 55.72	27.9 ± 5.5
nicotine 100μM	7	0.36 ± 0.04	41.58 ± 16.60	0.03 ± 0.01 (μM)**	166.88 ± 34.50	24.7 ± 4.7
oxotremorine:						
control	3	0.54 ± 0.11	8.71 ± 3.94	7.00 ± 3.95 (nM)	22.15 ± 0.76	22.0 ± 6.2
nicotine 1.0μM	4	0.49 ± 0.11	9.06 ± 4.90	0.48 ± 0.44 (nM)*	26.56 ± 3.73	23.4 ± 5.0

Significant difference from control: *p < 0.05; **p < 0.01

DISCUSSION

ACh is one of the main neurotransmitters in nervous system. It acts on its targets–mAChR and nAChR. Peripheral mAChR and nAChR are located in different tissues. The functional effects of ACh on the tissues basically depend on which type of cholinergic receptors, mAChR or nAChR, existing in these tissues. For example, ACh induces salivation through mAChR in salivary glands and induces neuromuscular transmission through nAChR in neuromuscular junction. But in CNS the distributions of mAChR and nAChR are not so clear cut as that in peripheral nervous system. Furthermore, the functions of brain cholinergic receptors, particularly nAChR, have not been fully studied. In central respiratory and cardiovascular center of medulla oblongata it seems that m– and nAChR are more or less independent and M– or N–antagonists can only prevent the effects of M– or N–agonists respectively (Liu et al., 1984; Huang & Liu 1985; Xie & Liu, 1985). But in cerebrum larger dose of M– or N–antagonists can prevent the effects of both M– and N–agonists. Some central cholinolytics, having both M– and N–antagonistic activities, showed much higher potentials than those of combined administration of M– and N–antagonists in prevention of AntiChE–induced epileptic seizures. It is expected that there will be close functional correlations between cerebral mAChR and nAChR (Liu, 1988).

Brain nAChR has been shown to be responsible for nicotine–induced convulsions(Miner et al., 1984;

Wang et al., 1990) and can be desensitized by repeated administration of nicotine (Miner&Collins,1988; Wang et al.,1990). In this experiment,the first subconvulsive dose of nicotine produced EEG desychronization, which was due to the activation of brain nAChR. After that the subsequent administration of nicotine was less effective. In acute or chronic nicotine tolerant animals,neither EEG seizures nor desynchronization were induced by larger doses of nicotine.The desensitization of brain nAChR resulted. It is interesting to note that the threshold doses of nicotine for producing EEG seizures in naive rats, mice and rabbits were the same (1.0 mg / kg iv). This indicates that the brain nAChR of the three species of animals has the same sensitivity to nicotine for producing EEG seizures. But the amount of nicotine needed for nAChR desensitization in these animals were different. Rat brain nAChR could be desensitized by two doses of nicotine (0.75 and 2.0 mg / kg iv at 10 min interval).For mouse brain nAChR desensitization, one injection of nicotine (0.5 mg / kg iv) was enough, while rabbit brain nAChR could be desensitized only after six injections of nicotine (0.1 mg / kg × 6 at 5 min interval).

The desensitization of brain nAChR does not mean that it can do nothing but only fails to response to ACh. It increases the sensitivity of mAChR to its agonists, as expressed in significant decreases of the threshold doses of M−agonists for producing EEG seizures or mAChR down−regulation. As shown in Fig.6−A, in acute nicotine tolerant rats the EEG seizure threshold doses of arecoline, pilocarpine, soman and physostigmine decreased by 66%,33%,41%,and 33% respectivly. Among them the facilitatory effect of nicotine on arecoline produced−EEG seizures was more potent than that on other M−agonist and AntiChE. It may be explained by the differences in pharmacological effects of these agents. Comparing to arecoline, pilocarpine has some noradrenergic effects on CNS (Mason, 1978), while AntiChE acts on both brain nAChR and mAChR. In addition, the facilitatory effect of nicotine on muscarinic seizures was more potent in acute tolerant rats than in chronic tolerant ones. (Fig.6−B) It may be explained by the facts that only brain nAChR desensitization is developed in acute nicotine tolerance, but up−regulation of brain nAChR is also developed in chronic nicotine tolerane, in addition to nAChR desensitization. (Schwartz & Keller,1983). In different species of animals, the facilitatory effect of acute nicotine pretreatment

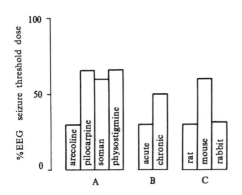

Fig.6. Nicotine induced decreases of EEG seizure threshold doses of M−agonists in various condition as compared to control. A: Various M−agonists in acute tolerant rats; B: Arecoline in acute and chronic tolerant rats; C: Arecoline in acute tolerant mice, rats and rabbits

on muscarinic seizures was more potent in rats and rabbits than that in mice (Fig.6—C).

Even if nicotine pretreatment was given, the increased seizure sensitivity to M—agonists was not observed in nicotine nontolerant animals. After three doses of nicotine (0.1 mg / kg × 3 iv at 5 min interval), rabbits remined sensitive to nicotine, just as in naive animals. 35 mg / kg of arecoline was necessary to produce EEG seizures. It is suggested that the hypersensitivity to M—agonists did not occur until nicotine tolerance developed. Similar results were obtained in chronic nicotine treatment. On the other hand, the time course for the development of nicotine tolerance revealed that mice became sensitive to nicotine normally 40 min after nicotine pretreatment (0.5 mg / kg iv). In this case, only arecoline 150 mg / kg, larger than EEG seizure threshold dose of 125 mg / kg, could produce EEG seizures. This indicates that the hypersensitivity of mAChR to its agonists subsided as tolerance to nicotine disappeared.

It has been reported that nicotine can not affect the binding parameters of [^{3}H]QNB binding to brain mAChR (Marks & Collins,1984). Some experiments revealed that M—agonist binding sites were not the same as M—antagonist binding sites(Birdsall et al.,1978; Hulme et al.,1978). In addition, M—antagonists have no intrinsic pharmacological efficacy and show higher affinity to mAChR than M—agonists. Therefore, the function of mAChR to agonists could not be detected by the binding profiles of M—antagonists. In this experiment, displacement of [^{3}H]QNB binding by M—agonists arecoline or oxotremorine was carried out to observe the affinity of mAChR to its agonists. It was found that the affinity of mAChR to its agonist increased in the presence of nicotine. However, had brain mAChR been labled by its agonist, more reasonable results might have been obtained. The facts that the affinity of mAChR to its agonists was increased in the presence of nicotine, at least, can partially explain the facilitatory effects of nicotine pretreatment on brain mAChR to its agonists.

It is proposed here that AntiChE can produce EEG seizure discharges through the inhibition of brain acetylcholinesterase and accumulation of endogenous ACh. The excitation of brain nAChR by ACh results in the initial tonic EEG seizure discharges and the subsequent desensitization of brain nAChR. The excitation of brain mAChR to ACh and the increased sensitivity of brain mAChR to ACh by brain nAChR desensitization, are responsible for the later long—lasting clonic EEG seizure discharges. N—antagonists can prevent the facilitatory effects of desensitized nAChR on brain mAChR to its agonists. It is suggested that the cholinolytics with both central nicotinic and muscarinic antagonistic effects are more potent against the EEG seizures induced by AntiChE.

CONCLUSION: Brain nAChR desensitization as a result of nAChR activation by repeated nicotine pretreatment selectively increases the sensitivity of brain mAChR high affinity binding sites to its agonists.

Acknowledgements:This research was supported by a grant from National Nature Science Foundation of China, No.3870879.

REFERENCES

Birdsall,N.J.M., Burgen,A.S.V., and Hulme,E.C. (1978) Mol. Pharmacol. 14,723—736
Freedman,S.B., Harleg,E.A., and Iversen,L.L. (1988) TIPS Suppl.,54—60
Huang, W.X., and Liu, C.G. (1985) Acta Physiologica Sinica 37, 191—198
Hulme,E.C.,Birdsall,N.J.M., and Burgen,A.S.V. (1978) Mol. Pharmacol. 14,737—750
Liu, C.G. (1988) S15.02 In: Programme and Abstracts 5th Southeast Asian and Western Pacific
    Regional Meeting of Pharmacologists, Beijing, China. P.120
Liu, C.G., Ma, X.Y., Huang, W.X., and Wang, Z.X. (1984) Information of the Chinese
    Pharmacological Society 4,42—43
Marks,M.J., and Collins,A.C. (1984) Pharmacol. Biochem. Behav. 22,283—291
Mason,S.T. (1978) Neuropharmacol. 17,1015—1021
Miner,L.L., and Collins,A.C. (1988) Pharmacol. Biochem. Behav. 29,375—380
Miner,L.L.,Marks,M.J., and Collins,A.C. (1984) J. Pharmacol. Exp. Ther. 231,545—554
Schwartz,R.P., and Keller,K.J. (1983) Sci. 220,214—216
Wang,H.,Cui,W.Y., He, X.P., and Liu,C.G. (1990) Chinese J. Pharmacol. Toxicol. 4,86
Wang, H., Cui, W.Y., He, X.P., and Liu, C.G. (1990) Chinese J. Applied Physiology. 6, 148
Xie, X.P., and Liu, C.G. (1985) Acta Pharmacologica Sinica 6, 87—90
Zhang,Y.N., and Liu,C.G. (1987) Chinese J. Pharmacol. Toxicol. 1,237
Zhang, Y.N. and Liu, C.G. (1988) O.27.07 and P2.11 In: Programme and Abstracts 5th Southeast
    Asian and Western Pacific Regional Meeting of Pharmacologists, Beijing, China. P. 210 and P.269

NICOTINE INDUCES OPIOID PEPTIDE GENE EXPRESSION IN HYPOTHALAMUS
AND ADRENAL MEDULLA OF RATS

V. Höllt and G. Horn

Department of Physiology, University of Munich, Pettenkoferstraße
12, D-8000 München 2, F.R.G.

SUMMARY: Two consecutive injections of nicotine (0.4mg/kg; s.c.;
4 pm and 9 am) resulted in an more than twofold increase in the
levels of mRNA coding for prodynorphin (PDYN) and for provaso-
pressin in the hypothalamus of rats measured 21 hour after the
first injection. In contrast, the levels of proenkephalin (PENK)
and proopiomelanocortin (POMC) mRNAs in the hypothalamus remained
unchanged. A much smaller increase (about 30%) in the level of
PDYN mRNA was observed when nicotine (4 mg/kg/day; s.c.) was
chronically applied via minipumps for 7 days. Moreover, PDYN mRNA
levels returned to normal when the same dose of nicotine was
continuously applied for 14 days indicating a complete desensitization to the effect of nicotine upon PDYN gene expression. Hypothalamic mRNA levels coding for the α_3-subunit of the neural
nicotinic receptors were unchanged suggesting no upregulation of
the biosynthesis of this particular receptor subunit by nicotine.
In the adrenal medulla, a pronounced increase in the PENK mRNA
levels was observed (more than twofold) after two injections of
nicotine. Moreover, PENK mRNA levels were still significantly
elevated (by about 60%) when nicotine was tonically infused by
minipumps (4 mg/kg/day) for 7 and 14 day indicating that the
nicotinic receptors in the medulla are not completely desensitized by these treatments. A more effective induction of the PENK
gene was obtained when nicotine was chronically applied in a
pulsatile fashion via minipumps (6 mg/kg/day for 7 days; in 6
pulses/day).

Several lines of evidence suggest that nicotine can affect the
release of opioid peptides. For instance, met-enkephalin and
other PENK-derived peptides are concomitantly secreted with catecholamines from adrenal medullary chromaffin cells in vitro and
in vitro in response to nicotine (Viveros et al. 1982). Moreover, nicotine has been shown to stimulate the release of vasopressin from the posterior pituitary (Castro de Souza and Rocha E
Silva, 1977). Since vasopressin and PDYN-derived peptides are
co-localized in the same hypothalamic neurones (Watson et al.
1982), a concomitant release of these peptides in response to
nicotine is likely. A persistent stimulation of release of the

peptides might result in an enhanced biosynthesis of the peptide
by activating the expression of its respective gene. It is, therefore, possible that a continuous administration of nicotine
causes specific alterations in the expression of opioid peptides
genes in the adrenal medulla and the hypothalamus of rats. In
fact, a marked induction of the levels of mRNA coding for PENK
was found in bovine adrenal medullary chromaffin cell after chronic exposure to nicotine (Eiden et al. 1984; Kley et al. 1987).
It is, however, not known whether or not the application of nicotine to rats results in an enhanced PENK gene expression in vivo.
In the present study we have investigated the effect of nicotine
administration to rats in vivo on the levels of PDYN mRNA in the
hypothalamus and on the levels of PENK mRNA in the adrenal medulla. It will be shown that the administration of nicotine causes
an induction of the opioid peptide gene expression in these tissues and that the mode of nicotine application (injection of
single doses; tonic or pulsatile infusion) markedly influences
the efficacy to nicotine to alter opioid peptide gene expression.

MATERIALS AND METHODS

Nicotine administration: Male Wistar rats (240-260 g) were treated with nicotine according to the following procedures: One
group received 0.4 mg/kg nicotine (s.c.) at 4 pm and the same
dose at 9 am at the next day. The animals were killed 4 hours
after the last injection. Control rats received saline. Other
rats were chronically treated with nicotine (4 mg/kg/day) or
saline for 7 or 14 days via s.c. implanted osmotic minipumps
(Alzet, 2001 or 2002). A further set of rats were treated with
nicotine in a pulsatile fashion (6 mg/kg/day in 6 pulses of 1
mg/kg at a flow rate of 1 μl/hour). Teflon tubings were alternately filled with 1 μl aliquots of a nicotine solution (0.25 μg
nicotine in saline) and 3 μl aliquots of silicone oil by suction
using an electronic pump. There was no significant distribution
of nicotine into the silicon oil as measured by using ^{3}H-nicoti-
ne. The filled teflon tubings (about 25 cm in length) were win-
ded up to a coil of 1 cm diameter, connected with the saline-
filled minipumps and s.c. implanted into rats. Control animals
were implanted with minipumps connected to teflon tubings in
which nicotine solution was replaced by saline.

Processing of the tissues: The rats were killed by decapitation and the trunk blood collected. The concentration of nicotine in the sera was measured by the Forschungslabor Dr. Adlkofer (Munich, F.R.G.) using gas chromatographic methods. After removing the brain from the skull, the hypothalami) were dissected out. In addition, the medullae were isolated from the adrenal glands. All tissues were weighed and immediately frozen on dry ice. Total RNA was extracted from the tissues using a LiCl-method (Auffray and Rougeon, 1980).

RNA measurements: RNA was quantified by spectrophotometry. RNA samples were denatured with glyoxal and 5 µg aliquots electrophoresed on a 1.2 % agarose gel and transferred to nylon filters. The filters were hybridized at 60°C to radiolabelled RNA probes complementary to rat PENK, PDYN, provasopressin (PVP), α_3-subunit of neuronal nicotinic receptors and to mouse POMC. The single stranded RNA probes were synthesized with RNA polymerases and α-^{32}P-UTP. The PDYN probe contains 74 bases complementary to nucleotide 513-587 of rat PDYN mRNA (Höllt et al. 1987). For preparation of the PENK probe a 938 kb SmaI/SacI restriction fragment of plasmid pRPE2 (a gift from Dr. Sabol, Bethesda, USA) was subcloned into plasmid pBS (Genofit, Heidelberg, FRG). Likewise, for preparation of the POMC probe a 150 bp HindIII restriction fragment of plasmid ME-150 (a gift from Dr. Roberts, New York) was subcloned into plasmid pBS. For preparation of the   vasopressin probe a PstI/DraI fragment of pRV8 which contains the vasopressin specific 3'end of the rat cDNA was subloned into pSP 65 (pSP2; Fehr et al. 1988; a generous gift from Dr. Schmale, Hamburg,FRG). The probe for the α_3-subunit of neural nicotinic receptors was made using a construct in which pCA48E DNA was subcloned into pSP65 (Goldman et al. 1987; a generous gift from Dr. Boulter, San Diego, USA). After hybridization, tzhe filters were washed, dried and exposed to x-ray film at -70°C. The autoradiograms were scanned using a laser densitometer (UltroScan XL, LKB).

Substances: (-)- Nicotine   (Serva, Heidelberg); ^{3}H-(-)nicotine, α^{32}P-UTP (Amersham, Braunschweig); enzymes and other chemicals (Boehringer, Mannheim).

RESULTS

Hypothalamic levels of mRNA coding for opioid peptides and vasopressin after two injections of nicotine into rats: An 0.4 mg/kg
dose of nicotine was s.c. injected into rats; thereafter, the
same dose was given 17 hours later and the animals killed 4 hours
after the second dose. Fig. 1 shows an analysis of RNA samples
isolated from the hypothalami of control and nicotine-treated
rats. The blots were consecutely hybridized to probes complementary to opioid peptides and vasopressin mRNAs. The mRNA species
which hybridize to the probes correspond in size to their corresponding mRNAs (PDYN = 2.6 kb; PENK = 1.45 kb; POMC = 1.2 kb;
PVP = .7 kb). When the staining of the bands was qunatified by
densitometry (Fig. 2) a significant increase in the levels of
PDYN mRNA and of mRNA coding for PVP of about 100% was observed
in the hypothalami of nicotine-treated rats. In contrast, the
levels of PENK and POMC mRNA were not significantly altered in
this tissue.

Hypothalamic levels of mRNA coding for PDYN and for the α_3-subu-
nit of nicotinic receptors after prolonged nicotine: Fig. 3 shows
an analysis of RNA samples obtained from hypothalami of control
rats and from rats which were chronically treated with nicotine
(4 mg/kg/day; s.c.) for 2 weeks. The blot was hybridized with a
PDYN cRNA (2.6 kb species) followed by hybridization with a probe
complementary to the α_3-subunit of neural nicotinic receptors
(3.5 kb species). As seen from the quantitative analysis in Fig.
4, there is no difference in the PDYN and α_3-subunit mRNA levels
between control rats and rats treated chronically with nicotine
for 2 weeks. In addition, as compared to the highly elevated
levels after the single injections, there is only a slight in-
crease(about 30%) when nicotine was infused by minipumps for 7
days. The serum levels of nicotine after 7 and 14 days minipump
adminstration were in the range between 50 and 170 ng/ml corresponding to those seen in heavy smokers.

PENK mRNA levels in the adrenal medulla after nicotine treatment:
Fig. 5 shows an analysis of RNA samples extracted form the adrenal medulla of the rats. The blots were hybridized with a probe

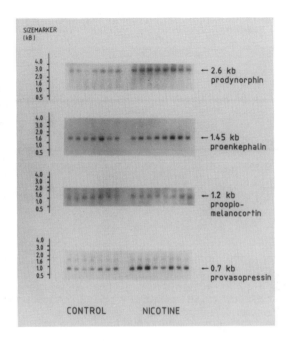

Fig.1   Analysis of RNA from hypothalami of control rats or of rats injected with two doses of 0.4 mg/kg nicotine. Each lane represents 5 μg RNA obtained from hypothalami of individual rats.

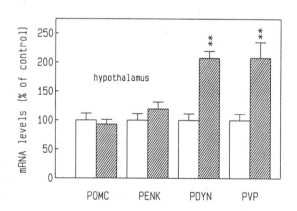

Fig.2   Quantification of mRNA levels coding for proopiomelanortin (POMC), proenkephalin (PENK), prodynorphin (PDYN) and provaso-pressin (PVP) by densitometry of the bands shown in Fig.1.
Open bars represent mean ± S.E.M. of control rats; hatched bars give the respective values for nicotine-treated rats (n = 7).

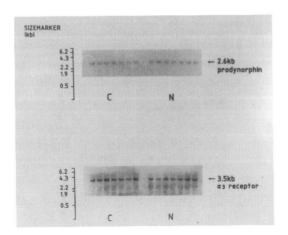

Fig.3   Analysis of RNA of hypothalami of rats chronically treated with nicotine (4mg/kg/day;s.c.) for 2 weeks. The blots were
hybridized with cRNAs complementary to prodynorphin and to the
α_3-subunit of neural nicotinic receptors. C = control; N =
nicotine.

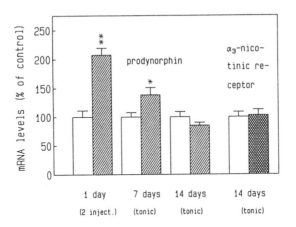

Fig.4   Quantification of hypothalamic mRNA levels coding for prodynorphin and for the α_3-subunit of nicotinic receptors by densitometry of RNA blots. Open bars represent mean ± S.E.M. of control rats; hatched bars give the respective values of nicotine-
treated rats (n = 7).

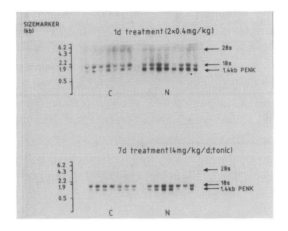

Fig. 5 Analysis of RNA from adrenal medullae of control rats or of rats which were treated with two injections (o.4 mg/kg) of nicotine or which were chronically infused with nicotine (4 mg/kg/day) for 7 days. The blots were hybridized with a proenkephalin (PENK) cRNA probe which cross-hybridizes to the 18 S ribosomal RNA under the conditions used. C = control; N = nicotine.

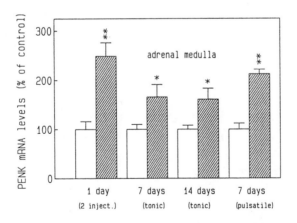

Fig.6 Quantification of of mRNA levels coding for proenkephalin (PENK) in the adrenal medullae of rats after nicotine treatment. See text for the description of the various modes of nicotine administration. Open bares: mean S.E.M. of control rats hatched bars give the values for nicotine-treated rats (n = 7).

complementary to PENK mRNA which migrates as molecular species of
1.45 kb. The probe cross-hybridizes also with the 18S ribosomal
RNA under the experimental conditions used. As can be seen by
the differential blackening of the PENK mRNA bands in Fig. 5 and
the quantitative densitometry of the data in Fig. 6, there is a
marked increase in the levels of PENK mRNA in the adrenal medullae of nicotine treated rats. The induction or PENK gene expression is highest after the single dose regimen, but still apparent
when nicotine was tonically infused for 7 or 14 days at a dose of
4 mg/kg/day. Interestingly, the increase in adrenal medullary
PENK mRNA levels is higher when nicotine was chronically applied
in a pulsatile fashion (6 mg/kg/day in 6 pulses; during each
pulse 1 mg/kg nicotine is delivered within 1 hour) (Fig. 6). By
this mode of adminstration high peak level of nicotine (up to 500
ng/ml) were measured.

DISCUSSION

The presented results show that administration of nicotine to
rats can cause a concomitant induction of the expression of the
PDYN and PVP gene in the hypothalamus. Within the hypothalamus
PDYN and PVP are synthesized by the the same magnocellular neurones of the supraoptic and paraventricular nuclei (Watson et al.
1982). Moreover, dynorphin and vasopressin are co-released from
the posterior pituitary of rats in response to various pharmacological and endocrinological stimuli (Höllt et al. 1981). Since
nicotine has been shown to release vasopressin from the pituitary
(Castro de Souza and Rocha E Silva, 1977) a concomitant release
of dynorphin and vasopressin is very likely. From our experiments
it appears that the injections of nicotine cause also a marked
increase in the biosynthesis of the peptides as indicated by the
large increase in the mRNAs coding for PDYN and PVP. In contrast,
the expression of the two other opioid peptide genes (POMC and
PENK) in the hypothalamus is not affected by the nicotine treatment indicating the specificity of the effect. Interestingly,
however, during continuous infusion of nicotine the inducing
effect of nicotine upon PDYN/PVP gene expression became smaller
(only 30% above control at 1 week) and disappeared after 2 week
treatment indicating that a pronounced tolerance develops to this

effect of nicotine. In a previous study we reported a marked increase (more than 3-fold) of the levels of PDYN mRNA in the hypothalamus of rats which were tonically infused with nicotine (4 mg/kg/day) for 7 days (Höllt and Horn, 1989). In three additional sets of experiments using the same mode of nicotine application we did not find such a large induction of the PDYN mRNA in the hypothalamus of rats. This might reflect a large variability in the time-course of desensitization in response to nicotine which has also been found by Villanueva and Rosecrans (personal communication; see this volume). After 2 week tonic infusion of nicotine (4 mg/kg/day) we consistently did not find any alterations of PDYN or PVP mRNA in the hypothalamus of rats indicating a complete tolerance to this effect of nicotine. Moreover, in preliminary experiments in which nicotine was administered to rats in a pulsatile fashion (6 mg/kg/day in 6 pulses) for 1 and 2 weeks, no significant increase in the PDYN and PVP levels in the hypothalamus could be found indicating that the development of tolerance is independent of the mode of drug application. Furthermore, the desensitization is not associated with any change of the biosynthesis of the α_3-subunit of the neural nicotinic receptors in the hypothalamus. This findings is interesting in view of the reported upregulation of nicotine binding sites in the brain of rodents chronically treated with nicotine (Marks et al. 1983; Nordberg et al. 1985). In addition, an increase in nicotinic cholinergic recognition sites in hypothalamic nuclei of rats after repetitive nicotine treatment has recently been reported (Kellar et al. 1989). From our experiments, however, we cannot exclude that the upregulation of nicotinic binding sites is due to an increase in the biosynthesis of other subunits (α_2, α_4) of neuronal nicotinic receptors or due to posttranslational processes (e.g. involving the assembling of the the channel forming subunits).

Short-term or chronic treatment of rats with nicotine results in a considerable induction of the PENK gene in the medully of the adrenal gland. As compared to the situation in the hypothalamus, there is no complete development of tolerance to the inducing effect of nicotine on PENK gene expression, since chronic infusion of nicotine for 1 or 2 weeks still results in a consistent enhancement of PENK mRNA levels in the adrenal medulla of the rats. The observation, however, that a higher induction of the

PENK gene is found when nicotine is chronically administered in a pulsatile fashion indicates that higher concentrations of nicotine are more effective in inducing PENK gene expression or that some resensitization occurs in the intervall phase when the blood nicotine levels declined. It is likely that nicotine causes its inducing effect on PENK gene expression in vivo by directly interacting with the nicotinic receptors localized on the chromaffin cells of the adrenal gland. We cannot, however, exclude a central action of nicotine which indirectly induces PENK gene expression via stimulation of the splanchnic nerve.

The direct induction of PENK gene expression by nicotine has been shown in bovine chromaffin cells in culture. In this preparation the continuous presence of nicotine causes a time-dependent increase in the PENK mRNA levels for several days (Kley et al. 1987). This findings indicate that there is no substantial tolerance to the inducing effect of nicotine in the chromaffin cells in vitro. In this preparation, nicotine has also been shown to induce the expression of the gene coding for tyrosine hydroxylase, the rate limiting enzyme in the biosynthesis of catecholamines. In fact, catecholamines and PENK-derived peptides are co-localized in the same vesicles of the chromaffin cells and concomitanly released in response to a variety of stimuli. It is possible that the application of nicotine in rats in vivo also results in a concomitant induction of tyrosine hydroxylase gene in the adrenal medulla. Experiments are presently carried out in our laboratory to study this question.

The mechanism whereby nicotine stimulates PENK gene expression in bovine chromaffin cells involves membrane depolarization followed by an influx of extracellular calcium (Kley et al. 1987). Activation of calcium dependent proteinkinases may activate transcription factors, such as the c-fos protooncogene which, in turn, may enhance PENK gene transcription after binding to a regulatory site of the PENK gene.

REFERENCES

Auffray, C., and Rougeon,F. (1980)
  Eur. J. Biochem. 107, 303-312.
Castro de Souza, B. and Rocha E Silva, M. (1977)
  J. Physiol. 265, 297-311.

Eiden, L.E., Giraud, P., Dave, J.R., Hotchkiss, A.J. and Affol
ter, H.U. (1984) Nature 312, 661-663.
Fehr,S., Schmale,H. and Richter,D. (1989)
Biochem. biophys. Res. Commun. 158, 555-561.
Goldman, D., Deneris, E., Luyten, W., Kohchar, A., Patrick, J.
and Heinemann, S. (1987) Cell 48, 965-973.
Höllt, V., Haarmann, I., Seizinger, B.R. and Herz, A. (1981)
Neuroendocrinology 33, 333-339.
Höllt, V., Haarmann, I., Millan, M.J. and Herz, A. (1987)
Neurosci. Lett. 73, 90-94.
Höllt, V. and Horn, G. (1989) In: Nicotinic Receptors in the CNS
(A. Nordberg, K. Fuxe, B. Homstedt and A. Sundwall, Eds),
Elsevier/North-Holland, Amsterdam, pp. 187-193.
Kellar, K.J., Giblin, B.A. and Lumpkin, M.D. (1989) In: Nicotinic
Receptors in the CNS (A. Nordberg, K. Fuxe, B. Holmstedt and A.
Sundwall, Eds), Elsevier/North-Holland, Amsterdam, pp. 209-216.
Kley, N., Loeffler, J.P., Pittius, C.W. and Höllt, V. (1987)
J. Biol. Chem. 262, 4083-4089.
Marks, M.J., Burch, J.B. and Collins, A.C. (1983)
J. Pharmacol. Exp. Ther. 226, 817-825.
Nordberg, A., Wahlstrom, G. Arnelo, U. and Larsson, C. (1985)
Drug Alcoh. Depend. 16, 9-16.
Viveros, O.H., Wilson, S.P. and Chang, K.J. (1982) In: Regulatory
Peptides (E. Costa and M. Trabucchi, Eds), Raven Press, New
York, pp. 217-224.
Watson, S.J., Akil, H., Fischli, W., Goldstein, A., Nilaver, G.
and von Wimmersma Greidanus, T.B. (1982) Science 216, 77-79.

THE MESOLIMBIC DOPAMINE SYSTEM AS A TARGET FOR NICOTINE

P.B.S. Clarke

Department of Pharmacology and Therapeutics, McGill University,
3655 Drummond St. Room 1325, Montreal, Canada H3G 1Y6

SUMMARY: Mesolimbic dopamine neurons appear to mediate the
locomotor stimulant and rewarding effects of d-amphetamine and
cocaine. In the rat brain, these neurons are known to possess
nicotinic receptors. In isolated tissues, nicotine directly
stimulates dopaminergic cell bodies, and can enhance dopamine
release via a local action as well. As expected, acute
administration of nicotine enhances mesolimbic dopaminergic
utilization in vivo, but whether chronic continuous exposure to
nicotine has the same effect is not clear. Activation of the
mesolimbic dopamine system appears to underlie the acute
behavioural stimulant effects of the drug and may contribute to
the maintenance of tobacco smoking.

When tobacco smokers are asked why they smoke, many reasons
are given. Cigarettes may be smoked for different effects,
depending on the individual and the context. Several
perceived effects of smoking appear to be linked to known
psychopharmacological effects of nicotine, particularly
increased subjective arousal, sedation/relaxation, and appetite
suppression (Clarke, 1987a; Surgeon General's Report, 1988).
Indeed, it is widely held that nicotine is the key to
dependence on tobacco, and that it is the central actions of
this drug which are paramount.

The diversity of nicotine's psychopharmacology probably
reflects, in large part, the widespread distribution of
nicotinic receptors in the CNS. Nicotinic receptor protein
itself has been mapped neuroanatomically with a number of
radioligands, most notably ^{3}H-nicotine and ^{3}H-acetylcholine
(Clarke, 1987b) and with antibodies (e.g. Swanson et al.,

1987). Recently, molecular biological studies have revealed that central nicotinic receptors comprise a distinct family: the number and relative prevalence of receptor subtypes is not yet clear, but certain pharmacological differences are already evident. In situ hybridization histochemical mapping of nAChR subunit-associated mRNA species reveal that certain putative nAChR subtypes are likely to be expressed widely in the brain, whilst others are not (Wada et al., 1989).

It might then appear that searching for a single neuropharmacological action of nicotine which could account for tobacco smoking would be akin to looking for a needle (or even several needles) in a haystack. However, in the last twenty years, evidence has accumulated to suggest that laboratory animals will avidly work for the opportunity to stimulate dopaminergic neurotransmission in the mesolimbic dopaminergic system, and it is this system that provides the entry point by which psychomotor stimulant drugs gain access to reward circuitry.

Thus, drugs such as d-amphetamine and cocaine appear to produce rewarding effects by increasing extracellular concentrations of mesolimbic dopamine (Pettit and Justice, 1989). The evidence, mostly obtained from laboratory rats, may be summarized as follows (see Wise and Rompre, 1989 for review): (1) appropriate doses of dopamine receptor antagonists appear to reduce the rewarding effects of intravenously self-administered d-amphetamine and cocaine, (2) animals tend to return to a place previously associated with d-amphetamine or cocaine (given iv), and acquisition of this "conditioned place preference" is reduced by dopamine receptor blockade, (3) depletion of mesolimbic dopamine reduces the rewarding effects of d-amphetamine and cocaine, as assessed in both intravenous self-administration and place preference paradigms, (4) intracerebral injection of d-amphetamine appears to be most rewarding when it is administered into the nucleus accumbens, a major terminal area of the mesolimbic dopamine system (Carr and White, 1986), and (5) the mesolimbic dopaminergic projection to the other major mesolimbic terminal area, the olfactory tubercle, does not appear to contribute to psychostimulant reward, insofar as dopamine depletion in this area did not alter the magnitude of d-amphetamine place preference (Clarke et al., 1990).

Parallel studies have shown that the locomotor stimulant effects of these drugs is also dependent on their ability to enhance dopaminergic tone in mesolimbic terminal areas (Kelly et al., 1975). It has been suggested, on the basis of intracranial drug administration experiments, that the olfactory tubercle it the important structure in mediating the locomotor stimulant effects of dopamine agonists (Cools, 1986). However, our own experiments, employing regionally selective depletion of mesolimbocortical dopamine, suggest the opposite: a critical role for the nucleus accumbens, with little obvious role for the olfactory tubercle (Clarke et al., 1988a).

In view of the central role of the mesolimbic dopaminergic system in the rewarding and stimulant effects of d-amphetamine and cocaine, several groups of researchers have examined the possibility that nicotine may also activate this system and that such an activation may underlie the drug's stimulant and rewarding effects. These studies, conducted in rodents, are reviewed below, and in greater detail elsewhere (Clarke, 1990).

BEHAVIOURAL EFFECTS OF NICOTINE

Several of the behavioural effects of d-amphetamine in rodents are known to be dependent on mesolimbic dopamine: (1) locomotor stimulation (Kelly et al., 1975; Clarke et al., 1988a), (2) reinforcing effects (Wise and Rompre, 1989), and (3) the ability to enhance the rewarding effect of electrical brain stimulation (Wise and Rompre, 1989). Although certain qualifications are necessary, nicotine can produce all three effects (see Clarke, 1987a for review). Thus, acute injection of nicotine reliably increases locomotor activity, at least in nicotine-experienced rats.   Nicotine is voluntarily self-administered, although the drug has not always been found to produce conditioned place preferences reliably.   Finally, nicotine has been found to enhance brain stimulation reward, at least in one behavioural paradigm in which d-amphetamine was also effective.   This behavioural profile, by analogy with that of d-amphetamine, is consistent with a stimulation of the mesolimbic dopaminergic system.

NICOTINIC RECEPTORS ARE ASSOCIATED WITH DOPAMINERGIC NEURONS

It appears that the predominant subtype(s) of brain nicotinic receptor activated by low ("smoking") doses of nicotine can be labelled with high affinity by tritiated agonists such as ^{3}H-nicotine and ^{3}H-ACh (Clarke, 1987b).   Such binding sites are present in moderate or high density in the brain areas containing dopamine cell bodies (the ventral tegmental area) and terminals (nucleus accumbens, olfactory tubercle).   Near-total destruction of mesolimbic (and nigrostriatal) dopamine neurons, effected with the neurochemically selective toxin 6-OHDA, reduced ^{3}H-nicotine binding in all these areas, suggesting that these nicotinic receptors were present on both cell bodies (and/or dendrites) and on terminals (Clarke and Pert, 1985).   This conclusion has been supported by the in situ hybridization histochemical demonstration of nAChR-associated mRNA in presumed dopamine cells of the ventral tegmental area (Wada et al., 1989).

The presence of nAChRs located on mesolimbic dopaminergic cell bodies and/or dendrites has recently been confirmed by electrophysiological experiments in vitro (Calabresi et al., 1989).   Nicotine and ACh both produced fast depolarizations which were inhibited by the nicotinic antagonists hexamethonium and kappa-bungarotoxin.   The persistence of nicotinic agonist effects in the presence of tetrodotoxin or cobalt suggests that these effects were indeed direct.   Responses to brief (1-2 min) application of nicotine or ACh desensitized and recovered slowly upon wash out of drug. The possibility of desensitization becomes important in considering in vivo data (see below).

Transmitter release studies have also confirmed the presence of nAChRs on terminals of mesolimbic dopaminergic neurons. In rat nucleus accumbens slices preloaded with ^{3}H-dopamine, low concentrations of nicotine (10^{-7} M and above) and other nicotinic agonists stimulated ^{3}H-dopamine release in a Ca^{++}-dependent manner (Rowell et al., 1987). Although prolonged or repeated applications of nicotine were not studied, analogous studies in rat striatal synaptosomes suggest that nicotine-induced dopamine release is susceptible to desensitization (Rapier et al., 1988).

NICOTINE ACUTELY STIMULATES MESOLIMBIC DOPAMINERGIC NEURONS IN VIVO

The in vitro data outlined above show that the mesolimbic dopamine system is a natural target for brief applications of nicotine. However, they also suggest that receptor desensitization may strongly limit the duration or size of this drug action. Nevertheless, there is ample evidence that acute systemic administration of the drug increases the activity of mesolimbic neurons.

Intravenous nicotine can increase mesolimbic cell firing in rats, but the effect is dependent on the presence of anaesthetic. Under general anaesthesia, little or no excitation occurred, although nicotine induced a bursting pattern of firing likely to result in increased transmitter release (Grenhoff et al., 1986; Mereu et al., 1987). However, in paralyzed, locally anaesthetized subjects, nicotine markedly increased cell firing. Clearly, analogous experiments in unstressed, conscious animals would be valuable.

The technique of intracerebral microdialysis permits
extracellular DA concentrations to be measured in freely moving
animals.   Acute systemic administration of nicotine increases
extracellular dopamine levels in nucleus accumbens, presumably
as a result of increased release (Imperato et al., 1986; Damsma
et al., 1989).   Acute administration of nicotine also
increases indirect measures of DA utilization in the rat
nucleus accumbens (Clarke et al., 1988b; Lapin et al., 1989;
Mitchell et al., 1989), and interestingly, cigarette smoke
exposure produces a similar effect which appears to be due to
nicotine (Fuxe et al., 1986).

TOLERANCE TO NICOTINE

Several mechanisms are likely to underlie tolerance to the
effects of nicotine.   In operational terms, it is useful to
distinguish two forms of tolerance: acute and chronic.   Acute
tolerance lasts for a few minutes or hours, is associated with
many of the central effects of nicotine, and may often reflect
receptor desensitization.   Chronic tolerance persists for days
or months; certain behavioural effects of nicotine (including
its locomotor stimulant and reinforcing effects) do not appear
to undergo chronic tolerance.   There is evidence for both
acute and chronic tolerance to certain effects of nicotine in
cigarette smokers (Surgeon General's Report, 1988).
Does the stimulant effect of nicotine on mesolimbic dopamine
utilization undergo tolerance in vivo ?   Chronic tolerance, if
it occurs, is not complete, since rats which have previously
received a number of daily doses of nicotine still show a
stimulant effect on this system (Clarke et al., 1988b).
Indeed, the weight of evidence suggests that chronic daily
injections of nicotine do not blunt subsequent mesolimbic
stimulation in response to acute nicotine challenge (Lapin et
al., 1989; Mitchell et al., 1989; Damsma et al., 1989).
Administration of nicotine in the form of chronic continuous

infusion, which might be expected to produce acute tolerance to nicotine, can in practice result in either tolerance or reverse-tolerance (see Clarke, 1990 for review).

MESOLIMBIC DOPAMINE AND THE LOCOMOTOR STIMULANT ACTION OF NICOTINE

The locomotor stimulant effect of nicotine is due to a direct central action and is more pronounced in rats that have received the drug on several previous occasions (Morrison and Stephenson, 1972; Clarke and Kumar, 1983a & b). In rats preexposed to nicotine through daily subcutaneous injections, we found that acute injection of nicotine increased locomotor activity in photocell cages, in a dose-dependent (0.1 - 0.4 mg/kg sc) and stereoselective (L>D) manner (Clarke et al., 1988b). A few days subsequent to locomotor activity tests, subjects were pretreated with an inhibitor of L-aromatic amino acid decarboxylase, challenged as before with nicotine and placed in the photocell cages for 30 min prior to sacrifice. Dopamine utilization was measured indirectly by L-DOPA/DA ratios in microdissected brain areas.

In parallel with the behavioural changes, we observed dose-dependent and stereoselective increases in dopamine utilization which were restricted to mesolimbic terminal regions at the doses tested. In order to test whether this was merely a spurious correlation, a further experiment was conducted. Rats received intra-accumbens infusions 6-hydroxydopamine, resulting in an 89% depletion of dopamine in mesolimbic terminal areas, compared to vehicle-infused controls. The lesion blocked the locomotor stimulant effect of nicotine when rats were tested 2 weeks after surgery. This study suggests that nicotine's ability to stimulate the mesolimbic dopamine system is sufficient to account for the drug's locomotor stimulant effect.

Microinjection experiments suggest that the locomotor stimulant effect of nicotine is mediated, possibly exclusively, through an action at the level of cell bodies within the ventral tegmental area (Pert and Clarke, 1987; Reavill and Stolerman, 1990).

Evidence has recently emerged that the rewarding effect of nicotine, self-administered intravenously, is also dependent upon the mesolimbic dopamine system (see Corrigall, this volume).

AREAS OF UNCERTAINTY

Animal studies have served to shape our notions of why nicotine appears to be so important in maintaining smoking behaviour (Clarke, 1987a; Surgeon General's Report, 1988). However, lest we conclude too readily that mesolimbic dopamine stimulation is important in the maintenance of the tobacco habit, it is worth reviewing what we do not know. First, neurotransmitter receptors often display pronounced species differences in their distribution, and we do not yet know whether mesolimbic dopaminergic neurons in human brain possess nicotinic receptors. Secondly, the effects of chronic continuous infusion of nicotine can be altogether different from the effects of long-term intermittent dosing in animals (see Clarke, 1990), and although cigarette smoking in humans provides twenty-four hour exposure to the drug, most animal studies employ acute injections. Thirdly, animal studies aimed at assessing the reinforcing effects of nicotine cannot readily allow for the subtle but important influences of behavioural context. Thus, for example, nicotine appears to be rather a weak reinforcer in the place preference paradigm (Clarke, 1987a), but robustly supports self-administration in animals (e.g. Corrigall and Coen, 1989). Lastly, no particularly effective treatment is available for the millions of tobacco smokers who would like to quit.

REFERENCES

Calabresi P., LAcey, M.G., and North, R.A. (1989) Br. J. Pharmac. 98, 135-140.

Carr, G.D., and White, N.M. (1986) Psychopharmac. (Berlin) 89, 340-346.

Clarke, P.B.S., and Kumar, R. (1983a) Br. J. Pharmac. 78, 329-337.

Clarke, P.B.S., and Kumar, R. (1983b) Br. J. Pharmac. 80, 587-594.

Clarke, P.B.S., and Pert, A. (1985) Brain Res. 348, 355-358.

Clarke, P.B.S., Jakubovic, A., and Fibiger, H.C. (1988a) Psychopharmac. (Berlin) 96, 511-520.

Clarke, P.B.S., Fu, D.S., Jakubovic, A., and Fibiger, H.C. (1988b) J. Pharmac. exp. Ther. 246, 701-708.

Clarke, P.B.S. (1987a) Psychopharmac. (Berlin) 92, 135-143.

Clarke, P.B.S. (1987b) Trends Pharmac Sci 8, 32-35.

Clarke, P.B.S., White, N.M., and Franklin, K.B.J. (1990) Behav. Brain Res. 36, 185-188.

Clarke, P.B.S. (1990) Biochem. Pharmac., in press

Cools, A.R. (1986) Psychopharmac. 88, 451-459.

Corrigall, W.A., and Coen, K.M. (1989) Psychopharmac. 99, 473-478.

Damsma, G., Day, J., and Fibiger, H.C. (1989) Eur J. Pharmac. 168, 363-368.

Fuxe, K., Andersson, K., Harfstrand, A., and Agnati, L.F. (1986) J. Neural Transm. 67, 15-29.

Grenhoff, J., Aston-Jones, G., and Svensson, T.H. (1986) Acta physiol. scand. 128, 351-358.

Imperato, A., Mulas, A., and Di Chiara, G. (1986) Nicotine preferentially stimulated dopamine release in the limbic system of freely moving rats. Eur. J. Pharmac. 132, 337-338.

Kelly, P.H., Seviour, P.W., and Iversen, S.D. (1975) Pharmac. Biochem. Behav. 94, 507-522.

Lapin, E.P., Maker, H.S., Sershen, H., and Lajtha, A. (1989) Eur. J. Pharmac. 160, 53-59.

Mereu, G., Yoon, K.-W.P., Boi, V., Gessa, G.L., Naes, L. and Westfall, T.C. (1987) Eur. J. Pharmac. 141, 395-399.

Mitchell, S.N., Brazell, M.P., Joseph, M.H., Alavijeh, M.S., and Gray, J.A. (1989) Eur. J. Pharmac. 167, 311-322.

Morrison, C.F., Stephenson, J.A. (1972) Br. J. Pharmac. 46, 151-156.

Pert, A., and Clarke, P.B.S. (1987) In: Tobacco Smoking and Nicotine (W.R. Martin, G.R. van Loon, E.T. Iwamoto and L. Davis, Eds), Plenum Press, New York, pp. 169-190

Pettit, H.O., and Justice, J.B. (1989) Pharmac. Biochem. Behav. 34, 899-904.

Rapier, C., Lunt, G.G., and Wonnacott, S. (1988) J. Neurochem. 50, 1123-1130.

Reavill, C., and Stolerman, I.P. (1990) Br. J. Pharmac. 99, 273-278.

Rowell, P.P., Carr, L.A., and Garner, A.C. (1987) J. Neurochem. 49, 1449-1454.

Swanson, L.W., Simmons, D.M., Whiting, P.J., and Lindstrom, J. (1987) J. Neurosci. 7, 3334-3342.

U.S. Surgeon General (1988) The Health Consequences of Smoking: Nicotine Addiction, U.S. Department of Health and Human Services.

Wada, E., Wada, K., Boulter, J., Deneris, E., Heinemann, S., Patrick, J., and Swanson, L.W. (1989) J. comp. Neurol. 284, 314-335.

Wise, R.A., and Rompre, P.-P. (1989) Ann. Rev. Psychol. 40, 191-225.

PRESYNAPTIC ACTIONS OF NICOTINE IN THE CNS

S. Wonnacott & A.L. Drasdo

Department of Biochemistry, University of Bath, Bath BA2 7AY, UK

Nicotine stimulates the release of neurotransmitters in the CNS.
One locus of its action is directly on the nerve terminal,
through presynaptic nicotinic acetylcholine receptors. The
nicotinic stimulation of dopamine release from striatal nerve
terminals has been particularly widely studied. Nicotine acts in
a dose-dependent manner (EC_{50}=4µM) to elicit Ca^{2+}-dependent
dopamine release. The pharmacological profile of this action
favours a ganglionic type of nicotinic receptor, but molecular
biological techniques have revealed several subtypes of nicotinic
receptors in the CNS. The pharmacological specificity of
nicotine-evoked transmitter release is consistent with the
receptor class identified by high affinity [^{3}H]nicotine binding.
This correlation is supported by the loss of such binding sites
following nerve degeneration after lesion experiments or in
degenerative diseases, and subcellular fractionation experiments
indicate that a high proportion of [^{3}H]nicotine binding sites are
associated with isolated nerve terminals. Thus presynaptic
nicotinic receptors may constitute a major target of nicotine in
the brain.

Nicotine stimulates the release of neurotransmitters and hormones
in the CNS (Balfour, 1982; Fuxe et al., 1990). One locus where
nicotine acts to achieve this effect is at the nerve terminal.
This is indicated by the failure of tetrodotoxin (which blocks
nerve conduction) to prevent nicotine-evoked transmitter release
in brain slices (Giorguieff et al., 1977) and intact preparations
(Xu & Kato, 1988). A direct presynaptic action is confirmed by
the ability of isolated nerve terminals (synaptosomes) to respond
to nicotine (Rapier et al., 1988). All of these examples relate
to dopamine release, either in the striatum or nucleus accumbens.
The dopaminergic system is the best characterised transmitter
system with respect to nicotine, and this interaction may
underlie some of the locomotory effects of nicotine, as well as
the reward pathway thought to contribute to the development of

dependence on psychostimulant drugs (see Clarke, this volume). The striatum is very amenable for neurochemical studies, and this chapter will focus on the pharmacological characterisation of the presynaptic actions of nicotine on striatal dopaminergic nerve terminals. The evidence indicates that nicotine acts through presynaptic nicotinic receptors, and these functional receptors will be compared with putative receptor subtypes characterised by ligand binding assays.

NICOTINE-EVOKED DOPAMINE RELEASE FROM STRIATAL NERVE TERMINALS: PHARMACOLOGICAL CHARACTERISATION

Nicotine acts in a stereoselective and dose-dependent manner to elicit [^{3}H]dopamine release from striatal synaptosomes (Rapier et al., 1988). The EC_{50} for this response is 4 µM. The pharmacological profile of nicotine-stimulated striatal dopamine release is consistent with its mediation by a nicotinic receptor mechanism. Thus other nicotinic agonists such as cytisine and DMPP can also evoke [^{3}H]dopamine release, and the endogenous ligand acetylcholine is equipotent with nicotine (Rapier et al., 1990).

Nicotinic agonist-evoked release from striatal synaptosomes is blocked by low micromolar concentrations of mecamylamine and dihydroβerythroidine (Rapier et al., 1990, see Fig. 1). Comparison of a number of novel neurotoxins with nicotinic blocking activity was made, in order to better define the presynaptic receptor. The potent neuromuscular antagonist αbungarotoxin had little effect on nicotine-evoked release, in contrast to neosurugatoxin (Rapier et al., 1985). This shellfish toxin is a ganglionic blocker with little activity at muscle nicotinic receptors (Hayashi et al., 1984; Wonnacott, 1987). Another toxin with specificity for neuronal nicotinic receptors is neuronal bungarotoxin (nBgt), a minor component of the venom from Bungarus multicinctus (Chiappinelli, 1985). This toxin has the advantage that it is now commercially available, in contrast to neosurugatoxin which is very scarce. nBgt inhibited nicotine

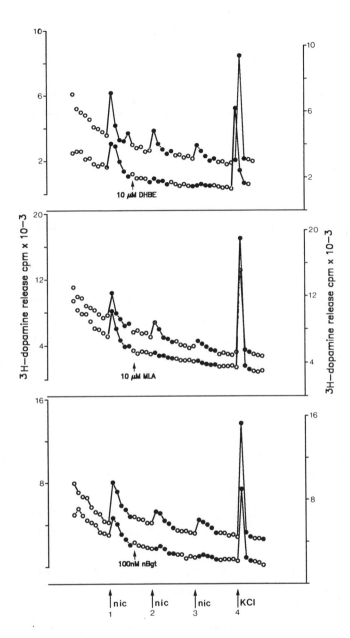

Fig. 1 Antagonism of nicotine-evoked release of
[³H]dopamine from striatal synaptosomes

Transmitter release was monitored in superfusate fractions (750
μl; 3 min). Control (upper traces) and test chambers (lower
traces) were stimulated in parallel as indicated, with nicotine
(3 μM) or KCl (20 mM). Antagonists were introduced into the
perfusing buffer after S1.

evoked [^{3}H]dopamine release by 50% when tested at 100nM (Fig.1); αbungarotoxin at the same concentration was without any significant effect. We have recently examined methyllycaconitine (MLA), a product of <u>Delphinium</u> <u>brownii</u>, in the same preparation. At 1μM, this compound had no effect, but 10 μM MLA inhibited nicotine-evoked release by 80% (Fig. 1).

NICOTINIC RECEPTOR SUBTYPES IN THE BRAIN

Two types of ligand binding site displaying a nicotinic pharmacology can be distinguished in the brain. Sites that bind the neuromuscular antagonist αbungarotoxin are distinct from those labelled by tritiated agonists (nicotine, methylcarbamyl-choline, acetylcholine in the presence of atropine). The different anatomical distributions of these binding sites confirms their separate identity (Clarke <u>et</u> <u>al</u>., 1985). This is illustrated in Fig. 2, in which we have compared the distributions of binding sites for [^{3}H]nicotine, [^{125}I]α-bungarotoxin and the muscarinic ligand [^{3}H]quinuclidinyl benzylate (QNB) in sections of rat brain taken at the level of the striatum. In these photographs of the autoradiographs, labelled areas appear white. Thus [^{3}H]nicotine labels the striatum, whereas [^{125}I]αbungarotoxin shows little binding to this brain region. [^{3}H]QNB also labels the striatum, reflecting the considerable cholinergic activity in this region. In the cortex, however, labelling patterns for all three ligands are distinct.

Molecular genetics techniques have provided further insights into the nature of nicotinic receptors in the brain. Several closely related agonist-binding and non-agonist-binding subunits have been recognised (Deneris <u>et</u> <u>al</u>., 1989), in addition to less similar αbungarotoxin binding proteins (Schoepfer <u>et</u> <u>al</u>., 1990). Presently three subunit combinations have been shown to result in nicotinic receptors when expressed in <u>Xenopus</u> oocytes. These are the α2, α3 and α4 agonist-binding subunits, in combination with the β2 (non-α) subunit (Deneris <u>et</u> <u>al</u>., 1989).

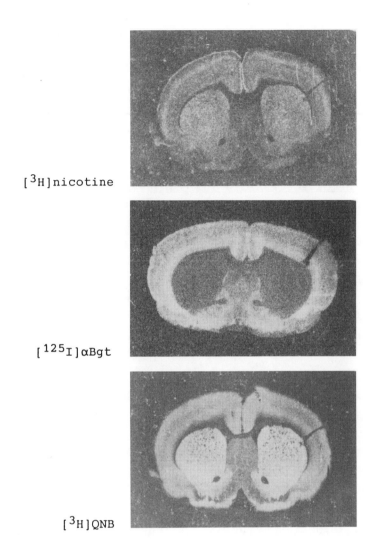

[^{3}H]nicotine

[^{125}I]αBgt

[^{3}H]QNB

Fig. 2   Autoradiographic distributions of cholinergic ligand binding sites

Photomicrographs of rat striatal sections (10 μm) labelled with [^{3}H]nicotine (3.5 μM; top), [^{125}I]αbungarotoxin (1 nM; centre) and [^{3}H]QNB (2 nM; bottom).

The question arises as to which nicotinic receptor subtype corresponds to the presynaptic receptor present on dopamine-containing nerve terminals in the striatum. This question is not easy to address. The pharmacological profiles of the three nicotinic receptor candidates identified by gene cloning are not yet fully defined. They appear to differ little in their sensitivities to different drugs (Luetje et al., 1990), and radioligands are not available to identify each of the subtypes. Evidence suggests that [^{3}H]nicotine binding corresponds to the α4β2 receptor (Whiting et al., 1987). The α3β2 configuration may bind nBgt, which can be iodinated, but there is overlap with other receptors (see below). It is presently unclear if the α2β2 complex is identified by any radioligand; the α2 subunit is present in very low amounts in the brain which has handicapped its analysis. However, the absence of the α2 gene product in the substantia nigra and caudate (Wada et al., 1989) discount it as a candidate subunit of the presynaptic receptor discussed here.

COMPARISON OF PRESYNAPTIC NICOTINIC RECEPTORS WITH NICOTINIC RADIOLIGAND BINDING SITES

Table I compares the characteristics of nicotine-evoked [^{3}H]dopamine release with the pharmacological properties of nicotinic binding sites labelled by [^{125}I]αbungarotoxin and [^{3}H]nicotine. Although both binding sites, by definition, are sensitive to nicotine, the K_i value for nicotine at the αbungarotoxin site (5 μM; Rapier et al., 1990) is closest to the EC_{50} for nicotine-evoked dopamine release (4μM; Rapier et al., 1988). Nicotine appears to be three orders of magnitude more potent at the [^{3}H]nicotine binding site. However, the latter is considered to represent the high affinity, desensitized form of the receptor (the inevitable consequence of the long incubation with nicotine in the binding assay), and is therefore not comparable with the functional receptor, in terms of agonist sensitivity.

A more discriminating measure is the stereoselectivity, in

favour of (-)nicotine, shown by different preparations. The [^{3}H]nicotine binding site and nicotine-evoked [^{3}H]dopamine release display a marked, hundredfold preference for the natural enantiomer (Wonnacott, 1986; Rapier et al., 1988), whereas the [^{125}I]αbungarotoxin binding site shows only a fivefold preference for (-)- versus (+)nicotine. This correspondence between the [^{3}H]nicotine binding site and the presynaptic receptor is supported by the comparison of competitive antagonists. Thus both measures are insensitive to αbungarotoxin but inhibited by nanomolar concentrations of neosurugatoxin (Rapier et al., 1985; 1990). Dihydroβerythroidine does not discriminate between the ligand binding sites (Rapier et al., 1990), but its K_i value of 1.3 μM at the [^{3}H]nicotine site is consistent with the 80% inhibition of nicotine-stimulated [^{3}H]dopamine release by 10 μM dihydroβerythroidine (see Fig. 1). Methyllycaconitine (MLA) is unusual in that it is a more potent inhibitor of [^{125}I]αbungarotoxin binding (K_i 1.4 nM) than of [^{3}H]nicotine binding (K_i 3.7 μM) (MacAllan et al., 1988). Again its potency at the latter site accords with the sensitivity of nicotine-evoked [^{3}H]dopamine release to MLA (Fig. 1).

Table I   Comparison of the pharmacological specificity of nicotinic ligand binding sites and presynaptic receptors

| | Binding Sites | | Presynaptic receptor |
	[^{125}I]αBgtx	[^{3}H]nicotine	nictotine-evoked [^{3}H]dopamine release
(-) Nicotine	+	++	+
Stereoselectivity	+	++	++
Dihydroβerythroidine	+	+	+
Neosurugatoxin	-	++	++
αBungarotoxin	++	-	-
nBungarotoxin	+	-?	+
Methyllycaconitine	++	+	+
Mecamylamine	-	-	+

for drugs, + denotes μM potency; ++ denotes nM potency

The remaining two agents documented in Table I do not appear to fit this correlation. The failure of mecamylamine to inhibit either radioligand binding site while blocking nicotine-evoked [^{3}H]dopamine release is explained by the non-competitive nature of its antagonism; mecamylamine is now recognised to be a nicotinic channel blocker. The sensitivity of neuronal nicotinic receptors to such agents is supported by the antagonism of nicotine-evoked [^{3}H]dopamine release by the well established nicotinic channel blocker, histrionicotoxin (Rapier et al., 1987).

Neuronal bungarotoxin (nBgt) competes for binding to [^{125}I]αbungarotoxin sites in muscle and neuronal tissues, but with affinities that are 100 and 10 times lower, respectively, than αbungarotoxin (Chiappinelli, 1985). However, nBgt is rather ineffective as an antagonist at muscle nicotinic receptors (Loring & Zigmond, 1988), but it is a potent antagonist of nicotinic responses in autonomic ganglia (Chiappinelli, 1985; Loring & Zigmond, 1988). In contrast, in this preparation αbungarotoxin is without effect, and αbungarotoxin binding sites are located extrasynaptically (Jacob & Berg, 1983). In the CNS, nBgt has been reported to inhibit nicotinic responses in the cortex (Vidal & Changeux, 1989), cerebellum (de la Garza et al., 1989) and caudate-putamen (Schulz & Zigmond, 1989), but a nBgt-insensitive response to nicotine has recently been described in medial habenula (Mulle & Changeux, 1990).

These somewhat confusing results can be reconciled by the differential sensitivity of various receptor subtypes in the brain to nBgt (Luetje et al., 1990). Thus the most sensitive receptor is the α3β2 configuration: when expressed in oocytes this receptor is almost totally blocked by 10 nM nBgt. The α4β2 combination of subunits is markedly less sensitive, requiring 1 μM nBgt for partial blockade in oocytes, whereas α2β2 is insensitive. Thus nBgt is the only pharmacological probe currently available that discriminates between these closely related subtypes of neuronal nicotinic receptor.

Does the partial block of nicotine-evoked [^{3}H]dopamine release by 100nM nBgt observed in the present study (Fig 1) accord with either the α3β2 or α4β2 receptor? This sensitivity falls between the two subtypes, according to the study carried out in the artificial expression system (Luetje et al., 1990). A presynaptic nicotinic receptor in cortex required 1.4 µM nBgt for blockade (Vidal & Changeux, 1989), consistent with α4β2, but studies on striatal slices reported complete blockade of dopamine release with 100 nM nBgt (Schulz & Zigmond, 1989), favouring α3β2.

The α4β2 receptor subtype corresponds to the high affinity [^{3}H]nicotine binding protein (Whiting et al., 1987). The binding of this radioligand is reputedly insensitive to nBgt in chick brain (Wolf et al., 1988); in rat brain we have observed a $K_i > 1µM$ against [^{3}H]nicotine binding. However, the conditions under which binding studies are carried out may not be optimal for nBgt binding.

Thus the consensus of information summarised in Table I is consistent with a correlation between the presynaptic nicotinic receptor on dopaminergic striatal nerve terminals and high affinity [^{3}H]nicotine binding site. But the data from nBgt, while inconclusive, raises the possibility of other nicotinic receptors, not recognised by conventional radioligands, modulating the release of [^{3}H]dopamine.

EVIDENCE FOR THE PRESYNAPTIC LOCALISATION OF [^{3}H]NICOTINE SITES

Binding assays are typically carried out on whole brain membranes, thus providing no information about the synaptic localisation of receptor sites. Molecular biology studies identify subtypes at the site of protein synthesis, i.e. in the cell bodies, which may not accord with the ultimate destination of the functional receptor in the neuron. Nevertheless, several lines of evidence point to the localisation of a high proportion of [^{3}H]nicotine binding sites on presynaptic nerve terminals.

Nerve degeneration following experimental lesion or in degenerative diseases is accompanied by loss of [^{3}H]nicotine binding. With respect to the striatum, 6-hydroxydopamine lesions of the nigro-striatal projection resulted in losses of 30% of [^{3}H]nicotine binding sites in the terminal regions (Clarke & Pert, 1985). Furthermore, postmortem examination of brain tissue from Parkinson patients revealed a 50% deficit in [^{3}H]nicotine binding in the caudate nucleus (Perry et al., 1989).

Subcellular fractionation experiments have demonstrated that [^{3}H]nicotine binding sites are enriched in synaptosome fractions, rather than in bulk plasma membrane fractions (Wonnacott & Thorne, 1990). This contrasts with the distribution of muscarinic binding sites labelled with [^{3}H]QNB, which are predominantly postsynaptic on cell bodies and dendrites. Further evidence for the common identity of presynaptic nicotinic receptors mediating striatal dopamine release and [^{3}H]nicotine binding sites comes from the parallel effects of chronic agonist treatment on these two indices (Rowell & Wonnacott, 1990). Chronic infusion of a low dose of anatoxin-a to rats resulted in a 32% increase in the number of [^{3}H]nicotine sites on striatal nerve terminals, and a 42% increase in nicotine-evoked [^{3}H]dopamine release from the same preparation.

CONCLUSIONS

This paper attempts to correlate functional nicotinic receptors on dopamine nerve terminals with a subtype of nicotinic receptor characterised by high affinity [^{3}H]nicotine binding. The evidence indicates that a high proportion, perhaps the majority, of these sites is presynaptic. However, it should be emphasised that there are other subtypes of putative nicotinic receptor present in the CNS which are not identified by [^{3}H]nicotine.

Acknowledgements Research carried out in the authors' laboratory was supported by grants from The Medical Research Council of Great Britain, The Wellcome Trust and RJ Reynolds Tobacco Co. ALD is supported by a Postgraduate Training Award from the SERC.

REFERENCES

Balfour, D.J.K. (1982) The effects of nicotine on brain neurotransmitter systems **Pharmac.Ther.** 16, 269-282

Chiappinelli, V.A. (1985) Actions of snake venom toxins on neuronal nicotinic receptors and other neuronal receptors **Pharmac.Ther.**, 31, 1-32

Clarke, P.B.S. and Pert, A. (1985) Autoradiographic evidence for nicotine receptors on nigrostriatal and mesolimbic dopaminergic neurons **Brain Res.** 348, 355-358

Clarke, P.B.S., Schwartz, R.D., Paul, S.M., Pert, C.B. and Pert, A. (1985) Nicotinic binding in rat brain: autoradiographic comparison of [³H]acetylcholine, [³H]nicotine and [¹²⁵I]alphabungarotoxin **J.Neurosci.** 5, 1307-1315

de la Garza, R., Freedman, R. and Hoffer, B.J. (1989) Kappa-bungarotoxin blockade of nicotine electrophysiological actions in cerebellar neurons **Neurosci.Lett.** 99, 95-100

Deneris, E.S., Boulter, J., Connolly, J., Wade, E., Wada, K., Goldman, D., Swanson, L.W., Patrick, J. and Heinemann, S. (1989) Genes encoding neuronal nicotinic acetylcholine receptors. **Clin.Chem.** 35, 731-737

Fuxe, K., Andersson, K., Harfstrand, A., Eneroth, P., Perez de la Mora, M. and Agnati, L.F. (1990) Effects of nicotine on synaptic transmission in the brain. In **Nicotine Psychopharmacology** (ed. S.Wonnacott, M.A.H. Russell & I.P. Stolerman) Oxford University Press pp 194-225

Giorguieff, M.F., Le Floc, H.M.L., Glowinski, J. and Besson, M.J. (1977) Involvement of cholinergic presynaptic receptors of nicotinic and muscarinic types in the control of the spontaneous release of dopamine from striatal dopaminergic terminals in the rat **J.Pharmacol. Exp. Ther.** 200, 535-544

Hayashi, E., Isogai, M., Kagawa, Y., Takayanagi, N. and Yamada, S. (1984) Neosurugatoxin, a specific antagonist of nicotinic acetylcholine receptors **J. Neurochem.** 42, 1491-1494

Jacob, M.H. & Berg, D. (1983) The ultrastructural localisation of αbungarotoxin binding sites in relation to synapses on chick ciliary ganglion neurons **J.Neurosci.** 3, 260-271

Loring, R.H. and Zigmond, R.E. (1988) Characterisation of neuronal nicotinic receptors by snake venom neurotoxins **Trends Neurosci.** 11, 73-78

Luetje, C.W., Wada, K., Rogers, S., Abramson, S.N., Tzuji, K., Heinemann, S. and Patrick, J. (1990) Neurotoxins distinguish between different neuronal nicotinic acetylcholine receptors. **J. Neurochem.** in press

MacAllan, D.R.E., Lunt, G.G., Wonnacott, S., Swanson, K.L., Rapoport, H. and Albuquerque, E.X. (1988) Methyllycaconitine and anatoxin-a differentiate between nicotinic receptors in vertebrate and invertebrate nervous systems. **FEBS Lett.** 226, 357-363

Mulle, C. & Changeux, J.-P (1990) A novel type of nicotinic receptor in the rat central nervous system characterised by patch-clamp techniques. **J.Neurosci.** 10, 169-175

Perry, E.K., Smith, C.J., Perry, R.H., Johnson, M. and Fairbairn, A.F. (1989) Nicotinic (^{3}H-nicotine) receptor binding in human brain: characterisation and involvement in cholinergic neuropathology **Neurosci. Res. Comm.** in press

Rapier, C., Harrison, R., Lunt, G.G. and Wonnacott, S. (1985) Neosurugatoxin blocks nicotinic acetylcholine receptors in the brain.**Neurochem. Int.** 7, 389-396

Rapier, C., Wonnacott, S., Lunt, G.G. and Albuquerque, E.X. (1987) The neurotoxin histrionicotoxin interacts with the putative ion channel of the nicotinic acetylcholine receptors in the central nervous system **FEBS Lett.** 212, 292-296

Rapier, C., Lunt, G.G. and Wonnacott, S. (1988) Stereoselective nicotine-induced release of dopamine from striatal synaptosomes: concentration dependence and repetitive stimulation **J. Neurochem.** 50, 1123-1130

Rapier, C., Lunt, G.G. and Wonnacott, S. (1990) Nicotinic modulation of [^{3}H]dopamine release from striatal synaptosomes: pharmacological characterisation **J. Neurochem.** 54, 937-945

Rowell, P.P. and Wonnacott, S. (1990) Evidence for functional activity of up-regulated nicotine binding sites in rat striatal synaptosomes **J. Neurochem.** in press

Schoepfer, R., Conroy, W.G., Whiting, P., Gore, M. and Lindstrom, J. (1990) Brain alpha-bungarotoxin-binding protein cDNAs and mABs reveal subtypes of this branch of the ligand-gated ion channel gene superfamily **Neuron.** in press

Schulz, D.W. and Zigmond, R.E. (1989) Neuronal bungarotoxin antagonises nicotinic function in rat caudate-putamen **Neurosci. Lett.** 98, 310-316

Wada, E., Wada, K., Boulter, J., Deneris, E., Heinemann, S., Patrick, J. and Swanson, L.W. (1989) Distribution of alpha2, alpha3, alpha4 and beta2 neuronal nicotinic receptor subunit mRNAs in the central nervous system: a hybridisation histochemical study in the rat **J.Comp.Neurol.** 284, 314-335

Whiting, P., Esch, F., Shimasaki, S. and Lindstrom, J. (1987) Neuronal nicotinic acetylcholine receptor ß-subunit is coded for by the cDNA clone α4 **FEBS Lett.** 219, 459-463

Vidal, C. and Changeux, J.P. (1989) Pharmacological profile of nicotinic acetylcholine receptors in the rat prefrontal cortex: an electrophysiological study in a slice preparation. **Neuroscience** 29, 261-270

Wolf, K.M., Ciarlegio, A. and Chiappinelli, V.A. (1988) K-Bungarotoxin: binding of a neuronal nicotinic receptor antagonist to chick optic lobe and skeletal muscle **Brain Res.** 439, 249-258

Wonnacott, S. (1986) α-Bungarotoxin binds to low-affinity nicotine binding sites in rat brain **J.Neurochem.** 47, 1706-1712

Wonnacott, S. (1987) Neurotoxin probes for neuronal nicotinic receptors. In **Neurotoxins and their pharmacological implications**. Jenner, P. New York: Raven Press pp 209-231

Wonnacott, S. and Thorne, B. (1990) Separation of pre- and post-synaptic receptors on Percoll gradients **Biochem.Soc.Trans.** 18, in press

Xu, M. and Kato, T. (1988) Brain Dialysis: changes in the activities of dopamine neurons in rat striatum by perfusion of acetylcholine agonists under freely moving conditions **Neurochem. Int.** 12, 539-545

MECHANISMS UNDERLYING THE PROTECTIVE EFFECTS OF CHRONIC NICOTINE TREATMENT AGAINST DEGENERATION OF CENTRAL DOPAMINE NEURONS BY MECHANICAL LESIONS

K. Fuxe, A.M. Janson, J. Grenhoff , K. Andersson, J. Kåhrström , B. Andbjer, Ch. Owman , T. Svensson and L.F. Agnati

Department of Histology and Neurobiology, and *Pharmacology, Karolinska Institute, Box 60400, 104 01 Stockholm, Sweden; **Department of Medical Cell Research, University of Lund, Lund, Sweden and ***Department of Human Physiology, University of Modena, Modena, Italy

SUMMARY: In a series of studies on the effects of chronic nicotine treatment via minipumps on retrograde and anterograde degenerative processes in nigrostriatal dopamine (DA) neurons following a partial hemitransection, morphological, biochemical, functional and neurophysiological evidence has been obtained of protective actions of chronic nicotine treatment against the degeneration of nigrostriatal DA neurons in the male rat.
    1. By means of tyrosine hydroxylase (TH) immunocytochemistry in combination with image analysis it was demonstrated that chronic nicotine treatment via minipumps ((-)nicotine hydrogen (+)tartrate: 0.125 mg/kg/h for 2 weeks) significantly counteracted the disappearance of TH immunoreactive (IR) nerve cell body, dendrite and terminal profiles of the nigrostriatal DA neurons produced by a partial di-mesencephalic hemitransection. The most marked protection was observed against the degeneration of the TH dendritic profiles in the substantia nigra.
    2. By means of DA fluorescence histochemistry in combination with quantitative histofluorimetry as well as of biochemical analysis of DA stores it was demonstrated that this type of chronic nicotine treatment also protected against the disappearance of DA stores within the substantia nigra and within nucleus caudatus putamen and the nucleus accumbens induced by a partial di-mesencephalic hemitransection. By the use of the TH inhibition method using α-methyl-($\pm$)-p-tyrosine methyl ester (α-MT) evidence was also obtained that chronic nicotine treatment pre-

erentially and substantially reduces forebrain DA utilization on
the lesioned side.
    3. By means of analysis of local cerebral blood flow and glu-
cose utilization within the neostriatum by computer assisted
autoradiography using the radioligands 14C-deoxyglucose and 14C-
iodoantipyrin it was demonstrated that the partial hemitransection at the di-mesencephalic level produced a substantial reduction in both glucose utilization and in blood flow in the neostriatum of the lesioned side. It was demonstrated that this
reduction of both glucose utilization and of cerebral blood flow
in the neostriatum was counteracted by the chronic nicotine
treatment using minipumps (same dose treatment schedule as
above).
    4. By means of the standard single cell recording method the
extracellular activity of nigral DA neurons was analyzed two
weeks following partial hemitransection at the meso-diencephalic
level. It was found that the chronic nicotine treament as
described above produced a significantly lower burst firing in
the nicotine treated hemitransected animals compared with the
hemitransected group treated with saline. No differences in firing rate or regularity of firing was observed between the two
groups.
    It is suggested that the mechanism underlying the protective
action of nicotine in this mechanical lesion model is represented a desensitization of the excitatory nicotinic cholinoceptors located on the nigral DA nerve cells and on the DA forebrain terminals. Desensitization of the nicotinic cholinoceptors
leads to the demonstrated reduction of burst firing, and thus to
a preferential and marked reduction of DA utilization in the
surviving DA nerve terminals. In this way reduced energy demands
develop in the surviving nigrostriatal DA neurons as well as a
reduced calcium ion influx, which is known to have neurotoxic
activity. Desensitization of the nicotinic cholinergic receptors
per se will also lead to a reduced calcium influx in view of the
closure of the ligand gated kation channel in the nicotinic
cholinoceptors. In conclusion, chronic nicotine treatment, via
desensitization of the excitatory nicotinic cholinoceptors, may
enhance repair mechanisms in central neuronal systems, possessing large numbers of nicotinic receptors, such as the monoamin-
ergic and cholinergic neurons. There is an urgent need to
further evaluate protective actions of chronic nicotine treatment in animal models of Parkinson's disease and Alzheimer's
disease in view of its potential therapeutic role (see also
Janson et al., this symposium).

INTRODUCTION

A negative association exists between smoking and Parkinson's
disease independent of other associated factors (Kessler &
Diamond, 1971; Bauman et al., 1980; Godwin-Austen et al., 1982;

see also Godwin-Austen, this symposium). In a series of studies we have therefore, analyzed if chronic nicotine treatment can exert protective actions against the degeneration of nigrostriatal DA neurons in the rat following partial hemitransections at the meso-diencephalic junction using morphological, biochemical, functional and electrophysiological techniques (Janson et al., 1986, 1988; Fuxe et al., 1990; Grenhoff et al., 1990; Owman et al., 1989a,b). The present article will review this work involving mechanical lesions, while the article by Janson et al. (this symposium) will review modulating actions of chronic nicotine treatment on MPTP-induced lesions of the nigrostriatal DA neurons in the black mouse, which emphasizes important differences in the protective activity of nicotine against mechanical versus neurotoxic injuries of the nigrostriatal DA system. This review article emphasizes consistant and substantial protection of chronic nicotine treatment against mechanically induced degeneration of nigrostriatal DA neurons probably based on desensitization of excitatory nicotinic cholinoceptors located on the DA neurons, leading to reduced energy demands as well as a reduction in calcium influx.

MORPHOMETRIC STUDIES ON NIGROSTRIATAL DOPAMINE NEURONS AFTER PARTIAL HEMITRANSECTION AND THE PROTECTIVE ACTION OF CHRONIC NICOTINE TREATMENT USING MINIPUMPS

For details on TH immunocytochemistry and on image analysis of TH IR cell bodies, dendrites and nerve terminals, see Janson et al. (1986, 1988) and Agnati et al. (1988). For details on the partial hemitransection, see Agnati et al. (1983) and Janson et al. (1988). This lesion predominantly produces a marked disappearance of DA nerve terminals within the neostriatum and within the anterior parts of the nucleus accumbens and tuberculum olfactorium. In contrast, the cholecystokinin (CCK) costoring DA nerve terminals of the nucleus accumbens and tuberculum olfactorium are not significantly effected by the lesion.

The nicotine treatment protocol was as follows: immediately following the lesions the rats received four intraperitoneal (i.p.)
injection of (-)nicotine(+)tartrate in a dose of 0.5 mg/kg with
30 min time intervals. At the time of the lesion Alzet minipumps
model 2002 were implanted subcutaneously. Nicotine was delivered
at a rate of 0.125 mg/kg/h. The rats were decapitated after 14
days of treatment. The serum nicotine and cotinine levels
obtained by the chronic nicotine treatment are shown in Table I.
The serum nicotine levels obtained in the sham-operated and partially hemitransected rats were not significantly different and
were in the order of 60-80 ng/ml. In contrast, the serum cotinine levels were significantly increased in the partially hemitransected animals. This result may be related to a reduced
clearance in the partially hemitransected rats (see Agnati et
al., 1985).

TABLE I

SERUM NICOTINE AND COTININE LEVELS IN SHAM-OPERATED AND
PARTIALLY HEMITRANSECTED RATS AFTER CHRONIC NICOTINE TREATMENT

Treatment	Serum nicotine	Serum cotinine
Sham + saline	not detectable	not detectable
Sham + nicotine	84.9±17.6	252.6±10.8
Partial hemitran- section + saline	not detectable ns	not detectable ***
Partial hemitran- section + saline	65.0±5.8	412.2±46.4

Means ± S.E.M. are given in ng/ml, n=12. Statistical analysis
with two-tailed Mann-Whitney U-test, ns = no significant differences between the groups, *** p< 0.002.

We have demonstrated (Janson et al., 1986, 1988) that the chronic nicotine treatment as described above, produces a significant protection against the disappearance of TH IR nerve cell bodies, dendrites and terminals in the nigrostriatal DA neurons following the partial hemitransection. The results obtained are illustrated in Fig. 1. A marked and highly significant protection against the disappearance of TH IR dendritic profiles is observed in the sampled field in both the rostral and caudal part of the substantia nigra (medial part) as evaluated two weeks following a partial mes-diencephalic hemitransection. The protective action of chronic nicotine treatment against the disappearance of TH IR nerve cell body profiles and nerve terminal profiles in the substantia nigra and nucleus caudatus putamen, respectively, were substantially less pronounced than the protective activity reported in Fig. 1 against the disappearance of TH IR dendritic profiles. It seems possible that the increased presence of especially the TH IR dendrites in the substantia nigra may by itself lead to an increased trophic support of the remaining DA cells, since these dendrites may produce trophic factors. In this way they may
increase survival of the DA cells as well as their ability to compensate for the loss of a large number of nigral DA nerve cells by e.g. enhancement of the collateral sprouting of DA nerve terminals within the neostriatum.

It was suggested based on these protective actions of chronic nicotine treatment that the mechanism involved was a desensitization of excitatory nicotinic cholinergic receptors leading to a reduction in the firing rate of nigral DA nerve cells and thus to reduced energy demands and increased DA nerve cell survival (see Janson et al., 1986, 1988).

In support of this hypothesis it was also demonstrated that chronic nicotine treatment leads to an enhancement of the apomorphine-induced rotational behaviour in the partially hemitransected rat (Janson et al., 1988). Furthermore, the higher the serum nicotine levels, the higher the degree of ipsilateral rotational behaviour induced by apomorphine. These results indi-

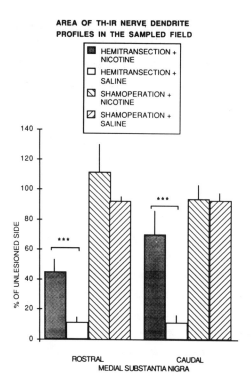

AREA OF TH-IR NERVE DENDRITE
PROFILES IN THE SAMPLED FIELD

Fig. 1. Effects of chronic nicotine treatment on the lesion-
induced decrease of the area of TH IR nerve dendrite profiles in
the sampled field of the medial substantia nigra at the rostral
and caudal levels. All values are expressed as a percentage of
the respective intact side. Means ± S.E.M., n= 4 or 5. The absolute field area (expressed in µm2) for profiles on the respective intact side are at the rostral level 59,800±4800 (mean ±
S.E.M.) for the lesioned nicotine-treated animals, 78,800±12,600
for the lesioned saline-treated animals, 70,600±13,000 for the
sham-operated nicotine-treated animals and 66,300±5600 for the
sham-operated saline-treated animals. The corresponding values
for the caudal level are: 56700±8000, 45,100±4000, 49,000±5300
and 50,600±4500. Two-tailed Mann-Whitney U-test. *** p<0.001.

cate a further reduction of the DAergic function of the surviving DA cells of the lesioned side, which is in fact predicted by
the above hypothesis, stating that the chronic nicotine treatment leads to a reduction in the firing activity of the surviving DA cells on the lesioned side.

CATECHOLAMINE FLUORESCENCE HISTOCHEMICAL AND BIOCHEMICAL STUDIES
ON DOPAMINE LEVELS AND UTILIZATION WITHIN THE SUBSTANTIA NIGRA
AND SURVIVING FOREBRAIN DOPAMINE NERVE TERMINAL SYSTEMS AFTER A
PARTIAL DI-MESENCEPHALIC HEMITRANSECTION AND THE PROTECTIVE
ACTIONS OF CHRONIC NICOTINE TREATMENT USING MINIPUMPS

For details on the DA fluorescence method and quantitative
histofluorimetrical analysis, see Andersson et al. (1985). For
details on the HPLC analysis in combination with electrochemical
detection, see Jonsson et al. (1980). The TH inhibition method
was used to study DA utilization and the inhibitor used was α-
MT. The higher the degree of DA depletion after TH inhibition
the higher the degree of DA utilization (see Andén et al.,
1969). Partial di-mesencephalic hemitransections and the chronic
nicotine treatment were performed as described above. The overall serum nicotine levels obtained was 64.6±2.7 ng/ml.

It was demonstrated (see Fuxe et al. 1990) that partial hemitransection produced a marked and highly significant disappearance of DA stores in various types of neostriatal DA nerve terminals as well as in the diffuse types of DA nerve terminals
within the nucleus accumbens and tuberculum olfactorium. The
results obtained in the neostriatum are shown in Fig. 2. A
marked depletion of DA stores is observed both in the marginal
zone of the nucleus caudatus representing DA terminals of the
islandic type as well as in the medial and central part of the
nucleus caudatus. On the intact side no effects on DA stores
were observed.

In Fig. 3 it is shown that the chronic nicotine treatment
produces a highly significant and substantial protection of the
neostriatal DA stores. The effects of chronic nicotine treatment
in the partially hemitransected rats are highly significantly
different from the effects of chronic nicotine treatment in the
sham-operated animals. The chronic nicotine treatment does not
influence DA stores within the DA nerve terminals of the neostriatum on the intact side in the hemitransected animals. These
results give further support for the view that chronic nicotine

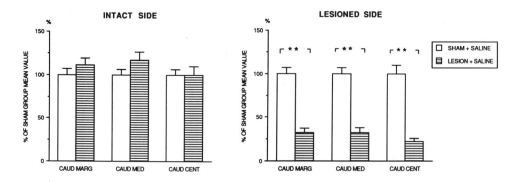

Fig. 2. Effects of a partial hemitransection on DA stores in discrete neostriatal DA nerve terminal systems of the intact side and lesioned side. Means±S.E.M., n=11-12. The values are expressed as a percentage of the sham-operated group mean values. Level: A8700 according to the König and Klippel Atlas. Mann-Whitney U-test applying the Bonferoni's procedure. ** p<0.01. CAUD MARG= marginal zone of the nuc. caudatus; CAUD MED= medial part of the nuc. caudatus; CAUD CENT= central part of the nuc. caudatus.** p<0.01.

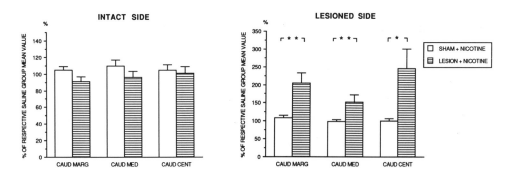

Fig. 3. Effects of chronic nicotine treatment on DA fluorescence in discrete forebrain DA nerve terminal systems in the caudate putamen of the non-operated (intact) side and lesioned side of the male rat. The animals were sham-operated or hemitransected and subsequently treated with nicotine (0.125 mg/kg/h) during 14 days. On the day of the experiment the rats were given saline i.p. 2 h before decapitation. The sham-operated + nicotine treated group is expressed as a percentage of the sham-operated + saline treated group mean values and the hemitransected + nicotine treated group is expressed as a percentage of hemitransected + saline treated group mean values. Means±S.E.M., n= 10-12 in all groups except for all experiments on the lesioned side in the hemitransected + nicotine treated group, where n= 6. Statistical analysis according to the two-tailed Mann-Whitney U-test applying the Benferoni's procedure.

treatment can protect against the degeneration of DA nerve terminal systems induced by mechanical lesions of the nigrostriatal
DA neurons. This represents the simplest explanation for the
ability of chronic nicotine treatment to increase DA stores on
the lesioned side of the hemitransected rats, while chronic
nicotine treatment in sham-operated rats does not influence the
DA stores within the neostriatum. The same is true also for the
nucleus accumbens and the olfactory tubercle (see Fuxe et al.,
1990). Thus, results obtained both with TH immunocytochemistry
and DA fluorescence histochemistry indicate that chronic nicotine treatment protects against the degeneration of DA nerve
terminal systems of the neostriatum following mechanical lesion.

The biochemical analysis of DA stores within the substantia
nigra also lend support to the results obtained in the morphological analysis in this case of TH IR profiles in the substantia nigra (see Fuxe et al., 1990). Thus, chronic nicotine treatment significantly increased DA stores on the lesioned side following a partial hemitransection, while chronic nicotine treatment in sham-operated animals did not produce any significant
effects.

To test the hypothesis of the existence of reduced activity
in the nigrostriatal DA neurons upon chronic nicotine treatment
on the lesioned side of partially hemitransected rats studies
were also performed on DA utilization within the forebrain DA
nerve terminal systems (see Fuxe et al., 1990). As seen in Fig.
4, chronic nicotine treatment of hemitransected rats produced a
highly significant reduction in the depletion of DA stores produced by the TH inhibitor in the various DA nerve terminal systems of the neostriatum compared with the effects of chronic
nicotine treatment on the lesioned side of sham-operated rats
(Fig. 4). Thus, it seems as if chronic nicotine treatment preferentially and markedly reduces DA utilization in surviving DA
nerve terminals in the neostriatum of the lesioned side after a
partial meso-diencephalic hemitransection. Similar results were
obtained within the nucleus accumbens and tuberculum olfactorium

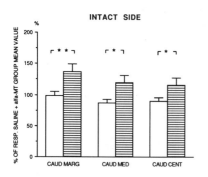 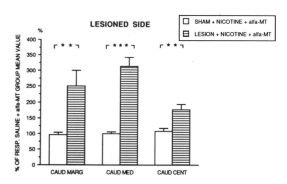

Fig. 4. Effects of chronic nicotine treatment on α-MT-induced
disappearance of DA fluorescence in discrete forebrain DA nerve
terminal systems in the caudate putamen of the non-operated
(intact) side and lesioned side of the male rat. The animals
were sham-operated or hemitransected and subsequently treated
with nicotine (0.125 mg/kg/h) during 14 days. On the day of the
experiment the rats were given α-MT (250 mg/kg i.p.) 2 h before
decapitation. The percent amount remaining in the sham-operated
+ nicotine + αMT treated group is expressed as a percentage of
that of the sham-operated + saline + αMT treated group mean
value. By analogy the percent amount remaining in the hemitransected + nicotine + αMT treated group is expressed in the same
way as a percentage of the hemitransected + saline + αMT treated
group mean values. Means ± S.E.M., n= 12 in all groups except
for CAUD marg, CAUD med, CAUD cent on the lesioned side in the
hemitransected + nicotine + αMT treated group, where n=9.
Statistical analysis according to the two-tailed Mann-Whitney U-
test applying the Bonferoni's procedure. * p<0.05, ** p<0.01.

(see Fuxe et al., 1990). Also on the intact side of partially
hemitransected rats the chronic nicotine treatment produced a
significant reduction in DA utilization compared with the
effects of chronic nicotine treatment in sham-operated rats
(Fig. 5). This reduction of DA utilization of the intact side of
the hemitransected rats was weak compared with that obtained of
the denervated side. This action of chronic nicotine treatment
on the intact side of the partially hemitransected rats may be
related to a possible balancing action, existing on the intact
side of partially hemitransected rats, and aiming to maintain
symmetry between the two sides of the brain. Furthermore, the
analysis of the substantia nigra demonstrated that the activa-

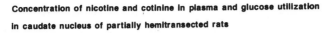

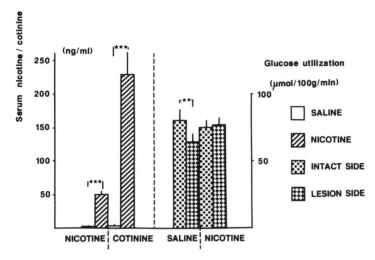

Fig. 5. Effects of chronic nicotine treatment via minipumps on neostriatal glucose utilization of partially hemitransected rats on day 14 after operation. The nicotine and cotinine serum levels are shown to the left and glucose utilization to the right. Means ± S.D., n= 5. Student's paired t-test was used for the analysis of the data obtained on glucose utilization and two-tailed Mann-Whitney U-test for the serum data. ** p<0.01, *** p<0.001.

tion of DA utilization observed in the substantia nigra on the partially hemitransected side was abolished by the chronic nicotine treatment (Fuxe et al., 1990).

Taken together, these results give further evidence for our hypothesis (Janson et al., 1988) that a protective action of chronic nicotine treatment on nigrostriatal DA neurons following a partial hemitransection is related to a desensitization of excitatory nicotinic cholinergic receptors, since a marked reduction of DA utilization was present, leading to reduced energy demands. The marked reduction of DA utilization in the surviving DA terminals may be related at least in part to the fact that desensitization of the nicotinic cholinoceptors leads to reduced firing rates in the nigral DA neurons. It is also important to emphasize that the desensitization of the nicotinic cholinergic receptors leads to reduced sodium and calcium influx (Revah & Changeux, 1988) which should not only reduce the energy cost to maintain ion homeostasis but also reduce calcium influx and thus reduce neurotoxicity (see Griffiths et al., 1983).

Finally, it should be pointed out that corticosterone does not appear to be involved in the production of the desensitization of the nicotinic cholinergic receptors, since the serum corticosterone levels were not significantly influenced by the chronic nicotine treatment (see Fuxe et al., 1990). It has been shown that adrenocortical hormones produce a reduction in the nicotine sensitivity in physiological and behavioural tests (see Pauly et al., 1988). We have also recently been able to demonstrate that acute and chronic corticosterone treatment selectively reduces the affinity of cholinergic receptors in the subcortical limbic forebrain (see von Euler et al., 1990).

STUDIES ON STRIATAL GLUCOSE UTILIZATION AND BLOOD FLOW USING COMPUTER ASSISTED AUTORADIOGRAPHY IN PARTIALLY HEMITRANSECTED RATS AND THE EFFECTS OF CHRONIC NICOTINE TREATMENT VIA MINIPUMPS

For methodological details on studies on glucose utilization and on blood flow using 14C-deoxyglucose and 14C-iodoantipyrin, see Owman and Diemer (1985). Partial hemitransections and the chronic nicotine treatment were performed as described above. It was demonstrated by Owman and colleagues (1989a,b) that the partial hemitransection induced a reduction of glucose utilization and of blood flow in the neostriatum. These results are illustrated in Fig. 5., where also the serum nicotine and cotinine levels are shown. Following the chronic nicotine treatment the asymmetry in striatal glucose utilization was abolished, mainly related to an increase in glucose utilization on the lesioned side. Preliminary studies on the striatal blood flow (see Owman et al., 1989a) indicated that the chronic nicotine treatment exert similar effects also on striatal blood flow. Thus, the lesion-induced reduction of striatal blood flow was counteracted by the chronic nicotine treatment. These results provide functional evidence for a restoration of function in the neostriatum following the chronic nicotine treatment. This action may be related at least in part to an increased survival of striatal DA nerve terminals, which may produce trophic factors and improve the biochemical machinery of the striatal nerve cells. It should also be considered that the chronic nicotine treatment via effects also on other types of striatal nicotinic receptors not related to the DAergic systems may have positive consequences for the metabolic performance of the striatal nerve cells.

ELECTROPHYSIOLOGICAL STUDIES ON NIGROSTRIATAL DOPAMINE NEURONS FOLLOWING PARTIAL MESO-DIENCEPHALIC HEMITRANSECTIONS AND EFFECTS OF CHRONIC NICOTINE TREATMENT VIA MINIPUMPS

For details on the extracellular, single cell-recording techniques employed in this study, see Grenhoff et al. (1988). Nigral cells were identified as DA cells on the basis of a typical triphasic action potential shape (> 2 msec), rate between 1-10 Hz and a typical firing pattern (see Grenhoff & Svensson, 1988). A negative current was passed through the recording electrode producing a blue dye spot at the recording site after the experiment, making it possible to locate the position of the electrode. Interspike interval histograms were analyzed in special programs allowing the analysis of neuronal firing rate, burst firing and regularity. The burst firing was defined as the ratio between spikes in bursts and total number of spikes of an interspike interval histogram. The regularity was measured by determination of the variation coefficient. It should be emphasized that the partial meso-diencephalic hemitransection and the chronic nicotine treatment was performed in an identical way to that described above. Recordings of the nigral DA neurons were made on day 14 after the lesion. Markedly fewer active cells were observed in the substantia nigra of the hemitransected animals compared to sham-operated rats, probably related to the disappearance of DA cells, but could also be related to changes in the number of spontaneously active cells. The recorded cells of hemitransected rats of saline or nicotine treated rats showed no differences in firing rates or in regularity of firing. Of substantial interest (Grenhoff et al., 1990) was the fact that burst firing was significantly lower in the nicotine treated hemitransected rats compared with the saline treated hemitransected rats. The demonstrated reduced burst firing could elegantly explain the marked reduction of DA utilization demonstrated after chronic nicotine treatment in the partially hemitransected rats. It also gives important new evidence for the hypothesis that nicotine treatment increases DA nerve cell survival by reducing energy demands via reducing the activity of the DA nerve cells. The reduction of burst firing is probably produced then by the desensitization of the cholinergic nicotinic receptors leading to a reduction of sodium influx and calcium influx, since the ion channel of the nicotinic receptor becomes closed (see Changeux et al. 1984). It should also be considered, however that depolarization blockade may develop for short time periods due to the excessive activation of the neurons (Grace & Bunney, 1986). It has also recently been demonstrated that lesions of nigrostriatal DA neurons lead to an increased tendence of the DA cells of the substantia nigra to develop depolarization blockade following treatment with antipsychotic drugs (Hollerman & Grace, 1989). Nevertheless, it must be pointed out that these periods of depolarization blockade must be short, since otherwise this leakage of ions over the nerve cell membrane will lead to increased energy demands (to maintain ion homeostasis) and obviously via the depolarization the voltage sensitive calcium channels will remain open and increase calcium influx and thus neurotoxicity. The major action of chronic nicotine treatment is therefore the desensitization of the nicotinic cholinergic receptors, leading to a reduction in the sodium and calcium influx, which probably underlies the observation of Grenhoff and colleagues (1990) that chronic nicotine treatment in this model can produce a significant reduction of burst firing in the DA cells of the substantia nigra of the lesioned side.

Acknowledgements: This work has been supported by Forschungsrat Rauchen und Gesundheit, FRG.

REFERENCES

Agnati, L.F., Fuxe, K., Calza, L., Benfenati, F., Cavicchioli, L., Toffano, G. and Goldstein, M. (1983) Acta Physiol. Scand. 199, 347-363.

Agnati, L.F., Fuxe, K., Toffano, G., Calza, L., Zini, I., Giardino, L., Mascagni, F. and Goldstein, M. (1985) In: Quantitative Neuroanatomy in Transmitter Research (L.F. Agnati and K. Fuxe, Eds), Macmillan Press, London, pp. 145-156

Agnati, L.F., Fuxe, K., Zoli, M., Zini, I., Härfstrand, A. Toffano, G. and Goldstein, M. (1988) Neuroscience 26, 461-478.

Andén, N.-E., Corrodi, H. and Fuxe, K. (1969) In: Metabolism of Amines in the Brain (G. Hooper, Ed), Macmillan Press, London, pp. 38-47.

Andersson, K., Fuxe, K. and Agnati, L.F. (1985) Acta Physiol. Scand. 123, 411-426.

Bauman, R.J., Jameson H.D., McKean, H.E., Haack, D.G., and Weisberg, L.M. (1980) Neurology 30, 839-843.

Changeux, J.-P., Devillers-Thiéry, A. and Chemouilli, P. (1984) Science 225, 1335-1345.

von Euler, G., Fuxe, K., Finnman, U.-B., and Agnati, L.F. (1990) Brain Res, in press.

Fuxe, K., Janson, A.-M., Jansson, A., Andersson, K., Eneroth, P., and Agnati, L.F. (1990) Arch. Pharmacol. 341, 171-181.

Grace, A.A., and Bunney, B.S. (1986) J. Pharmacol. Exp. Ther. 238, 1092-1100.

Grenhoff, J., and Svensson, T. (1988) Life Sci., 42, 2003-2009.

Grenhoff, J., Ugedo, L., and Svensson, T. (1988) Acta Physiol. Scand. 134, 127-132.

Grenhoff, J., Janson, A.-M., Svensson, T.H., and Fuxe, K. (1990) Brain Res. in press.

Godwin-Austen, R.B., Lee, P.N., Marmot, M.G., and Stern, G.M. (1982) J. Neurol. Neurosurg. Psychiatry 45, 577-581.

Griffiths, T., Evans, M.C. and Meldrum, B.S. (1983) In: Excitotoxins (K. Fuxe, P. Roberts and R. Schwarcz, Eds), MacMillan Press, London, pp. 331-342.

Hollerman, J.R., and Grace, A.A. (1989) Neurosci. Lett., 96, 82-88.

Janson, A.M., Fuxe, K., Kitayama, I., Härfstrand, A. and Agnati, L.F. (1986) Neurosci. Lett., Suppl. 26, S88.

Janson, A.M., Fuxe, K., Agnati, L.F., Kitayama, I., Härfstrand, A., Andersson, K. and Goldstein, M. (1988) Brain Res. 455, 332-345.

Jonsson, G., Hallman, H., Mefford, I. and Adams, R.N. (1980) In: Central Adrenaline Neurons: Basic Aspects and Their Role in Cardiovascular Disease (K. Fuxe, M. Goldstein, B. Hökfelt and T. Hökfelt, Eds), Pergamon, New York, pp. 59–71.

Kessler, II., and Diamond, K.L. (1971) Am. J. Epidemiol. 94, 16–25.

Owman, Ch., and Diemer, N.H. (1985) In: Quantitative Neuroanatomy in Transmitter Research (L.F. Agnati and K. Fuxe, Eds), Macmillan Press, London, pp. 71–87.

Owman, Ch., Fuxe, K., Janson, A.M. and Kåhrström, J. (1989a) Prog. Brain Res. 79, 267–276.

Owman, Ch., Fuxe, K., Janson, A.M. and Kåhrström, J. (1989b) Neurosci. Lett. 102, 279–283.

Pauly, J.R., Ullman, E.A., and Collins, A.C. (1988) Physiol. Behav. 44, 109–116.

Revah, R., and Changeux, J.-P. (1988) In: Transport through membranes: Carriers, channels and pumps (A. Pullman, Ed), Kluwer Acad Publ., Amsterdam, pp. 321–335.

THE EFFECT OF CHRONIC NICOTINE TREATMENT ON 1-METHYL-4-PHENYL-1,2,3,6-TETRA-HYDROPYRIDINE-INDUCED DEGENERATION OF NIGROSTRIATAL DOPAMINE NEURONS IN THE BLACK MOUSE.

A.M. Janson*, K. Fuxe*, L.F. Agnati#, E. Sundström† & M. Goldstein¤

Dept of Histology & Neurobiology*, Karolinska Institutet, Box 60400, S-104 01 Stockholm, Sweden; Dept of Human Physiology#, Univ. of Modena, I-41100 Modena, Italy; Dept of Geriatrics†, Huddinge Hospital, S-141 86 Huddinge, Sweden. Dept of Psychiatry¤, New York Univ. Medical Center, New York, 100 16 NY, USA.

SUMMARY Protective activity of (-)nicotine on 1-methyl-4-phenyl-1,2,3,6-tetra-hydropyridine-induced (MPTP, 50 mg/kg sc) neurotoxicity is demonstrated in the mesostriatal dopamine (DA) system in the male black mouse with an acute inter-mittent (-)nicotine treatment (0.5 mg/kg x 4, starting immediately following the MPTP injection with a 30-minute time interval) together with a two-week continuous administration of (-)nicotine via Alzet minipumps implanted sub-cutaneously the day before MPTP is given. In contrast, when omitting the acute (-)nicotine treatment a dose-dependent increase in the MPTP-induced neurotoxi-city is seen.

INTRODUCTION

Protective actions of (-)nicotine after partial hemitransection of the meso-striatal DA system in the male Sprague-Dawley rat have been shown (Janson et al. 1986, 1988a, Fuxe et al., 1990). These results support an involvement of nicotine in the negative correlation between smoking and Parkinson's disease (Baron, 1986). Indications of anti-Parkinsonian activity of (-)nicotine have also been obtained by studying if chronic (-)nicotine treatment can protect nigrostriatal DA neurons from the neurotoxic actions of MPTP, known to cause chronic parkinsonism in humans (Davis et al., 1979, Langston et al., 1983), using a sensitive black mouse, C57Bl/6 (Perry et al., 1985). This protection by (-)nicotine after MPTP (Janson et al., 1988b) has now been further studied.

MATERIALS AND METHODS

Male C57 Bl/6 mice, (22 g b.wt., 10 weeks old, Alab, Sollentuna, Sweden) were treated with (-)nicotine hydrogen(+)tartrate (BDH Chemicals, Poole, UK) for two weeks ((-)nicotine dose 0.125mg/kg/h), via Alzet minipumps (model 2002, SMA, London, UK) implanted subcutaneously under halothane anaesthesia the day before MPTP-HCl (RBI, Natick, MA, USA, 50 mg/kg sc) was given. Starting immediately following the injection of MPTP (-)nicotine (0.5 mg/kg ip) was given four times with a 30-minute time interval.

The animals were sacrificed and the brains taken to immunohistochemical and biochemical analysis. For details on the protocol, see Janson et al., 1988b. Coronal neostriatal 20μm cryotome sections were taken to immunohistochemical analysis applying the avidin-biotin procedure (Vector Laboratories, Burlingame, CA, USA), after incubation with a polyclonal tyrosine hydroxylase (TH) antibody (Markey et al., 1980) diluted 1/1500. The TH immunoreactivity (IR) was then semi-quantitatively analyzed using an IBAS (Kontron, München, FRG) linked to a video camera (Bosch, FRG) to determine the total area of the TH IR striatal nerve terminal profiles (FA) and their staining intensity, grey value (GV), as well as the total IR (T-IR) using the formula $T\text{-}IR=GV{\times}FA$. Endogenous DA levels were determined in the substantia nigra in separate experiments using HPLC with electrochemical detection (Jonsson et al., 1980).

Similar experiments giving (-)nicotine for two weeks via mini-pumps in six different doses (0.037 mg/kg/h to 3 mg/kg/h, see Table I) were carried out o-mitting the acute injections of (-)nicotine after the lesion with MPTP (40 mg/kg sc). TH immunohistochemistry in the substantia nigra and HPLC determinations of neostriatal DA, 3,4-dihydroxyphenylacetic acid (DOPAC) and homovanillic acid

Table I
Serum nicotine and cotinine levels in the C57/Bl 6 mouse after MPTP and 14 days of treatment with various doses of (-)nicotine.

Treatment		serum nicotine (ng/ml)		serum cotinine (ng/ml)	
A saline+saline		0		0	
B saline+nicotine 3mg/kg/h		259.8±32.5		1652.6±249.2	
C MPTP+saline		0		0	
D MPTP+nicotine 0.037 mg/kg/h		4.0±0.26		12.6±1.2	
E MPTP+nicotine 0.11 mg/kg/h		7.2±1.0		21.3±3.3	★
F MPTP+nicotine 0.33 mg/kg/h	$p<0.01$	22.7±2.6	$p<0.01$	101.1±6.0	
G MPTP+nicotine 1 mg/kg/h		68.4±3.9		400.3±33.8	
H MPTP+nicotine 3 mg/kg/h		307±23.9		2567±152	

Means±SEM, n=6. Statistical analysis of the MPTP+nicotine treated groups (D-H) with the Jonckheere-Terpstra test and comparison of the saline+nicotine 3 mg/kg/h (B) and the MPTP+nicotine 3 mg/kg/h (H) with two-tailed Mann Whitney U-test. *=$p<0.05$.

(HVA) were performed in these studies, where also the serum nicotine and cotinine levels were determined in blood samples (trunk blood collected at time of
decapitation) using capillary column gas chomatography (Curvall et al., 1982)
(detection limit: 0.2 ng/ml for nicotine and 1.3 ng/ml for cotinine).

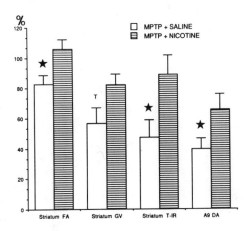

Fig. 1.   Means±SEM, n =4-6 for the immunohistochemical parameters (neostriatal
FA, GV, and T-IR, see methods) and n =11-12 for the analysis of DA levels in
the substantia nigra (A9). All values are expressed as a percentage of the respective control group not receiving MPTP. The absolute values (mean±SEM) for
the groups treated only with saline or nicotine are given in the order saline/
nicotine (not significantly different from eachother): FA:2.21±0.70/2.20±0.58
mm²; GV:26.8±1.8/28.7±1.8; T-IR:61±6/63±5; A9 DA (ng/g of tissue wet weight):
1109±117/948±86. Two-tailed Mann-Whitney U-test. *=p<0.05; T=0.05<p<0.10.

RESULTS

Effects of chronic (-)nicotine treatment combined with acute intermittent injections of (-)nicotine immediately after MPTP:

MPTP produced a disappearance of TH IR in the neostriatum and a depletion of
nigral DA stores (Fig. 1). The (-)nicotine treatment significantly antagonized
the MPTP-induced effects on TH IR within the neostriatum, as seen from both the
FA and the T-IR values, as well as on the nigral DA stores (Fig. 1).

Effects of chronic (-)nicotine treatment without acute intermittent injections
of (-)nicotine immediately after MPTP:

The serum nicotine and cotinine levels after 14 days of chronic treatment with
six different doses of (-)nicotine are seen in Table I and shown to be clearly

dose-related. The serum cotinine levels were significantly increased in the MPTP treated group compared with the saline + nicotine treated group.

A dose dependent enhancement by chronic (-)nicotine of the MPTP-induced depletion of DA stores in the neostriatum and of the disappearance of TH IR nerve cells in the substantia nigra was seen. The group receiving (-)nicotine (3mg/kg/h) alone (without MPTP) was not significantly different from the saline control group (see Fig. 2 and legend to Fig. 3).

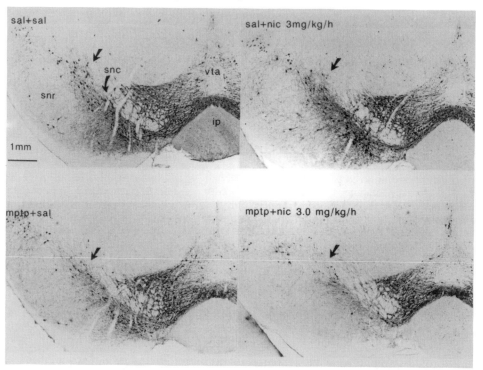

Fig. 2.   Substantia nigra illustrating the TH IR nerve cell bodies and dendrites. Upper row shows saline (left) and nicotine (3 mg/kg/h, right) treated animals (control - no MPTP, not different from eachother). Lower row shows MPTP lesioned animals with saline (left) and nicotine treatment (3 mg/kg/h, right). Straight arrows point at the central substantia nigra, where an enhancement of the MPTP-induced disappearance of TH IR nerve cells is seen after nicotine treatment. Abbreviations: ip= interpeduncular nucleus; vta=ventral tegmental area; snr=substantia nigra, reticular part; snc=substantia nigra, compact part.

DISCUSSION

A protective activity of (-)nicotine on MPTP-induced neurotoxicity in the mesostriatal DA system in the black mouse was demonstrated with acute intermittent

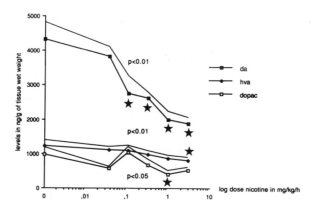

Fig. 3.   HPLC analysis of DA, HVA and DOPAC levels in the neostriatum after
MPTP and with increasing doses of nicotine (log scale). Means±SEM (dotted
line), n =4-7. The absolute values (in ng/g of tissue wet weight) for the
groups treated only with saline or nicotine were (not significantly different
from eachother): DA - saline:7382±865, nicotine (3 mg/kg/h): 6615±847; DOPAC -
saline: 768±24, nicotine: 1099±186; HVA - saline: 1280±80, nicotine 1402±121.
The significancies marked in the figure are found with the Jonckheere-Terpstra
test. Comparisons with the respective saline values are marked with asterisks
in the vicinity of the respective dose of DA, DOPAC and HVA and refer to the
treatment vs control test ⋆ = p<0.05.

treatment with (-)nicotine (0.5 mg/kg x 4) together with a low dose of (-)nico-
tine (0.125 mg/kg/h) for 14 days (Janson et al., 1988b). The neurotoxicity of
MPTP depends on the MAO-B catalyzed oxidation of MPTP to MPDP[+] and MPP[+] (Markey et al., 1984, Heikkila et al., 1985). The uptake of MPP[+] in DA neurons is
linked to the DA reuptake mechanism and can be antagonized by DA uptake inhibitors (Javitch et al., 1985, Sundström and Jonsson, 1985). When (-)nicotine was
administered acutely immediately following the MPTP injection, an enhancement
of striatal DA release (Andersson et al., 1981) could take place and thus compete with the MPP[+] up-take and lead to a reduced neurotoxic effect. These
effects may lead to an increased survival of the nigrostriatal DA neurons.

   In contrast, when omitting the acute (-)nicotine treatment and giving various doses of (-)nicotine a dose dependent increased neurotoxicity could be
seen. MPTP treatment furthermore leads to an increase in serum cotinine levels
vs saline treated mice after chronic (-)nicotine treatment, which may reflect a
reduced renal clearance (cf. Fuxe et al., 1990). However, it can not be excluded that the MPTP treatment leads to an altered cotinine metabolism.

   A possible explanation for the enhancement by chronic (-)nicotine treatment
of MPTP-induced neurotoxicity is a failure of the nicotinic cholinoceptors to

desensitize due to a detergent-like action of MPTP on the membrane (Hallman et al., 1985). In agreement with this view the MPTP-induced increase in DA utilization was not counteracted by the chronic (-)nicotine treatment (cf. Fuxe et al., 1990). In view of the failure of the nicotinic cholinoceptors to desensitize, the chronic (-)nicotine treatment still activates these receptors and thus increases the influx of Na^+ and Ca^{2+} (Revah and Changeux, 1988), leading to a possible increase in neurotoxicity (Griffiths et al, 1983). It should be considered that the decrease in mitochondrial oxidative phosphorylation seen after MPTP also leads to a release of mitochondrial Ca^{2+} and to increased intracytoplasmatic Ca^{2+} levels (Frei and Richter, 1986, Kass et al., 1988).

The mechanisms behind the dose dependent (-)nicotine-induced enhancement of the degeneration caused by MPTP of nigrostriatal DA neurons as well as its protective activity, when given in acute intermittent doses together with MPTP, remain to be further studied.

Acknowledgements: We are grateful to Mrs Ulla Altamimi, Mrs Beth Andbjer and Mrs Ulla-Britt Finnman for skilful technical assistance. This work has been supported by grants from the Forschungsrat Rauchen und Gesundheit, FRG and from the Swedish Medical Society.

REFERENCES

Andersson K, Fuxe K, Agnati LF, Eneroth P (1981) Effects of acute central and peripheral administration of nicotine on ascending dopamine pathways in the male rat brain. Evidence of nicotine induced increases of dopamine turnover in various telencephalic dopamine nerve terminal systems. Med Biol 59:170-176
Baron JA (1986) Cigarette smoking and Parkinson's disease. Neurology 36:1490-1496
Curvall M, Kazemi-Vala E, Enzell CR (1982) Simultaneous determination of nicotine and cotinine in plasma using capillary column gas chromatography with nitrogen-sensitive detection. J Chromatogr 232:283-293
Davis GC, Williams AC, Markey SP, Ebert MH, Caine ED, Reichert CM, Kopin IJ (1979) Chronic parkinsonism secondary to intravenous injection of meperidine analogues. Psychiatry Res 1:249-254
Frei B, Richter C (1986) N-methyl-4-phenylpyridine (MMP^+) together with 6-hydroxydopamine or dopamine stimulates Ca^{2+} release from mitochondria. FEBS Lett 198:99-102
Fuxe K, Janson AM, Jansson A, Andersson, K, Eneroth P, Agnati LF (1990) Chronic nicotine treatment increases dopamine levels and reduces dopamine utilization in substantia nigra and in surviving forebrain dopamine nerve terminal systems after a partial di-mesencephalic hemitransection. Naunyn-Schmiedeberg's Arch Pharmacol 341:171-181
Griffiths T, Evans MC, Meldrum BS (1983) Temporal lobe epilepsy, excitotoxins and the mechanism of selective neuronal loss. In: Fuxe K, Roberts P, Schwarcz R (eds) Excitotoxins, Wenner-Gren International Symposium Series, vol 39. Mac Millan, London, pp 331-342

Hallman H, Lange J, Olson L, Strömberg I, Jonsson G (1985) Neurochemical and histochemical characterization of neurotoxic effects of 1-methyl-4-phenyl-1,2,3,6-tetrahydropyridine on brain catecholamine neurones in the mouse. J Neurochem 44:117-127

Heikkila RE, Hess A, Duvoisin, RC (1985) Dopaminergic neurotoxicity of 1-methyl-4-phenyl-1,2,3,6-tetrahydropyridine (MPTP) in the mouse: relationships between monoamine oxidase, MPTP metaboilism and neurotoxicity. Life Sci 36: 231-236

Hollander M, Wolfe, DA (1973) Non-parametric statistical methods. Wiley, New York

Janson AM, Fuxe K, Kitayama I, Härfstrand A, Agnati LF (1986) Morphometric studies on the protective action of nicotine on the substantia nigra dopamine nerve cells after partial hemitransection in the male rat. Neurosci Lett (suppl 26) S88

Janson AM, Fuxe K, Agnati LF, Kitayama I, Härfstrand A, Andersson K, Goldstein, M (1988a) Chronic nicotine treatment counteracts the disappearance of tyrosinehydroxylase immunoreactive nerve cell bodies, dendrites and terminals in the mesostriatal dopamine system of the male rat after partial hemitransection. Brain Res 455:332-345

Janson AM, Fuxe K, Sundström E, Agnati LF, Goldstein, M (1988b) Chronic nicotine treatment partly protects against the 1-methyl-4-phenyl-1,2,3,6-tetrahydropyridine-induced degeneration of nigrostriatal dopamine neurons in the black mouse. Acta Physiol Scand 132:589-591

Javitch JA, D'Amato RJ, Strittmatter SM, Snyder SH (1985) Parkinsonism-inducing neurotoxin N-methyl-4-phenyl-1,2,3,6-tetrahydropyridine: uptake of the metabolite N-methyl-4-phenylpyridine by dopamine neurons explains selective toxicity. Proc Natl Acad Sci USA 82:2173-2177

Jonsson G, Hallman H, Mefford I, Adams, RN (1980) The use of liquid chromatography with electrochemical detection for the determination of adrenaline and other biogenic monoamines in the CNS. In: Fuxe K, Goldstein M, Hökfelt B, Hökfelt T (eds) Central adrenaline neurons: Basic aspects and their role in cardiovascular disease. Pergamon Press, New York, pp. 59-71

Kass GE, Wright JM, Nicotera P, Orrenius S (1988) The mechanism of 1-methyl-4-phenyl-1,2,3,6-tetrahydropyridine toxicity: role of intracellular calcium. Arch Biophys Biochem 260:789-797

Langston JW, Ballard P, Tetrud JW, Irwin I (1983) Chronic parkinsonism in humans due to a product of meperidine-analog synthesis. Science 219:979-980

Markey KA, Kondon S, Schenkman L, Goldstein M (1980) Purification and characterization of tyrosine hydroxylase from a clonal chromocytoma cell line. Mol Pharmacol 17:79-85

Markey SP, Johannessen JN, Chiueh CC, Burns RS, Herkenham MA (1984) Intraneuronal generation of a pyridinum metabolite may cause drug-induced parkinsonism. Nature 311:464-467

Perry TL, Yong VW, Clavier RM, Jones K, Wright JM, Foulks JG, Wall RA (1985) Partial protection from the dopaminergic neurotoxin N-methyl-4-phenyl-1-2-3-6-tetrahydropyridine by four different antioxidants in the mouse. Neurosci Lett 60:109-114

Revah R, Changeux J-P (1988) Functional organization of the acetylcholine receptor: a model of ligand gated ion channel. In: Pullman A (ed) Transport through membranes: Carriers, channels and pumps. Kluwer Academic Publishers, Amsterdam, pp 321-335

Sundström E, Jonsson G (1985) Pharmacological interference with the neurotoxic action of 1-methyl-4-phenyl-1,2,3,6-tetrahydropyridine (MPTP) on central catecholamine neurons in the mouse. Eur J Pharmacol 110:293-299

EFFECTS OF POSTNATAL EXPOSURE TO CIGARETTE SMOKE ON HYPOTHALAMIC
CATECHOLAMINE NERVE TERMINAL SYSTEMS AND NEUROENDOCRINE FUNCTION
IN WEANLING AND ADULT MALE RATS

K. Andersson[1,2], A. Jansson[1], B. Bjelke[1], P. Eneroth[3] & K. Fuxe[1]

[1] Department of Histology & Neurobiology, Karolinska Institute,
   Box 60 400, 104 01 Stockholm, Sweden.
[2] Department of Internal Medicine F-62, Huddinge Hospital, 141 86
   Huddinge, Sweden.
[3] Unit for Applied Biochemistry, F-62, Huddinge Hospital. 141 86
   Huddinge, Sweden.

Male rats were exposed to the smoke from 2 cigarettes (Kentucky
reference IR-1 type) every morning from the day 1 after birth for
a period of 5, 10 or 20 days. The rats were decapitated 24 h, 1
week or 7 months after last exposure. Serum LH levels were
increased 24 h after a 10 and 20 day exposure to cigarette smoke.
In adult life (6 months of age) following a 20 day postnatal
exposure to cigarette smoke increased serum levels of
corticosterone and prolactin were observed. Postnatal exposure to
cigarette smoke produced increased catecholamine (CA) utilization
in the medial palisade zone (MPZ) of the median eminence (ME) as
well as a reduction in CA utilization in the paraventricular
hypothalamic nucleus (PA) 24 h after a 20 day exposure to
cigarette smoke. These changes were not present in adult life. In
conclusion, postnatal exposure to cigarette smoke, most likely
produced by its nicotine component, produces hypersecretion of
prolactin and corticosterone in the adult but not in the weanling
male rat. The hypothalamic CA nerve terminal systems do not seem
to mediate this effect. Thus, postnatal exposure to cigarette
smoke seem to produce functional consequences for neuroendocrine
regulation mainly after maturation of the animal.

INTRODUCTION

Previous studies on adult male rats have shown that acute and

chronic exposure to cigarette smoke produce increases in dopamine

(DA) utilization in the medial and lateral palisade zones (MPZ and

LPZ) of the median eminence, which also were found to be

associated with inhibition of prolaction and LH serum levels
(Andersson et al. 1985a,c; Fuxe et al. 1987). Furthermore, acute
exposure to cigarette smoke produces increases in noradrenaline
(NA) utilization within discrete hypothalamic catecholamine (CA)
nerve terminal systems. These effects were shown to be
counteracted by mecamylamine pretreatment and thus the effects of
cigarette smoke are most likely produced by its nicotine component
(Andersson 1985). In paraventricular and subependymal NA nerve
terminal systems this effect is subjected to a tolerance
development following chronic exposure to cigarette smoke as well
as a withdrawal reaction, which may be associated with changes in
corticosterone secretion (Andersson et al. 1989).

It has previously been demonstrated that prenatal nicotine
exposure leads to increases in regional NA- but not DA-levels and
turnover as evaluated at various time-intervals during the
postnatal period (Slotkin et al. 1987a). We have also demonstrated
that pre- and postnatal nicotine treatment produces a selective
and permanent activation in DA nerve terminal systems of the
external layer of the median eminence (ME), probably related to
alterations in DA neuronal differentiation (Jansson et al. 1989).
Furthermore, prenatal nicotine treatment interferes with the
development of the gonadal axis, sexual dimorphism of saccharin
preference (Lichtensteiger and Schlumpf 1985) and an enhancement
of the rate avoidance acquisition takes place in offsprings of
dams exposed to cigarette smoke (Bertolini et al. 1982). Prenatal
treatment with nicotine increases 3H-nicotine binding sites in the
fetal and postnatal rat brain (Hagino and Lee 1985; Slotkin et al.
1987b) and decreases the number of dopaminergic receptor binding
sites, associated with an increase in their receptor affinity, in
striatum in adult male rats (Fung and Lau 1989).
The present article reviews our present work on postnatal exposure
to cigarette smoke (2 cig/day) on neuroendocrine regulation and CA
in the weanling and adult rat.

MATERIAL and METHODS

Male rats were exposed to the smoke from 2 cigarettes (Kentucky reference IR-1 type) every morning from the day 1 after birth for a period of 5, 10 or 20 days. The rats were decapitated 24 h, 1 week or 7 months after last exposure. CA stores were analysed in discrete hypothalamic nerve terminal systems using the Falck-Hillarp methodology in combination with quantitative histofluorimetry. CA utilization was determined by studying the decline in CA stores following a-methyl-dl-p-tyrosine methyl ester (aMT) treatment (Andersson et al. 1985b). Within the hypothalamus the CA fluorescence represents NA nerve terminals in the parvo- and magnocellular parts of the paraventricular hypothalamic nucleus (PA FP and PA FM), in the anterior and posterior periventricular hypothalamic regions (PV I and PV II), in the dorsomedial hypothalamic nucleus (DM) an in the subependymal layer (SEL) of the median eminence (ME). In the lateral palisade zone (LPZ) of the ME CA flourescence represents DA nerve terminals and in the medial palisade zone (MPZ) there is a mixture of DA and NA nerve terminals where DA nerve terminals predominate (Andersson et al. 1985b).

Serum levels of TSH, prolactin, LH and corticosterone were determined by using standard RIA techniques (Jansson et al. 1989). Statistical analysis according to the Mann-Whitney U-test (Hollander and Wolfe 1973).

RESULTS

Effects of postnatal exposure to cigarette smoke: Postnatal exposure to cigarette smoke produced increased CA utilization in the MPZ of the ME as well as a reduction in CA utilization in the PA 24 h after a 20 day exposure to cigarette smoke (Table I). These changes were not present in adult life (data not shown).

Serum LH levels were increased 24 h after a 10 and 20 day exposure to cigarette smoke (Table II). In adult life (6 months of age) following a 20 day postnatal exposure to cigarette smoke increased serum levels of corticosterone and prolactin were observed (Table II).

TABLE I. Effects of a 3-week postnatal exposure to cigarette smoke on CA levels and aMT-induced CA disappearance in discrete NA and DA nerve terminal systems in the hypothalamus and the median eminence of the male rat as evaluated on day 21, 24 h after the last exposure

Treatment	SEL	MPZ	LPZ	PA FP	PA FM
Air control	100±12	100± 8	100± 9	100±4	100±12
	(91±10)	(120±10)	(340±32)	(124±4)	(145±17)
Cigarette smoke	104± 9	124±12	88± 5	86±8	85± 4
Air control + aMT 2h	67± 8	47± 5	36± 4	67±4	67± 4
Cigarette smoke + aMT 2h	71± 7	27± 2	40± 4	83±3	92± 4

Means ± s.e.m. are given as a percentage of respective control group mean values. n = 6-8 rats. Below the respective 100 % values, the absolute concentration of CA are given in nmol/g tissue wet weight. Statistical analysis according to the Mann-Whitney U-test. * = p < 0.05. For abbreviations see Material and Methods.

TABLE II. Effects of a daily 10-day or a 3-week postnatal exposure to cigarette smoke on TSH, prolactin, LH and corticosterone serum levels of the male rat evaluated 24 h or 7 months after the last exposure.

Treatment	TSH	Prolactin	LH	Corticosterone
11 day old				
Air control	100± 8	n.d.	100± 24	n.d.
	(1290±98)		(225± 55)	
Cigarette smoke	99± 6	n.d.	(301± 76)	n.d.
21 day old				
Air control	100± 7	100±43	100± 10	100±24
	(772±53)	(2.8± 1.2)	(118± 12)	(168±40)
Cigarette smoke	108±12	138±45	226± 48	92±22
7 month old				
Control		100±34	100± 14	100±12
		(7.6± 2.5)	(726±100)	(170±21)
Cigarette smoke		258±53	82± 16	281±42

Means ± s.e.m. are given as a percentage of respective control group mean values. n = 10 to 18 rats. Below the 100 % values, the absolute serum hormone levels are given in pg/ml (TSH and LH), ng/ml (prolactin) or nmol(1 (corticosterone). Statistical analysis according to the Mann-Whithey U-test. ** = p < 0.01. n.d. = not detectable.

DISCUSSION

Previous studies in the rat have demonstrated that chronic prenatal nicotine treatment produces increases in NA- but not DA- levels and turnover in certain central CA pathways (Slotkin et al. 1987a). Acute nicotine treatment of the fetus results in increases in midbrain DA levels (Lichtensteiger et al. 1979). Chronic prenatal nicotine treatment produces in persistent alterations in the functional state of forebrain DA neurons (Ribary and Lichtensteiger 1989). Following pre- and postnatal nicotine treatment increases in DA utilization in the MPZ and LPZ of the ME was found one week after cessation of pre- and postnatal treatment with nicotine as well as in adulthood in the female diestrous rat (Jansson et al. 1989). An indication for such a change was also observed in the male rat. This permanent and apparently sex- specific change may be related to a change in the differentiation of the tubero-infundibular DA neurons (Jansson et al. 1989).

In the present study we have investigated the effects of exposure to cigarette smoke on hypothalamic CA nerve terminal systems and on neuroendocrine function. In the adult male rat the effects produced by cigarette smoke exposure were similar to the changes produced by nicotine injections on hypothalamic CA nerve terminal systems and on pituitary hormone secretion (Fuxe et al. 1989). Furthermore, mecamylamine pretreatment counteracts the effects produced by cigarette smoke exposure and thus the effects are most likely produced by the nicotine component of the smoke (Andersson 1985). In the present study we can report that postnatal exposure to cigarette smoke produces an increase in LH serum levels present 24 h after a 10- and 20-day exposure but not after a 5-day postnatal exposure and this change in serum LH levels is no longer observed 1 week and 6 months following a 3- week postnatal exposure. These results point to the possibility that infants exposed to cigarette smoke may show episodes of hypersecretion of LH, which may lead to increased serum concentrations of gonadal steroids and thus to an altered influence of gonadal steroids on brain function leading e.g. to changes in puberty development and sexual differentiation.

In spite of a selective and substantial increase in DA utilization in the MPZ seen 24 h after the last day of postnatal exposure there was no alteration in the serum prolactin levels at this time interval. The increase in DA utilization may be a withdrawal phenomenom, in view of the fact that such a reaction has been observed in the adult rats 48 and 72 h after withdrawal from exposure to cigarette smoke (Andersson et al. 1989). In the adult male rat a substantial increase in serum prolaction and corticosterone levels was shown after postnatal treatment. These findings demonstrate that after postnatal exposure to cigarette smoke abnormalities in prolactin and corticosterone secretion may develop after the completed maturation processes of the CNS and/or the anterior pituitary gland. The mechanism underlying these changes are unknown but it indicates that such permanent increases in corticosterone levels may lead to alterations in the glucocorticoid receptor activity in the CNS and thus to altered stress responses and to alterations in learning, especially in fear-motivated behaviours (Fuxe et al. 1989). Moreover, both prolaction and corticosterone are known to play a major role in the regulation of the immune system (Holaday et al. 1988).

Conclusion: The present results demonstrate that postnatal exposure to cigarette smoke, most likely by its nicotine component, produces hypersecretion of prolactin and corticosterone in the adult male rat, processes in which hypothalamic CA nerve terminal systems do not seem to participate in a major way. Postnatal exposure to cigarette smoke thus seems to produce functional consequences for neuroendocrine regulation in the adult rat.

Acknowledgements: We are grateful to Mrs Ulla Altamimi, Mrs Beth Andbjer, Mrs Ulla-Britt Finnman and Mr Lars Rosen for skilful technical assistance. This work has been supported by a grant (1223) from the Council for Tobacco Research, New York, U.S.A., a grant from Svenska Tobaks AB and grants (12x-08275 and 12P-08127) from the Swedish Medical Research Council and the Karolinska Institute Founds.

REFERENCES

Andersson K. (1985) Acta Physiol. Scand. 125, 445-452.
Andersson K., Eneroth P., Fuxe K., Jansson A. and Härfstrand A.
    (1989) Arch. Pharmacol. 339, 387-396.
Andersson K., Eneroth P., Fuxe K., Mascagni F. and Agnati L.F.
    (1985a) Neuroendocrinology 41, 462-466.
Andersson K., Fuxe K. and Agnati L.F. (1985b) Acta Physiol.
    Scand. 123, 411-426.
Andersson K., Fuxe K., Eneroth P., Mascagni F. and Agnati L.F.
    (1985c) Acta Physiol. Scand. 124, 277-285.
Bertolini A., Bernardi M., Genedani S. (1982) Neurobehav. Toxi-
    col.Teratol. 4, 545-548.
Fung K.Y. and Lau Y.-S (1989) Pharmacol.Biochem.Behav. 33, 1-6.
Fuxe K., Andersson K., Eneroth P., Härfstrand A. and Agnati L.F.
    (1989) Psychoneuroendocrinology 14, 19-41.
Fuxe K., Andersson K. Eneroth P., Härfstrad A. Nordberg A. and
    Agnati L.F. (1987) In: Tobacco Smoking and Nicotine (W.R.
    Martin, G.R. van Loon, E.T. Iwamoto and L. Davis, Eds.),
    Plenum Press, New York, pp. 225-262.
Hagino L. and Lee J.W. (1985) Int. J. Dev. Neurosci. 3, 567-571
Holady J.W., Bryant H.U., Kenner J.R. and Bernton E.W. (1988)
    Progress in Neuroendocrine Immunology 1, 6-13.
Hollander M. and Wolfe D.A. (1973) Nonparametrical Statistical
    Methods, John Wiley and Sons, New York, U.S.A.
Jansson A., Andersson K., Fuxe K., Bjelke B. and Eneroth P.
    (1989) J.Neuroendocrinology 1, 455-464.
Lichtensteiger W., Felix D. Hefti F. and Schlumpf M (1979)
    In: Electrophysiological Effects of Nicotine (A. Remong and
    C. Izard, Eds.), Elsevier/North-Holland Biomedical Press,
    Amsterdam, Holland, pp. 15-30.
Lichtensteiger W. and Schlumpf M. (1985) Pharmacol.Biochem.
    Behav. 23, 439-444.
Ribary U and Lichtensteiger W (1989) J. Pharmacol. Exp. Therap.
    248, 786-792.
Slotkin T.A., Cho H. and Whitmore W.L. (1987a) Brain Res. Bull.
    18, 601-611.
Slotkin T.A., Orband-Miller L. and Queen K.L. (1987b) J. Pharma-
    col. Exp. Ther. 242, 232-237.

NICOTINE INFUSION IN MAN STIMULATES PLASMA-VASOPRESSIN, -ACTH AND -CORTISOL IN A DOSE DEPENDENT MANNER

J. Stalke, O. Hader, J. Hensen, V. Bähr, G. Scherer[*], and W. Oelkers

Division of Endocrinology, Department of Medicine, Klinikum Steglitz, Freie Universität Berlin, Germany
[*]Analytisch-Biologisches Forschungslabor Prof. Adlkofer, München, Germany

Summary: We examined the response of plasma-vasopressin (AVP), -ACTH and -cortisol to nicotine infusion in man. Intravenous nicotine infusions led to dose-dependent increments of plasma nicotine, -AVP, -ACTH and -cortisol. The threshold for the stimulation of all parameters measured was between 0.5 and 1.0 μg/kg b.w./min of nicotine infusion. The occurrence of nausea was not a prerequisite for the stimulation of AVP release. The activation of all variables measured in this study at a similar threshold dosage of nicotine suggests a common basic mechanism at a nicotinergic receptor site. It is also possible that stimulation of AVP and ACTH by the same threshold dose of nicotine reflects corticotropin (ACTH) release by AVP.

INTRODUCTION

Nicotine has an antidiuretic effect, which is supposed to be mediated by vasopressin (Burn et al. 1945, Husain et al. 1975). According to Goldsmith et al. (1988), who studied effects of a nicotine chewing gum, vasopressin release is stimulated by nicotine only if nausea occurs. Rowe et al. (1980) infused nicotine into healthy subjects in order to obtain nicotine plasma levels similar to those during smoking. While smoking stimulated vasopressin release, nicotine infusion failed to do so. The authors concluded that vasopressin stimulation due to smoking was mediated by an airway-specific mechanism.

Nicotine elevates plasma cortisol levels when applied peripherally (Hökfelt 1961, Wilkins et al. 1982), an effect which is mediated by ACTH release from the adenohypophysis (Chiodera et al. 1984).

Thus, the present study was performed to investigate the response of plasma-vasopressin (AVP), -ACTH and -cortisol to nicotine infusion in man.

SUBJECTS AND METHODS

Seven male, healthy, non-smoking subjects (age range 25-38 yr, weight range 62-80 kg) were each studied three times, receiving nicotine infusions of 0.5 , 1.0 and 2.0 μg/kg body weight/min respectively over 20 minutes on different days. Blood was drawn twice at baseline and 5, 10, 15, 20, 25, 30, 40, 55 and 70 minutes after starting the infusion. ACTH was measured in unextracted plasma using a commercial IRMA (Nichols Institute, San Juan Capistrano, CA). Plasma cortisol was measured using a commercial kit (Sorin Biomedica, Saluggia, Italia). AVP in plasma was measured radioimmunologically by the method of Morton et al. (1985) after extraction. Nicotine in plasma was measured by gas-chromatography after extraction (Feyerabend et al. 1979). Statistical calculations (ANOVA followed by dependent t-test) were performed using "Stats + "statistics program (Copyright 1987, StatSoft, Inc.). All data are expressed as the mean $\pm$ SEM.

RESULTS

Infusion of nicotine led to a dose-dependent increase of plasma nicotine to peak levels of about 5, 10 and 20 ng/ml respectively in the three sets of experiments (Fig. 1, $p < 0.001$ at 20' vs 0' for all dosages). The lowest dose of nicotine caused mild lightheadedness and a feeling of arousal in three subjects. Infusion of the medium dose (1 μg/kg/min) led to lightheadedness in six subjects and to mild nausea in one. 2.0 μg/kg b.w./min of nicotine led to nausea in two subjects and to lightheadedness in the others.

Changes of AVP, ACTH, and cortisol during nicotine infusion are given in Fig 2. No changes in hormone levels were observed with the lowest dose of nicotine. Infusion of the

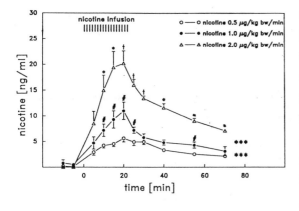

medium dose resulted in a clear-cut increase of all hormones measured (AVP: p < 0.05 at 30'; ACTH: p < 0.05 at 20', 30' and 40'; cortisol: p < 0.05 at 30' ,40' ,55' and 70' vs. 0' each). The increment of AVP, ACTH and cortisol was even higher with 2.0 µg/kg b.w./min (AVP: p < 0.05 at 20', 25', 30', 40', 55' and 70'; ACTH: p < 0.05 at 20', 25', 30', 40' and 55'; cortisol: p < 0.05 at 20', 25', 55', 70' and p < 0.001 at 30' and 40' vs. 0' each).

Fig 1. Plasma-Nicotine (x ± SEM) after nicotine infusion (* p < 0.05 vs. 1.0 ug/kg/min; # p < 0.05 vs. 0.5 ug/kg/min; t p < 0.001 vs. 1.0 ug/kg/min; *** p < 0.001 within curves).

DISCUSSION

In this study, nicotine levels and time course studies after infusion of nicotine are well in accordance with the findings of others (Rosenberg et al. 1980). Plasma concentrations were similar to those observed after smoking. Peak levels of nicotine were reached at the end of infusion, with rapidly decreasing blood levels after the end of infusion.

Intravenous nicotine infusions led to dose-dependent increments of plasma-AVP, -ACTH and -cortisol. The threshold for the stimulation of all parameters measured was between 0.5 and 1.0 µg/kg b.w./min of nicotine infusion. Rowe et al. (1980) infused 2 mg nicotine bitartrate, which is 33% nicotine, over five minutes representing a total dose of 0.66 mg, which is comparable to the medium dose used in our study. No effect on plasma vasopressin was observed. In the same study, test cigarettes containing 0.6 and 1.2 mg of total nicotine induced an increases in plasma vasopressin, and the authors concluded that some component in the vapor phase of smoke might be responsible for the vasopressin release. Unfortunately there are no data on nicotine levels in plasma after infusion of nicotine and after smoking in the paper of Rowe et al. It is known that the technique of

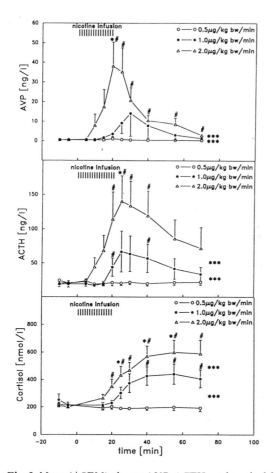

inhalation of cigarette smoke determines blood nicotine levels (Herning et al. 1983). Thus it is possible that the blood levels of nicotine achieved after smoking the low nicotine cigarette were higher than those after intravenous infusion, reaching the stimulatory threshold level observed in our study. The occurrence of nausea, which is a potent stimulus for AVP release (Rowe et al. 1979) was not a prerequisite for the stimulation of AVP release in our study.

Our data show that nicotine in the absence of nausea induces vasopressin release and that this effect is dose related.

Fig. 2. Mean (± SEM) plasma AVP, ACTH- and cortisol during nicotine infusion (* p < 0.05 versus 1.0 ug/kg/min, # p < 0.05 versus 0.5 ug/kg/min, *** p < 0.001 within curves).

To our knowledge, this is the first report of the response of plasma nicotine, -AVP, -ACTH and -cortisol to intravenous nicotine administration in the same subjects.

The activation of all variables measured in this study at a similar threshold dosage of nicotine suggests a common basic mechanism at a nicotinergic receptor site. Recently Weidenfeld et al. (1989) reported that nicotine activates the hypothalamo-hypophyseal-adrenal axis by a central mechanism which requires the integrity of the paraventricular nucleus. This effect was attributed to activation of central nicotinic cholinergic receptors.

It is also possible that stimulation of AVP and ACTH by the same threshold dose of nicotine reflects corticotropin (ACTH) release by AVP. A clear answer to this question requires further studies.

REFERENCES

Burn, J.H., Truelove, L.H., and Burn, I. (1945) Brit.Med.J. 1 403-406.

Chiodera, P., Camellini, L., Rossi, G., Maffei, M.L., Volpi, R., and Coiro, V. (1984) Horm.metabol.Res. 16 501-502 .

Feyerabend, C., and Russell, M.A.H. (1979) J.Pharm.Pharmacol. 31 73-76

Goldsmith, S.R., Katz, A., and Crooks, P.A. (1988) Clin Pharmacol Ther 44 478-481.

Herning, R.I., Jones, R.T., Benowitz, N.L., and Mines, A.H. (1983) Clin Pharmacol Ther 33 84-90

Hökfelt, B. (1961) Acta.med.Scand. 170 123-124.

Husain M.K., Frantz, A.G., Ciarochi, F., and Robinson, A.G. (1975) J Clin Endocrinol Metab 41 1113-1117.

Morton, J.J., Connell, J.M.C., Hughes, M.J., Inglis, G.C., and Wallace, E.C.H. (1985) Clin.Endocrinol 23 129

Rosenberg, J., Benowitz, N.L., Jacob, P., and Wilson, K.M. (1980) Clin.Pharmacol.Ther. 28 517-522.

Rowe, J.W., Shelton, R.L., Helderman, J.H., Vestal, R.E., and Robertson, G.L. (1979) Kidney International 16 729-735.

Rowe, J.W., Kilgore, A., and Robertson, G.L. (1980) J.Clin.Endocrinol.Metab. 51 170-172.

Weidenfeld, J., Bodoff, M., Saphier, D., and Brenner, T. (1989) Neuroendocrinology 50 132-138.

Wilkins, J.N., Carlson, H.E., Van Vunakis, H., Hill, M.A., Gritz, E., and Jarvik, M.E. (1982) Psychopharmacology 78 305- 308.

Acknowledgements

We wish to thank Ms. P. Exner and Ms. B. Faust for their technical assistance.

This study was supported by a grant of the Research Council on Smoking and Health, Hamburg, FRG.

A SINGLE DOSE OF NICOTINE IS SUFFICIENT TO INCREASE TYROSINE HYDROXYLASE ACTIVITY IN NORADRENERGIC NEURONES

Katherine M. Smith, Michael H. Joseph and Jeffrey A. Gray

MRC Brain, Behaviour and Psychiatry Research Group,
Department of Psychology, Institute of Psychiatry,
Denmark Hill, London SE5 8AF, England.

SUMMARY: Nicotine (0.8mg/kg in saline) was given to rats in one or two daily doses by subcutaneous injection. Tyrosine hydroxylase activities were increased relative to saline controls in noradrenergic cell bodies in the locus coeruleus 7 days later and in noradrenergic terminals in the hypothalamus and hippcampus after 21 days, a pattern consistent with enzyme induction. These increases were of the same magnitude as when nicotine was given by daily injection for the full 7 or 21 days, implying that as little as a single dose of nicotine is as effective as chronic nicotine administration in increasing tyrosine hydroxylase activity in noradrenergic neurones.

INTRODUCTION

It is well established that the cholinergic agonist nicotine releases catecholamines from slices and synaptosomal preparations of noradrenergically innervated areas, including the hypothalamus, hippocampus and cerebellum, of rat brain in vitro (Balfour, 1982). We have recently reported that nicotine given to rats increases catecholamine synthesis, in a regionally selective manner, in the hypothalamus and hippocampus; in the latter we have demonstrated increased noradrenaline release with in vivo techniques (Joseph et al, 1990; Mitchell et al, 1989 and 1990). We have also shown (Smith and Joseph, 1989) that repeated daily injections of nicotine for between 3 and 28 days increase the activity of tyrosine hydroxylase, the rate limiting enzyme in catecholamine biosynthesis, in noradrenergic neurones. This involves an early response in the cell bodies in the locus coeruleus, followed by increases in enzyme activity in the terminals (cerebellum, hypothalamus, hippocampus and frontal cortex) over a time course which seems to be directly related to the distance of the terminals from their cell bodies. In addition, the increase in tyrosine hydroxylase activity in the terminals appears to be independent of continued nicotine administration after 7 days of treatment (Smith and Joseph, 1990). We have now determined whether a more limited exposure to nicotine results in a similar increase in tyrosine hydroxylase activity in these neurones over a comparable time scale.

MATERIALS AND METHODS

Dosing: (-)-Nicotine di-(+)tartrate (0.8mg/kg as free base, neutralised in 0.9% sterile saline), was given in a single subcutaneous, interscapular injection for one day or on two consecutive days to male Sprague-Dawley rats, initial weight 200-250g. Controls received isotonic saline (1ml/kg) alone for one or two days.

Dissection: Animals were killed on the seventh day after the first injection for cell bodies, or on the 21st day after the first injection for the terminals. The locus coeruleus, hypothalamus and hippocampus were dissected on ice from 1.5 or 2.0mm coronal slices then frozen in liquid nitrogen.

Assay: Brain samples were thawed and assayed for tyrosine hydroxylase activity using the assay described in Joseph et al (1990) in which side chain tritiated tyrosine was used as substrate and 6-methyltetrahydropterin as cofactor with ascorbic acid as its reductant. The product, ^{3}H-DOPA, accumulated in the presence of the decarboxylase inhibitor NSD 1015, was isolated on alumina columns then quantified by liquid scintillation counting.

RESULTS

A single injection of nicotine on the first day or on each of the first two days resulted in tyrosine hydroxylase activities in the locus coeruleus 7 days later of 232% and 250% of saline controls respectively (Fig. 1). This is very comparable with the increase to 223% of controls after 7 days of repeated single daily injection with 0.8mg/kg nicotine (Smith and Joseph, 1989). 21 days after a single dose of nicotine, tyrosine hydroxylase activities had increased to 157% of controls in the hippocampus and to 136% in the hypothalamus. These effects are somewhat smaller in magnitude than those at 21 days resulting from nicotine injection for the first 7 days only (182% in the hippocampus and 147% in the hypothalamus), which are very similar to those following 21 days of continued nicotine administration (180% and 147% in the hippocampus and hypothalamus respectively).

DISCUSSION

We have shown previously that chronic nicotine increases tyrosine hydroxylase activity in a sequential manner from the cell bodies to the terminals of noradrenergic neurones (Fig. 2). This effect is analogous in its time course and regional selectivity to that of a single dose of reserpine, which has been attributed to enzyme induction (Zigmond, 1979). A similar dose of reserpine has been demonstrated to increase tyrosine hydroxylase mRNA (Faucon-Biguet et al, 1986), confirming that this time course represents induction of the enzyme in the cell bodies; this is presumably followed by axonal transport to the terminals. The results of the present study imply that a single exposure to nicotine is as effective as chronic nicotine in promoting the

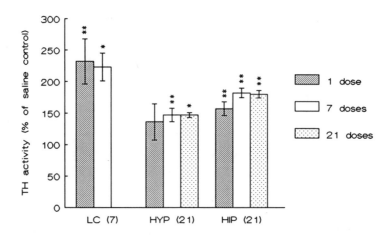

Figure 1. Effects of single or multiple injections of nicotine on tyrosine hydroxylase activity in noradrenergic neurones. One injection and 7 daily injections of nicotine (0.8mg/kg) are compared in the locus coeruleus (LC) after 7 days; 1, 7 or 21 daily injections are compared in the hypothalamus (HYP) and hippocampus (HIP) after 21 days. Results as % of respective saline controls; * P< 0.02, ** P< 0.01; n=6 for each group. Error bars represent +/- 1 S.E.M. Typical saline control activities were, in nmols DOPA/g wet weight tissue/hour, 73.3 +/- 11.0 (LC); 207.3 +/- 22.7 (HYP); 8.8 +/- 0.2 (HIP).

increases in tyrosine hydroxylase activity in noradrenergic cell bodies; this response too may be a result of enzyme induction. The increased activity in the terminals appears to be substantially independent of continued nicotine administration. We have also observed (Fig. 2; Smith and Joseph, 1989) that these increases in tyrosine hydroxylase activity are not sustained, even with chronic nicotine administration. A desensitisation following an initial exposure to the drug might account for the transient nature of the response.

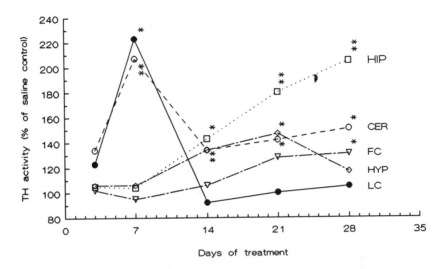

Figure 2. Effect of chronic nicotine (0.8mg/kg by subcutaneous injection), compared with saline injected controls, on tyrosine hydroxylase activity in the locus coeruleus (LC), cerebellum (CER), hypothalamus (HYP), hippocampus (HIP) and frontal cortex (FC). * P< 0.02, ** P< 0.01; n=7 for each group.

Acknowledgements: KMS is the holder of a Wellcome Trust Prize Studentship. We also thank R J Reynolds Tobacco Company and the MRC for financial support.

REFERENCES

Balfour, D.J.K. (1982) The effects of chronic nicotine on brain neurotransmitter systems. Pharmacol. Ther. 16, 269-282.
Faucon-Biguet, N., Buda, M., Lamoroux, A., Samolyk, D. and Mallet., J. (1986) Time course of the changes in tyrosine hydroxylase mRNA in rat brain and adrenal medulla after a single injection of reserpine. EMBO J. 5, 287-291.
Joseph, M.H., Peters, S.L., Prior, A., Mitchell, S.N., Brazell, M.P. and Gray, J.A. (1990) Chronic nicotine administration increases tyrosine hydroxylase selectively in the rat

hippocampus. Neurochem. Int. <u>16</u>, 269-273.

Mitchell, S.N., Brazell, M.P., Joseph, M.H., Alavijeh, M.S. and Gray, J.A. (1989) Regionally specific effects of acute and chronic nicotine on rates of catecholamine and 5-hydroxytryptamine synthesis in rat brain. Eur. J. Pharmacol. <u>167</u>, 311-322.

Mitchell, S.N., Brazell, M.P. and Gray, J.A. (1990) Nicotine-stimulated noradrenaline release in the hippocampus of freely moving rats. Br. J. Pharmacol. <u>99</u>, 266P.

Smith, K.M. and Joseph, M.H. (1989) Regionally selective effects of chronic nicotine on rat brain tyrosine hydroxylase activity: A time course study. Br. J. Pharmacol. <u>98</u>, 696P.

Smith, K.M. and Joseph, M.H. (1990) Contrasting effects of chronic nicotine on tyrosine hydroxylase activity in noradrenergic and dopaminergic neurones. Br. J. Pharmacol. <u>99</u>, 75P.

Zigmond, R.E. (1979) Tyrosine hydroxylase activity in noradrenergic neurones of the locus coeruleus increases after reserpine administration: Sequential increase in cell bodies and nerve terminals. J. Neurochem. <u>32</u>, 23-29.

THE EFFECT OF ACUTE AND REPEATED NICOTINE INJECTIONS ON BRAIN DOPAMINE ACTIVATION: COMPARISONS WITH MORPHINE AND AMPHETAMINE

P. Vezina, D. Hervé, J. Glowinski and J.-P. Tassin

Chaire de Neuropharmacologie, INSERM U. 114, Collège de France, 11, place Marcelin-Berthelot, 75231 Paris Cedex 05, France

SUMMARY: Nicotine was first compared to morphine in its ability to activate brain dopamine (DA) turnover in cortical and subcortical DA terminal fields after acute and repeated injections. In a second series of experiments, the effects of repeated injections of nicotine and amphetamine on the regulation of mu-opiate binding sites in subcortical DA terminal fields were compared. Results suggest that while nicotine may resemble morphine and amphetamine in its ability to activate the ascending mesencephalic DA systems acutely, it differs importantly from these two drugs in the way it activates these systems after repeated injections.

Acute systemic injections of morphine, amphetamine and nicotine elicit increased locomotion in rats and repeating these injections produces an enhanced or sensitized effect (Babbini & Davis, 1972; Ksir et al., 1987; Robinson & Becker, 1986). Different lines of evidence suggest that mesolimbic DA activation underlies the acute locomotor effects of these drugs and that enhanced activity in this system is responsible for sensitization to the locomotor effects of morphine and amphetamine (Clarke et al., 1988; Kalivas, 1985; Robinson & Becker, 1986). Relatively little is known, however, about the effects on DA activation of repeated injections of nicotine.

We have found that acute injections of (-)-nicotine bitartrate (0.4-0.8 mg/kg, base, s.c.) produce substantial increases (+41-49%) in DOPAC/DA in the nucleus accumbens (N.Acc.) and more moderate increases (+26-37%) in the antero-medial striatum. Repeating these injections reduced (to 15-26% in the N.Acc.) or abolished (in the antero-medial striatum) these increases in DA turnover, extending similar findings reported by others (Grenhoff & Svensson, in press). Interestingly, the opposite effect was obtained in the medial prefrontal cortex (mPFC). Small increases in DOPAC/DA produced by acute injections (+18%) gave way to larger increases (+31%) after repeated injections.

In comparison, acute morphine hydrochloride (10.0 mg/kg for repeated injections and 5.0 mg/kg at test, base, s.c.) produced large increases in DOPAC/DA in the N.Acc. (+59%) and the mPFC (+61%). Unlike with nicotine, however, repeating these injections substantially enhanced the effect of morphine in the N.Acc. (+96%) but slightly reduced it in the mPFC (+52%).

These findings suggest that nicotine differs from morphine (and amphetamine) in the manner in which it produces enhanced locomotor effects following repeated injection and possibly in its relation to DA as a mediator of reward. Indeed, repeated nicotine reduces the acute activating effects of this drug in subcortical DA terminal fields and increases them in the mPFC while repeated morphine increases these effects in subcortical sites and produces small decreases or no changes in the mPFC. These results are, furthermore, consistent with the view that a dynamic or functional interaction exists between the effects of DA action in the mPFC and the N.Acc. (Tassin et al., in press).

Results from a second series of experiments suggest that the differential effects of these drugs on DA activation may also have consequences on the regulation of mu-opiate receptors in subcortical DA terminal fields. Preliminary results obtained by quantitative autoradiography indicate that mu-opiate binding sites in these areas are upregulated (up to +48%) following repeated injections of d-amphetamine sulfate (0.75 mg/kg, base, i.p.) but not of nicotine (0.4 mg/kg, base, s.c.). These effects, opposite to those obtained (down-regulation of mu-opiate receptors) when the ascending DA projections are lesioned or after chronic DA receptor blockade (Trovero et al., in press), may be related to the ability of amphetamine (and morphine) but not of nicotine to produce sensitization of subcortical DA function after repeated injection.

It remains to be determined how decreased nicotine-induced DA activation in the N.Acc. can be reconciled with a sensitized locomotor response to this drug. Nicotine may thus elicit at least some of its effects on locomotor behavior via actions on brain neurotransmitter systems other than DA, a possibility not inconsistent with the reported quantitative (and qualitative?) differences between nicotine- and morphine- or amphetamine-induced locomotion. Conversely, or in addition, such differences may reflect the differential effect of these drugs on the state of equilibrium between DA activity in the mPFC and the N.Acc. (Tassin et al., in press).

ACKNOWLEDGEMENTS: Supported by NSERC, INSERM and Philip Morris Europe.

REFERENCES

Babbini, M. and Davis, W.M. (1972) Br. J. Pharmacol. 46, 213-224.
Clarke, P.B.S., Fu, D.S., Jakubovic, A. and Fibiger, H.C. (1988) J. Pharmacol. Exp. Ther. 246, 701-708.
Grenhoff, J. and Svensson, T.H. (in press) Naunyn-Schmiedeberg's Arch. Pharmacol.
Kalivas, P.W. (1985) J. Pharmacol. Exp. Ther. 235, 544-550.
Ksir, C., Hakan, R.L. and Kellar, K.J. (1987) Psychopharmacology 92, 25-29.
Robinson, T.E. and Becker, J.B. (1986) Brain Res. Rev. 11, 157-198.
Tassin, J.-P., Hervé, D., Vezina, P., Trovero, F., Blanc, G. and Glowinski, J. (in press) In: The Mesolimbic Dopaminergic System: From Motivation to Action (P. Willner, Ed.), Wiley.
Trovero, F., Hervé, D., Desban, M., Glowinski, J. and Tassin, J.-P. (in press) Neurosci.

PRELIMINARY STUDIES ON THE IN VIVO DESENSITIZATION OF CENTRAL
NICOTINIC RECEPTORS BY (-)-NICOTINE

Heidi F. Villanueva, Shawali Arezo and John A. Rosecrans

Department of Pharmacology and Toxicology, Va. Commonwealth Univ.-
Medical College of Va., MCV Box 613, Richmond, VA 23298-0613, USA

SUMMARY: Preliminary evidence of in vivo desensitization of
nicotinic receptors by nicotine is presented via two methods.
Nicotine partially attenuated the memory deficits produced by the
cholinergic neurotoxin AF64A on a radial arm maze task in rats.
Nicotine also acted as a nicotinic antagonist by decreasing drug-
appropriate responding in a standard drug discrimination procedure.

INTRODUCTION

Studies on animal tissues have revealed that nicotinic agonists
cause a desensitization of the nicotinic receptors after an
extended exposure time (Beanie et al., 1985). The present
experiments provide in vivo evidence for desensitization of the
nicotinic receptor via nicotine. Experiment 1 examined the
ability of (-)-nicotine to protect cholinergic receptors from an
acetylcholinergic neurotoxin, AF64A (ethylcholine aziridinium ion).
AF64A has been proposed as a selective presynaptic neurotoxin which
acts primarily by inhibiting the high affinity choline transport
(HAChT) system in cholinergic terminals. (Fisher et al., 1983).
AF64A is proposed to compete with endogenous choline and thus be
bound or transported into the cholinergic terminals, acting as a
specific neurotoxin. Intraventricular (IVT) or intracerebral (IC)
administrations result in long-term reduction of cholinergic
markers with no decrease in other neurochemical parameters (Glick
et al., 1973; Sandberg et al., 1984). Studies utilizing a radial
arm maze (RAM) memory task have demonstrated that IVT

administration of AF64A produces deficits in the rat 35 or 80 days after treatment (Walsh et al., 1984).

Experiment 2 utilized a standard two-lever operant drug discrimination procedure. Numerous studies in this laboratory have investigated nicotine as a discriminative stimulus (DS) and subsequent antagonism of this DS by nicotinic antagonists (Rosecrans et al., 1989). The present study investigated the ability of nicotine to antagonize the nicotinic DS, presumably via desensitization of the nicotinic receptor.

MATERIALS AND METHODS

Experiment 1: Rats were trained to perform a short-term memory task in an eight arm radial maze. Following training, rats were implanted with 14 day Alzet osmotic mini pumps containing (-)-nicotine (1.5 mg/day) or saline. Seven days after pump implantation, rats were injected IVT (bilaterally) with 6 ng AF64A or with distilled water. All animals were tested for short-term memory deficits seven days after IVT injections. Following behavioral testing, animals were sacrificed by microwave irradiation and levels of acetylcholine and choline were measured in the hippocampus and striatum via HPLC.

Experiment 2: Rats were trained to discriminate 0.4 mg/kg (-)-nicotine (s.c.) from saline on a VI-15 second schedule of food reinforcement. Following discrimination training, animals were injected with 0.8 mg/kg nicotine (s.c.) 15, 30, 60, 90, 120, 150 or 180 minutes prior to discrimination testing. Five minutes prior to testing, animals were injected with 0.4 mg/kg nicotine. Percentage of drug-appropriate responding was recording during 2-minute test sessions to determine whether pre-treatment with nicotine antagonized the discrimination of the training dose of nicotine.

RESULTS

Experiment 1: As seen in Figure 1, animals receiving distilled water displayed no memory deficits, while those receiving AF64A following pre-treatment with saline displayed significant deficits. Animals infused with nicotine prior to receiving AF64A showed moderate memory deficits, suggesting that nicotine may have partially desensitized acetylcholinergic receptors, thus preserving some short-term memory ability. Levels of ACh and choline in the hippocampus and striatum (Table I) were not significantly different in the three treatment groups, however the animals pre-treated with nicotine did show higher levels of Ach and choline in the hippocampus than did those animals pre-treated with saline prior to AF64A administration.

Figure 1: RAM Errors

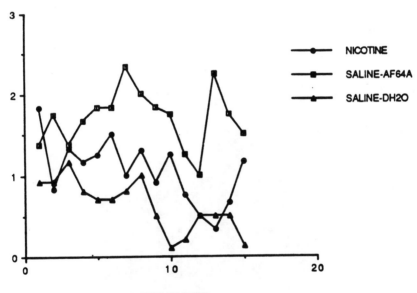

2 DAY BLOCKS

Mean number of errors on radial arm maze memory task following AF64A administration. Each data point is an average of 2 days for each group. Nicotine animals were pre-treated with 1.5 mg/day of nicotine (sc) for 7 days prior to AF64A administration (ivt); saline animals were pre-treated with saline (sc) for 7 days prior to AF64A or distilled water administration (ivt).

Table I: Mean and SEM for brain levels of acetylcholine and
           choline expressed as nm/mg protein. N = 6/group.

| | Hippocampus | | Striatum | |
	ACh	Ch	ACh	Ch
Saline+	.27	.32	.59	.26
dist H2O	(.06)	(.11)	(.11)	(.05)
Saline+	.13	.10	.42	.16
AF64A	(.04)	(.04)	(.05)	(.04)
Nicotine	.25	.21	.40	.21
+ AF64A	(.11)	(.08)	(.11)	(.08)

Experiment 2: When nicotine administration was preceded by an
additional 0.8 mg/kg dose of nicotine, responding on the drug-
appropriate lever was decreased in a sub-population of rats. The
peak effect of the desensitization varied for individual rats and
ranged from 15 minutes to 3 hours.   Figure 2 shows the mean
percentage of responding on the drug-appropriate lever at each time
tested.

Figure 2: Nicotinic Desensitization

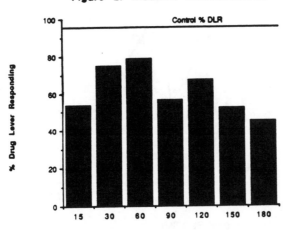

Mean percentage of responses on the drug-appropriate
lever following administration of 0.8 mg/kg nicotine 15-180 minutes
prior to administration of 0.4 mg/kg nicotine.   All animals were
trained to discriminate 0.4 mg/kg nicotine from saline. Control
%DLR represents responding following 0.4 mg/kg nicotine alone.

DISCUSSION

The results from these preliminary studies provides _in vivo_ evidence that nicotine desensitizes the nicotinic receptors. Nicotine partially attenuates the neurotoxic effects of the presynaptic neurotoxin AF64A and antagonizes the nicotinic discriminative cue in a drug discrimination procedure. The ability of nicotine to attenuate memory deficits adds further evidence to the present data which suggests that the memory deficits associated with Alzheimer's disease are linked to a loss of cholinergic innervation to parts of the cortex and hippocampus (Coyle et al., 1983). This information may contribute to the pharmacological treatment of Alzheimer's disease in the future. Also, the individual differences observed in Experiment 2 may provide an explanation as to the individual differences among smokers. It is proposed that smoking behavior is linked to desensitization of the nicotinic receptors, and an individual's smoking behaviors are thus determined by the rate of desensitization of those receptors.

REFERENCES

Beanie, L., Bianchi, C., Nilsson, L., Nordberg, A., Romanelli, L. & Sivilotti, L. (1985). Naunyn Schmiedebergs Archives of Pharmacology, 331, 293-296

Fisher, A., Mantione, C., Grauer, E., Levy, A. & Hanin, I. (1983). Behavioral Model and the Analysiss of Drug Action, (Spiegelstein and Levy, eds.) Amsterdam: Elsevier, 333-342.

Glick, S. Mittag, T. & Green, J. (1973). Neuropharmacology, 12, 291-296.

Sandberg, K. Hanin, I., Fisher, A. & Coyle, J. (1984). Behavioral Neuroscience, 98, 162-166.

Walsh, T., Tilson, H., DeHaven, D., Mailman, R., Fisher, A. & Hanin, I. (1984). Brain Research, 321, 91-102.

Coyle, J. Price, D. & DeLong, M. (1983). Science, 219, 1184-1190.

Rosecrans, J., Stimler, C., Hendry, J. & Meltzer, L. (1989). Progress in Brain Research, 79, 239-250.

A COMPARISON OF THE CHARACTERISTICS OF NICOTINE RECEPTORS IN YOUNG AND AGED RAT CORTEX.

R. Loiacono, S. Harrison, P. Allen, J. Stephenson and F. Mitchelson.

School of Pharmacology, Victorian College of Pharmacy, 381 Royal Parade, Parkville, 3052, Victoria, Australia.

SUMMARY: There is no difference in the affinity and numbers of binding sites for $[^3H]-(-)$-nicotine in homogenates of cerebral cortex from young adult (2-6 months) or aged (18-24 months) rats. Preliminary autoradiographic studies however indicate a greater degree of binding of $[^3H]-(-)$-nicotine to coronal sections of aged rat whole brain. There is no difference in the affinity of d-tubocurarine in displacing $[^3H]-(-)$-nicotine from its binding site in homogenates of rat cortex from young adult or aged rats. Activation of a presynaptic population of nicotine receptors on cortical cholinergic nerves by nicotine enhances $[^3H]$acetylcholine release. At higher frequencies of stimulation, where activation of presynaptic nicotine receptors by acetylcholine may be occurring, d-tubocurarine inhibits $[^3H]$acetylcholine release.

INTRODUCTION

The characterization of nicotine receptors within the normal mammalian CNS has received much attention (see review by Wonnacott, 1987). Further, examination of these sites during pathological changes observed in patients with Alzheimer's disease have revealed losses in the numbers of nicotine receptors within the cerebral cortex (Flynn and Mash, 1986). The aim of the present study was to examine changes in the nicotine receptor population in the mammalian CNS during the aging process by comparing the characteristics and role of nicotine receptors in brain from young adult (2-6 months) and aged (18-24 months) rats.

METHODS

Radioligand binding studies: Homogenates of rat cerebral cortex were incubated in Hepes buffer, pH 7.4 at 22°C for 20 min with [^{3}H]-(-)-nicotine (0.1 - 100 nM).   [^{3}H]-(-)-Nicotine was displaced from its binding site in a concentration dependent manner by d-tubocurarine (1 μM - 7 mM). Non specific binding was defined by 1 mM carbachol.

Autoradiography: Slide mounted coronal sections (14 μm) of rat brain were incubated at 22°C for 30 min in 50 mM Tris buffer, pH 7.4, containing 8 mM CaCl$_2$ and [^{3}H]-(-)-nicotine (1 - 50 nM) and exposed to [^{3}H]-Ultrofilm for ca 10-13 weeks. Non specific binding was defined by 1 mM carbachol.

[^{3}H]-Acetylcholine release studies: Slices of rat cerebral cortex were incubated for 30 min with 0.1 μM [^{3}H]-choline. [^{3}H]-Acetylcholine release in response to electrical field stimulation (60 pulses at 0.5 Hz or 240 pulses at 2 Hz) was deduced from the efflux of radioactivity in to the Krebs-Henseleit solution perfusing the slices. Each experiment used two periods of stimulation; the release of radioactivity evoked in the second period (drug present unless control) was expressed as a percentage of the first (no drug present; %S2/S1).

RESULTS

[^{3}H]-(-)-Nicotine binds with high affinity to a single population of sites in homogenates from both young adult (2-6 months) and aged (18-24 months) rat cortex (Table I). There was no difference in the affinity and number of binding sites for [^{3}H]-(-)-nicotine in young adult or aged rats.

d-Tubocurarine displaces [^{3}H]-(-)-nicotine from its binding site according to a two site model in homogenates of cortex from either young adult or aged rats. There was no difference in the potency of d-tubocurarine in displacing [^{3}H]-(-)-nicotine from its binding site in homogenates of rat cortex from young adult or aged rats (Table II).

Table I.   The characteristics of $[^3H]$-(-)-nicotine binding to homogenates of young adult (2-6 months) and aged (18-24 months) rat cerebral cortex.   Data derived from 5-8 rats.

	K_D (nM)	B_{max} (fmol/mg prot)
young adult	5.49 (4.87-6.19)	81.3 ± 4.9
aged	5.98 (1.54-8.62)	90.8 ± 5.9

95% confidence limits are given for K_D values; s.e.m are given for B_{max} values.

In contrast to homogenate binding data preliminary autoradiographic studies comparing the localization of $[^3H]$-(-)-nicotine in young adult and aged rat whole brain indicates a greater degree of binding of $[^3H]$-(-)-nicotine to coronal sections of aged rat whole brain than to young adult whole brain.

Nicotine (3 µM) enhanced the release of $[^3H]$-acetylcholine evoked by electrical field stimulation (60 pulses at 0.5 Hz) in young adult cerebral cortex slices (Fig. 1); while at higher frequencies of stimulation (250 pulses at 2 Hz), d-tubocurarine inhibits the release of $[^3H]$-acetylcholine evoked by electrical field stimulation in young adult cerebral cortex slices (Fig. 1).

Table II.   The characteristics of d-tubocurarine in displacing $[^3H]$-(-)-nicotine from homogenates of young adult (2-6 months) and aged (18-24 months) rat cerebral cortex.   Data is derived from 5-6 rats.

	K_I (high) (µM)	K_I (low) (µM)
young adult	16.6 (7.4-38)	781 (14-42.8mM)
aged	7.9 (1.6-40.7)	520 (16-15.6mM)

95% confidence limits are give for K_I values

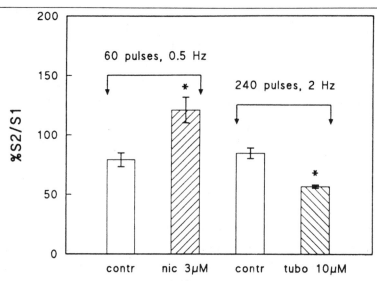

Figure 1.   The effects of nicotine (3 μM) and d-tubocurarine
(10 μM) on [^{3}H]-acetylcholine release evoked by 60 pulses at
0.5 Hz and 240 pulses at 2 Hz, respectively, in young adult
cerebral cortex slices.   Data derived from 4-8 rats.   *
indicates significance at the P<0.05 level.

DISCUSSION

Previous studies have demonstrated that the number of binding
sites for [^{3}H]-(-)-nicotine appear to decrease in frontal
cortex of patients suffering from Alzheimer's disease (Flynn
and Mash 1986).   Age related reductions in these binding sites
in normal human frontal cortex (Flynn and Mash, 1986), and age
related increases in normal human thalamus (Nordberg et al
1988) have been reported.   In contrast, the present study
indicates no change in either affinity or number of binding
sites for [^{3}H]-(-)-nicotine in young adult (2-6 months) or
aged (18-24 months) rat cerebral cortex homogenates.   Yamada
et al., (1986) reported a progressive increase in the numbers
of binding sites for [^{3}H]-(-)-nicotine in rat foetal
forebrain, reaching maximum levels at 14-28 days of age; these
being not different from adult (6-8 months) rat.   A single
population of binding sites for [^{3}H]-(-)-nicotine were

detected in either young adult or aged rats. d-Tubocurarine displaces $[^3H]$-(-)-nicotine from its binding site according to a two site model in either young adult or aged rats. This may be due to the differential affinity of d-tubocurarine for the $[^3H]$-(-)-nicotine binding site and an allosteric site linked to the $[^3H]$-(-)-nicotine binding site. Further, the potency of d-tubocurarine in displacing $[^3H]$-(-)-nicotine was not different in young adult or aged rats.

It is not clear why preliminary autoradiographic studies indicate an increase in binding sites for $[^3H]$-(-)-nicotine in aged rats.

The beneficial effects of nicotine in Alzheimer's disease may in part be due to the activation of a population of nicotine receptors associated with cortical cholinergic nerve endings. In the present study nicotine was able to enhance acetylcholine release evoked by electrical stimulation in young adult rat cortical slices. At higher frequencies of stimulation, d-tubocurarine was able to inhibit evoked acetylcholine release. Under these conditions d-tubocurarine may be disrupting a positive feedback loop on transmitter release mediated through the activation of presynaptic nicotine receptors by endogenous acetylcholine.

REFERENCES

Flynn, D. and Mash, D. (1986) Characterization of $[^3H]$-(-)-nicotine binding in human cerebral cortex: comparison between Alzheimer's disease and the normal. J. Neurochem. 47, 1948-54.

Nordberg, A., Adem, A., Nilsson, L., Romanelli, L. and Zhang. X. (1988) Heterogenous cholinergic nicotinic receptors in the CNS. In: Nicotinic Acetylcholine Receptors in the Nervous System (Clementi et al Eds) Springer-Verlag Berlin pp. 331-350.

Wonnacott, S. (1987) Brain nicotine binding sites. Human Toxicol. 6, 343-53.

Yamada, S., Kagawa, Y., Isogai, M., Takayanagi, N. and Hayashi E. (1986) Ontogenesis of nicotinic acetylcholine receptors and presynaptic cholinergic neurons in mammalian brain. Life Sci 38, 637-44.

THYMOPOIETIN POTENTLY AND SPECIFICALLY INTERACTS WITH THE NICOTINIC α-BUNGAROTOXIN RECEPTOR IN NEURONAL TISSUE

M. Quik, R. Afar and G. Goldstein[1]

Department of Pharmacology, McGill University, 3655 Drummond St., Montreal, Quebec H3G 1Y6, Canada and [1]Immunobiology Research Institute, Box 999, Route 22 East, Annandale, New Jersey, USA 08801-0999

SUMMARY: Thymopoietin (TPO), a polypeptide linked to immune function, potently and specifically interacted at the α-bungarotoxin (BGT) receptor (IC50 = 3-10 nM) in neuronal membranes. As well, TPO up-regulated the α-BGT sites in neuronal cells in culture. In a neuronal cell line (PC12), TPO enhanced process formation suggesting that it may be a neurotrophic factor. Since TPO has been identified in nervous tissue, these findings suggest that the polypeptide may represent an endogenous ligand for the neuronal α-BGT site, possibly with a trophic role in neuronal function. Moreover, the present work provides further evidence for a link between the nervous and immune system.

EVIDENCE FOR MULTIPLE NICOTINIC RECEPTORS IN NEURONAL TISSUE

Extensive evidence suggests that there exists at least two populations of neuronal nicotinic acetylcholine (Ach) receptors (Lindstrom et al., 1987; Berg and Halvorsen, 1988; Quik and Geertsen, 1988; Steinbach and Ifune, 1989). These include (1) the nicotinic receptor which binds [3H]Ach or [3H]nicotine with high affinity, but does not bind α-BGT. (2) The other site is the α-BGT site, which binds the α-toxin with high affinity and nicotinic receptor ligands with a lower affinity.

Evidence for at least these two nicotinic receptors stems
from the results of numerous studies including: **functional
studies**, which showed that α-BGT did not block nicotinic receptor mediated sensitivity in neuronal tissue (Oswald and
Freeman, 1981; Quik and Geertsen, 1988); receptor **localization
experiments** which indicated that the labeling pattern of
nicotinic receptor ligands such as [^{3}H]Ach and [^{3}H]nicotine did
not correlate well with the distribution of [^{125}I]α-BGT sites
(Clarke et al., 1985); receptor **purification studies**, which
demonstrated that the nicotinic receptor is composed of two
different subunits designated α and β (Lindstrom et al., 1987;
Steinbach and Ifune, 1989), while the brain α-BGT binding site
consists of more than 2 different subunits; and studies involving receptor **regulation**.

FUNCTIONAL ROLE OF THE NICOTINIC RECEPTOR POPULATIONS

With regard to function, the nicotinic ACh receptor which
binds nicotinic ligands with high affinity, appears to be involved in synaptic events, while the functional role of the
nicotinic α-BGT site in neuronal tissue is currently not known.
However, the observation that the toxin binds with such
specificity to a site with the characteristics of a nicotinic
receptor, infers that the α-BGT site is of significance.

The toxin binding site may have developed to mediate other
functions related to the nicotinic system. The α-BGT receptor
has been implicated in trophic functions in neuronal tissues.
Freeman (1977) showed that the α-BGT site appeared to exert a
trophic influence on incoming neurons. Studies in rat brain
indicated that the ontogenesis of toxin binding sites precedes
the development of other components of the cholinergic system.
This could implicate a role for the toxin binding sites in the
development, maintenance and/or maturation of synaptic contacts
(Fiedler et al., 1987). As well, evidence has suggested that
the α-BGT receptors may be involved in hormonal regulation.

ENDOGENOUS LIGANDS FOR THE α-BGT SITES IN ADDITION TO ACH

In addition to the notion that the α-BGT site may mediate functions distinct from those conventionally associated with the nicotinic receptor, it is also possible that the toxin binding site may be the receptor for other ligands, as well as Ach. The α-toxin binding site may have evolved in a fashion such that both Ach and another ligand could interact at the receptor. Very recent work (Quik et al., 1989) has shown that a novel polypeptide TPO potently inhibited α-BGT binding in brain.

TPO is a 49 amino acid polypeptide isolated from thymus, which is involved in immune mediated responses (Goldstein, 1974, 1987; Audhya et al., 1987). Earlier work had implicated a role for TPO in myasthenia gravis, an autoimmune disorder characterized by muscle weakness (Goldstein and Schlesinger, 1975). Subsequent studies to assess whether TPO exerted a direct effect at the nicotinic receptor showed that the polypeptide resulted in a marked inhibition of $[^{125}I]$α-BGT binding in electroplax (Venkatasubramanian et al., 1986). Studies to determine the functional effect of TPO at the neuromuscular junction indicated that the polypeptide may be involved in nicotinic receptor desensitization (Revah et al., 1987). Surprisingly, it did not appear to result in a direct block of neuromuscular transmission, leaving its role in muscle function somewhat unclear.

INTERACTION OF TPO AT THE α-BGT IN NEURONAL TISSUES

Since TPO interacted at the electroplax nicotinic α-BGT receptor, the question arose whether this polypeptide might also interact at the toxin binding site in neuronal tissues. Subsequent studies showed that TPO potently and specifically inhibited $[^{125}I]$α-BGT binding (IC50 = 3 nM) to brain membranes (Quik et al., 1989). Previous work had shown that nicotinic receptor ligands inhibited α-BGT binding (Oswald and Freeman, 1981). Thus, toxin binding could be affected by two classes of agents, the polypeptide TPO at nM concentrations and nicotinic receptor ligands at higher concentrations. Brown et al. (1986) have demonstrated the presence of TPO-like immunoreactivity in mouse spinal cord and brain extracts. Since TPO has been identified in the nervous tissue, these results suggest that TPO could be an endogenous ligand for the nicotinic α-BGT site.

Subsequent work was then done to determine whether TPO could modulate nicotinic receptors and/or the nicotinic α-BGT sites. Long term exposure of chromaffin cells in culture to TPO resulted in an increase in the number of α-BGT sites (Quik et al., 1990), without affecting the functional nicotinic receptor. Interestingly, this TPO induced response could be reversed in the presence of nicotine. These studies show that the α-BGT sites can be regulated both by the novel polypeptide TPO and by nicotinic cholinergic ligands.

Experiments were then done to assess possible functional consequences of TPO in nervous tissue. PC12 pheochromocytoma cells in culture were used as these have been studied extensively as a model system for neuronal cell function. Initial studies showed that TPO potently (4 nM) interacted at the α-BGT site in competition binding experiments (Cohen et al., 1989). Subsequent work which involved exposure of the PC12 cells to the thymic polypeptide for several days, showed that TPO enhanced process formation in the cells in culture in a time and concentration dependent manner. These results thus suggest that TPO, a polypeptide associated with the immune system, may be a trophic agent for neuronal cells. To date only a few other trophic factors, including NGF and fibroblast growth factor, have been shown to result in neurite formation in PC12 cells. α-BGT produced qualitatively similar results as TPO in inducing neurite formation in the cells in culture, although this effect occurred over a longer time course and using higher concentrations. Thus, these studies may point to a trophic role for the α-BGT/TPO site.

CONCLUSION: The present studies suggest that TPO may be an endogenous ligand for the nicotinic α-BGT receptor in neuronal cells. Furthermore, the demonstration that TPO induces neurite outgrowth in PC12 cells, indicates the peptide may have a growth or trophic function. These studies also provide further evidence for a link between the nervous and immune system, in analogy to the previously identified link between the gastrointestinal and nervous system.

Acknowledgements: The authors thank the FRSQ and MRC (Canada) for financial support.

REFERENCES

Audhya, T., Schlesinger, D.H., and Goldstein, G. (1987) Proc. Nat. Acad. Sci 84, 3545-3549.
Berg, D.K., and Halvorsen, S.W. (1988) Nature 334, 384-385.
Brown, R.H., Schweitzer, T., Audhya, T., Goldstein, G., and Dichter, M.A. (1986) Brain Res. 381, 237-243.
Clarke, P.B.S., Schwartz, R.D., Paul, S.M. Pert, C.B., and Pert, A. (1985) J. Neurosci. 5, 1307.
Cohen, R., Audhya, T., Goldstein, G., and Quik, M. (1989) Neurosci. Abstr. 15, 280.
Fiedler, E.P., Marks, M.J., and Collins, A.C. (1987) J. Neurochem. 49, 983-990.
Freeman, J.A. (1977) Nature 269, 218-222.
Goldstein, G. (1974) Nature 247, 11-14.
Goldstein, G. (1987) In Immune regulation by characterized polypeptides (G. Goldstein et al., Eds) Alan R. Liss, Inc., New York, pp.51-59.
Goldstein, G., and D.H. Schlesinger (1975) Lancet 2, 256-262.
Lindstrom, J., Schoepfer, R., and Whiting, P. (1987) Neurobiol. 1, 281-337.
Oswald, R.E., and Freeman, J.A. (1981) Nuisance 6, 1-14.
Quik, M., and Geertsen, S. (1988) Can. J. Physiol. Pharmacol. 66, 971-979.
Quik, M., Afar, R., Audhya, T., and Goldstein, G. (1989) J. Neurochem. 53, 1320-1323.
Quik, M., Afar, R., Geertsen, S., Audhya, T. Goldstein, G., and Trifaro, J.M. (1990) Mol. Pharmacol. 37, 90.
Revah, F., Mulle, C., Pinset, C., Audhya, T., Goldstein, G. and Changeux, J.P. (1987) Proc. Nat. Acad. Sci. 84, 3477-3481.
Steinbach, J.H., and Ifune, C. (1989) Trends Neurosci. 12, 3-6.
Venkatasubramanian, K., Audhya, T., and Goldstein, G. (1986) Proc. Nat. Acad. Sci. U.S.A. 83, 3171-3174.

LOW CONCENTRATIONS OF NICOTINE INCREASE THE FIRING RATE OF NEURONS
OF THE RAT VENTRAL TEGMENTAL AREA IN VITRO.

Mark S. Brodie

Neuroscience Research, D-47H, AP10, Abbott Park, IL 60064, U.S.A.

SUMMARY: Dopamine neurons of the ventral tegmental area are
thought to be involved in the mediation of the rewarding effects of
drugs of abuse, like nicotine. Low concentrations of nicotine (50
nM to 1 μM) produce excitation of VTA neurons recorded in a brain
slice preparation. At these concentrations, the response to
nicotine does not desensitize with prolonged administration.

INTRODUCTION

Nicotine-containing agents are used throughout the world as stimulants. Despite this widespread usage, the reinforcing properties of
nicotine (NIC) which underlie its high abuse potential are not
known. Recent studies provide substantial evidence that the rewarding properties of numerous drugs of abuse are mediated by central
dopamine systems (see Wise, 1987). The pathways most strongly implicated in these reward processes are the mesolimbic and mesocort-
ical systems, both arising from the ventral tegmental area of Tsai
(VTA) (Albanese and Minciacchi, 1983; Beckstead et al., 1979).
Disruption of dopaminergic transmission in these pathways has been
reported to decrease the rewarding efficacy of many drugs of abuse
(Roberts, et al., 1977; Roberts and Koob, 1982; Spiraki, et al.,

1983; Wise 1978). By studying the action of NIC on brain reward centers, it will be possible to understand and possibly control the reinforcing effects of this drug.

NIC was reported to excite dopaminergic neurons of the substantia nigra zona compacta recorded extracellularly in vivo (Mereu, et al., 1987; Clarke, et al., 1985; Lichtensteiner, et al., 1982). Furthermore, NIC has been shown to depolarize neurons of the VTA maintained in a brain slice preparation (Calabresi, et al., 1989). However, the concentrations used in that study were quite high, exceeding the plasma concentration of NIC in human smokers by 1000 fold. If the brain slice preparation is to be used to assess the effects of various agents on brain tissue, it is important to assess those effects at relevant concentrations. Furthermore, it has been reported that a pronounced desensitization to NIC occurs shortly after exposure of dopamine neurons to concentrations of NIC of 10 to 100 μM (Calabresi et al., 1989). The following study was performed to assess the effects of low concentrations of NIC (50 nM to 5 μM) on VTA neuronal activity and to determine whether these lower concentrations of NIC produce tachyphylaxis.

MATERIALS AND METHODS

Brain slices from Sprague-Dawley rats (100-200 gm) containing the VTA were prepared as previously described (Brodie and Dunwiddie, 1987). Animals used in this study were treated in strict accordance with the NIH Guide for the Care and Use of Laboratory Animals. Coronal sections (400 μm) containing the VTA were cut and placed in the recording chamber. The slice was covered with medium and a superfusion system then maintained the flow of medium at 2 ml/min and at a temperature of 35 °C. The composition of the artificial cerebrospinal fluid (aCSF) in these experiments was (in mM): NaCl 125, KCl 2.5, NaH_2PO_4 1.25, $MgSO_4$ 2, $NaHCO_3$ 25, glucose 11, $CaCl_2$ 2; the aCSF was saturated with 95% O_2/ 5% CO_2. The small volume chamber (about 300 μl) used in this study permitted the rapid infusion and washout of drug solutions.

Extracellular recording electrodes (6-10 MΩ) were made from 1.5 mm diameter glass tubing. One hour or more after preparation of the slice, an electrode was lowered into the VTA under visual guidance. Frequency of firing was determined using a window discriminator and ratemeter, the output of which was fed to a chart recorder and to an IBM-PC-based data acquisition system.

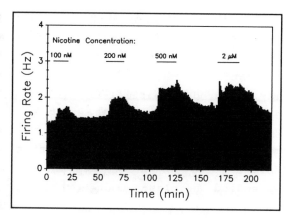

Fig. 1. Excitation of a VTA neuron by nicotine. Firing rate (computed over 20 sec intervals) was plotted as a function of time. Horizontal lines represent the duration of application of nicotine.

RESULTS

Cells chosen for study conformed to the characteristics of dopamine-type neurons of the mesencephalon which have been previously established (Brodie and Dunwiddie, 1987; Wang 1981). These characteristics include a distinctive waveform, a slow (0.5 to 4 Hz) firing rate, and a regular firing pattern. In all cells tested (n=32), NIC produced a clear excitation of VTA neurons (Fig. 1). This excitation occurred at concentrations as low as 50 nM, with an EC50 of about 170 nM. The maximum increase was about 80% of control firing rate (Fig. 2).

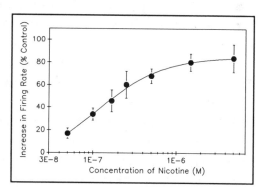

Prolonged administration of low concentrations of NIC to VTA neurons did not produce apparent desensitization to NIC. In fact,

Fig. 2. Concentration-response curve for nicotine-induced excitation of VTA neurons. Each point on the line represents the mean (±S.E.M.) of from 4 to 12 neurons.

long administrations (20-30 minutes) produced an extended excitation which subsided upon washout of NIC (Fig. 3). Prolonged administration of higher con-
centrations of NIC (5-100
μM) did produce partial desensitization in some cells,
but the firing rate in the
presence of NIC was generally higher than the control
rate even after 20 minutes
of NIC exposure; i.e., desensitization was not complete. Nicotine-induced excitation was blocked by hexamethonium (100-500 μM) and
mecamylamine (100-500nM).

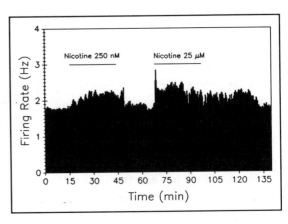

Fig. 3. Prolonged administration of high and low concentrations of nicotine. Firing rate (20
sec bins) plotted vs. time. Note that
in this paradigm, 25 μM nicotine
produces some desensitization.

DISCUSSION

As has been demonstrated by others, NIC excites dopaminergic
neurons of the ventral tegmental area. The results presented here
indicate two important points regarding this effect of NIC. First,
low concentrations of NIC profoundly excite VTA neurons in vitro.
Previous studies in vitro have used much higher concentrations (100
-1000 fold higher); these higher concentrations grossly exceed the
NIC levels which might be estimated to occur in brain during smoking. The concentrations used in the present study are similar to
those in the plasma of human smokers (Russell and Feyerabend,1978).

The second important point demonstrated in the present study is
that low concentrations of NIC did not produce observable desensitization. Many other studies report a slow desensitization during
prolonged application of NIC, so that the magnitude of the NIC
effect is reduced with time. This is clearly not the case in the
VTA when the concentration of NIC is kept below 1 μM (Figure 3).
The primary difference from previous studies is the lower concen-

trations used in these experiments. In a recent study of substantia nigra and VTA neurons recorded intracellularly in a brain slice preparation, desensitization to NIC was reported with concentrations of 10 to 100 μM (Calabresi, et al., 1989). A comparison with Figure 2 above indicates that these are supermaximal concentrations for the effect on firing rate. In the present study, some desensitization was seen when high, supermaximal concentrations of NIC were used (Figure 3).

Nanomolar concentrations of NIC produced non-desensitizing activation of VTA neurons. Presumably, at concentrations found in smokers' plasma (150-300 nM, Russell and Feyerabend, 1978), there is no desensitization to the rewarding effects of NIC. It follows that there should be no desensitization to NIC at pharmacological concentrations in brain areas involved in reward. From this viewpoint, the data presented in this report support the hypothesis that the VTA is involved in the rewarding effects of NIC.

REFERENCES

Albanese, A. and Minciacchi, D. (1983) J. Comp. Neurol. 216, 406-420.
Beckstead, R.M., Domesick, V.B. and Nauta, W.J.H. (1979) Brain Research 175, 191-217.
Brodie, M.S., and Dunwiddie, T.V. (1987) Brain Research 425, 106-113.
Calabresi P., Lacey, M.G., and North, R.A. (1989) Br J Pharmacol 98, 135-40.
Clarke, P.B.S., Hommer, D.W., Pert, A. and Skirboll, L.R. (1985) Br. J. Pharmacol. 85, 827-835.
Lichtensteiner, W., Hefti, F., Felix, D., Huwyler, T., Melamed, E. and Schlumpf, M. (1982) Neuropharmacology 21, 963-968.
Mereu, G., Yoon, K.W., Boi, V., Gessa, G.L., Naes, L., and Westfall, T.C. (1987) Eur J Pharmacol 141, 395-9.
Roberts, D.C.S., Corcoran, M.E., and Fibiger, H.C. (1977) Pharmacol. Biochem. Behav. 6, 615-620.
Roberts, D.C.S. and Koob, G.F. (1982) Pharmacol. Biochem. Behav. 17, 901-904.
Russell, M.A.H. and Feyerabend, C., Cigarette smoking: a dependence on high-nicotine boli. (1978) Drug Metab. Rev. 8, 29-57.
Spyraki, C., Fibiger, H.C., and Phillips, A.G. (1983) Psychopharm. 79, 278-283.
Wang, R.Y. (1981) Brain Research Reviews 3, 123-140.
Wise, R.A. (1978) Brain Research 152, 215-247.
Wise, R.A. (1987) Pharmac. Ther. 35, 227-263.

EFFECT OF NICOTINE AND SOME OTHER HEDONIC AGENTS ON SPONTANEOUS
ACTIVITY OF HIPPOCAMPUS NEURONES IN THE RAT

I. Jurna and H. Barkemeyer

Institut für Pharmakologie und Toxikologie der Universität des
Saarlandes, 6650 Homburg/Saar

The effects of nicotine and some other hedonic agents (ethanol,
morphine, caffeine, DL-amphetamine) on spontaneous activity of
hippocampus neurones in the rat have been studied. It was found
that only nicotine and ethanol exclusively increase the number
of impulses discharged from single neurones, while morphine,
caffeine, and DL-amphetamine increase or reduce the activity.

Neuronal Activity in the limbic system including the hippocampus
is associated with emotions and affective behaviour (Kupfermann,
1985) and may play a role in the development of hedonic drug
effects. It was therefore aimed to determine the effect of
nicotine on spontaneous activity of neurones in   the rat
hippocampus and to compare it with that of some other drugs
possessing hedonic properties, i.e. ethanol, morphine, caffeine,
and DL-amphetamine. Nicotine has frequently been reported to
excite hippocampal pyramidal cells (Storm-Mathisen, 1977), but
the doses employed were high, i.e. exceeding 1 mg/kg. Therefore,
doses were employed that corresponded to those computed from the
amount of nicotine retrieved in plasma after absorption from
cigarette smoke (Rotenberg et al., 1980).

METHODS
The experiments were carried out on rats of both sexes (Sprague-
Dawley/SIV; 250-300 g body weight) under urethane anaesthesia.
They received an intraperitoneal injection of urethane (1.2
g/kg) to induce and maintain anaesthesia for surgery and the
experiment properly. At the end of surgery, an additional
subcutaneous injection of urethane (120 mg/kg) was given. This
treatment warranted a sleeping time for more than 6 h. The
animals breathed spontaneously. Body temperature was monitored

in the rectum and kept between 37.5 and 38°C by radiant heat. A
cannula was inserted into a tail-vein for intravenous (i.v.)
injection. A hole was drilled into the skull on both sides at
sites where the electrodes were inserted. The co-ordinates for
recording from the hippocampus were AP: 2.5-3 mm; L: 2.8-3 mm;
V: 2.3-3.5 mm (according to the atlas of Fifková and Marsala,
1967). Activity was recorded extracellularly from single
neurones in the hippocampus with a tungsten microelectrode (tip
diameter 1 µm; resistance 10 MΩ) using a stereotaxic device. The
activity of single neurones was amplified, passed through a
window discriminator and evaluated with a computerized program.
The frequency of impulses discharged from the neurones was
integrated over periods of 1 min duration. When 5-6 recordings
showed stable baseline activity, the drugs were administered
intravenously and their effects were determined at 10 min
intervals till 60 min after one injection (morphine, caffeine,
DL-amphetamine), the last injection of a block of four
injections (nicotine) and after the beginning of an infusion
(ethanol). The data of 7 to 12 experiments were pooled for
statistical evaluation. Each drug was tested only once in each
animal.

RESULTS

Nicotine was given in blocks of four subsequent injections made
at intervals of 10 min. A block of injections lasted 30 min.
Recordings were made during this period immediately before each
injection and until 60 min after the last injection. The doses
employed were 0.01, 0.03, 0.05, and 0.1 mg/kg, the total
cumulative doses being 0.04, 0.12, 0.2 and 0.4 mg/kg. These
injections always caused an immediate and dose-dependent
increase in the number of impulse discharges, except at the
lowest dose which was ineffective (Fig. 1). Doses of 0.05 and
0.1 mg/kg produced nearly the same increase (i.e. by about 200%
of the controls), while the effect of the higher dose lasted
longer. It was significant until the end of the experiment at 60

min after the last injection of nicotine.

Ethanol was infused for a period of 20 min at concentrations of 0.1, 0.2, and 0.3%. This yielded total doses of 100, 200, and 300 mg/kg delivered at the end of infusion. The two high concentrations increased the activity in hippocampus neurones by about 90%, while the low concentration was ineffective (Fig. 2). The effect of the two high concentrations was significant till the end of the experiment at 40 min after stopping the infusion.

Morphine administered by single injections of 0.5, 1, and 2 mg/kg either increased or reduced spontaneous impulse discharges. The proportion of neurones being excited or depressed by morphine was the same at all doses (Fig. 3). Naloxone (0.2 mg/kg) antagonized the depressant but not the excitatory effect of morphine.

Caffeine injected at a dose of 4 mg/kg increased the activity in some hippocampus neurones and reduced it in others. As shown by the curves (Fig. 4), there may be a dose-response relationship of the two types of effects.

Likewise, DL-amphetamine administered at doses of 0.1 and 0.2 mg/kg either excited or depressed the neurones (Fig. 5). The maximum increase was produced by a dose of 0.05 mg/kg, so the dose-response relationship of the excitatory effect of DL-amphetamine follows a bell-shaped curve.

DISCUSSION

The results show that nicotine and other hedonic agents differ in their effects on spontaneous activity of hippocampus neurones in the rat. Nicotine and ethanol exclusively increase the number of impulses discharged from single neurones, while morphine, caffeine, and DL-amphetamine increase or reduce the activity. It is possible that these three latter drugs simultaneously excite and inhibit one and the same neurone by direct and indirect actions, and that one of these effects prevails out of as yet unknown reasons. This may account for the absence of a linear dose-dependence of the excitatory and inhibitory effects. It is conceivable that the difference in the effects of the agents tested reflect part of a pattern of neurones activity in the limbic system to which the hedonic action is related and that is particular to each drug. This assumption is supported by results recently obtained when studying the effects of nicotine and the other agents on spontaneous activity in other areas of the limbic system. The results summarized in Table 1 were obtained in a similar way as described above. The total number of neurones studied was 567. It is evident that the pattern of the effects caused by each drug on neuronal activity in various areas of the limbic system is different or, in other words, each drug gives its characteristic fingerprint. Only nicotine produced excitation in all areas. It is reasonable to assume that the difference in the actions of the hedonic agents in various limbic areas accounts for the well known disparity in their hedonic effects.

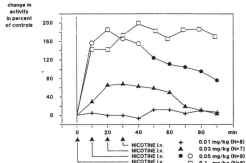

Figure 1   Time course of the effect of nicotine on spontaneous activity of single hippocampal neurones. Nicotine was injected in four subsequent doses as indicated by the arrows. Ordinate: increase in number of impulse discharges expressed in per cent of controls. Abscissa: time in min after the first injection. The points on the curves present the mean value of the number of determinations indicated in brackets beside the doses. Open symbols: values differ significantly from controls (P < 0.05 or less). Filled symbols: values do not differ significantly from the controls.

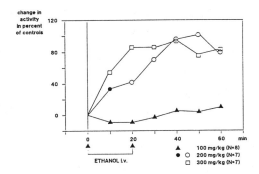

<u>Figure 2</u> Time course of the effect of ethanol on spontaneous activity of single hippocampal neurones. Ethanol was administered by infusion of three different concentrations giving the final concentrations of 100, 200, and 300 mg/kg at the end of a 20 min infusion. The begin and the end of infusion is indicated by arrows. Other details as in legend to Figure 1.

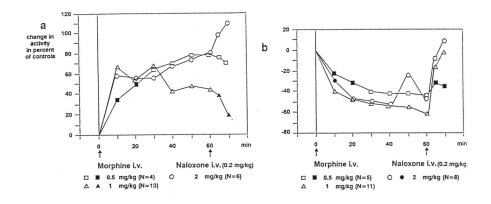

<u>Figure 3</u> Time course of the effects of morphine on spontaneous activity of single hippocampal neurones. The neurones of one group were excited by three different doses of morphine (a), those of the other group were inhibited (b). Naloxone was injected as indicated by the arrow. Other details as in legend to Figure 1.

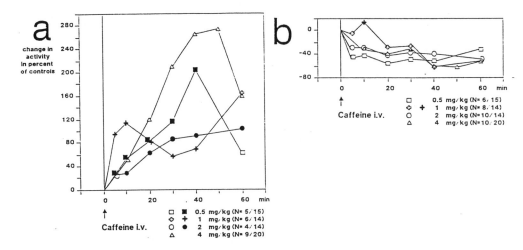

Figure 4 Time course of the effects of caffeine on spontaneous activity of single hippocampal neurones. The neurones of one group were excited by four different doses of caffeine (a), those of the other group were inhibited (b). Other details as in legend to Figure 1.

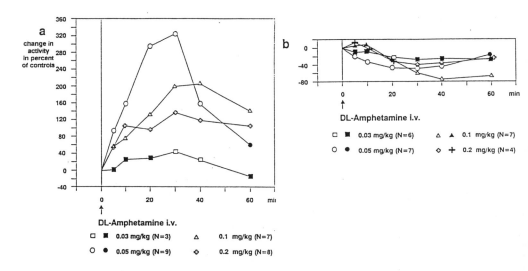

Figure 5 Time course of the effects of DL-amphetamine on spontaneous activity of single hippocampal neurones. The neurones of one group were excited by four different doses of DL-amphetamine (a), those of the other group were inhibited (b). Other details as in legend to Figure 1.

area \\ agent	hippocampus	accumbens septi	lateral amygdala	basal amygdala
nicotine	+	+	+	+
ethanol	+	+/-	+	+/-
morphine	+/-	+/-	-	-
caffeine	+/-	+	+	+
DL-ampheta-mine	+/-	+/-	+	+

+   increase in activity

-   decrease in activity

+/- increase in some neurones,
     decrease in others

Table 1   Changes in spontaneous activity of neurones in different areas of the limbic system caused by hedonic agents

REFERENCES
Fifková, E. and Marsala, J., Stereotaxic atlases for the cat, rabbit and rat. In: J. Bures, H.M. Petrán and J. Zachar (Eds.) Electrophysiological Methods in Biological Research, Academic Press New York and London, pp. 653-731 (1967)
Kupfermann, I., Hypothalamus and limbic system I: peptidergic neurons, homeostasis, and emotional behaviour. In: E.R. Kandel and J.H. Schwartz (Eds.) Principles of Neural Science, Elsevier, New York, Amsterdam and Oxford, pp. 609-625 (1985)
Rotenberg, K.S., R.P. Miller and J. Adir, Pharmacokinetics of nicotine in rats after single-cigarette smoke inhalation. J. Pharm. Sci. 69, 1087-1090 (1980)
Storm-Mathisen, J., Localization of transmitter candidates in the brain: the hippocampal formation as a model. Progr. Neurobiol. 8, 119-181 (1977)

V.

Behavioral Effects of Nicotine in Animals

THE ROLE OF THE FOREBRAIN CHOLINERGIC PROJECTION SYSTEM IN PERFORMANCE IN THE RADIAL-ARM MAZE IN MEMORY-IMPAIRED RATS.

H. Hodges, J.A. Gray, Y. Allen & J. Sinden

MRC Brain, Behaviour and Psychiatry Research Group, Department of Psychology, De Crespigny Park, Denmark Hill, London SE5 8AF, UK.

SUMMARY: After chronic administration of alcohol for 6 months, or ibotenate lesions of the cholinergic projection nuclei to the neocortex and hippocampus, rats showed enduring deficits in 4 different components of memory tested in the radial-arm maze: reference and working spatial and associative memory. Cell suspension transplants of foetal cholinergic-rich (C-R), but not cholinergic-poor (C-P), brain tissue, placed into neocortex and/or hippocampus, substantially reversed deficits in all 4 memory components over a period of 7-13 weeks after grafting. Both alcohol-treated (ALC) and lesioned (LES) animals showed reduced cortical and hippocampal levels of choline acetyltransferase (ChAT) activity. However, in rats with C-R but not C-P transplants, ChAT activity did not differ from control level. Rats were tested before grafting to determine their reaction to acute systemic administration of muscarinic and nicotinic receptor agonists and antagonists. Both ALC and LES animals showed increased sensitivity to the beneficial effects on cognitive performance of cholinergic agonists, and to the deleterious effects of antagonists. Effects of both muscarinic and nicotinic agents were more marked for working than reference memory. Nicotine was particularly effective in improving spatial working memory in LES rats. However, in LES rats tested after grafting with C-R tissue, this compound had the surprising effect of making performance worse, though it continued to reduce errors in animals with sham, or C-P grafts.

The forebrain cholinergic projection system (FCPS) has been assigned a key role in processes of memory and/or attention on the basis of converging evidence from human post mortem (PM) studies, for loss of cholinergic cells and markers in patients with severe memory deficits (Perry et al., 1978, Arendt et al., 1983); from effects of lesions to the FCPS in animals (Dunnett, 1985, Arendt et al., 1989); and from the marked impairments produced in man and animals by cholinergic antagonists, together with some evidence for improvements with agonists, in tests of learning, memory or attention (Warburton and Wesnes, 1984, Sahakian et al., 1989, Ridley et al., 1986). However none of this evidence is wholly convincing. PM studies are correlational, and memory deficits may reflect widespread damage that occurs in neurodegenerative diseases, rather than cholinergic dysfunction alone (Kopelman 1986). Lesion studies suffer from the lack of a neurotoxin specific to cholinergic cells, whilst pharmacological enhancement of cholinergic transmission has not proved clinically useful in demented patients. The case for a role for acetylcholine (ACh) in memory and/or attention has been strengthened by recent findings that in impaired aged (Dunnett et al., 1988), alcohol-treated (Arendt et al., 1988) or lesioned (Dunnett et al., 1985) animals, with low cortical and/or hippocampal levels of choline acetyl transferase (ChAT) activity, cholinergic-rich, but not cholinergic-poor, grafts of foetal neural tissue have promoted recovery of cognitive function. Thus transplants may be used not only to investigate cholinergic involvement in aspects of memory, but may also lead to the development of future treatment strategies for memory loss. We therefore investigated the effects of cholinergic-rich (C-R) transplants in two groups of rats which displayed long-lasting deficits in radial maze performance, following 28 weeks of alcohol ingestion (ALC), or ibotenate lesions (LES) to the nucleus basalis (NBM), medial septal and diagonal band (MS-DB) brain regions, at the source of the cortical and hippocampal branches of the FCPS, respectively. In addition we examined effects of nicotinic and muscarinic receptor agonists and antagonists in LES and ALC rats, as a method of challenging the functional integrity of cholinergic transmission, and to look for possibly different cognitive effects of these agents. We used the Jarrard (1986) version of the 8-arm radial maze to assess both long-term ('reference') and short-term ('working') memory in both spatial ('place') and associative ('cue') tasks.

MATERIALS AND METHODS

Behavioural training and testing: Male Sprague Dawley rats (80 for each experiment) were trained in 2 tasks (place and cue) to find sucrose at the end of 4 arms in 2 8-arm radial mazes, placed one above the other (upper 112 cm and lower 62 cm above floor level). For the place task (upper maze, with good visibility for cues around the room) the sucrose was always at the end of the same 4 arms, so the rat had to learn their location. For the cue task (lower maze) each arm contained a distinctively textured insert (e.g. sandpaper, carpet); 4 inserts were always rewarded but were moved to different arms after each trial, so that the rat had to learn the association between cue and reward, regardless of cue position. Within each task reference memory errors (first entry/trial into arms that were never rewarded) and working memory errors (re-entry into any arm within a trial) provided measures of failure in long and short term memory, respectively. ALC rats (Experiment 1) were trained for 12 weeks (40 trials/task), commencing a month after withdrawal form alcohol, to assess treatment effects on acquisition. Effects of cholinergic drugs were then tested on asymptotic performance, followed by grafting, and testing of transplant effects for 12 weeks. LES rats (Experiment 2) were pre-trained (60 trials/ task) and effects of lesion tested on asymptotic performance for 10 weeks, followed by testing with cholinergic drugs. Transplants were then given, and the time course of their effects followed for 14 weeks, before re-assessment of effects of cholinergic drugs.

Alcohol treatment and lesions: Rats were given alcohol (20% v/v in drinking water, sweetened with 8.75 mg/ml sucrose) for 23 h/day, and water for 1 h/day, for 28 weeks. Mean consumption was 20 ml/day (6.7-8.5 g/kg/day) and blood alcohol concentrations (BACs), measured enzymatically in the dark period during week 24, averaged 101 mg% (range 24-167 mg%). Treatment was initiated and withdrawn slowly by 2% changes in alcohol concentration /day; no withdrawal symptoms were seen. Controls received an isocaloric diet, including sucrose. For lesions ibotenic acid (10 ug/ul) was infused by cannula under stereotaxic control into 2 sites in NBM and 3 in MS-DB areas bilaterally, in anaesthetised rats; controls received vehicle (see Hodges et al., 1989).

Drug treatments: Groups of ALC, LES and control rats (n = 8-16) were injected i.p. (1 ml/kg) with a cholinergic agonist or antagonist agent 15 min before testing for comparison with baseline (BL) days, when saline vehicle was given, as follows:- (doses in mg/kg) nicotinic agents: nicotine (0.05, 0.1), mecamylamine (1.0, 2.0), hexamethonium (1.0); muscarinic agents: arecoline (0.5, 1.0), scopolamine (0.05, 0.1), n-methylscopolamine (0.1).

Transplants: C-R tissue was dissected from foetal basal forebrain (BF) or ventral mesencephalon (VM) at embryonic (E) day 15 or 18, and C-P tissue from hippocampus (HC) at E-16. Tissue was dissociated and suspended in trypsin (promoting survival of cholinergic, but not noadrenergic cells; Bjorklund et al., 1986), and implanted into 2 cortical and/or 2 hippocampal sites bilaterally (see Arendt et al., 1989, Hodges et al., 1990). Controls were sham operated. In Exp. 1 there were 5 groups (n = 8-9): those receiving C-R grafts (groups BF and VM), control C-P grafts (HC), and sham operated ALC and non-treated (CON) controls; all transplants were made into both cortex and hippocampus. In Exp. 2 C-R BF grafts were placed in cortex (AC), hippocampus (AH), or both sites (AC+H), control C-P hippocampal tissue grafted at both sites (HC+H), whilst LES and nonlesioned (CON) controls were sham operated, making 6 groups (n = 8-9) in all. For PM studies brains were bisected; 3-5 half brains/group were reserved for histology, in which alternating 30 um sections through graft and lesion sites were stained for nissl, cholinesterase (AChE), tyrosine hydroxylase (TH) and dopamine-B-hydroxylase (DBH). The remaining brains were used to assess ChAT activity (Fonnum, 1975), or amines and metabolites by HPLC (see Arendt et al., 1989).

RESULTS

Effects of alcohol treatment and lesions: Alcohol treatment significantly increased all 4 types of error during acquisition in the radial maze (F [2,74] = 45.07, p< .0001). Rats were divided retrospectively into groups below (LO ALC) and above (HI ALC) the median BAC of 100 mg%. Fig. 1 shows that place task error rates were higher in the HI ALC than the LO ALC group, and that the difference became more marked as training progressed; similar results were obtained in the cue

Effects of Alcohol on radial maze acquisition, Place Task

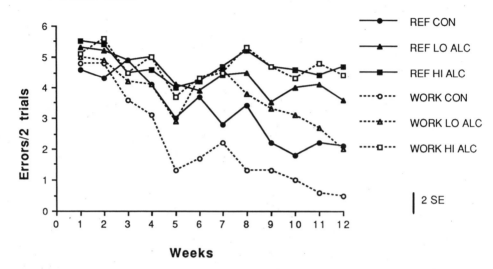

Fig. 1: Mean Place task reference (REF) and working (WORK) memory errors/2 trials in controls (CON) and alcohol-treated rats with BACs below (LO ALC) or above (HI ALC) 100 mg%. Training began 1 month after withrawal from alcohol.

Effects of lesion on radial maze performance, Place Task

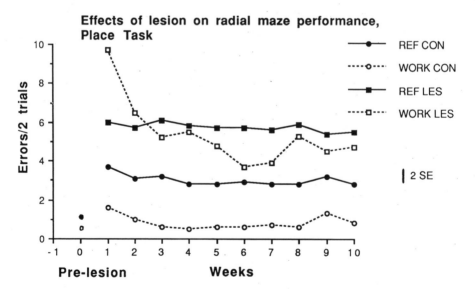

Fig. 2: Mean Place task reference (REF) and working (WORK) memory errors/2 trials in pre-trained control (CON) and lesioned (LES) rats over 10 weeks of post-lesion testing. Data from Hodges et al. (1990).

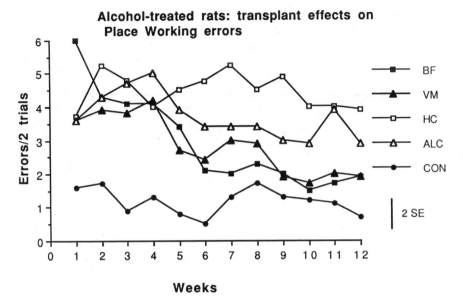

Fig. 3: Mean Place task working memory errors/2 trials in controls (CON) and alcohol-treated rats receiving cholinergic-rich (BF, VM), control (HC), or sham (ALC) grafts.

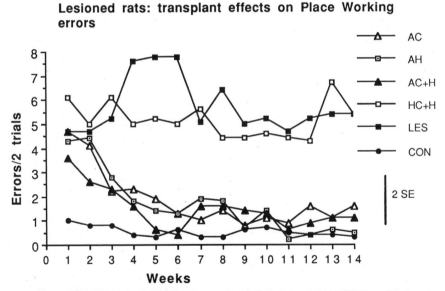

Fig. 4: Mean Place task working memory errors/2 trials in controls (CON) and lesioned rats receiving cholinergic-rich (AC, AH, AC+H), control (HC+H) or sham (LES) grafts. Data from Hodges et al., (1990).

task. LES rats (see Fig. 2 for place task errors) were even more impaired in all 4 aspects of memory than ALC rats (F [1,78]=107.76, p< .0001), with high stable error rates throughout post lesion testing.

Effects of transplants in alcohol-treated and lesioned rats: In Exp. 1 rats receiving C-R grafts (groups BF and VM) showed a marked trend of improvement over trials in all 4 aspects of memory, reaching control (CON) levels by weeks 7-9 (8-10 weeks after transplantation). As Fig. 3 shows for place working memory, no consistent improvement was shown in the control transplant and alcohol groups (HC, ALC), so that as well as a significant difference between groups (F [4,42] = 12.53, p< .0001), there was a marked interaction between groups and the linear trend of trials (F [4,42] = 10.07, p< .0001). In Exp. 2 the 3 groups of lesioned rats with C-R transplants to cortex, hippocampus and both areas (AC, AH, AC+H) also showed substantial improvement over trials, particularly in working memory (Fig. 4), relative to lesion and control transplant groups (LES, HC+H), with a time course of recovery comparable to that found in ALC rats. Thus there were marked differences between groups (F [5,44] = 24.59, p< .0001), and an interaction between groups and the linear trend of trials (F [5.44] = 3.75, p< .01). However there were no interactions between site of transplant and type of memory error; all 3 transplant groups performed similarly. Table I shows ChAT activity in the groups from Experiments 1 and 2. It can be seen that activity was significantly reduced in ALC and LES groups, whereas in rats with C-R transplants it was elevated to, or above, control level, commensurate with site of graft. HPLC assays of groups in Exp. 1 (data not shown) found significant reduction in forebrain noradrenaline (NA) content in all ALC animals, regardless of type of transplant, and significant increases in cortical dopamine (DA) in all groups receiving transplants, of any type.

Table I: Post-transplant choline acetyltransferase activity in alcohol-treated and lesioned rats (Mean activity/group as p moles/min/mg wet weight, in 3 brain regions: frontal cortex: FC, dorsolateral cortex: IC, hippocampus: HC; difference from control (CON): * p< .05, ** p< .025, *** p< .01)

GROUPS	FC	IC	H
Experiment 1:			
VM	16.3 ± 7.6	10.5 ± 3.2	15.5 ± 4.2
BF	17.9 ± 4.7*	12.6 ± 3.5*	17.5 ± 4.5
HC	11.3 ± 1.8*	9.6 ± 1.5*	13.6 ± 0.9**
ALC	10.0 ± 1.6**	8.9 ± 0.7*	14.6 ± 1.1*
CON	14.4 ± 1.8	11.0 ± 1.5	17.1 ± 0.9
Experiment 2:	FC	IC	H
AC	11.7 ± 2.0	7.7 ± 1.3	12.3 ± 3.2**
AH	7.2 ± 2.3***	5.4 ± 1.2***	15.2 ± 2.2
AC+H	11.6 ± 2.2	9.2 ± 2.2*	14.7 ± 0.9
HC+H	6.9 ± 1.7***	4.9 ± 1.0***	10.7 ± 1.5***
LES	6.6 ± 1.3***	5.4 ± 1.4***	11.1 ± 0.9***
CON	11.0 ± 1.6	7.9 ± 1.2	14.5 ± 1.1

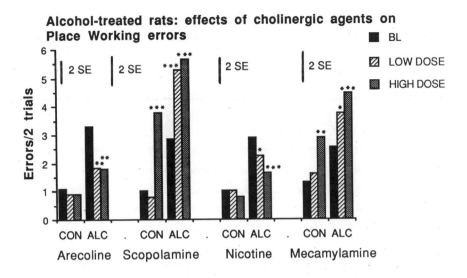

Fig. 5: Mean Place task working memory errors/2 trials in control and alcohol-treated rats (LO and HI ALC scores pooled) after treatment with cholinergic drugs, (low & high doses as in text). Difference from BL: *p< .05, **p< .025, ***p< .01.

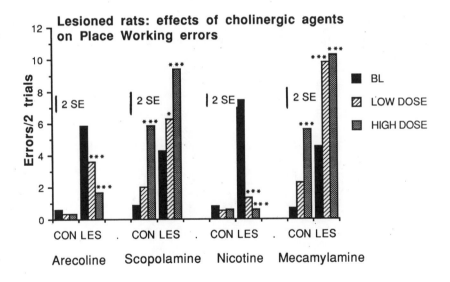

Fig. 6: Mean Place task working memory errors/2 trials in control and lesioned rats after treatment with cholinergic drugs. Legend as for Fig. 5.

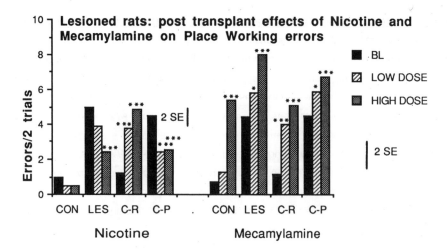

Fig. 7: Mean Place task working memory errors/2 trials after nicotine and
mecamylamine (see text for doses) in controls, and lesioned rats receiving
cholinergic-rich (C-R: pooled scores for AC, AH & AC+H groups), -poor
(C-P) and sham (LES) grafts.

Difference from BL: * < .05, ** < .025, *** < .01

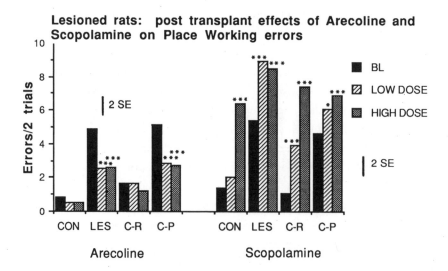

Fig. 8: Mean Place task working memory errors/2 trials after arecoline
and scopolamine in controls, and lesioned rats receiving grafts.
Legend as for Fig. 7.

<u>Effects of cholinergic receptor ligands in alcohol-treated and lesioned rats</u>: Rats in Exp. 1 and 2 showed a very similar pattern of response to cholinergic treatments (Figs. 5 and 6). Working memory error rates in impaired rats fell after treatment with arecoline and nicotine, but controls were not affected. Conversely working memory errors were substantially increased in response to scopolamine and mecamylamine, and this was found with the lower dose which did not significantly disrupt the performance of control rats. A similar pattern of results was found with the cue task (data not shown). In contrast to its marked ameliorating effect in LES rats, nicotine, given after transplantation (Fig. 7), impaired working memory in animals with C-R transplants, whereas arecoline was without effect in these rats (Fig. 8). Both agonists continued to improve performance in rats with sham (LES) or C-P transplants. The antagonists mecamylamine (Fig. 7) and scopolamine (Fig. 8) produced low dose impairment in formerly lesioned animals, regardless of transplant, and this effect was very marked in rats with C-R transplants.

DISCUSSION

The results show that both excitotoxic lesions to the FCPS, and chronic alcohol treatment produced long-lasting deficits in the radial maze, and reduced cortical and hippocampal ChAT activity, in line with the findings of Arendt et al. (1988, 1989). Several aspects of our results suggest that there may be a causal relationship between this cognitive impairment, and cholinergic damage. Firstly radial maze deficits were ameliorated by cholinergic-rich but not -poor foetal grafts, and these also elevated ChAT activity. Since C-R transplants from two different regions (BF and VM) were effective in ALC rats, our findings would be consistent with the participation of cholinergic cells, or other closely associated factors (Hefti, 1986) in functional recovery, since the two grafts differed with respect to other cell types. For instance TH staining was far more pronounced in VM than BF grafts. Moreover, though effects on other transmitters might contribute cognitive deficits, the recovery of function under transplants in Exp. 1 did not appear to be related to effects mediated by NA, which was reduced in all rats, regardless of transplant; by DA, which was abnormally elevated in all grafted rats, whether they showed behavioural recovery or not (this may reflect a reaction in the host cortex); or by serotonin (5-HT), which was not much affected by alcohol treatment or grafts.

Secondly ALC and LES animals showed a remarkably similar bi-directional response to the effects of cholinergic receptor ligands, which appeared to be of central origin, since the peripherally acting antagonists hexamethonium and n-methylscopolamine were without effect (data not shown). Both groups of rats demonstrated enhanced sensitivity to the cholinergic agents, since error rates were affected at low doses which did not alter performance in controls. Thus pharmacological challenge appears to be a useful method of testing the functional integrity of cholinergic transmission, in line with findings from lesion studies in animals (Ridley et al., 1986), and with the more marked effects of nicotine in Alzheimer patients, than in age-matched controls (Sahakian et al., 1989). Receptor binding studies indicate a loss of nicotinic binding sites in cortex and hippocampus of Alzheimer patients, which may involve pre- rather than post-synaptic receptors (Perry et al., 1987), and a less marked loss of pre-synaptic M2 muscarinic receptors, which is paralleled in lesioned rats (Mash et al., 1985). Thus behavioural sensitivity to low doses of cholinergic drugs may involve deficient pre-synaptic regulation of ACh release, so that post-synaptic drug action is untramelled. Even though tranplants promoted behavioural recovery, they did not normalise response to cholinergic agents. Indeed rats with C-R transplants showed greater impairment with the antagonists, than control or sham transplant groups. Paradoxicallly nicotine given after transplantation impaired performance in rats with C-R transplants, and thus may be a useful probe to assess transplant growth in the living animal. Since high doses of nicotine have been found to disrupt radial maze performance in normal rats (Mundy and Iwamoto, 1988), nicotine may have acted to produce a 'high dose' effect in these animals. If so, our findings suggest that an additive interaction may occur between host and transplant cholinergic function.

There were no interactions between our manipulations and place and cue tasks, suggesting that the different types of toxic damage, or types of receptor ligand did not exert differential effects on spatial or associative information processing. However drug and transplant effects were more marked on working than reference memory. Since working memory requires assimilation of recent information, it would be likely to engage attentional processes to a greater extent than reference memory, where attention is required only for registration of pertinent spatial/nonspatial cues. Thus pronounced effects on working memory would be consistent with a role for ACh in attention and rapid information processing (Warburton and Wesnes, 1984). Work with more explicit attentional paradigms (latent inhibition, Lubow, 1973; continuous nonmatching to sample, Pontecorvo, 1983) is in progress to assess the part played by ACh and nicotine in attention. The present experiments indicate that the FCPS is critically involved in the cognitive processes underlying learning and performance in the radial-arm maze, but further work is needed to clarify this role. They also lend support to our clinical observations (Sahakian et al., 1989) that nicotine may play a useful role in the treatment of cognitive dysfunction.

Acknowledgements: This work was supported by the Wellcome Trust, the UK Medical Research Council, and the R.J. Reynolds Tobacco Company. We wish to thank Tim Kershaw, Sanjay Patel and Peter Sowinski, Psychology Department, Institute of Psychiatry, for their assistance.

REFERENCES

Arendt,T., Bigl, V., Arendt, A., Tennestedt, A. (1983) Loss of neurons in the nucleus basalis of Meynert in Alzheimer's disease, paralysis agitans and Korsakoff's disease. Acta Neuropathol. 61, 101-108.

Arendt, T., Allen, Y., Sinden, J., Schugens, M.M., Marchbanks, R.M., Lantos, P., Gray, J.A. (1988) Cholinergic-rich brain transplants reverse alcohol-induced memory deficits. Nature 332, 448-450.

Arendt, T., Allen, Y., Marchbanks, R.M., Schugens, M.M., Sinden, J., Lantos, P.L., Gray, J.A. (1989) Cholinergic system and memory in the rat: effects of chronic ethanol, embryonic basal forebrain transplants, and excitotoxic lesions of the cholinergic basal forebrain projection system. Neurosci. 33, 435-462.

Bjorklund, A., Nornes, H., Gage, F.H. (1986) Cell suspension grafts of noradrenergic locus coeruleus neurons in rat hippocampus and spinal chord: reinnervation and transmitter turnover. Neurosci. 18, 685-698.

Dunnett, S.B. (1985) Comparative effects of cholinergic drugs and lesions of the nucleus basalis or fimbria fornix on delayed matching in rats. Psychopharmacol. 87, 357-363.

Dunnett, S.B., Toniolo, G., Fine, A., Ryan, C.N., Bjorklund, A., Iversen S.D. 1985) Transplantation of embryonic ventral forebrain neurons into the neocortex of rats with lesions of the nucleus basalis magnocellularis II: sensorimotor and learning impairments. Neurosci. 16, 787-797.

Dunnett, S.B., Badman, F., Rogers, D.C., Evenden, J.I., Iversen, S.D. (1988) Cholinergic grafts in the neocortex or hippocampus of aged rats: reduction of delay dependent deficits in the delayed nonmatching to position task. Exp. Neurol. 102, 57-64.

Fonnum, F. (1975) A rapid radiochemical method for the determination of choline acetyltransferase. J. Neurochem. 24, 407-409.

Hefti, F. (1986) Nerve growth factor promotes survival of septal cholinergic neurons after fimbrial transections. J. Neurosci. 6, 2115-2162.

Hodges, H., Thrasher, S., Gray, J.A. (1989) Improved radial maze performance induced by the benzodiazepine antagonist ZK 93 426 in cholinergic lesioned and alcohol-treated rats. Behav. Pharmacol. 1, 44-54.

Hodges, H., Allen, Y., Sinden, J., Lantos, P., Gray, J.A. (1990) Cholinergic-rich foetal transplants improve cognitive function in lesioned rats, but exacerbate response to cholinergic drugs. Prog. Brain Res. 82, 345-358.

Jarrard, L.E. (1986) In: The hippocampus (K.H. Pribram and R.L. Isaacson, Eds), Vol 4, Plenum Press, New York, pp 93-126.

Kopelman, M.D. (1986) The cholinergic neurotransmitter system in human memory and dementia: a review. Q.J.E.P. 38A, 535-573.

Lubow, R.E. (1973) Latent Inhibition. Psychol. Bull. 79, 398-407.

Mash D.C., Flynn, D.D., Potter, L.T. (1985) Loss of M2 muscarinic receptors in Alzheimer's disease and experimental cholinergic denervation. Science 228, 1115-1117.

Mundy, W.R., Iwamoto E.T. (1988) Nicotine impairs acquisition of radial maze performance in rats. Pharmacol. Biochem. Behav. 30, 119-122.

Perry, E.K., Tomlinson, B.E., Blessed, G., Bergman, K., Gibson, P.H., Perry, R.H. (1978) Correlation of cholinergic abnormalities with senile plaques and mental test scores in senile dementia. Brit. Med. J. 2, 1457-1459.

Perry, E.K., Perry, R.H., Smith, C.J., Dick, D.J., Candy, J.M., Edwardson, J.A., Fairburn, A., Blessed, G. (1987) Nicotinic receptor abnormalities in Alzheimer's and Parkinson's diseases. J. Neurol. Neurosurg. Psychiat. 50, 806-809.

Pontecorvo, M.J. (1983) Effects of proactive interference on rats' continuous nonmatching to sample. Animal Learning and Behav. 11, 356-365.

Ridley, R.M., Murray, T.K., Johnson, J.A., Baker, H.F. (1986) Learning impairment following lesions of the nucleus basalis of Meynert in the marmoset: modification by cholinergic drugs. Brain Res. 376, 108-116.

Sahakian, B., Jones, G., Levy, R., Gray, J.A., Warburton, D. (1989) The effects of nicotine on attention, information processing and short term memory in patients with senile dementia of the Alzheimer type. Brit. J. Psychiat. 154, 797-800.

Warburton, D.M., Wesnes, K. (1984) Drugs as research tools in psychology: cholinergic drugs and information processing. Neuropsychobiol. 11, 121-132.

ALTERATIONS IN SEPTOHIPPOCAMPAL CHOLINERGIC RECEPTORS AND RELATED
BEHAVIOR AFTER PRENATAL EXPOSURE TO NICOTINE.

Joseph Yanai, Chaim G. Pick, Yael Rogel-Fuchs and Eias A.
Zahalka. The Melvin A. and Eleanor Ross Laboratory for Studies in
Neural Birth Defects, Department of Anatomy and Embryology, The
Hebrew University-Hadassah Medical School, Box 1172, 91010
Jerusalem, Israel

SUMMARY: The present study applied the knowledge accumulated in
the model for phenobarbital neuroteratogenicity on nicotine. Mice
were exposed to nicotine prenatally by injecting the mother 1.5
mg/kg nicotine s.c. twice daily on gestation days 9-18. On age 50
days, the mice were tested in the eight-arm maze and their
hippocampi were assayed for muscarinic receptor binding using ^{3}H-
QNB as a ligand. Hippocampal muscarinic receptors B_{max} of
nicotine exposed mice was 58% higher than control (p < 0.05). On
the other hand, K_d was unaffected by prenatal nicotine exposure.
Nicotine-exposed mice made 26% more errors in the maze than
control (p < 0.01). The study suggests that nicotine administered
to the fetus alters septohippocampal chemistry and induces
deficits in hippocampus related behavior. The possible lineal
relationship between these two changes is the subject of our
current investigations.

Nicotine is a known teratogen drug both in human and animals.
Further studies established the nicotine as a neurobehavioral
teratogen as well (Brown and Fishman, 1984; Fried, 1989). In
animal model, prenatal exposure to nicotine induced long-lasting
alterations in various neurotransmitter innervation and in
several behavioral abilities (Brown and Fishman, 1984; Slotkin et
al., 1987; Lichtensteiger et al., 1988).

It appears that nicotine, like many other neuroteratogens,
alters numerous neural systems resulting in multiple behavioral
deficits; thus, making it difficult to ascertain the mechanism of
its neuroteratogenicity. In our model of phenobarbital
neuroteratogenicity we attempted to encounter this methodological
problem by studying alterations related to a specific brain
region and a specific biochemical process; i. e., the deficits in
the septohippocampal cholinergic pathways and their related

behaviors. Early exposure to phenobarbital induced in mice an increase in B_{max} of the hippocampal muscarinic receptors and in their second messenger, inositol phosphate (Yanai et al, 1990). On the other hand, marked presynaptic alterations could not yet be demonstrated.   Behavioral studies revealed concomitant deficits in hippocampus related behaviors, including eight-arm maze performance (Yanai and Pick 1988).

Previous studies already suggested that prenatal exposure to nicotine alters brain nicotine receptors (Slotkin et al., 1987) although results to the contrary were also published (Fung and Lau 1989). Consequently, it now became feasible to apply the early phenobarbital model on nicotine in order to gain an understanding on the mechanism by which this drug exerts its neuroteratogenic action. In the present study, pregnant mice were exposed to nicotine and the hippocampus related eight-arm maze performance of their offspring was studied. Parallel experiments ascertained the characteristics of the offspring hippocampal muscarinic receptors.

MATERIALS AND METHODS

General: Pregnant mice were exposed to nicotine on gestation days 9 to 18. The offspring and their respective controls were tested in the eight-arm maze beginning on age 50 days. On age 60 days, their hippocampi were assayed for muscarinic receptor binding.

Nicotine administration: Pregnant females, as evidenced by the existence of a vaginal plug, received subcutaneous injections of 1.5 mg/kg nicotine in saline solution twice daily on gestation days 9 to 18; control females received vehicle injections.

Eight-arm maze test: The test procedure in the radial eight-arm maze is described in detail elsewhere (Yanai and Pick, 1988). Mice were placed on a regimen of water deprivation that consisted of the administration of water for 30-minutes once a day. Then, they were introduced individually into the maze for 10 minutes of habituation, still without water reinforcement. The first 16 entries were recorded. In the following five testing days, the mice received water reinforcement, and were left in the maze until they had either entered all the eight arms or until they had made 16 entries. The number of trials needed to enter all arms through the 16 trials was recorded.

Muscarinic Receptor Binding: The saturation binding experiment was performed according to Yamamura and Snyder (1974). Briefly, to assay specific binding, 50μl of homogenized supernatant was incubated in varying concentrations of ^{3}H-QNB plus 0.05M Na-K phosphate buffer, pH=7.4. The identical experiment was performed with the addition of 10^{-6} atropine. At the final stage the contents passed through filters positioned over a vacuum, and the filters containing the samples were counted in a Scintillation Spectometer. Results were plotted according to Scatchard (1949). Protein was measured by the procedure of Bradford (1976).

RESULTS

Animals who were exposed to nicotine prenatally had a 58% higher specific ^{3}H-QNB binding in the hippocampus as compared to controls (p < 0.05, Table I). The alterations were not accompanied by significant changes in the dissociation constant (K_d).

Animals from both treatment groups improved their performance significantly (p < 0.01) in the eight-arm maze during the six days test period. That is, they needed fewer entries in order to visit all eight arms (Table II). However, the nicotine-exposed offspring needed more entries in order to visit all arms (p < 0.01) and the gap between groups increased up to 26% during the test period.

DISCUSSION

In the present study, prenatal exposure to nicotine induced in mice an increase in the number of hippocampal muscarinic cholinergic receptors and in the septohippocampal cholinergic related behavior, the performance in eight-arm maze.

Table I

The calculated muscarinic receptors B_{max} (pmol/mg protein, Mean±S.E.M.) and K_d (nM) in the hippocampus for 50 day old mice after prenatal exposure to nicotine.

TREATMENT	$\underline{B_{max}}$	$\underline{K_d}$
Control	1.12±0.21	1.42±0.41
Prenatal Nicotine	1.77±0.20	1.34±0.33

N = 48 animals (6 Scatchard plots) for the control group and 24 animals for each treatment group. p <0.05 for the differences in B_{max} between control and nicotine (ANOVA)

Table II

Number of entries needed to visit all eight arms in control and in animals after prenatal exposure to nicotine (mean±S.E.M).

Treatment (N)	DAYS					
	1	2	3	4	5	6
Control (12)	12.8±0.5	11.7±0.7	10.0±0.5	9.5±0.4	8.9±0.3	8.7±0.2
Nicotine (12)	13.9±0.6	13.2±0.8	12.3±0.8 *	11.7±0.6 *	11.2±0.6 **	10.7±0.6 *

Sample sizes are indicated in parentheses.
* P < 0,05,   ** P < 0.01, for the difference from control levels.

The deficits in eight-arm maze behavior indirectly suggests alterations in the septohippocampal cholinergic pathways, since the central role of the hippocampus and its afferents and efferents in this behavior is well established (Olton et al., 1978). Animals with hippocampal lesions performed poorly in the eight-arm maze, with no evidence of functional recovery (Olton, 1979). Thus the behavioral results are in line with, and confirm the alterations found in the muscarinic receptors.

In our phenobarbital model described above, the findings on alterations in the septohippocampal cholinergic pathways and their related behaviors made it possible to attempt to reverse the behavioral deficits by neuronal grafting. Transplantation of embryonic septal cholinergic cells, but not noradrenergic cells, into the early phenobarbital impaired hippocampus almost completely reversed the deficits in eight arm maze performance (Yanai and Pick, 1988) and the concomitant alterations in the muscarinic receptors number (manuscript in preparation). In studies on the regulating end, the septohippocampal cholinergic innervations were disinhibited by destroying the A10-septal dopaminergic pathways. The treatment ameliorated the impaired behavior of the drug-exposed animals (Yanai et al, 1989).

For the variables so far studied, nicotine, despite its great difference from phenobarbital in structure and action on the CNS, fit the model established in the phenobarbital studies. That is, it increased the B_{max} of the muscarinic receptors in the septohippocampal cholinergic pathways and caused deficits in hippocampal behavior. Whether this similarity of action between the two drugs remains true for other components of the septohippocampal cholinergic chemistry and whether these alterations can be corrected by neural transplantation and lesions of the regulation system remains the subject of our current investigation.

REFERENCES

Bradford, M.M. (1976) Analyt. Biochem. 72, 248-254.
Brown, M., Fishman, R.H.B. (1984) In: Neurobehavioral Teratology. (J. Yanai ED), Elsevier/Holland, pp. 3-54.
Fried, P.A. (1989) Neurotoxicol. 10, 577-584.
Fung, Y.K. and Lau Y.S. (1989) Pharmacol. Biochem. Behav. 33, 1-6.
Lichtensteiger, W., Ribary, U., Schlumpf, M., Odermatt, B., Widmer, R. (1988) Prog. Brain Res. 73, 137-157.
Olton, D.S. (1979). In: Functions of the septo-hippocampal system. Ciba Foundation Symposium, Elsevier, New York. pp. 327-349.
Olton, D.S., Walker, J.A., Gage, F.H. (1978) Brain Res. 139, 295-308.

Scatchard, G. (1949) Ann. N. Y. Acad. Sci. 51, 660-672.

Slotkin, T, Orband-Miller, L., Queen, K. (1987) J. Pharmacol. Exper. Ther. 242, 232-237.

Yamamura, H. I. and Snyder, S. H. (1974) Proc. Natl. Acad. Sci., USA. 71, 1725-1729.

Yanai, J., and C. G. Pick. Int. J. Dev. Neurosci. (1988) 6, 409-416.

Yanai, J., Laxer, U., Pick, C. G. and Trombka, D. Devl. Brain. Res. (1989) 48, 255-261.

Yanai, J., Newman, M.E.,Pick, C.G., Rogel-fuchs, Y., Trombka, D., Zahalka E.A. Paper presented at the 15th Ann.Meeting of the Behavioral Teratology Society (1990) Teratology, in press.

STUDIES ON THE ROLE OF MESOLIMBIC DOPAMINE IN BEHAVIOURAL RESPONSES TO CHRONIC NICOTINE

D.J.K. Balfour, M.E.M. Benwell and A.L. Vale

Department of Pharmacology and Clinical Pharmacology, University Medical School, Ninewells Hospital, Dundee, DD1 9SY, Scotland

SUMMARY: Three procedures, *in vivo* microdialysis in conscious freely moving rats; 6-hydroxydopamine lesions of the nucleus accumbens and operant avoidance conditioning, have been used to investigate the possible role of the mesolimbic dopamine (DA) system in the behavioural responses to chronic nicotine (0.4 mg/kg). The dialysis studies showed that pretreatment with nicotine for 5 days enhanced its ability to stimulate DA secretion in the nucleus accumbens and that this effect corresponded with an increased locomotor response to the drug. Since the lesion studies revealed that the stimulatory effects of chronic nicotine on locomotor activity are abolished by lesions of the mesolimbic system, the data appear consistent with the hypothesis that the increase in locomotor activity evoked by pretreatment with nicotine may be related to the enhanced mesolimbic DA response to nicotine. The results of the operant conditioning study showed that the effects of d-amphetamine (0.5 mg/kg) on the acquisition and performance of an unsignalled shock avoidance task were very similar to those of nicotine and that, as is the case for nicotine, amphetamine-withdrawal caused a disruption of avoidance performance. Since, in addition, d-amphetamine was able to attenuate the effects of nicotine-withdrawal, the data suggest that, in this avoidance model of nicotine dependence at least, the animals may become dependent upon increased levels of DA secretion evoked by the drugs in order to perform the task efficiently.

INTRODUCTION

There is good evidence to suggest that many of the behavioural responses to psychomotor stimulant drugs are related to their ability to stimulate dopamine (DA) secretion in the mesolimbic system of the brain. In particular, this system appears to mediate the locomotor stimulant properties of these agents and their ability to act as rewards in drug self-administration schedules (Wise and Bozarth 1987; Wise 1990). This observation has led Wise and Bozarth (1987) to suggest that the mesolimbic DA system forms a principal component of the reward systems of the brain and that the effects of amphetamine and cocaine on the

system are so powerful that they, in themselves, account for the development of
dependence upon these drugs. There is evidence that nicotine also stimulates DA
secretion in the mesolimbic system and that, as for the other psychostimulants,
this effect mediates the ability of the compound to act as a locomotor stimulant
(Imperato et al 1986; Clarke et al 1988) and a substrate in drug self-
administration schedules (Singer et al 1982). However, when compared with
amphetamine or cocaine, nicotine is a relatively weak reinforcer in drug self-
administration schedules (see Balfour 1984; Clarke 1987) and it seems likely that
other factors contribute to the rewarding properties of the drug. In particular
there is evidence to suggest that the locomotor stimulant response to nicotine is
enhanced in animals which have been pretreated with the drug prior to the test
day (Morrison and Stephenson 1972; Stolerman et al 1973) and that nicotine
dependence may develop more readily in aversive environments (Morrison 1974;
Gilbert 1979). The purpose of the studies reported here was to examine the
hypothesis that these effects of nicotine could also be attributed to its ability to
stimulate DA secretion.

MATERIALS AND METHODS

In vivo microdialysis studies: These experiments were performed on Sprague-
Dawley rats bred in the Animal Services Unit, Ninewells Hospital and Medical
School, from stock purchased from Interforna Ltd and weighed approximately 300g
at the beginning of the experiment. The animals were pretreated for 5 days with
daily injections of saline (1ml/kg SC) or nicotine (0.4mg/kg SC). Two hours after
the last injection, they were anaesthetised with avertin and dialysis loops
composed of 3mm of cuprophane tubing were located stereotaxically in the rostral
caudal plane in the nucleus accumbens and cemented into place. The dimensions
of the loop were 1.5mm in the vertical plane and 0.2mm in the lateral plane
with a cross section of approximately 0.5mm in the rostral caudal plane. Eighteen
hours later the animals were transferred to an activity box and the loops were
dialysed with a Ringer solution at a rate of 1.7μl per minute. Samples of
dialysate were collected every 20 minutes and analysed for DA and its principal
metabolites, dihydroxyphenylacetic acid (DOPAC) and homovanillic acid (HVA)
using HPLC with electrochemical detection. The activity of the rats in the box
was monitored using 4 infrared photobeams mounted along two sides of the box
(Vale and Balfour 1989). The biochemical and behavioural data collected during

the first 60 minutes of the session were used to establish the baseline activity. The animals were then given subcutaneous injections of nicotine or saline and the measurements were continued for a further 120 minutes.

Lesion studies: These experiments were also performed using male Sprague-Dawley rats bred in the Animal Services Unit at Ninewells Hospital which weighed approximately 300g at the beginning of the experiment. Lesions of the mesolimbic DA system were produced by injecting 6-hydroxydopamine bilaterally into the nucleus accumbens using a procedure similar to that described by Clarke et al (1988). Control animals received microinjections of the vehicle (0.05 percent ascorbic acid in saline). Starting 7 days after surgery, the rats were given 6 daily injections of nicotine (0.4mg/kg) to make them tolerant to the depressant effects on behaviour seen in animals treated acutely with nicotine (Clarke and Kumar 1983). The animals were then challenged at three day intervals with subcutaneous injections of saline, nicotine (0.1 and 0.4mg/kg) and d-amphetamine (0.5mg/kg), administered using a counter-balanced design. The locomotor activity of the rats was measured in an activity box (Vale and Balfour 1989) for 20 minutes starting 3 minutes after the injections of nicotine and 30 minutes after the injections of d- amphetamine. Twenty four hours after the last trial, the animals were killed by cervical dislocation and the concentrations of DA and DOPAC in the nucleus accumbens were measured using HPLC with electrochemical detection.

Shock avoidance studies: Male Wistar rats (Charles River UK Ltd) weighing approximately 250g at the beginning of the experiment were used. The rats were trained on an unsignalled shock avoidance task using the schedule described by Morrison (1974). In the first series of experiments the rats received subcutaneous injections of saline or nicotine (0.4mg/kg) three minutes prior to the beginning of each training session. In the second series the rats were given injections of saline or d-amphetamine (0.5mg/kg) 30 minutes prior to each training session. In both experiments the training was continued until the rats avoided 75 percent of the total number of shocks they could have received. At this point they were transferred to the test schedule in which shocks were delivered every 30 seconds. Each lever-pressing response delayed the next shock for 30 seconds. During the next three weeks the animals received the drug used in training prior to each test session except for day 4 of each week when the animals in each training groups were tested after an injection of saline, nicotine or d-amphetamine. In order to avoid sequence effects, these drugs were administered using a counter-balanced design.

Drugs: Nicotine hydrogen tartrate and d-amphetamine sulphate were purchased from Sigma. In each case the drug concentration has been expressed in terms of the free base and, where necessary, the pH of the drug solutions was corrected

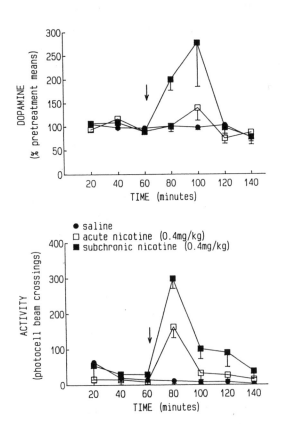

Figure 1: The effects of acute and subchronic nicotine administration on mesolimbic dopamine (DA) and locomotor activity. The concentrations of DA in the dialysate samples are expressed as the percentage ± SEM of the mean levels measured in the dialysates collected during the 60 minute control period prior to the injection of nicotine (0.4mg/kg) or saline (controls) at the point shown by the arrow. The activity data are expressed as means ± SEM of the number of photobeam crossings recorded during each 20 minute sampling period.

to 7 prior to its administration.

RESULTS

In our hands the acute administration of nicotine had no significant effect on the extracellular levels of DA in the nucleus accumbens (Fig. 1) although the drug did increase (P<0.05) the concentration of DOPAC (Fig. 2). After 60 minutes in the activity box the basal level of locomotor activity was very low and this was increased significantly (P<0.01) by the acute administration of nicotine (Fig. 1). Subchronic injections of nicotine not only increased extracellular levels of DOPAC (P<0.01) in the nucleus accumbens but also evoked significant increases (P<0.01) in the concentrations of DA and HVA (Fig. 2). The effects of subchronic injections of nicotine on locomotor activity were also greater (P<0.05) than those observed in animals treated acutely with the drug and approached those measured for rats treated acutely or subchronically with d-amphetamine (0.5mg/kg) (data not shown).

In agreement with the results of previous studies (Morrison and Stephenson 1972; Clarke and Kumar 1983), the subchronic administration of nicotine stimulated the locomotor of the rats (Fig. 3). The microinjection of 6-hydroxydopamine into the nucleus accumbens reduced (P<0.01) the DA concentration in the nucleus from 6.1 $\pm$ 0.8 to 0.8 $\pm$ 0.2 μg/g. The lesions resulted in a reduction in the locomotor activity of the rats which was independent of the treatment received prior to the trial although the effect in the rats given saline was modest when compared the reduction in activity observed for the rats given nicotine or d-amphetamine in which the lesions abolished the locomotor stimulant responses to the drugs.

The results of the shock avoidance experiments in which rats were trained and tested with nicotine showed that drug administration had no significant effects on either the number of shocks received or the number of lever-pressing responses made by the rats (Fig. 4). The withdrawal of nicotine, however, caused a significant reduction (P<0.05) in the number of lever-pressing responses and a significant increase (P<0.05) in the number of shocks received. The number of shocks received by the nicotine-withdrawn rats was also significantly higher (P<0.05) than the number received by the rats trained and tested with saline. The administration of d-amphetamine to rats trained with nicotine increased (P<0.01 when compared with rats trained and tested with either nicotine or saline) the number of lever-pressing responses made by the rats and attenuated the increase

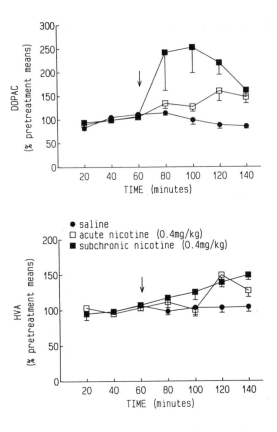

Figure 2: The effects of acute and subchronic nicotine administration on the recovery of dihydroxyphenylacetic acid (DOPAC) and homovanillic acid (HVA) in nucleus accumbens dialysates. The concentrations of DOPAC and HVA in the dialysate samples are expressed as the percentage ± SEM of the mean levelsmeasured in the dialysates collected during the 60 minute control period prior to the injection of nicotine (0.4mg/kg) or saline (controls) at the point shown by the arrow.

in the number of shocks received following nicotine-withdrawal (P<0.05) to the extent that the shocks received by the rats which were given d-amphetamine in place of nicotine was not significantly different to the number of shocks received by the rats trained and tested with nicotine. Rats trained and tested with d-amphetamine made more lever-pressing responses (P<0.01) and received fewer shocks (P<0.01) than control rats trained and tested with saline (Fig. 5). The withdrawal of d-amphetamine caused a reduction (P<0.01) in lever-pressing responses back to control levels and an increase in the number of shocks received which was significant (P<0.01) when compared with both the data for rats trained and tested with d-amphetamine and the rats trained and tested with saline. The acute administration of nicotine to these rats failed to attenuate the disruption of avoidance behaviour evoked by d-amphetamine-withdrawal.

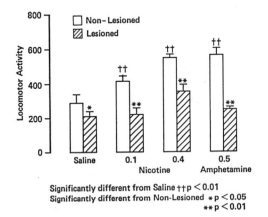

Significantly different from Saline ††p < 0.01
Significantly different from Non-Lesioned ∗p < 0.05
∗∗p < 0.01

Figure 3:   The effects of lesions of the mesolimbic dopamine system on locomotor responses to nicotine and d-amphetamine. Groups of rats (N=7 per group) were injected bilaterally in the nucleus accumbens with 6-hydroxydopamine or vehicle. Starting 7 days after surgery all the rats were given daily injections of nicotine (0.4mg/kg) for 6 days. Then, at three day intervals, locomotor responses were measure following subcutaneous injections of saline, nicotine (0.1 and 0.4mg/kg) and d-amphetamine (0.5mg/kg) using a counter-balanced design. The data are expressed as means ± SEM.

DISCUSSION

The results of the studies outlined above have shown that the response of the mesolimbic DA system to nicotine is enhanced in animals which have been pretreated with the drug for some days prior to the test day. The pretreatment procedure also results in an enhanced locomotor response to the drug and, since lesions of the mesolimbic system attenuate the effects of nicotine on locomotor activity, it is clearly tempting to suggest that the enhanced mesolimbic DA response may be associated with this effect on behaviour. The results of other studies, however, suggest that the enhanced locomotor response to nicotine, observed in animals treated chronically with nicotine is related to the development of tolerance to a depressant effect of the drug on locomotor

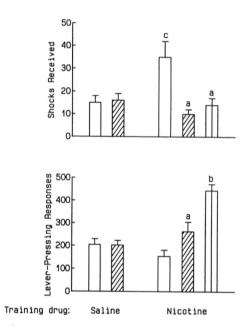

Figure 4: The avoidance performance of rats trained with nicotine or saline. The rats were trained with saline or 0.4mg/kg nicotine and then tested after injections of saline (open columns), 0.4mg/kg nicotine (hatched columns) or 0.5mg/kg d-amphetamine (striped columns). The bars represent the means $\pm$ SEM of the numbers of observations given in parenthesis. Significantly different from rats tested after saline; a=P<0.05; b=P<0.01. Significantly different from rats trained and tested with saline; c=P<0.05. (Taken from Balfour (1990) with permission).

activity (Clarke and Kumar 1983) and, therefore, further experiments are necessary to clarify this issue. Studies reviewed by Bozarth and Wise (1987) suggest that the increase in mesolimbic DA secretion evoked by psychostimulant drugs such as nicotine may contribute significantly to their rewarding properties and, thus, to the craving to continue drug taking behaviour. If this is the case, the data reported here imply that the putative rewarding properties of nicotine may be enhanced following a period of pretreatment. It is, however, important to

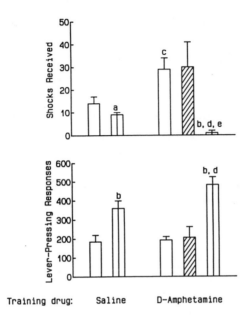

Figure 5: The avoidance performance of rats trained with d-amphetamine or saline. The rats were trained with saline or 0.5mg/kg d-amphetamine and then tested after injections of saline (open columns), 0.4mg/kg nicotine (hatched columns) or 0.5mg/kg d- amphetamine (striped columns). The bars represent the means ± SEM of the numbers of observations given in parenthesis. Significantly different from rats tested after saline; a=P<0.05; b=P<0.01. Significantly different from rats trained and tested after saline; c=P<0.05; d=P<0.01. Significantly different from rats trained with saline but tested after d-amphetamine; e=P<0.01. (Taken from Balfour (1990) with permission)

remember that the dialysis experiments were performed with rats tested in a novel environment and, therefore, that the stress associated with exposure to this environment may have contributed to the results obtained. In the case of nicotine this is a particularly important point because the results of other studies (Vale and Balfour 1989, Costall et al 1989), including the shock avoidance experiments reported above, suggest that nicotine does attenuate some of disruptive effects of stress on rat behaviour and that this effect may be mediated by its ability to stimulate DA secretion in the brain.

Since the effects of d-amphetamine and its withdrawal on unsignalled shock avoidance were very similar to those of nicotine, it seems reasonable to suggest that the increase in DA secretion evoked by nicotine may also underlie its effects on avoidance behaviour. If this is the case then the data further suggest that withdrawal of the drug may disrupt performance of the task because the rats have become dependent upon the enhanced levels of DA secretion in order to continue making the appropriate responses. The results to date have not implicated the mesolimbic system specifically in this effect and, indeed, further studies are necessary to establish the hypothesis with any certainty.

Acknowledgements: These studies were performed with financial assistance from the Wellcome Trust and the MRC (UK).

REFERENCES

Balfour, D.J.K. (1984) In: Nicotine and the Tobacco Smoking Habit - Section 114 of The International Encylopedia of Pharmacology and Therapeutics (D.J.K. Balfour ed) Pergamon Press, New York, pp. 101-112.
Clarke, P.B.S. (1987) Psychopharmacology 92, 135-143.
Clarke, P.B.S., Fu, D.S., Jakubovic, A. and Fibiger, H.C. (1988) J. Pharmacol. Exp. Ther. 246, 701-708.
Clarke, P.B.S. and Kumar, R. (1983) Br. J. Pharmacol. 80, 587-594.
Costall, B., Kelly, M.E., Naylor, R.J. and Onaivi, E.S. (1989) Pharmacol Biochem Behav 33, 197-203.
Gilbert, D.G. (1979) Psychol. Bull. 86, 643-661.
Imperato, A., Mulas, A. and Di Chiara G. (1986) Eur. J. Pharmacol. 132, 337-338
Morrison, C.F. (1974) Psychopharmacologia (Berl.) 38, 25-35.
Morrison, C.F. and Stephenson, J.A. (1972) Br. J. Pharmacol. 46, 151-156.
Singer, G., Wallace, M. and Hall, R. (1982) Pharmacol. Biochem. Behav. 17, 579-581.
Vale, A.L. and Balfour, D.J.K. (1989) Pharmacol. Biochem. Behav. 32, 857-860.
Wise, R.A. (1990) In: Psychotropic Drugs of Abuse - Section 130 of The International Encyclopedia of Pharmacology and Therapeutics (D.J.K. Balfour ed) Pergamon Press, New York pp 453-481
Wise, R.A. and Bozarth, M.A. (1987) Psychol. Rev. 94, 469-492.

REARING AND LEARNING BEHAVIORS OF RATS AFTER LONG-TERM
ADMINISTRATION OF NICOTINE

Kazuo Nagai, Hiroyuki Iso* and Sadao Miyata

Department of Pharmacology, and * Psychology, Hyogo College
of Medicine, Hyogo, Japan

Summary: The effects of long-term oral administration of
nicotine (N; 0.01%) on the rearing response and the complex
T-maze learning of WKY rats were investigated. N slightly
increased the locomotion, and increased the rearing response
more than the control (C) rats in the night. During daily 6-
trial T-maze learning for 4 days, N enhanced the learning in
both the number of errors and the latency measures.
Scopolamine (40mg/kg) administered on the 5th session
reduced the T-maze performance in the C rats, but not in the
N animals.

Nicotine (N) has been described as a psychotropic agent
(Nelson, 1976). Behavioral effects of N in animals may based
on via the central nervous system. Many studies have been
done to investigate the effects of N on the memory or the
learning in animals. However, if the procedures were
different from one to another, the results differed and
varied.

We studied the effects of chronic administration of N
on the rearing in the night period and on the T-maze
learning in the day time.

MATERIALS AND METHOD
Animals; Age matched male Wistar KYOTO rats (WKY) weighting
about 230-270g were used.

Apparatus; Animex was used to measure animals' activity in
the homecage. A square open-field made of plastic, sized
40x40cm was used for the activity cage or for a complex T-
maze. Ten infrared beams on each X and Y bank at 4cm above
the floor counted the horizontal movements, and the vertical
movement, rearing, of the rat was detected by 5 infrared
beams attached at 12 cm above the floor. Next, the same
apparatus was also used as a complex T-maze shown in Fig. 1.
Several number of clear plastic partitions were inserted,
and a 4-trap maze was constructed. Number of intrusions to
each trap were detected by a infrared beam system and
counted automatically. The latency, the time from the start
box to the goal, was also recorded in 0.1s.

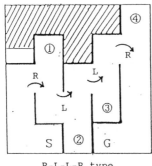

R-L-L-R type

Fig. 1. A complex T-maze pattern
(S:Start area, G:Goal, R:Right
turn, L:Left turn, 4 traps:
1,2,3,4)

Procedures; Animals were divided into two groups, N and
Control (n=10, each), and N (0.01%) or water was given in
their home cage over 30-60 days.
1) Activity and rearing test: After the start of N
administration, home cage activity at night was counted by
the Animex system. Rearing was counted in the open-field.
Food was always available in the home cage.
2) T-maze learning: After the activity and rearing test,
animals were deprived from the food and reduced to 80% of
their normal body weight over 8-10 days. Six trials were
conducted daily for 5 days. If animals intrude into one trap
continuously, responses after second one were neglected. Re-
intrusion into the initial trap after the other intrusion
was counted as a error. In the goal, 5 pellets (45mg

Results, Bio Serv. Co.) were given in each trial. Food was supplied after daily session to maintain their body-weight. Either water (C) or drugs (N) was always available in the homecage.

Drugs; Nicotine (free base) was purchased from Tokyo Kasei Co., and distilled at 1140 bp35. 0.01% N in tap water was given for rats orally. Scopolamine hydrobromide (Tokyo Kasei Co.) was dissolved in saline solution and injected intraperitoneally (40mg/kg).

Nicotine and cotinine determination; N and cotinine in urine and plasma were measured by the method of Watson with slight modification (Watson, 1976).

RESULTS

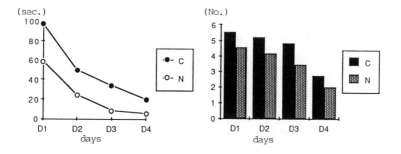

Fig.2. Mean latency (s) (Left panel) and mean number of errors
            (righ panel) during 4 days of T-maze learning (C:Control,
        N:Nicotine).

1) After administration of N, locomotive activity of the rats in the home cage were slightly increased at night. The rearing responses of N rats at night increased more than C animals from the 5th day of the N administration.

2) The N secretion in urine was highest at 3-4 days of administration and plasma levels also increased after 4th day. Cotinine, a metabolite of N, increased gradually both

in urine and plasma, and the level was kept constantly until 4th week.

3) The greatest number of rearings was recorded after 10-day of N administration.

4) N enhanced the T-maze learning in both the latency (Figure 2, left panel) and number of errors (Figure 2, right panel). A mixed-type ANOVA performed on the data of latency indicated that the main effects of group (G) and day (D) were significant [$F_{(1/18)}=9.03$, $p<.001$ and $F(3/54)=34.87$, $p<.001$, respectively], but G x D interaction was not. On the errors, the effects of G and D were also significant [$F_{(1/18)}=4.30$, $p<.05$ and $F(3/54)=19.54$, $p<.001$, respectively].

DISCUSSION

Numerous reports on the behavioral actions of N to animals were found as a psychoactive drugs (Bätting, 1981), and those actions were mediated by CNS (Clarke & Kumar, 1983; Pomerleau, 1986; Reavill & Stolerman, 1990). However, the procedures and animals used in such experiments were so different that the results differed and varied widely.

We used the WKY rats which was a conventional Wistar strain. Therefore, we would be able to compare the present results with those from spontaneously hypertensive rats (SHR) in the further reports.

N was given by way of chronic voluntary consumption (Flynn et al., 1989). Excreted N concentration in urine and plasma N level were measured. After the 7th day of N intake, these values reached to about 30nM in urine and 0.16 nM in plasma, respectively. Cotinine was also at the maximum concentration. The occurrence of the rearings of N rats was matched to the time when the excretion of N and cotinine were maximum. Number of rearings of N rats was about twice of the C animals.

In the T-maze learning, N enhanced learning in the two measures. There are three possible explanations about these data. First, N enhanced learning ability (memory construction). Second, N increased the motivation (hunger) of the animals. Third, N enhanced the memory storage. It is plausible the second one, since the performance on the first day was already different between N and C groups.

Application of scopolamine (40mg/kg) at the post-learning session showed different effects of N and C rats. C rats remained in the start area, but N animals ran to the goal with few errors. These results indicated that N competed to the amnesic effect of scopolamine in this T-maze situation. The development of reverse tolerance to effects of N in the CNS may correlate with these phenomena.

REFERENCES

Bättig, K (1981) Trends Pharmacol. Sci. 2, 145-147.
Clarke, P.D.S., and Kumar, R. (1983) Br. J. Pharmacol., 78, 329-337.
Flynn, F.W., Webster, M., and Ksir, C. (1989) Behavioral Neuroscience, 103, 356-364
Nelson, J.M. (1976) in Behavioral effect of nicotine (Bättig, K. ed.), Karger Verlag, Basel, 1-16.
Pomerleau, O.F. (1986) Psychopharmacology Bulletin, 22, 864-865.
Reavill, C., and Stolerman, I.P. (1990) Br. J. Pharmacol., 99, 239-278
Watson, I.D. (1977) Journal of Chromatography, 143, 203-206

REGULATION OF INTRAVENOUS NICOTINE SELF-ADMINISTRATION -- DOPAMINE
MECHANISMS

William A. Corrigall

Addiction Research Foundation, 33 Russell St., Toronto, Canada
M5S 2S1, and Department of Physiology, University of Toronto,
Toronto, Canada M5S 1A8

SUMMARY: Intravenous self-administration of nicotine has been
obtained in rats with a limited-access model.   Nicotine is a
powerful reinforcer in this situation. Dependency upon dose is
greatest at high and low ends of the dose range, but responding
is relatively insensitive to dose at mid-range values; this is
similar to the dependency of cigarette smoking behavior on
nicotine delivery. The role of dopamine in nicotine reinforcement
has been examined using this self-administration paradigm.
Selective D1 (SCH23390) and D2 (spiperone) dopamine antagonists
reduce nicotine self-administration in a dose-dependent fashion.
The same dose range of D1 and D2 antagonists produce compensatory
increases in cocaine self-administration.   These differences in
the effects of dopamine antagonists may stem from the different
regulation of nicotine as compared to cocaine. Lesions of the
mesolimbic dopamine system made by microinfusions of the
neurotoxin 6-hydroxydopamine into the nucleus accumbens attenuate
both nicotine self-administration and food-maintained behavior.
These data suggest that dopamine mechanisms, perhaps in the
mesolimbic system, play a role in nicotine reinforcement.

INTRODUCTION

We have recently demonstrated that nicotine is a reinforcer
in rodents when the drug is available for self-administration on
a limited-access schedule (Corrigall and Coen, 1989). Our intent
in developing this model was to have available a reliable paradigm
to study mechanisms of nicotine reinforcement. To this end we
have begun to examine the neurochemical mechanisms of nicotine
reinforcement. This report summarizes the main characteristics
of the self-administration model, and describes our findings which
suggest that dopamine plays a role in nicotine reinforcement.

Attention has focused recently on the possible involvement of the mesolimbic dopamine system in the behavioral effects of nicotine. Autoradiographic evidence indicates that there are nicotinic binding sites on mesolimbic neurons (Clarke & Pert, 1985), while electrophysiological studies have established that nicotine causes an excitation of neurons in the ventral tegmental area (Grenhoff et al., 1986; Mereu et al., 1987), apparently via a direct depolarizing action on presumptive dopaminergic cells themselves (Calabresi et al., 1989). Administration of nicotine systemically, as well as directly into the nucleus accumbens, has been shown to increase the extracellular dopamine concentration in that brain region (Imperato et al., 1986; Mifsud et al., 1989). In addition, dopamine turnover following acute nicotine treatment is greatest in the accumbens as compared to frontal cortex, olfactory tubercle, caudate-putamen, substantia nigra and ventral tegmental area (Lapin et al., 1989). Locomotor activity produced by systemic injections of nicotine is attenuated after 6-hydroxydopamine lesions of the nucleus accumbens (Clarke et al., 1988), whereas increases in locomotor activity occur after microinfusions of a nicotinic agonist into the ventral tegmental area (Museo & Wise, 1990).

The fact that nicotine appears to be able to modulate the dopamine system in general, and the mesolimbic pathway in particular, may have important implications for our understanding of the mechanisms of nicotine reinforcement, since cocaine reinforcement appears to depend on the ability of that drug to increase synaptic dopamine concentrations by blocking the reuptake into neuron terminals, possibly in the nucleus accumbens (Johanson & Fischman 1989). Recently it has been shown that nicotine, too, blocks dopamine reuptake in striatal tissue, and with a much lower IC_{50} than for release (Izenwasser et al., 1990). To investigate the possibility that dopaminergic mechanisms are important in the ability of nicotine to act as a reinforcer, we have examined the effects of dopamine receptor antagonists, and lesions of the mesolimbic dopamine projection to the nucleus accumbens, on nicotine self-administration.

METHODS

Experiments were carried out with male, Long-Evans rats which were drug-naive prior to the start of surgical or training procedures. Animal procedures were carried out in accordance with the guidelines of the Canadian Council on Animal Care.

Experimental methods have been described in detail in previous publications (Corrigall, 1990; Corrigall & Coen, 1989) and are only summarized here. Animals were trained for drug self-administration on a fixed-ratio 5 (FR5) schedule of reinforcement with a 1-minute time-out period (TO 1-min) following each infusion; session duration was 1 hour. A dose range of 0.003 to 0.06 mg/kg/infusion was used for nicotine self-administration. Control experiments were carried out using cocaine self-administration (0.03 to 1 mg/kg/infusion) with the same FR5 TO 1-min schedule of reinforcement.

The general pharmacological sensitivity of nicotine self-administration was examined by means of pre-session treatments with mecamylamine and hexamethonium, administered systemically. Pre-treatments were done acutely, with each dose of antagonist administered once. In addition, the long-lasting quaternary ganglionic blocker chlorisondamine was infused into the cerebral ventricles of animals previously trained to self-administer nicotine. For this experiment, animals were anesthetized, positioned in a stereotaxic frame, and infused bilaterally with 2.5 μg of chlorisondamine into each lateral ventricle. Following recovery from surgery, nicotine self-administration was retested.

To assess the consequences for nicotine reinforcement of blockade of dopamine receptors, dopamine antagonists were injected acutely prior to selected operant sessions. SCH23390 was used as the D1 antagonist and spiperone as the D2 antagonist; both compounds were tested over a dose range of 0-30 μg/kg. Doses of dopamine antagonists were chosen in part for their ability to alter nicotine-induced locomotor activity (Corrigall & Coen, 1990). SCH23390 was administered 30 min before the start of operant sessions, while spiperone was given 60 min prior to the start of the sessions.

To examine the role of the ventral tegmental-to-nucleus accumbens projection in nicotine self-administration, 6-hydroxydopamine (6-OHDA) was injected bilaterally into the nucleus accumbens region of animals trained to self-administer nicotine (Corrigall et al, 1990). Lesions were done with established procedures (Clarke et al., 1988). Following recovery from surgery, animals were re-tested for nicotine self-administration for a 3-week period. After completion of behavioral testing, brains were removed and assayed for dopamine.

RESULTS

Acquisition of nicotine self-administration at 0.03 mg/kg/ infusion requires 3-4 weeks. After this time, substantial responding occurs on the lever that delivers nicotine, and the number of infusions received typically ranges between 10-20 in the 1-hour sessions. Responding on an inactive lever virtually does not occur (Fig. 1).

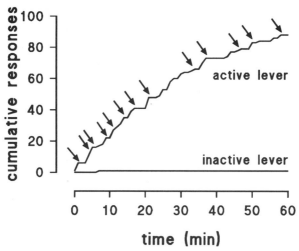

Figure 1. An example of nicotine self-administration by one subject. Nicotine infusions occurred at the times indicated by the arrows. Notice that responding on the inactive lever is essentially absent; this subject pressed the inactive lever only once, whereas approximately 100 responses were made on the active lever, to receive 13 infusions of nicotine.

Self-administration of nicotine does not change very much over a 10-fold range of unit doses (0.003 to 0.03 mg/kg/infusion; Fig. 2). Above this plateau, responding begins to decrease; at the 0.06 mg/kg unit dose, the number of infusions is halved compared to its value at 0.03 mg/kg, whereas the intake in the session remains the same as at 0.03 mg/kg/infusion. Therefore, nicotine acts as a reinforcer over a range of doses, but the degree of responding does not seem to be particularly sensitive to dose (or total intake) until very high or very low values are reached.

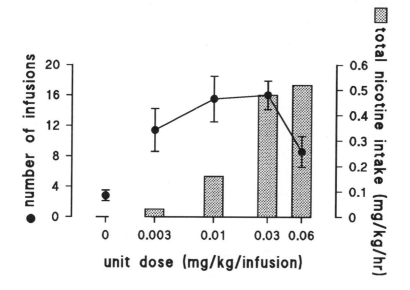

Figure 2. Dose-effect curve for nicotine self-administration (n=8). Points indicate mean values; bars represent ± one standard error of the mean. Modified from Corrigall and Coen (1989), with permission.

Acute treatment of animals with hexamethonium (0-3 mg/kg) before self-administration sessions had no effect on the number of nicotine infusions received (Corrigall and Coen, 1989). In contrast, treatment with mecamylamine (0-3 mg/kg) decreased the number of infusions obtained. In addition, a single microinfusion of the quaternary ganglionic blocker chlorisondamine into the cerebral ventricles caused a sustained reduction in nicotine self-administration; the number of infusions obtained by the subjects after chlorisondamine was comparable to saline extinction.

Both of the dopamine antagonists tested had similar effects on the number of nicotine infusions received in treatment sessions. Over a dose range of 0-30 μg/kg, the D1 antagonist SCH23390 and the D2 antagonist spiperone each produced a dose-related decrease in the number of infusions obtained. Analysis of the time course during treatment sessions showed that both antagonists produced response-decreasing effects, usually during the second half of operant sessions (except for the highest dose of SCH23390, which resulted in a complete cessation of responding for the first half of the session, and severely attenuated responding in the remainder).

In contrast, cocaine self-administration, with the same schedule of reinforcement, was affected in the opposite direction by pre-treatment with these dopamine antagonists. For example, cocaine at a dose of 1 mg/kg/infusion produces a comparable rate of responding to that of nicotine at a unit dose of 0.03 mg/kg. Treatment with the same dose range of SCH23390 and spiperone that decreased nicotine self-administration, produced increases in cocaine self-administration throughout the test sessions, except in the case of the highest dose of SCH23390 which again caused complete cessation of responding for the first half of the cocaine self-administration session, and reduced responding thereafter, a result comparable to its effect on nicotine self-administration.

Microinfusions of 6-OHDA into the nucleus accumbens resulted in a marked reduction in nicotine self-administration when testing began 3-4 days post-surgery, whereas sham-treated animals were not affected. Testing was continued for a 3-week period, and during this time self-administration of nicotine by the sham-treated group remained unchanged or increased slightly from pre-treatment values, whereas self-administration by the 6-OHDA treated group was reduced to approximately 25% of pre-lesion values (the number of infusions obtained by the lesioned animals was comparable to saline extinction). Food responding, tested intermittently in these same subjects, was also reduced substantially in the dopamine-lesioned as compared to the sham-treated rats. Analysis of brain tissue *post mortem* showed that dopamine was depleted by 93% in the nucleus accumbens, whereas depletion in the caudate-putamen was only 25%.

DISCUSSION

When nicotine is available on a limited-access FR5 schedule of reinforcement, substantial responding occurs specifically on the manipulandum which is programmed to deliver drug. Nicotine self-administration occurs over a dose range of 0.003 to 0.06 mg/kg/infusion, with, however, a different dose-dependency than other drugs such as cocaine. Whereas rats adjust the degree of cocaine self-administration in response to dose changes, intravenous nicotine seems to be regulated less strongly as a function of dose -- responding shows the greatest sensitivity to changes in dose at the extremes of the range. The regulation of intravenous nicotine self-administration therefore seems to be similar to the compensation that occurs in cigarette smoking as a function of changes in nicotine yield.

The reinforcing effects of nicotine occur at sites within the central nervous system. Part of the evidence for this comes from the demonstration that self-administration is decreased following acute peripheral injections of mecamylamine but not hexamethonium (Corrigall & Coen, 1989). More direct proof, however, derives from the demonstration that microinfusion of the ganglionic blocker chlorisondamine into the ventricles produces a sustained reduction in nicotine self-administration (Corrigall et al., 1990). The effect of chlorisondamine is specific to nicotine and not the result of a generalized effect on behavior, since cocaine self-administration is not affected.

The alteration in nicotine self-administration produced by D1 and D2 antagonists is clearly suggestive of a role for dopamine in nicotine reinforcement. However, given that dopamine antagonists produced decreases in nicotine self-administration,

and that dopaminergic manipulations are capable of producing motor
deficits, it is important to exclude the latter as a possible
mechanism. Response patterns after treatment with the antagonists
provide the means to do so.

The response patterns for nicotine after treatment with
SCH23390 or spiperone show that animals were not unable to
respond; rather, comparison of antagonist- and vehicle-treatment
sessions demonstrate that response patterns were similar at the
beginning of the sessions, and diverged thereafter. In other
words, effects on nicotine self-administration generally developed
well after treatment sessions had begun, after subjects had
sampled nicotine. This pattern of effect is not consistent with
motor impairment as the mechanism for the effects on nicotine
self-administration. Only after the highest dose of SCH23390 (30
μg/kg) do response patterns suggest that motor impairment is
responsible for the effect of this antagonist on either nicotine
or cocaine self-administration, since after treatment with this
dose animals appear to be unable to respond for the first part of
the session.

The absence of compensatory increases in nicotine self-
administration after treatment with dopamine antagonists presents,
at first glance, a problem in interpretating the nicotine data as
being an effect on reinforcement systems. From research with
other drugs, such as cocaine, we have come to expect that a
reduction in reinforcing value will lead to a compensatory
increase in responding. However, one must recall that nicotine
is regulated differently than cocaine, and that its self-
administration shows a large "plateau" in the infusions-versus-
dose curve. Therefore, the decreases in nicotine self-
administration that occur after treatment with dopamine
antagonists may still represent a direct effect on nicotine
reinforcement, its manifestation is simply different because of
the fundamental difference in the regulation of nicotine. This
is consistent with the observations that systemic mecamylamine and
central chlorisondamine each decrease nicotine self-
administration without showing any compensatory increase in

responding. Therefore, one might expect a decrease in nicotine self-administration to occur if a substrate of its reinforcement system were blocked.

Observations following neurochemically-selective destruction of part of the mesolimbic dopamine projection provide further support for the idea that dopamine mechanisms are involved in nicotine reinforcement. Microinfusion of 6-OHDA into the nucleus accumbens produced a long-lasting decrease in nicotine self-administration. Were the lesion-produced reductions in responding a consequence of motor impairment? Two facts suggest that this is not so. First, 6-OHDA lesioned rats were capable of responding for food at rates greater than their pre-lesion response rates for nicotine; i.e., animals were capable of responding at a higher rate for nicotine than they did. Second, lesions produced a dopamine depletion in the caudate-putamen of approximately 25%, substantially less than the depletions of the nigrostriatal pathway usually associated with motor deficits. Nonetheless, the fact remains that the lesion did reduce responding for each of nicotine and food, and it is not possible to exclude completely that effects other than a direct one on reinforcement are responsible for the reduction in nicotine self-administration after the 6-OHDA lesion. Alternatively, it may be that the accumbens is a critical site for a variety of reinforcing stimuli.

In conclusion, intravenous nicotine self-administration represents drug reinforcement with a novel form of regulation. Manipulations of the dopamine system alter nicotine self-administration, and it is probable that these effects arise directly from a reduction in the reinforcing properties of the nicotine, mediated possibly by the ventral tegmentum-to-nucleus accumbens projection. However, at present it is not possible to state unequivocally that the mesolimbic dopamine system underlies nicotine's reinforcing properties. Further experiments, such as using chlorisondamine to block the effects of nicotine in specific dopaminergic regions of the brain will help to establish whether dopaminergic cells participate directly in nicotine reinforcement.

Acknowledgements: The author is grateful to Drs. P.B.S. Clarke and K.B.J. Franklin for their collaboration in the 6-OHDA lesion experiments, and to K.M. Coen for her expert technical assistance. Research supported by the Addiction Research Foundation of Ontario.

REFERENCES

Calabresi, P., Lacey, M.G., and North, R.A. (1989) Br. J. Pharmacol. 98, 135-140.

Clarke, P.B.S., and Pert, A. (1985) Brain Res. 348, 355-358.

Clarke, P.B.S., Fu, D.S., Jakubovic, A., and Fibiger, H.C. (1988) J. Pharmacol. Exp. Ther. 246, 701-708.

Corrigall, W.A. (1990) In: Animal Models of Drug Addiction. Neuromethods, vol 21 (A.A. Boulton, G.B. Baker, and P.H. Wu, Eds), Humana Press, Clifton, NJ, in press

Corrigall, W.A., and Coen, K.M. (1989) Psychopharmacology 99, 473-478.

Corrigall, W.A., and Coen, K.M. (1990), in preparation.

Corrigall, W.A., Franklin, K.B.J., and Clarke, P.B.S. (1990) in preparation

Grenhoff, J., Aston-Jones, G., and Svensson, T.H. (1986) Acta Physiol. Scand. 128, 351-358.

Imperato, A., Mulus, A., and Di Chiara, G. (1986) Eur. J. Pharmacol. 132, 337-338.

Izenwasser, S., Jacocks, H.M., and Cox, B.M. (1990) NIDA Res. Monograph, in press.

Johanson, C.E., and Fischman, M.W. (1989) Pharmacol. Rev. 41, 3-52

Lapin, E.P., Maker, H.S., Sershen, H., and Lajtha, A. (1989) Eur. J. Pharmacol. 160, 53-59.

Mereu, G., Yoon, K.-W.P., Boi, V., Gessa, G.L., Naes, L., Westfall, T.C. (1987) Eur. J. Pharmacol. 141, 395-399.

Mifsud, J.C., Hernandez, L., and Hoebel, B.G. (1989) Brain Res. 478, 365-367.

Museo, E., and Wise, R.A. (1990) Pharmacol. Biochem. Behav. 35, 735-737.

Effects of Nicotine on Biological Systems
Advances in Pharmacological Sciences
© Birkhäuser Verlag Basel

NICOTINE-SEEKING BEHAVIOR IN RHESUS MONKEYS

T. YANAGITA, K. ANDO, Y. WAKASA and K. TAKADA

Preclinical Research Laboratories, Central Institute for
Experimental Animals, Nogawa, Kawasaki 213, JAPAN

SUMMARY: Nicotine-seeking behavior was studied in rhesus
monkeys mainly in intravenous self-administration
experiments on nicotine and partly in cigarette-smoking
experiments. As a result, the nicotine-seeking behavior by
the intravenous route was observed mostly in the daytime
with a relatively stable daily pattern, but by the smoking
route such behavior was unstable from day to day. The
intensity of the intravenous drug-seeking behavior for
nicotine observed in a progressive ratio experiment was
found to be quite strong but weaker than that for morphine
or cocaine. Pretreatment with frequent intravenous doses of
nicotine for 4 weeks did not enhance the intensity. This
result demonstrates marked difference between physical
dependence on opiates such as morphine or codeine and on
nicotine. For analysis of nicotine-seeking behavior, serum
nicotine level at the time of lever pressing for nicotine
was determined. The result may indicate that the nicotine-
seeking behavior is triggered by lowering of the serum
nicotine level, which is proportional to the maintained dose
level. Also analyzed were the intravenous threshold doses of
nicotine in self-administration experiments and in drug
discrimination experiments. The threshold dose for
reinforcement in the former experiments and that for
discrimination in the latter were quite close each other.
This result may support the observation that nicotine-

seeking behavior is developed as a consequence of its
subjective effects.

INTRODUCTION

Nicotine is known to have psychic dependence potential. In
animal experiments, this is demonstrated as the reinforcing
effect of nicotine in self-administration of the drug. The
characteristics and intensity of nicotine-seeking behavior
and the relation of the behavior with both serum nicotine
level and discrimination behavior were studied in the rhesus
monkey, mainly in intravenous self-administration
experiments on nicotine and partly in cigarette smoking
experiments and nicotine discrimination experiments.

METHODS, RESULTS, AND DISCUSSION

1. Characteristics of nicotine-seeking behaviors:

a. General methods: Male and/or female rhesus monkeys
weighing 4 to 6 kg were restrained in individual cages by
free-jointed metal arms and metal harnesses. When they
adapted to life in the cage, an intravenous catheter was
implanted into the jugular vein under pentobarbital
anesthesia and the catheter was connected via tubing passing
through the restraining arm to an automatic injector located
outside the cage. When the monkey pressed a lever switch
located inside the cage a certain number of times, the
injector was activated and a preset dose of nicotine was
delivered to the monkey.

b. Continuous self-administration: Under the schedule of
this experiment, the monkeys were tested with saline to
begin with, and when infrequent intake of saline was

confirmed, the saline was replaced with nicotine and
initiation and maintenance of nicotine self-administration
was observed continuously around the clock for 4 weeks or
longer. Intravenous self-administration of nicotine at unit
doses of 20, 50, and/or 200 μg/kg were observed in 6
monkeys previously experienced with self-administration of
stimulants such as cocaine or lefetamine (SPA) and in 7
naive monkeys. All 6 experienced monkeys and 4 out of 7
naive monkeys initiated and maintained self-administration
of nicotine. For the most part, the animals exhibited the
nicotine-seeking behavior during the daytime in a relatively

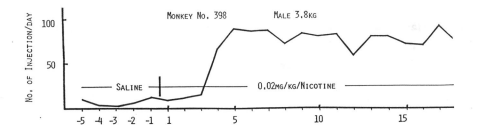

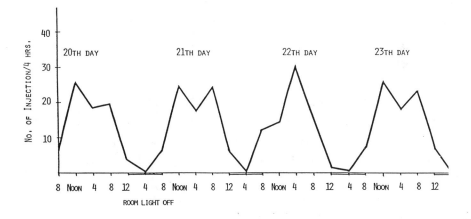

Fig. 1.
     A case of intravenous self-administration of nicotine in a rhesus
monkey. The monkey initiated and maintained self-administration of
nicotine when became available. Upper: Daily number of injections.
Lower: Daily pattern of intake (Yanagita et al., 1983).

stable daily pattern [Fig. 1]. Their daily intake in averaged over the observation period reached up to about 100 infusions/day at a unit dose of 20 μg/kg and 20 infusions/day at 200 μg/kg [Table I]. No marked general behavioral changes were observed in these monkeys (Yanagita et al., 1983).

c. Self-smoking behavior: Self-smoking experiments in rhesus monkeys were conducted by having them smoke cigarette via a metal pipe leading from an individual cage to an automatic cigarette dispensing machine. When the monkey sucked air from the pipe, a cigarette would be lit which, upon burning down a certain amount, would be replaced automatically with a new one. For training to shape air-sucking behavior, a sugar solution was used as reinforcer. When the behavior was

Table I. CONTINUOUS INTRAVENOUS SELF-ADMINISTRATION OF NICOTINE

	SUBJECT	PREVIOUS DRUG	INITIATION	APPROX. DAILY DOSE (μg/kg/inj. X No.)	PATTERN OF INTAKE
EXPERIENCED	1[1)	SPA	+	200 X 15	CONSTANT
	2[1)	SPA	+	20 X 100	CONSTANT
	3[1)	SPA	+	20 X 50	CONSTANT
	4[2)	SPA	+	20 X 35	IRREGULAR
	5[1)	COCAINE	+	50 X 40	CONSTANT
	6[1)	COCAINE	+	20 X 100	CONSTANT
NAIVE	7	NONE	+	200 X 20	CONSTANT
	8	NONE	+	200 X 20	CONSTANT
	9	NONE	+	20 X 80	CONSTANT
	10	NONE	+	20 X 50	CONSTANT
	11	NONE	−		
	12	NONE	−		
	13	NONE	−		

1) Time interval longer than 1 month before nicotine study
2) Study followed immediately after 9 day's SPA intake

established cigarette smoke was presented to them through the pype. The self-smoking experiment then began in earnest, and nicotine-seeking behavior was again observed without sugar reinforcement mostly during the daytime, and the serum nicotine levels of some monkeys reached up to 40 ng/ml, which is comparable to the levels in human smokers. However, their daily numbers of puffs were relatively erratic (1,000~3,000 puffs/day). Thus, application of this method for behavioral studies as a model of human smoking was found to be somewhat limited because of the unstable behavioral baseline. In order to confirm whether the monkeys' self-smoking behavior developed in order to obtain

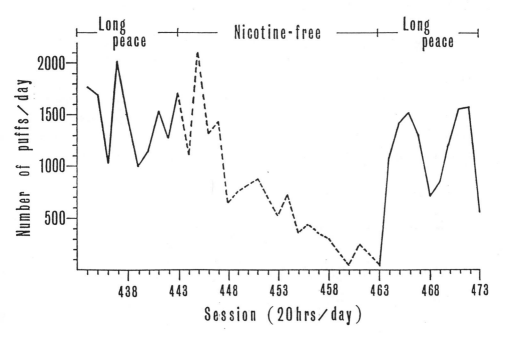

Fig. 2.   Self-smoking experiment in a rhesus monkey.   When regular cigarettes were replaced with nicotine-free cigarettes the number of puffs/day markedly decreased (Ando and Yanagita, 1981).

nicotine, nicotine-free cigarettes were given in place of
regular cigarettes during a self-smoking period. As a
result, their smoking behavior tended to extinguish as
indicated by a marked decrease of the number of puffs per
day [Fig. 2] (Ando & Yanagita, 1981). It thus became clear
that the monkeys' smoking behavior derives from the
nicotine-seeking behavior. However, it is unclear why
nicotine-seeking behavior is stable by the intravenous route
yet unstable when ingested by smoking. It may be possible
to explain this unstable behavior in terms of the
unpleasantness of the cigarette smoke to the monkeys as a
stimulus.

2. Intensity of nicotine-seeking behavior:

a. Self-administration in progressive ratio schedule: As a
method to measure the intensity of drug-seeking behavior, a
progressive ratio schedule can be used with self-
administration of the drug: the number of lever presses
requisite for a drug administration is progressively
increased at each drug intake until reaching the breaking
point at which lever pressing is abandoned (final ratio).

Table II. RESULTS OF PROGRESSIVE RATIO TESTS (Yanagita et al.)

Drug	Unit dose (mg/kg/infusion)	Final ratio (max)	Influence of pretreatment on final ratio
Morphine	0.5	12,800	Marked enhance
Codeine	1.0	12,800	Marked enhance
Cocaine	1.0	12,800	Decrease
Alcohol	8,000	6,400	Slight enhance
Diazepam	1.0	3,200	Tend to decrease
Nicotine	0.25	2,800	Decrease

Under this procedure the final ratios obtained with several
dependence-producing drugs are shown in Table II (Yanagita,
1975 and 1987, Yanagita et al. 1983). The final ratios for
morphine or codeine tested under the physically dependent
state as well as for cocaine reached up to 12,000 lever
presses for one dose. The ratio obtained with nicotine 250
μg/kg/infusion reached up to 2,800 [Fig. 3]. These values
can be regarded to be significantly high although much lower
than for opiates or cocaine.

b. Influence of nicotine pretreatment: In the case of
opiates such as morphine or codeine, it is well known that
pretreatment with the drugs prior to conducting the
progressive ratio test markedly enhances the drug-seeking
behavior, but nicotine pretreatment at hourly doses of 250
μg/kg for 4 weeks was not found to enhance the nicotine-
seeking behavior [Fig. 3]. This result demonstrates a marked
difference between physical dependence on the opiates and
that on nicotine.

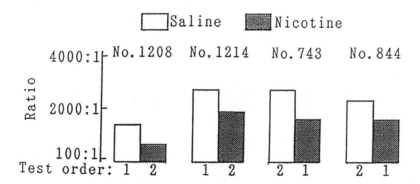

Fig. 3.   Average final ratios for nicotine obtained in progressive ratio
tests in intravenous self-administration in 4 monkeys. Tests were
conducted following 4-weeks' pretreatment with hourly intravenous
infusion of nicotine 0.25mg/kg/inj. (shaded column) and saline 0.25
ml/kg/inj. (blank column).   Nicotine pretreatment did not increase
but rather decreased the final ratios.

3. Induction of nicotine-seeking behavior and serum
concentration: Intravenous self-administration experiments
were conducted at 3 unit doses of nicotine (30, 120, and 240
μg/kg/infusion) in 4 monkeys each in random order. During
each 10- to 14-day self-administration period at a fixed
unit dose, the serum concentration of nicotine was
determined 4 times in each monkey immediately following a
lever press and before infusion to find the concentration at
which the drug-seeking behavior was triggered. As a result,
these concentration levels were 17.6, 37.3, and 60.1 ng/ml,
respectively. These levels were found to be considerably
different depending on the daily ingestion level of
nicotine; the daily ingestion doses on the average were 0.9
mg/kg/day at 30 μg/kg, 2.3 mg/kg/day at 120 μg/kg, and 4.0
mg/kg/day at 240 μg/kg. Thus it can be said that the
higher the daily ingestion level of nicotine, the higher the
serum concentration at which nicotine was sought. It may
mean that the nicotine-seeking behavior is triggered not
when the serum nicotine level lowers beyond a fixed level
(absolute concentration) but when the lowering exceeds a
certain value which is proportional to the maintained level
(relative concentration).

4. Nicotine-seeking behavior and nicotine discrimination
behavior: Since drug-seeking behavior is generally believed
to be initiated and maintained as a result of pleasurable
subjective effects, the dose relationship between the
reinforcing effect and discriminative effect of nicotine was
examined by observing the intravenous threshold doses for
reinforcement in the self-administration experiment and for
discrimination in a drug discrimination experiment in 2
monkeys each. To briefly explain the experimental method for
a drug discrimination experiment, a rhesus monkey was

of orange juice was placed in the front of the monkey, and
nicotine or saline was infused through intravenously
implanted catheter. Following infusion of nicotine or saline
the stimulus light was turned on and when the monkey pressed
an appropriate lever, the orange juice was presented as a
reinforcer. The unit doses tested for nicotine
reinforcement were 15, 30, and 60 µg/kg and those for
nicotine discrimination were 12.5, 25, and 50 µg/kg. As a
result, the threshold dose for the reinforcement test was 15
to 30 µg/kg and that for the discrimination test was 25
µg/kg [Table III]. This result may support the observation
that nicotine-seeking behavior is developed as a consequence
of its subjective effects.

Table III THRESHOLD DOSES FOR REINFORCEMENT AND DISCRIMINATION

Monkey No.	896	1089
Self-administration		
15 µg/kg, i. v.	+	−
30	+	+
60	+	+

Monkey No.	885	914
Drug Discrimination		
12.5 µg/kg, i. v.	−	−
25	+	+
50	+	+

Acknowledgements: We are grateful for the generous financial aid provided by Tobacco Research Foundation of Japan in supporting our studies from which most of the data presented in this paper were obtained.

REFERENCES

Ando, K. and Yanagita, T. (1981) Psychopharmacology <u>72</u> , 117-127.
Yanagita, T. (1975) Pharmacological Reviews <u>27</u> , 503-509.
Yanagita, T. (1987) In: Methods of assessing the reinforcing properties of abused drugs (Bozarth, M.A. Ed), Springer-Verlag, New York, pp.189-198
Yanagita, T., Ando, K., Kato, K. and Takada, K. (1983) Psychopharmacol. Bull. <u>19</u> , 409-412.

REGULATION OF VARIOUS RESPONSES TO NICOTINE BY DIFFERENT NICOTINIC RECEPTOR SUBTYPES

ALLAN C. COLLINS and MICHAEL J. MARKS

Institute for Behavioral Genetics and Department of Psychology, University of Colorado, Boulder, Colorado 80309, U.S.A.

SUMMARY: Nicotine administration affects many behavioral and physiological systems, and for many responses both stimulant and depressant effects may be observed. Presumably these actions are initiated by the binding of nicotine to appropriate receptor systems. Results presented in this paper suggest that the responses to nicotine are regulated by two distinct systems that may be related to two different nicotinic receptor systems.

INTRODUCTION

Nicotine elicits a broad array of physiological and behavioral effects in humans and in animals. The use of nicotine, in tobacco, by humans generally results in alterations in physiological measures such as blood pressure, heart rate, gastrointestinal motility, and respiration rate and in a number of behavioral measures where both stimulant and depressant effects have been reported (reviewed in: Russell, 1976; Henningfield, 1984). Animals administered nicotine exhibit changes in blood pressure, heart rate, respiratory rate and body temperature as well as effects on a variety of behaviors including locomotor activities, operant responding and conditioned avoidance, aggression, food and water intake, and level of arousal (reviewed in Clarke, 1987).

Whenever multiple and sometimes opposing drug effects are detected the possibility that more than one receptor exists for the drug must be considered. Indeed, it is likely that multiple forms of receptors for many drugs, hormones and neurotransmitters is the rule, rather than the exception. Recent evidence suggests that brain tissue contains more than one nicotinic receptor variant. Before 1980, only one brain nicotinic receptor had been described. Many studies had demonstrated that brain tissue contains a protein that binds the snake neurotoxin, α-bungarotoxin, with high affinity, but several lines of evidence suggested that the brain α-bungarotoxin-binding protein is not identical to the muscle nicotinic receptor (reviewed in: Lindstrom et al., 1987). In 1980, Romano and Goldstein reported that rat brain contained a high affinity binding site for $[^3H]$-nicotine. Numerous studies since that time have demonstrated that α-bungarotoxin and nicotine bind to different sites in brain. We (Marks et al., 1986) and Wonnacott (1986) have argued that the $[^3H]$-nicotine binding site is a high affinity nicotinic receptor and that the α-bungarotoxin binding site based primarily on the affinity of the compound for these two binding sites.

Variation in response to nicotine is axiomatic. For nearly every response, both stimulant and depressant effects can result, depending upon dose and time after administration. These differences in response may be due to differences in the test procedures used or to differences in dose, but genetic factors are also important. Both rat (Battig et al., 1976) and mouse (Marks et al., 1983; 1989a) strains differ in sensitivity to nicotine. These differences in response to nicotine are not due to nicotine metabolism differences (Hatchell and Collins, 1980; Petersen et al., 1984) which suggests that genetic factors regulate tissue sensitivity to nicotine.

Although not used very often, genetic strategies can be very useful in answering questions about mechanism of drug action. For example, if mouse strains that differ in sensitivity to nicotine also differ in some aspect of nicotinic receptor biochemistry evidence will have been obtained that the receptor difference explains the sensitivity difference.

Pharmacological evidence, primarily obtained from studies with antagonists, is also useful in determining mechanism or site of action. In addition, antagonist studies can be useful in identifying receptor subtypes. For example, the observation that responses to nicotine elicited by application of the drug to autonomic ganglia are most effectively inhibited by hexamethonium whereas nicotine's effects on skeletal muscle are most effectively inhibited by decamethonium was useful in establishing that the ganglionic and skeletal muscle nicotinic receptors are different. Given that brain and peripheral nervous system nicotinic receptors are not identical, it is not surprising that much of the pharmacology of brain nicotinic receptors is not known. However, Freund et al. (1990) have reported that hexamethonium failed to block nicotine effects in hippocampal slices obtained from mouse brain whereas mecamylamine was very effective in this regard which suggests that a study of mecamylamine's effects on nicotine actions might provide insight into those receptor systems that regulate response to nicotine.

The studies reported here used two approaches, genetic correlation and pharmacological antagonism, to attempt to identify which brain nicotinic receptors control several behavioral and physiological responses to nicotine.

MATERIALS AND METHODS

Animals: The genetic studies used mice from 19 inbred strains: A/J, AKR/J, BUB/BnJ, C3H/2J, C57BL/6J, C57BL/10J, C57BR/cdJ, C57L/J, C58/J, CBA/J, DBA/1J, DBA/2J, LP/J, P/J, RIIIS/J, SJL/J, ST/bJ and SWR/J. The antagonism studies used mice from the C3H/2 and DBA/2 strains. All mice were 60-90 days of age at the time of testing.

Measurement of Nicotine Sensitivity: Varying doses of nicotine were injected intraperitoneally and the effects of this injection, or injection with an isotonic saline solution, were measured using a multifactorial test battery (Marks et al., 1985). Respiration rate was measured using a Columbus Instruments Respiration Rate Monitor starting 1 min after injection. At 3 min after injection the acoustic startle response was measured using a Columbus Instruments Reponder Startle reflex Monitor. Locomotor and rearing activities were measured for a 3 min time period in a symmetrical Y-maze starting 5 min after injection. Heart rate was measured 9 min after injection with an E & M Physiograph and body temperature was measured at 15 min after injection using a Bailey Instruments rectal probe. These time points were chosen from time course analyses of nicotine actions. In other studies, animals were injected with higher doses of nicotine and whether this dose did, or did not, elicit a seizure was recorded.

Dose-response curves for each of the nicotine effects were constructed for each strain. The resulting data were analyzed by calculating "ED" values: respiration, ED_{260} (the dose required to elevate respiratory rate to 260 breaths/min); startle response, slopes of the dose-response curves were calculated; Y-maze crosses and rears, ED_{50} (the dose that reduced these activities by 50%); heart rate, ED_{-100} (the dose that reduced heart rate by 100 beats/min); body temperature, ED_{-2} (the dose that reduced body temperature by 2°); seizures, ED_{50} (the dose that elicited seizures in 50% of the animals).

Antagonism Studies: Mecamylamine antagonism of these same nicotine effects was determined by pretreating C3H and DBA/2 mice with saline or varying doses of mecamylamine 10 min before challenge with nicotine (Collins et al., 1986). A 1.5 mg/kg dose was used in the DBA/2 strain and a 2.0 mg/kg dose was used in the C3H mice for the test battery responses and seizure antagonism was monitored following challenge with 3.75 mg/kg (C3H) or 6.75 mg/kg (DBA) doses of nicotine. The mecamylamine dose that inhibited each of the responses to nicotine by 50%, the ED_{50} value, was determined.

Measurement of Nicotinic Receptors: The binding of $[^{3}H]$- nicotine and α-$[^{125}I]$-bungarotoxin to membranes prepared from eight brain regions (cortex, midbrain, hindbrain, cerebellum, hippocampus, striatum, hypothalamus, and colliculli) were measured using the methods described previously (Marks et al., 1986). Complete saturation curves were constructed for each of the 19 inbred mouse strains in each of these regions. The total number of binding sites (B_{max}) and affinity (K_D) were calculated using Scatchard analysis of the binding curves.

RESULTS

The 19 mouse strains differed dramatically in their relative sensitivities to nicotine as is evident from the data presented in table I, below. The strains varied in sensitivity for all of the nicotine effects; maximal differences in sensitivity, as measured by the "ED" values, were 3-4 fold.

TABLE I

SENSITIVITY OF FIVE INBRED MOUSE STRAINS TO NICOTINE

STRAIN	Y-MAZE CROSS (ED_{50})	Y-MAZE REAR (ED_{50})	TEMP (ED_{-2})	SEIZURES (ED_{50})
C3H/2	1.78 ± .33	1.50 ± .10	1.32 ± .09	3.13 ± .09
DBA/2	0.97 ± .31	0.80 ± .06	0.89 ± .19	5.21 ± .12
C57BL/6	0.51 ± .18	0.45 ± .18	0.80 ± .16	5.30 ± .26
ST/b	0.93 ± .21	0.64 ± .27	1.47 ± .23	2.34 ± .09
BUB	1.89 ± .33	1.32 ± .36	2.53 ± .08	4.52 ± .06

The "ED" values are reported as the mean ± S.E.M. in mg/kg.

The data from the 19 inbred strains were subjected to factor analysis; two major factors were extracted. One of these loaded heavily on the two Y-maze tests and body temperature. A lesser loading was seen for heart rate. The second factor loaded heavily on the two seizure measures. Both factors loaded on the respiratory rate and acoustic startle tests. The most parsimonious explanation for this result is that two major genes regulate nicotine effects.

The effects of pretreatment with mecamylamine on the various responses to nicotine were determined in two of the mouse strains, the DBA/2 and C3H/2. The ED_{50} values for mecamylamine inhibition of nicotine responses are presented in table II. Mecamylamine pretreatment inhibited all of the nicotine effects; the missing DBA/2 values arose because the challenge dose of nicotine did not have a replicable effect on the respiratory rate and acoustic startle response tests in this mouse strain. The ED_{50} values for mecamylamine inhibition of nicotine response fell into two ranges, one near 1.0 mg/kg and the other near 0.1 mg/kg.

TABLE II

MECAMYLAMINE INHIBITION OF NICOTINE EFFECTS

TEST	DBA	C3H
Respiration rate	---	0.96 ± 0.22
Y-Maze crosses	1.36 ± 0.09	2.29 ± 0.55
Y-maze rears	1.39 ± 0.11	1.92 ± 0.40
Heart rate	0.88 ± 0.17	0.89 ± 0.09
Body temperature	0.96 ± 0.19	0.85 ± 0.48
Enhanced startle	---	0.09 ± 0.01
Clonic seizure	0.09 ± 0.01	0.07 ± 0.02

All data are presented as the mean ± S.E.M in mg/kg.

The mouse strains also differed from one another with respect to receptor binding in most, but not all, brain regions. Very little variance was detected in cortex and cerebellum, for example. Table III presents the maximal (B_{max}) binding of the two ligands in two brain regions from the five strains. As can be seen from the representative data presented in this table, the mouse strains vary by as much as a factor of two in number of nicotinic receptors in these two brain regions. No significant differences in K_D values were detected.

Table IV presents the results of a correlational analysis of the potential relationship between nicotine effects on the Y-maze activity-temperature measures and seizures and "overall" brain [^{3}H]-nicotine and α-[^{125}I]-bungarotoxin binding. This "overall" measure represents a weighted average of the binding in the brain regions that showed variance among the strains for each ligand. As is evident from this table, nicotine binding is highly correlated to the activity-temperature measures while α-bungarotoxin binding is highly correlated to the seizure measures. The correlations betwen the two binding sites were very low.

TABLE III

BINDING OF NICOTINE (NIC) AND BUNGAROTOXIN (BTX) IN TWO BRAIN REGIONS

STRAIN	HYPOTHALAMUS		HIPPOCAMPUS	
	NIC	BTX	NIC	BTX
C3H/2	90.6 ± 7.7	47.1 ± 2.3	53.0 ± 4.2	80.8 ± 2.4
DBA/2	78.9 ± 3.4	34.1 ± 1.5	32.8 ± 1.3	54.0 ± 1.7
C57BL/6	67.5 ± 10.1	38.8 ± 5.7	50.1 ± 2.6	77.5 ± 3.2
ST/b	69.2 ± 20.1	50.7 ± 3.4	30.4 ± 2.4	94.9 ± 3.1
BUB	52.8 ± 2.8	38.1 ± 2.0	43.6 ± 2.8	76.2 ± 3.1

All binding values are presented as the mean ± S.E.M. in fmoles/mg protein.

TABLE IV

CORRELATIONS AMONG RESPONSES TO NICOTINE AND BRAIN NICOTINIC RECEPTORS

	Nicotine Binding	Activity-Temperature	BTX Binding	Seizures
Nicotine	---	-0.62	+0.31	-0.27
Act-Temp		---	+0.10	+0.15
BTX			---	-0.63

The correlations presented in table V were calculated from averaged data for both the behavioral and biochemical measures. Different outcomes might be obtained if binding in specific brain regions is compared with specific nicotine effects.

DISCUSSION

The results presented here provide evidence that all of the responses to nicotine may not be mediated by the same receptors. The genetic experiments suggest that the seizure measures are regulated by the α-bungarotoxin binding site. Since seizures are elicited by high doses of nicotine it is not entirely surprising that the best correlation between response and receptor number was detected for the low affinity, for nicotine, α-bungarotoxin binding site. However, if two traits are highly correlated it is not necessarily the case that a causal relationship exists between the two. Nonetheless, we suspect that an examination of the function of the α-bungarotoxin binding site in a brain region that regulates nicotine-induced seizures, such as the hippocampus, may prove valuable in resolving the controversy over the function of this binding site.

The observation that mecamylamine blocks nicotine-induced seizures at lower doses than are required to antagonize all of the other responses, other than the acoustic startle response, also argues that seizures are mediated by a different mechanism than are the other responses. However, it is not clear which of the nicotinic receptors are affected by mecamylamine.

In the last few years molecular biological methods have identified a wide variety of nicotinic receptor subtypes. At this stage, we know virtually nothing about the function of these subtypes. We suspect that the strategies outlined in this report (genetic correlation and antagonism studies) will prove useful in determining the functions of these binding sites.

CONCLUSION: The results presented here indicate that more than one functional form of the nicotinic receptor family exists. Our results suggest that the site labeled with [^{3}H]-nicotine regulates nicotine effects on locomotor activities and body temperature and the α-bungarotoxin binding site regulates nicotine-induced seizures. The regulation of other responses is more complex which may mean that more than one receptor subtype is invloved in regulating these effects of nicotine.

ACKNOWLEDGMENTS: This work was supported by DA-03194 and DA-00116.

REFERENCES

Battig, K., Driscoll, P., Schlatter, J., and Uster, H.J. (1976) Pharmacol. Biochem. Behav.4, 435-439.
Clarke, P.B.S. (1987) Psychopharmacology. 92, 135-143.

Freund, R.K., Jungschaffer, D.A., and Collins, A.C. (1990) Brain Research. 511, 187-191.

Hatchell, P.C. and Collins, A.C. (1980) Psychopharmacology (Berlin) 71, 45-49.

Henningfield, J.E. (1984) Adv. Behav. Pharmacol. 4, 131-210.

Lindstrom, J., Schoepfer, R., and Whiting, P. (1987) Molecular Neurobiology 1, 281-337.

Marks, M.J., Burch, J.B. and Collins, A.C. (1983) J. Pharmacol. Exp. Ther. 226, 291-302.

Marks, M.J., Romm, E., Bealer, S.M. and Collins, A.C. (1985) Pharmacol. Biochem. Behav. 23, 325-330.

Marks, M.J., Stitzel, J.A., Romm, E., Wehner, J.M., and Collins, A.C. (1986) Molecular Pharmacology 30, 427-436.

Marks, M.J., Romm, E., Campbell, S.M. and Collins, A.C. (1989a) Pharmacol. Biochem. Behav. 33, 679-689.

Marks, M.J., Stitzel, J.A. and Collins, A.C. (1989b) Pharmacol. Biochem. Behav. 33, 667-678.

Petersen, D.R., Norris, K.J., and Thompson, J.A. (1984) Drug Metab. Dis. 12, 725-731.

Romano, C., Goldstein, A. and Jewell (1981) Psychopharmacology (Berlin) 74, 310-315.

Russell, M.A.H. (1976) In: Research Advances in Alcohol and Drug Problems (R.J. Gibbins, Y. Israel, H. Kalant, R.E. Popham, W. Schmidt, R.G. Smart, Eds), Wiley, New York, pp. 1-47.

Wonnacott, S. (1986) J. Neurochemistry 47, 1706-1712.

INFLUENCE OF NEONATAL NICOTINE EXPOSURE ON DEVELOPMENT OF NICOTINIC RECEPTOR SUBTYPES IN BRAIN AND BEHAVIOUR IN ADULT MICE

Xiao Zhang[1], Per Eriksson[2], Anders Fredriksson[3], and Agneta Nordberg[1]

[1]Department of Pharmacology, [2]Department of Zoophysiology, and [3]Department of Toxicology, Uppsala University, Uppsala, Sweden.

SUMMARY: Neonatal exposure of nicotine influences the development of nicotinic receptor subtypes in brain. The displacement curves of [3H]nicotine/(-)nicotine in adult mice cortex indicate both high and low affinity nicotinic binding sites for saline treated animals, while only high affinity nicotinic binding site for neonatally nicotine treated animals. Neonatal nicotine exposure also changes the behavioural effect of nicotine in adult mice.

INTRODUCTION

Receptor binding studies indicate that there are nicotinic receptor subtypes with different affinity to nicotine ligands in adult rodent brains (Larsson and Nordberg, 1985; Wonnacott, 1987; Zhang et al., 1987). These observations were supported by molecular biological studies showing a whole family of genes coding for nicotinic receptor subunits in rodent brain (Patrick et al., 1989). The number of nicotinic receptors gradually increases after birth and reaches adult level at age of one month in rodent brain (Larsson et al., 1985; Slotkin et al., 1987; Zhang et al., 1990). Studies have shown that during the development only one population of nicotinic receptors with high affinity is present in the cortex of 1-day and 5-day old mice, while in adult mice cortex both high and low affinity nicotine binding sites can be detected (Zhang et al., 1990). Several studies indicate that repeated exposure of nicotine to mice or rats can alter the number of nicotine binding sites for various nicotinic ligands in the brain and affect the behaviour of the animals (Falkeborn et al.,1983; Schwartz and Kellar, 1983; Nordberg et al., 1985; Marks et al., 1986; Larsson et al., 1986; Nordberg et al., 1989). The aim of this study is to investigate the influence of neonatal nicotine exposure on development of nicotinic receptor subtypes, as well as behaviour in adult mice.

METHODS

Male NMRI mice received (-)nicotine (nicotine-bi-d-tartrate, Sigma) (66μg base/kg bw, s.c.) twice a day (8 am and 5 pm) between 10-16 postnatal day. The control group received saline in the same way. The mice were kept until adult age of 4 months, when they after behavioural tests were killed and their brain cortices were used for binding assays.

[^{3}H]nicotine (5 nM) (62 Ci/mmole, Amersham) was incubated with cortical membranes and different concentrations of (-)nicotine (10^{-10}M – 10^{-4}M) in 50 mM Tris-HCl buffer (pH 8.0) at $4°C$ for 40 min. The samples were then filtered through Whatman GF/C glass filters. The filters were washed three times with assay buffer and the radioactivity was counted. The specific binding was determined in the presence of 10^{-3} M unlabeled (-)nicotine.

Spontaneous behaviour and induced nicotine behaviour were tested in mice at the age of four months. Motor activity, observed as total activity was measured in an automated device, Rat-O-Matic® (ADEA Elektronik AB, Uppsala, Sweden) as earlier described (Eriksson, et al., 1990).

The data obtained from binding assays were computer analyzed using EBDA/LIGAND program. For statistical evaluation of the behavioural data, ANOVA with a solit-plot-design was used (Kirk, 1968)

RESULTS AND DISCUSSION

The total number of [^{3}H]nicotine binding sites in adult mouse cortex was not changed in neonatally nicotine treated mice compared to saline treated mice (see table I, R$_H$ and R$_L$). The computer-assisted analysis of the displacement curves for [^{3}H]nicotine/(-)nicotine indicated that for saline treated mice the data were best fitted to a two-site model (F=5.65; P=0.014), while the data for neonatally nicotine treated mice were best fitted to an one-site model (F=0.66; P=0.531) with absence of low affinity binding component (Table I). An interconversion of low-affinity nicotinic binding sites to high-affinity binding sites in brain has been reported following subchronic nicotine treatment of adult rats (Romanelli, et al., 1988). The present study shows that neonatal treatment with nicotine can prevent/interrupt the development of nicotinic receptor subtypes in brain.

When the neonatally nicotine exposed mice were subjected to spontaneous and nicotine induced behavioural tests, there was no significant change in the spontaneous behaviour in the adult mice (Fig. I). However, nicotine-induced behaviour revealed that

Table I: The computer-assisted analysis with EBDA/LIGAND program of the displacement by (-)nicotine for [^{3}H]nicotine binding to the cortices of 4-month-old mice.

	K$_H$	K$_L$	R$_H$	R$_L$
	(nM)		(pmoles/g protein)	
NaCl	3.60 ± 1.10	123 ± 42	22.2 ± 6.1	5.5 ± 1.4
			(80.2%)	(19.8%)
(-)Nicotine	2.97 ± 0.36		28.0 ± 1.0	

Each value is the mean ± SE of three independent experiments.

neonatally nicotine exposed mice displayed a hypoactive condition whereas the saline-treated mice displayed a hyperactive condition, when receiving 40 µg or 80 µg nicotine /kg bw (Fig. I). A hyperactive condition is earlier known to be induced in naïve adult mice receiving the same amount of nicotine as in the present study (Nordberg and Berg, 1985).

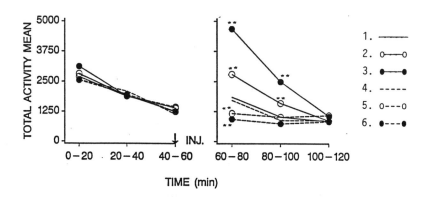

Fig.1: Spontaneous (A) and nicotine induced (B) behaviour in 4-month-old mice. Mice received 66 µg nicotine base/kg bw (group 4,5,6) or 10 ml saline/kg bw (control group 1,2,3) between the 10-16 postnatal day. At 4-month of age the motor activity was measured over 3 x 20 min period in a Rat-O-Matic®. The nicotine induced behaviour was studied by using two different doses of nicotine; 40 µg/kg bw for groups 2 and 5; 80 µg/kg bw for groups 3 and 6; 10 ml/kg bw saline for groups 1 and 4. The statistical difference from control is indicated by ** $p < 0.01$.

CONCLUSION: The present study shows that nicotine, when given to mice during their rapid brain growth period, can influence the development of nicotinic receptor subtypes in brain to consist of only high affinity nicotinic binding sites in the mature adult animal. An early exposure to nicotine also results in an opposite behaviour reaction to nicotine in the adult animal compared to saline treated mice, resulting in hypoactivity and hyperactivity , respectively.

Since the nicotine dose used in this study was low it might

have a physiological implication for the maturation of the human brain when exposed to nicotine during the perinatal period.

Acknowledgement: This study was financially supported by grants from the Swedish Medical Research Council, The Swedish Tobacco Company, The Bank of Sweden Tercentenary Foundation.

REFERENCES

Eriksson, P., Nilsson-Håkansson, L., Nordberg, A., Aspberg, A., and Fredriksson, A., (1990) NeuroToxicol. in press.

Falkeborn, Y., Larsson, C. and Nordberg, A.(1983) Drug Alcohol Depend. 8, 51-60

Kirk R.E. (1968) California, Belmont, Brooks/Cole.

Larsson, C., Nilsson, L., Halén, A., and Nordberg, A. (1986) Drug Alcohol Depend. 17, 37-45

Larsson, C. and Nordberg, A. (1985) J. Neurochem. 45, 24-31

Larsson, C., Nordberg, A., Falkeborn, Y. and Lundberg, P-Å. (1985) Int. J. Devel. Neurosci.3, 667-671.

Marks,M.J., Stitzel, J.A., Romm, E., Wehner, J.M. and Collins, A.C. (1986) Mol. Pharmacol. 30, 427-436

Nordberg, A. and Bergh, C. (1985) Acta Pharmacol. et Toxicol. 56, 337-341

Nordberg, A., Romanelli, L., Sundwall, A., Bianchi, C., and Beani, L. (1989) Br. J. Pharmacol. 98, 71-78.

Nordberg, A., Wahlström, G., Arnelo, U. and Larsson, C. (1985) Drug Alcohol Depend. 16, 9-17.

Patrick, J., Boulter, J., Deneris, E., Wada, K. Wada, E., Connolly, J., Swanson, L. and Heinemann, S., (1989) In: Progress in Brain Research, Vol. 79 (Nordberg, A., Fuxe, K., Holmstedt, B. and Sundwall, A., eds) Elsevier, Amsterdam, pp. 22-33

Romanelli, L., Öhman, B., Adem, A. and Nordberg, A. (1988) Eur. J. Pharmacol. 148,289-291

Schwartz, R.D. and Kellar,K.J.(1983) Science 220, 214-216

Slotkin, T.A., Orband-Miller, L. and Queen, K.L. (1987) J. Pharmacol. Expt. Ther. 242, 232-237.

Wonnacott, S. (1987) Human Toxicol. 6, 343-353

Zhang, X., Stjernlöf P., Adem, A. and Nordberg, A. (1987) Eur. J. Pharmacol. 135, 457-458

Zhang, X., Wahlström, G. and Nordberg, A (1990) Int. J. Devl. Neurosci. in press.

BEHAVIORAL INTERACTION BETWEEN NICOTINE AND CAFFEINE

Caroline Cohen (1), Hans Welzl, Karl Bättig

Swiss Federal Institute of Technology, Department of Behavioral
Sciences, Turnerstrasse 1, 8092 Zurich, Switzerland
(1) Present Address: NIDA, Addiction Research Center, P.O. Box
5180, Baltimore MD 21224, USA.

SUMMARY: The interactive effect of caffeine and nicotine on
spontaneous locomotion in a tunnel maze was determined in
nicotine-naive and nicotine-tolerant rats. Caffeine was found to
regulate the effect of nicotine on locomotor activity depending
upon previous nicotine exposure and level of activity.

INTRODUCTION

To date, only few data exist on the combined effects of
caffeine and nicotine on behavior (Lee et al. 1987, White 1988).
To investigate the interactive effect of caffeine and nicotine on
locomotor activity, nicotine-naive and nicotine-tolerant rats
were tested after caffeine and/or nicotine injections in a
hexagonal tunnel maze.

METHODS

The hexagonal tunnel maze consisted of an outer hexagonal
ring, an inner hexagonal ring and an open field in the center,
all connected through short radial arms.

Experiment 1: Male Wistar rats were injected subcutaneously with saline or nicotine (0.4 mg/kg) on 12 consecutive days (pretreatment: Days 1-12). On Days 13-14, the animals were exposed to the maze for a 6-min trial. On Day 15, the animals were injected intraperitoneally with saline, nicotine (0.2 mg/kg), caffeine (8 mg/kg), or nicotine (0.2 mg/kg) + caffeine (8 mg/kg). The injected rat was then placed into a holding cage for 15 min before being placed into the tunnel maze system for a 30-min trial.

Experiment 2: Male Wistar rats were daily exposed to the tunnel maze after subcutaneously injection of saline or nicotine (0.2 mg/kg). They were then sequentially exposed to caffeine dose of 4/8/16/32 mg/kg.

STATISTICAL ANALYSIS

The effect of treatments on locomotor activity was evaluated by performing repeated measures analysis of variance with Tukey post hoc tests.

RESULTS

Experiment 1 (Fig. 1): In nicotine-naive rats, nicotine decreased locomotor activity when compared with saline during the first 4 test intervals (all P's<0.05). Caffeine administered together with nicotine alleviated the decrease in locomotor activity induced by nicotine in nicotine-naive rats (second, third, and fourth intervals, P's<0.05). In nicotine-tolerant rats, caffeine plus nicotine increased locomotor activity when compared to saline for the third, fourth, and fifth interval (all P's<0.05).

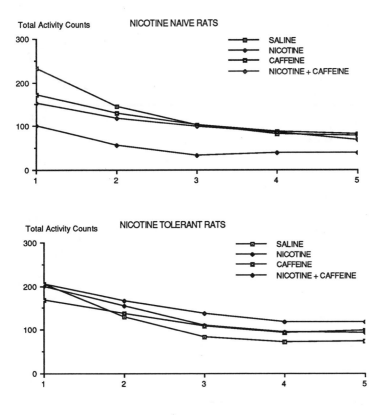

Fig. 1: Intrasession habituation for the successive 6-minute time intervals is plotted for the 4 treatments in nicotine naive rats (top panel) and in nicotine tolerant rats (bottom panel).

Experiment 2 (Fig. 2): Intersession habituation was not observed in nicotine treated rats. Caffeine (32 mg/kg) reduced the activity produced by nicotine in nicotine-treated rats (p<0.001) whereas caffeine (8 mg/kg) increased locomotor activity in saline-treated rats (p<0.01).

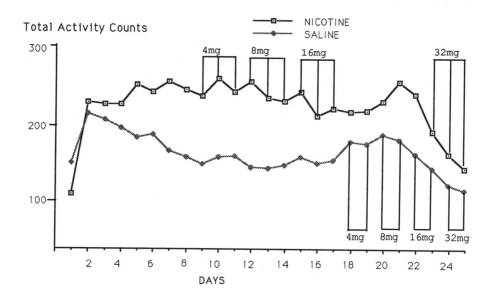

Fig. 2: Locomotor effects of caffeine (4/8/16/32 mg/kg) in nicotine-treated rats and saline-treated rats (6-min trial).

CONCLUSION

Our results suggest that caffeine regulates the effects of nicotine on locomotor activity in nicotine-naive and nicotine-tolerant rats. In nicotine-naive rats, the interaction of the effects of nicotine and caffeine results in an abolition of the locomotor depressant effects of nicotine by caffeine. In nicotine-tolerant rats, the locomotor effects of nicotine and caffeine depend upon the level of activity induced by nicotine alone.

REFERENCES

Lee, E.H., Tsai M.J., Tang Y.P., Chai C.Y. Differential biochemical mechanisms mediate locomotor stimulation effects by caffeine and nicotine in rats. Pharmacol. Biochem. Behav. 26:427-430; 1987.

White, J.M. Behavioral interactions between nicotine and caffeine. Pharmacol. Biochem. Behav. 29:63-66; 1988.

EFFECTS OF PRE- AND POSTNATAL INJECTIONS OF "SMOKING DOSES" OF
NICOTINE, OR VEHICLE ALONE, ON THE MATERNAL BEHAVIOR AND SECOND-
GENERATION ADULT BEHAVIOR OF ROMAN HIGH- AND LOW-AVOIDANCE RATS

P. Driscoll, C. Cohen, P. Fackelman, H.P. Lipp and K. Bättig

Labor für Verhaltensbiologie, ETHZ, Turnerstrasse 1, 8092-Zürich,
Switzerland, and Anatomisches Institut der Universität Zürich,
Winterthurerstrasse 190, 8057-Zürich, Switzerland

SUMMARY: Genetically-influenced (line-specific) effects of pre-
and postnatal injections of physiological saline (stressor alone)
or "smoking doses" of nicotine were found on several traits of
maternal behavior in 2 psychogenetically-selected lines of rats,
over 2 generations. The pre- and postnatal administration of
nicotine was also found to alter two-way avoidance acquisition
and hippocampal morphology in the line bred for superior shuttle
box performance, however in a functionally-opposed fashion.

Roman high- and low-avoidance (RHA/Verh and RLA/Verh) rats are

selected and bred for the rapid vs non-acquisition of two-way

active avoidance behavior, respectively. Based on neurochemical,

hormonal and behavioral evidence, the RLA/Verh line is considered

to be the more anxious one (Driscoll & Bättig, 1982). Differences

in the sensitivity to nicotine between the lines have long been

recognized in connection with locomotor tasks (Bättig et al,

1976). As part of a series of studies measuring the effects of

pre- and postnatal stress, nicotine and alcohol on the behavior

of RHA/Verh and RLA/Verh rats, the present studies involved the

interaction of pre- and postnatal nicotine administration and

injection stress (e.g. Grimm & Frieder, 1987) on 2 generations of

maternal behavior of females, as well as on the 2-way avoidance

behavior of adult males derived from the affected litters.

MATERIALS AND METHODS

Experiment (Exp.) 1: Using a time-sampling method like that

described previously (Driscoll et al, 1979), we compared several
aspects of maternal behavior in 40 RHA/Verh and RLA/Verh litters,
making 7 daily observations of 30 sec each on each cage, between
10:00 and 22:00 hr (the lights were on from 9:00 to 23:00 hr).
Most of the behaviors observed are indicated in Fig. 1. The
mothers of the litters were either undisturbed (true control) or
injected 3-4 times nightly with physiological saline (NACL) or
with 0.3 mg/kg nicotine (NIC), starting two weeks before birth
and continuing until two weeks after birth.

Exp. 2: Maternal behavior was also compared in 30 RHA/Verh and
RLA/Verh litters of the next generation. The mothers used were
derived either from control litters, or had been exposed through
their mothers, pre- and postnatally, to the multiple, nightly
NACL or NIC injections. They were otherwise experimentally naive.

Exp. 3: 5 month-old males, also derived from the Exp. 1 litters,
were tested for two-way avoidance behavior in a shuttle box. As
associations had been previously suggested to exist between a
role for the hippocampus and the behavioral effects of NIC in
these rat lines (Fitzgerald et al, 1986), as well as the extent
of the hippocampal intra/infrapyramidal mossy fiber projection
(IIP-MF) and a predisposition to acquire the avoidance response
in RHA/Verh vs RLA/Verh rats, the hippocampi of the RHA/Verh rats
involved were examined microscopically with Timm stain and all
component parts of the CA3/CA4 were measured (Schwegler & Lipp,
1983).

RESULTS

Exps. 1 and 2: There were no differences in the number of pups
born, or in pre-weaning mortality, among any of the conditions.
Basic differences in maternal behavior between the RHA/Verh and
RLA/Verh lines are shown in Fig. 1, based on observations of the
control rats over 3 generations. Some of the important results of
Exps. 1 and 2 can be seen in Table I. In addition to the data

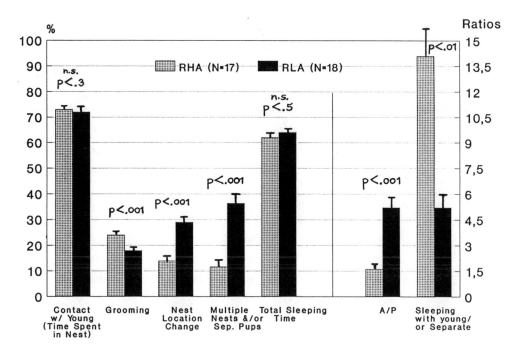

Fig. 1: Maternal behavior comparison between control RHA/Verh and RLA/Verh mothers during postnatal days 1-16, expressed as percent of observations or as ratios. S.E.M.s are shown. A/P = active/passive nursing position (see Fümm & Driscoll, 1981).

shown, RHA/Verh mothers also showed a significant reduction in the A/P ratio after NIC in Exp. 1, and the RLA/Verh rats showed a significant reduction in the "sleeping with pups/sleeping away from pups" ratio with NACL in both Exps. 1 and 2. It should be noted that effects of the stressor (NACL) alone were seen only in RLA/Verh rats, that none of those changes occurred with NIC, and that the results often encompassed both generations. RHA/Verh rats, on the other hand, showed no NACL effects. A significant NIC effect seen in both lines of rats, over both generations, was an increase in the total observed sleeping time.

Exp. 3: The most striking result with the shuttle box testing was that the RHA/Verh rats which had been exposed to NIC in utero and postnatally showed an attenuated acquisition of avoidance (20% vs

RLA/Verh groups (n)	Contact with young	Nest location changes	Multiple nests &/- or separated pups
Control (12)	70 (3.0) a,b	32 (1.1) A,B	40 (3.7) A,B
Exp. 1: NACL (7)	57 (3.1) B	23 (2.7) b	24 (4.0) b
Exp. 1: NIC (7)	70 (3.0) b	20 (3.7) b	26 (4.8) a
Exp. 2: NACL (6)	58 (4.2) A	22 (3.8) a	25 (9.8)
Exp. 2: NIC (6)	72 (2.4) a	27 (3.3)	36 (4.4)
Kruskal-Wallis:	p<.01	p<.02	p<.05
RHA/Verh groups (n)			
Control (11)	72 (1.4)	13 (1.8) B,C	9 (1.6) B
Exp. 1: NACL (7)	74 (3.5)	17 (2.2)	15 (4.1)
Exp. 1: NIC (6)	76 (2.8)	25 (2.0) b,c	19 (3.6) b
Exp. 2: NACL (4)	69 (1.8)	17 (5.2) +	13 (3.2) +
Exp. 2: NIC (5)	75 (2.0)	20 (3.5) b	21 (4.4) b
Kruskal-Wallis:	p<.8 (n.s.)	p<.05	p<.2 (n.s.)

Table I: Some results of Exps. 1 and 2, expressed as percent of observations during postnatal days 1-16 (with S.E.M.s). Further results may be found in the text. Control mothers were combined here, as no significant within-line differences were seen between the generations. The vertical statistical comparisons represent within-group calculations made with the Mann-Whitney U, 2-tailed test: A vs a = p<.05, B vs b = p<.02, C vs c = p<.001.

40/45% during trials 1-10, and 45% vs 80/65% during trials 11-20) compared to non-handled controls and NACL rats, respectively. It was also found that pre- and postnatal NIC reduced the extent of IIP-MF (1.74 vs 2.16 %vol of CA3/CA4; p<.05).

DISCUSSION

As the moderate change in IIP-MF seen in Exp. 3 was actually in the opposite direction to that which would have been expected according to the shuttle box results (Schwegler & Lipp, 1983), it was concluded that the behavioral effects of NIC seen were most likely mediated elsewhere. One good possibility would be the mesolimbic/nigrostriatal dopaminergic pathways, where profound differences in metabolism between RHA/Verh and RLA/Verh rats have reflected their divergent, inherent locomotor and emotional characteristics (D'Angio et al, 1988; Driscoll et al, 1990). NIC has often also been shown to have an affinity for these pathways.

Exps. 1 and 2 demonstrated genetic differences in the effects of injection stress vs NIC, both on an acute and pre-/postnatal basis. Whereas, in the RHA/Verh rats, the NACL condition could be considered to be what would normally constitute a "vehicle control", upon which the éffects of NIC were superimposed, NACL alone produced the most dramatic changes in the RLA/Verh rats. Two implications of these results should be briefly considered. First of all, many studies using "non-selected" stocks of rats can be assumed to be, in fact, using animals similar to either of the Roman lines, i.e. largely stress-resistant or stress-suscept-ible ones. Exps. 1 and 2 have clearly shown that conclusions about "rats" based on results with only one such stock would, at best, be limited in value. In addition, the results with the RLA/Verh mothers emphasize the importance of including additional controls (i.e. non-injected rats) in types of studies in which "control injections" might in themselves act as stressors, perhaps through causing permanent alterations in the hypothalamic control of the hypothalamic-pituitary-adrenal axis (Peters,1982).

REFERENCES

Bättig, K., Driscoll, P., Schlatter, K., and Uster, H.J. (1976) Pharmacol. Biochem. Behav. 4, 435-439.
D'Angio, M., Serrano, A., Driscoll, P., and Scatton, B. (1988) Brain Res. 451, 237-247.
Driscoll, P., and Bättig, K. (1982) In: Genetics of the Brain (I. Lieblich, Ed), Elsevier Biomedical Press, Amsterdam, pp 95-123
Driscoll, P., Fümm, H., and Bättig, K. (1979) Experientia 35, 786-788.
Driscoll, P., Dedek, J., D'Angio, M., Claustre, Y., and Scatton, B. (1990) In: Farm Animals in Biomedical Research (Advances in Animal Breeding and Genetics, Vol. 5) (V. Pliska and G. Stranzinger, Eds), Verlag Paul Parey, Hamburg, pp 97-107
Fitzgerald, R.E., Driscoll, P., and Bättig, K. (1986) Behav. Brain Res. 20, 95-96.
Fümm, H., and Driscoll, P. (1981) Anim. Behav. 29, 1267-1269.
Grimm, V.E., and Frieder, B. (1987) Int. J. Neurosci. 35, 67-72.
Peters, D.A.V. (1982) Pharmacol. Biochem. Behav. 17, 721-725.
Schwegler, H., and Lipp, H.P. (1983) Behav. Brain Res. 7, 1-38.

VI.
Behavioral Effects of Nicotine in Man

THE PLEASURES OF NICOTINE

David M. Warburton

Department of Psychology, University of Reading, Building 3, Earley Gate, Whiteknights, Reading RG6 2AL. England

SUMMARY: Nicotine use is described as pleasurable, but the effects are rather subtle. Mood and performance tests show that there is both mild pleasurable stimulation and mild pleasurable relaxation. It is this unique combination which makes the pleasure of nicotine use distinctive.

INTRODUCTION

In this chapter, I will present evidence on the nature of the pleasurable effects of nicotine. The first point to be made is that the psychopharmacological effects of nicotine at smoking doses are very subtle, in comparison with the prototypical euphoriants, heroin and cocaine. Heroin and cocaine users report a strong pleasurable thrill which some users describe in sexual terms (Wikler, 1953).

Anyone who has experienced nicotine would agree that the pleasure from this substance is not comparable to those described for opiates and cocaine. Indeed, anyone who has experienced both alcohol and nicotine would agree that the pleasurable effects of alcohol and nicotine are dissimilar, rather smokers report mild mood effects, which have proved hard to quantify.

In our own work, we have tried to measure the pleasurable effects of nicotine that are obtained by smoking. It has seemed to us that the concept of "pleasure" covers two separable experiences, "pleasurable stimulation" and "pleasurable relaxation". We tested 139 subjects for their recall of their experience of different substances and activities. In this way, we obtained ratings of the pleasurable-stimulation and pleasurable-relaxation of a set of substances and activities, including tobacco. These ratings enabled us to derive a comparative ranking of the substances and activities on these two kinds of pleasure.

For our purposes, the most interesting comparisons were those of tobacco use with other substances and activities. Alcohol, amphetamines, amyl nitrite, cocaine, glue, heroin, marijuana and sex were significantly more stimulating than tobacco. Sleeping tablets and tranquilizers were significantly less stimulating than tobacco, while there were no statistically reliable differences between tobacco, coffee or chocolate, in terms of pleasurable stimulation.

On the pleasurable relaxation dimension, alcohol, heroin, sex, sleeping tablets and tranquilizers were significantly more relaxing than tobacco. Amphetamine, amyl nitrite, cocaine, coffee and glue were significantly less relaxing than tobacco, while there were no statistically reliable differences between tobacco and chocolate in terms of pleasurable relaxation.

In a related study, Kozlowski et al (1989) has also done a retrospective comparison of pleasure from cigarettes with other substances. He asked some problem users to compare smoking with their own problem substance. Of the problem users of alcohol, 57.5 percent said that their pleasure from cigarettes was less than for alcohol, 28.7 percent said it was similar and only 13.8 percent rated smoking as giving more pleasure than alcohol. Of the cocaine users, only 2.7 percent said their pleasure from cigarettes was greater, 1.4 percent said their pleasure was similar, 96 percent said their pleasure was less strong. These data show clearly that smokers do not experience cigarettes as strongly pleasurable, like cocaine and alcohol, but only as mild pleasurable effects.

In psychopharmacology, mood effects of drugs are assessed

with visual analogue scales with two adjectives at each end, such as happy-sad, calm-excited, etc. Using this sort of test, we have found that a prescribed anxiolytic, such as diazepam, or an over-the-counter anxiolytic, such as alcohol, produces greater calmness at the expense of decreased alertness and clear-headedness.

In our studies of the puff-by-puff mood effects of cigarettes (Warburton, Revell and Walters, 1988) subjects completed a set of Bond-Lader mood scales after each puff. For all cigarettes, there were significant increases in calmness on the calm-excited scale, when measured after successive puffs. Subjects also reported that they became more tranquil, more sociable, more friendly, more contented, more relaxed and happier over successive puffs and these changes were highly significant. When a second cigarette was given 30 min later, (i.e. minimal deprivation), the mood changes increased until they were above the level that was achieved at the end of the first cigarette. The mood changes were correlated with plasma nicotine so that, as plasma nicotine increased throughout the cigarette, the mood changes increased. Non-nicotine cigarettes produced no improvements.

The changes on the "relaxation" set of scales were curvilinear, while the puff-by-puff changes on the "contentedness" were linear. These differences argue against the subjects merely giving "expected" answers but were making different assessments of each mood state. In addition, it suggests that there must be different biochemical mechanisms underlying the separate mood effects.

It was interesting to compare the effects of nicotine with those of alcohol and diazepam. All three produced very similar calming effects, although the time course of action of alcohol and diazepam was much longer. A cigarette was equivalent to 10 mg of diazepam or 0.5 grams per kilogram of alcohol, which would give blood alcohol concentrations of 50 milligram percent in males in terms of calmness. However, nicotine has these effects without impairing alertness.

In a previous paper, Pomerleau, Turk and Fertig (1984) found that anxiety, as measured by the Profile of Mood States Questionnaire, could be generated using anagrams. The anxiety was markedly decreased in all subjects more after smoking their own nicotine-containing cigarettes than after non-nicotine cigarettes, showing the beneficial effects of nicotine on mood. An important part of the study was that subjects were only deprived of cigarettes for half an hour prior to the study and so the results could not have been due to a reversal of nicotine withdrawal.

Murphree (1979) performed a study to determine whether smokers have significant tolerance to some of the effects of nicotine. As part of this study, Murphree asked his subjects to fill in a mood adjective checklist before and after nicotine infusions. Both smokers and non-smokers felt more energetic, happier and calmer. They felt less tired and less tense.

The finding that nicotine has pleasurable stimulation and pleasurable relaxing effects concurs with the results of an early smoking motives questionnaire of Ikard, Green and Horn (1969). Positive affect smoking had included the components stimulation, pleasurable relaxation and manipulation. The calming effects of smoking seem to be the experience of many smokers (McKennell, 1973). Surveys have found that 80% of smokers say that they smoke more when they are worried, 75% say that they light up a cigarette when they are angry and 60% feel that smoking cheers them up (Russell, Peto and Patel, 1974; Warburton and Wesnes, 1978). These opinions suggest that smokers feel that smoking improves their mood and that this is an important reason for smoking.

INFORMATION PROCESSING

Information processing is a term that was adopted from computing and refers to the processes that operate when a person is perceiving, attending, memorising, conceptualizing, imagining, judging and reasoning. Clearly, the normal person is continuously engaged in these activities throughout their waking day. Surveys have indicated that the majority of smokers believe that smoking helps them process information more efficiently, by aiding thought and concentration (Russell, Peto and Patel, 1974; Warburton and Wesnes, 1978).

Current research has focussed on the extent to which nicotine and smoking can improve attention and memory and whether the effects are attributable to a reversal of withdrawal, because most early studies tested people after nicotine abstinence. The belief of current smokers that smoking does improve processing agues against the idea that there is tolerance to the effects.

ATTENTION: One type of test for examining attention is vigilance. Vigilance tasks are highly monotonous and attention has to be directed to an input for long periods of time and the subject is required to detect and respond to brief, infrequent changes in the input. The original vigilance task was the Mackworth Clock (Mackworth, 1950), in which subjects watched a clock face with the clock hand rotating once a minute. Once every 30 seconds on average the hand stopped briefly. Performance was assessed in terms of the detection rate, i.e. the proportion of signals correctly detected. During a typical vigilance session, the detection rate decreased, a change called the vigilance decrement.

In a study of nicotine using the Mackworth Clock, smoking cigarettes at 20 min. intervals helped smokers who were previously deprived for 10 hours maintain their detection of experimental targets in the 80 min. vigilance task (Warburton and Wesnes, 1978). In contrast, detection dropped for a group of non-smokers from a baseline of 0.76 to 0.70 and from a baseline of 0.82 to 0.65 for a group of deprived smokers who were not allowed to smoke. There was no significant difference between deprived smokers and non-smokers in their performance, i.e. no evidence of a withdrawal effect. This suggests that smoking was not merely reversing deprivation but was enhancing performance.

In order to investigate the importance of tolerance, we designed an experiment (Wesnes, Warburton and Matz, 1983), which studied the effects of nicotine tablets (0, 1 or 2 mg) on non-smokers, light smokers (less than 5 per day) and heavy smokers (more than 15 per day) who were all undergraduates. During the task the subjects took nicotine or placebo tablets at 20, 40 and 60 min. The unbuffered nicotine tablets were held in the mouth for five minutes and then swallowed.

Nicotine reduced the vigilance decrement which occurred over time in the placebo condition (Wesnes, Warburton and Matz, 1983). The nicotine tablets produced the same effects in non-smokers, light smokers (less than 10 per day) and heavy smokers (more than 15 per day). These data give no evidence for tolerance.

Another attention task involves rapid visual information processing, in which a series of digits are presented at a rapid rate by computer and subjects are required to detect certain specified three-digit sequences. Measures of both the speed and the accuracy of detection are made in order to assess the ability of the subjects to sustain their attention.

In our initial studies (Wesnes and Warburton, 1984a), we found that smoking after 10 hours deprivation produced improved performance in terms of both speed and accuracy. Of course, these results could merely be a reversal of a withdrawal effect. Accordingly, we tested the same smokers after minimal deprivation of one hour and after 10 hours deprivation. There was no difference in the improved performance when the subjects were smoking after one hour or 10 hours deprivation, i.e. there was no evidence of desensitization to the effects.

In order to investigate the importance of tolerance, we tested doses of 0.5 mg, 1.0 mg and 1.5 mg of nicotine which were given to non-smokers. The taste of the nicotine was masked by capsaicin which was placed on both the active and placebo tablets.

Then the subjects performed the rapid visual information

processing task (Wesnes and Warburton, 1984b). The highest dose of nicotine (1.5 mg) produced a performance improvement in non-smokers which closely resembled that produced by smoking in smokers. The 1.0 mg dose had a lesser effect and 0.5 mg was without effect. This finding provides strong evidence that nicotine plays the major role in the improvements in focused attention tasks that are produced by smoking. The use of capsaicin as an irritancy control argues against the performance improvements being due to peripherally-mediated arousal. The improvement in the performance of non-smokers argues against the improvement being only a reversal of withdrawal.

In a study at the Institute of Psychiatry in London, the effects of subcutaneous doses of nicotine on information processing performance of patients with senile dementia of the Alzheimer type, were examined (Sahakian et al. 1989). Doses of nicotine produced a dose-related improvement in performance in the detection of signals in the rapid visual information processing task and they approached the performance of the normal elderly. The equivalent data for reaction time show that nicotine doses produce improvements in reaction times in the patients with senile dementia of the Alzheimer type in comparison with the baseline and placebo conditions. A much larger study by the same group have replicated this finding and a comparison of smokers with non-smokers found no differences in the effect of nicotine, which was in accord with the results with healthy volunteers.

The evidence from this section is consistent. Nicotine improves attentional processing capacity and the results do not depend on deprivation from nicotine and so the effects cannot be attributed to the reversal of withdrawal.

MEMORY: Information storage consists of input of the information, registration of the information in immediate, shorter term memory and consolidation of the information in longer term memory. However, information storage can only be demonstrated by the effective retrieval of the information from storage. Thus, it is important to distinguish a drug's effect on storage from an effect on retrieval. One method of doing this is by using a state dependent design and studies have shown that nicotine has no effect on retrieval (Warburton, Wesnes, Shergold and James, 1986).

First, I will consider the effects of smoking and, by implication, of nicotine on information input as shown by immediate memory.

Peeke and Peeke (1984), studied the effects of smoking on immediate memory in two hour deprived smokers. Recall of a 50-word list was tested immediately and after intervals of 10 and 45 minutes. Pretrial smoking of their own brand resulted in improved recall immediately after learning. In contrast, another study compared a low (0.40 mg) and high (1.38 mg) level of nicotine cigarette in light and heavy smokers using pretrial smoking. The high nicotine cigarette resulted in improved immediate recall. The low nicotine cigarette was less effective.

In our own work, we have been studying the effects of smoking and nicotine tablets on memory. Subjects were presented with a list of 32 words in sets of four words at a time. Deprived smokers were improved on the task, but, more importantly, non-smokers were improved after they had received nicotine tablets.

Of the delayed recall studies, Mangan and Golding (1983) studied the effects on memory of one hour deprived smokers and those smoking a single cigarette after one hour of deprivation but immediately after acquisition of a paired-associate learning task. Subjects were tested for retention of the memorized material at intervals of 30 minutes, one day, one week and one month. Nonsmokers showed superior recall compared with all smokers at the 30-minute retest. After one month subjects who smoked a 0.8 mg and 1.3 mg nicotine cigarette were better than non-smokers.

In a series of studies, Peeke and Peeke (1984) tested the effects of smoking one cigarette on verbal memory and attention in four experiments using smokers who were deprived for two hours, i.e. minimally deprived. In one study, subjects were tested when they smoked their own brand of cigarette prior to learning, after learning and not smoking. Recall of a 50-word list was tested immediately and after intervals of 10 and 45 min. Pre-learning smoking resulted in improved recall but there was no effect of smoking after learning, as Mangan and Golding (1983) had found.

The results of the memory studies suggest that it can be improved after minimal deprivation and by nicotine in non-smokers. Animal studies have demonstrated that information storage is improved in animals given a 'smoking dose' of nicotine, approximately 0.1 milligram per kilogram. In these sorts of doses, nicotine facilitated the acquisition of a shuttle-box avoidance, maze learning, and two-choice visual discrimination behaviour, when administered prior to training (e.g. Bovet et al, 1966). Post-trial nicotine treatment has been shown to result in facilitated retention of a variety of tasks (e.g, Battig, 1970). The finding of post-trial facilitation in animals confirms the conclusion from the human studies, that nicotine is affecting memory storage processes. Of course, these studies also demonstrate that the facilitation effects cannot be due to withdrawal effects, because the animals had been exposed to very little nicotine in their life.

PLEASURES OF NICOTINE

Previously, I have outlined a functional view of nicotine to explain nicotine use (Warburton, 1987). The functional approach regards a person's use of nicotine to control their psychological state. In this view, different smokers can smoke for different reasons and the same smoker may smoke for different effects on different occasions.

Other substances, such as alcohol and diazepam produce pleasurable relaxation and can be regarded as drugs of escape from problems. In contrast, caffeine produces pleasurable

stimulation and is used to help performance, but there is
increased tension. Figure 1 shows the changes in performance
and mood with alcohol, caffeine and diazepam, in comparison with
smoking from data that we described in earlier sections.

Percentage Changes In Performance And Mood

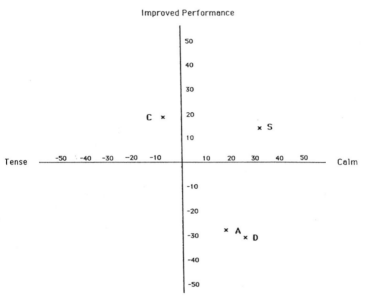

C= Caffeine
S= Smoking
A= Alcohol
D= Diazepam

Nicotine is unique in having both calming, as well as
stimulating effects. These effects occur in smokers, who were
only deprived of nicotine for a short time and nicotine produces
improved mood and performance in non-smokers. The improved
performance is important, because it can produce a different
sort of pleasure, the pleasure of coping more efficiently with
situations and being more successful. The experience of this
enhanced efficiency will reduce the person's anticipatory stress

about tasks.   In addition, the person will be better able to concentrate and ignore distressing and distracting thoughts.

In brief, the pleasure of nicotine for smokers comes from the enhanced mastery over all aspects of their lives and the way it enables them to confront everyday problems.

REFERENCES

Battig, K. (1970). Psychopharm. 18, 68-76.

Bovet, D, Bovet-Nitti, F,Oliverio, A. (1966). Effects of nicotine on avoidance conditioning of inbred strains of mice. Psychopharm. 10, 1-5.

Ikard, F F, Green, D E, Horn, D A. (1969) Int. J. Addict. 4, 649-59.

Kozlowski, L T, Wilkinson, D A, Skinner, W, Kent, C, Franklin, T, Pope M. (1989) J.A.M.A. 261, 898-901.

Mackworth, N H. (1950). Researches on the Measurement of Human Performance. Medical Research Council Special Report, (No. 268). London: HMSO.

Mangan, G L, Golding, J F. (1983). J. Psychol. 115, 65-77.

McKennell, A C. (1973). A Comparison of Two Smoking Typologies. Research Paper No. 12. London: Tobacco Research Council.

Murphree, H B. (1979). In: Electrophysiological Effects of Nicotine, (A Remond and C. Izard, eds), Elsevier - North Holland Biomedical Press, Amsterdam, pp 227-244.

Peeke, S C, Peeke, H V S. (1984). Psychopharm. 84, 205-216.

Pomerleau, O F, Turk, D C, Fertig, J B. (1984). Addict. Behav. 9, 265-271.

Russell, M A H, Peto, J, Patel, U A. (1974). J. Roy. Stat. Soc. A. 137, 313-333.

Sahakian, B, Jones, G, Levy, R, Gray, J, Warburton, D M. (1989) Brit. J. Psychiat. 154, 797-800.

Warburton, D M. (1987). The functions of smoking. In: Tobacco smoke and Nicotine. (W R Martin, G R Van Loon, E T Iwamoto, and D L Davis, eds.) Plenum Press, New York pp. 51-62.

Warburton, D M, Revell, A, Walters, A C. (1988). In: The Pharmacology of Nicotine. (M J Rand, K Thurau eds.), IRL Press, Oxford, pp. 359-373.

Warburton, D M, Wesnes, K. (1978). In: Smoking Behaviour (R E Thornton, ed) Churchill-Livingstone, Edinburgh, 19-43.

Warburton, DM, Wesnes K, Shergold K and James M (1986). Facilitation of learning and state dependency with nicotine. Psychopharm. 89, 55-59.

Wesnes, K, Warburton, D M. (1984a). Psychopharm. 82, 338-342.

Wesnes, K, Warburton, D M. (1984b). Psychopharm. 82, 147-150.

Wesnes, K, Warburton, D M, Matz, B. (1983). Neuropsychobiol. 9, 41-44.

Wikler, A. (1953) Opiate Addiction: Springfield, Ill. C.C. Thomas.

NICOTINE DEPRIVATION AND NICOTINE REINSTATEMENT: EFFECTS UPON
A BRIEF SUSTAINED ATTENTION TASK

A.C. Parrott and G. Roberts

Department of Psychology, Polytechnic of East London, London E15
4LZ.

SUMMARY: Twenty regular smokers (+15 cigarettes/day), were
tested on a brief letter cancellation task, over four successive
days. On one of the test days subjects were nicotine deprived
for +12 hours, while on the other days they were not nicotine
deprived. The first letter cancellation test was given prior
to smoking. Then, the following one cigarette, a second letter
cancellation test was given. Performance was significantly
impaired by nicotine deprivation, when assessed both by response
time (p<.05) and target detection (p<.05). Following cigarette
smoking, the performance of the previously deprived subjects
returned to baseline, while performance remained basically
unchanged in the other conditions. There was no evidence of
performance differences between high, mid and low frequency
letter targets, either during nicotine deprivation, or nicotine
reinstatement.   Thus,   while   sustained   attention   was
significantly affected by nicotine, there was no evidence of
changed attentional selectivity.

INTRODUCTION

Nicotine has been shown to affect various indices if human
performance, including memory, psychomotor control, and
sustained attention (reviewed in: Warburton and Wesnes, 1984;
Wesnes and Warburton, 1983; Surgeon General, 1988). Nicotine
deprivation has been shown to impair performance (Tarriere and
Hartman, 1964), while nicotine given to deprived subjects, has
been shown to improve performance (Williams, 1980; also reviews
by: Wesnes and Warburton, 1983; Surgeon General, 1988). While

investigations into the effects of nicotine upon human
performance, of necessity, use nicotine deprived subjects, the
interpretation of any performance change following nicotine
reinstatement is open to question. Does the improved
performance reflect an improvement above the normal baseline,
or does it represent a level of performance found prior to the
deprivation? In order to answer this question, the effects of
nicotine deprivation, and of nicotine reinstatement, need to be
assessed within the same experimental paradigm. This does seem
to have been undertaken before, and was the main aim of the
present study. A secondary aim was to assess whether nicotine
deprivation would be evident using only a brief attentional
task, since previous research demonstrating performance
decrements following nicotine deprivation had used prolonged
tasks of 1-6 hours duration (Tarriere and Hartmann, 1964; also
above reviews).

METHODS

Twenty nurses and medical students, in the age range 20-27
years, comprised the unpaid volunteers for the study. All were
regular smokers (+15 cigarettes/day).

Twelve matched versions of the letter cancellation task were
used. Each had 105 letter targets, scattered amongst the 1350
pseudo-random letters on each sheet. Each sheet contained three
stimulus targets: 60 high frequency, 30 mid frequency, and 15
low frequency. The stimulus letters were varied systematically
across response sheets. Different sheets were given at all
sessions. Pre-trial task training/practice was given on two
separate days.

Subjects were tested on four successive days. Days one and four
comprised baseline days. Then either on day 2 (half the
subjects) or day 3 (other half of the subjects), the volunteers

agreed to abstain from smoking before testing (12+ hours abstinence). Subjects were warned that non-smoking compliance would be randomly checked, using an expired breath CO monitor. On the other three days smoking was permitted up to 60 minutes before testing. As far as possible, subjects were tested at the same time each day, in the morning.

The first test session was given prior to smoking. This was followed by one cigarette, then a second letter cancellation test. The data for days 2 and 3 were analysed by 2x2 split-plot factorial analysis of variance, with the following terms: drug condition (deprived, non-deprived), subject group (group 1, group 2), and session (first, second). The last term also comprised the drug x group interaction. The pre and post cigarette data were analysed separately.

RESULTS

Performance was significantly impaired by nicotine deprivation, when assessed both by response time (p<.05) and target detection (p<.05: Table 1). In addition to these significant drug effects, there was a significant session (ie. group x drug) effect for letter cancellation error (p<.05). Group 2 (nicotine deprived on day 3) showed a greater change in letter cancellation error, than group 1 (nicotine deprived on day 2). It is not known whether this reflects a random (chance) factor, or a genuine performance effect. There was no evidence of a session effect with letter cancellation response time. On the second test following cigarette smoking, the performance of the previously deprived subjects returned to baseline, while performance remained basically unchanged in the other conditions.

Table 1. Group mean letter cancellation scores across the four test
sessions

	Session 1 non-deprived	Session 2/3 non-deprived	Session 3/2 deprived	Session 4 non-deprived
Reaction Time				
pre-cigarette	384	359	400*	372
post-cigarette	355	365	360	363
pre-post diff	29	-6	40*	9
Total Hits				
pre-cigarette	89.7	93.3	88.5*	93.0
post-cigarette	92.9	93.6	95.1	94.3
pre-post diff	3.2	0.3	6.6*	1.3

* $p < .05$ ANOVA drug effect (across sessions 2/3).

There was no evidence of a speed accuracy tradeoff; the
correlation between target detection change and response time
change was almost zero (Spearman r=.002, non-significant).
Throughout the experiment, there was no evidence of performance
differences between the high, mid and low frequency targets.
All showed reduced target detection under nicotine deprivation,
and near baseline values on nicotine reinstatement.

DISCUSSION

Nicotine deprivation led to an impairment in sustained
attention, while reinstatement of nicotine led to performance
close to baseline. In contrast, there was little evidence of
performance change following cigarette smoking, in the non-
deprived subjects. Several studies have shown that nicotine
deprivation can impair vigilance (Tarriere and Hartmann, 1964;
reviews by: Surgeon General, 1988; Wesnes and Warburton, 1983).
These studies have invariably involved classic vigilance types
of task, with overall duration of hour or more. The present
findings demonstrate that the effects of nicotine deprivation
can be measured on a brief task, as long as that task is
sensitive (Williams, 1980). The present findings also confirmed

the importance of using nicotine deprived subjects, when attempting to assess the performance effects of nicotine. For instance, Parrott et al (this symposium), compared three research groups investigating the performance effects of nicotine chewing gum. Two groups used fully nicotine deprived subjects (+12 hours Snyder/ Henningfield; +12 hours Parrott/Craig), and they each demonstrated performance effects with the nicotine gum. In contract, one group used only briefly deprived subjects (+2 hours Michel/Hasenfrantz), and found no performance change (although EEG changes were evident).

There was no evidence of altered attentional selectivity. High, medium and low probability targets were affected in similar ways, during nicotine deprivation, and nicotine reinstatement. It has been suggested that under high arousal, for instance as induced by loud noise, attention may narrow, or focus upon high probability events (eg. Hockey, 1970, 1973; see also: Mangan and Golding, 1984, p.168). There was however no evidence of any changes in attentional strategy, either within the present data, or in the other paper being presented here (Parrott et al, this symposium).

REFERENCES

Hockey GRJ (1970). Effect of loud noise on attentional selectivity. Quarterly Journal of Experimental Psychology, 22, 28-36.
Hockey GRJ (1973). Changes in information-selection patterns in multi-source monitoring as a function of induced arousal shifts. Journal of Experimental Psychology, 101, 35-42.
Mangan GL, Golding JF (1984). The psychopharmacology of smoking. Cambridge UK, Cambridge University Press.
Parrott AC, Craig D, Haines M, Winder G (in press). Nicotine polacrilex gum and sustained attention. Proceedings of this symposium.
Surgeon General (1988). Nicotine addiction: the health consequences of smoking. Washington, US Government Printing Office.
Tarriere HC, Hartmann F (1964). Investigations into the effects of tobacco smoke on a visual vigilance task. Proceedings of the 2nd International Congress on Ergonomics, p525-530.

Warburton DM, Wesnes K (1984).   Drugs as research tools in
    psychology: cholinergic drugs and information processing.
    Neuropsychobiology, 11, 121-132.
Wesnes K, Warburton DM (1983).   Smoking, nicotine, and human
    performance.   Pharmacology and Therapeutics, 21, 189-208.
Williams GD (1980).   Effect of cigarette smoking on immediate
    memory and performance.   British Journal of Psychology, 71,
    83-90.

NEUROELECTRIC CORRELATES OF SMOKING BEHAVIOR

V. Knott

Clinical Neurophysiological Services and Departments of
Psychiatry and Psychology, University of Ottawa, Royal
Ottawa Hospital, 1145 Carling Avenue, Ottawa, Ontario, K1Z
7K4, Canada

SUMMARY: Quantitative electroencephalographic (EEG)
assessments of smoking behavior appear to be dependent upon
cigarette familiarity, with 'preferred' habitual cigarettes
resulting in different EEG profiles relative to 'novel'
cigarettes. Neuroelectric profiles emerge gradually through
the smoking of a cigarette and the presence and onset of EEG
changes varies with tar-nicotine yield. Smoking-induced EEG
changes also appear to be variable across the scalp and are
also dependent upon cigarette yield.

Since many of the current biological hypotheses of tobacco
use concern relationships between neurotransmitter systems-
events at the level of the individual neuron - and
psychological/behavioral states, human non-invasive
psychophysiological measurement strategies may prove
particularly useful in understanding the intermediate
effects of smoking occurring between neuron and behavior
(Muller & Muller, 1965). The scalp-recorded
electroencephalographic (EEG) signal is considered an
expression of perceptual/cognitive/emotional
processes and as such, it's quantification provides an
empiral correlate of brain states and the effects of
substances thereon. As observed in Fig. 1, the EEG consists
of an oscillating signal which varies in amplitude (uV, uV^2

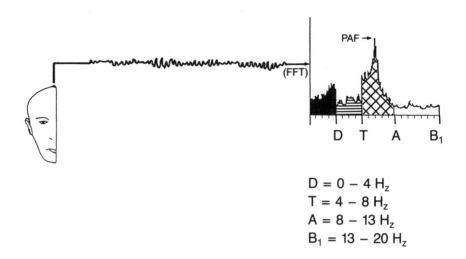

D = 0 – 4 H$_z$
T = 4 – 8 H$_z$
A = 8 – 13 H$_z$
B$_1$ = 13 – 20 H$_z$

Fig. 1   EEG and (FFT) Spectrum Analysis

or % power), frequency (cycles per second; Hz) and phase.
Quantitative (fast fourier transform: FFT) spectral
analysis of electrical activity recorded from the intact
skull of a conscious, relaxed subject, reveals continuous
spectra of frequencies starting below 1Hz and gradually
becoming imperceptible beyond 40-50 Hz (Johnson, 1980).
Typically, these electrical potentials are divided into at
least four main frequency bands including delta
(frequencies below 4 Hz), theta (frequencies between 4-8
Hz), alpha (waves between 8-13 Hz) and beta (frequencies
above 13 Hz). Spectra derived from posterior regions of the
scalp while subjects are relaxing with eyes closed
ordinarily show a major peak at 10-11 Hz (dominant or peak
alpha frequency; PAF). EEG may be particularly useful in
examining 'arousal-control' theories of tobacco use (e.g.,
Gilbert, 1971) as variations in frequency and voltage
characteristics of the EEG are believed to index an
individual's state on a behavioral-arousal continuum (Fig.
2) ranging from deep sleep through to normal relaxed

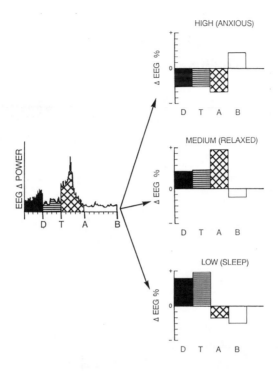

Fig. 2   EEG and Behavioral Arousal States

awakefulness and extreme emotional states including
anxiety, fear and rage, etc., Routenberg, 1968). Although
reports of EEG-behavioral dissociations (Sannita, 1983) or
lack of EEG behavioral associations (Daniel, 1966) have
challenged the notion that EEG reflects the level of
behavioral arousal, one might still argue that the degree of
synchronization of the EEG or its opposite,
desynchronization, can be useful to indicate when a 'change'
has taken place in brain state.

The 'pharmaco-EEG' approach to psychotropic effects may
be of particular interest in interpreting the tobacco-
related psycho-functional changes in the CNS as one might
compare tobacco-EEG profiles with acute single dose EEG
profiles of a variety of psychotropics to determine whether

they produce a common or different CNS effect and thereby
derive indications of tobacco's potential psychoactive
properties (Herrmann, 1982). The EEG profiles of major
psychoactive compounds with acute single behavioral
threshold dosages are similar for compounds with similar
psychoactive effects. Thus, as shown in Fig. 3, and as most
recently and succinctly described by Saletu (1987), drugs

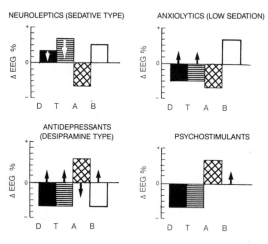

Fig. 3 Pharmaco-EEG Profiles

with similar psychotropic action   (all neuroleptics or all
antidepressants or all anxiolytics or all psychostimulants)
have been shown to produce similar EEG alterations (Hermann
& Schaerer, 1986; Fink, 1980; Itil, 1982). As indicated,
low potency, sedative neuroleptics induce mainly an increase
in theta and delta, a decrease in alpha and a slight
increase in fast beta activities. Antidepressants of the
desipramine type enhance mainly alpha activity while
anxiolytics increase beta and attenuate alpha.
Psychostimulants, in contrast to the high behavioral-EEG
arousal profile observed in Figure 2, do not diminish but
augment alpha and occasionally lower beta.

Attempts to characterize the spectral EEG profile of
acute cigarette smoking have met with considerable intra-
study variability (Church, 1989; Conrin, 1980; Edwards &
Warburton, 1983) and although profile discrepancies may in
part be due to differences in subject samples and/or
procedural/recording/analytical methods, it is within reason
to contend that variations in smoking administration may
have contributed to study differences. Relative to Church's
(1989) recommendation to restrict the unit of acute
treatment to one cigarette, many studies have adopted
smoking procedures (e.g., 2-3 cigarettes in 5-10 min) which
have essentially 'overdosed' the smoker thus producing
aversive states which potentially confound the 'true'
smoking EEG profile. Studies such as Golding's (1987),
which have administered only a single cigarette (in this
study: 1.3 mg nicotine; 14 mg Tar), have, as shown in a
<u>schematic representation</u> of his tabled data (Fig. 4),
reported power reductions in theta and increments in beta.

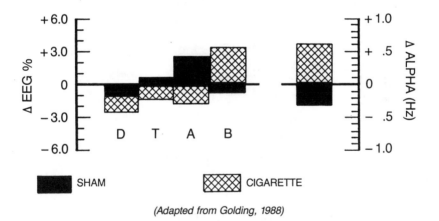

(Adapted from Golding, 1988)

Fig. 4 Acute EEG Effects of Single Cigarette
(Novel, non-preferred cigarette)

Alpha was also reduced but this effect reached significance
only in 'eyes-open' recording conditions. Although this
profile is markedly similar to the 'high' EEG-behavioral
arousal profile observed above in Fig. 2, it is not typical
of psychostimulant-EEG profiles (Fig. 3) which evidence
increases, not decreases, in alpha activity. Alpha
attenuation has been a characteristic EEG response in the
majority of studies which have administered a 'novel'
standard cigarette to smokers and, as such, one might argue
that the response effect may reflect the impact, not of
nicotine, but of emotional/cognitive reactivity to non-
preferred, unacceptable cigarettes.

The smoking of a single 'preferred' medium yield
cigarette (mean nicotine yield = 1.1 mg; mean tar yield =
19.5 mg) from the smoker's own chosen brand does in fact
result in a psychostimulant-EEG profile as reported by Knott
(1988). As observed in Fig. 5, this profile is
characterized by power reductions in slow wave activity

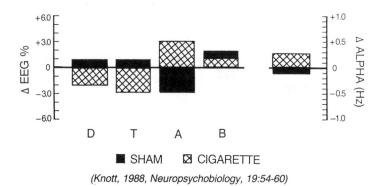

(Knott, 1988, Neuropsychobiology, 19:54-60)

Fig. 5 Acute EEG Effects of Medium T-N Yield Cigarette
(Preferred, single cigarette)

(delta and theta) and increases in both alpha power and PAF.
Similar psychostimulant-EEG profiles have been recently

reported by Knott (1989) following the smoking of a single
'preferred' low yield cigarette (mean nicotine yield = .30
mg; tar yield = 3.0 mg).

Smokers exert 'finger-tip' control over volume, duration
and interval of puffs in order to titrate the timing and
degree of smoke impact (Pomerleau & Pomerleau, 1984). As
shown in Fig. 6, attempts to examine the puff-by-puff EEG

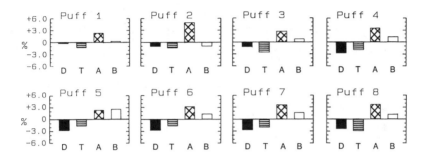

Fig. 6 Dynamic EEG Effects of Medium T-N Cigarettes
(Preferred, single cigarette)

dynamics of smoking behavior have shown, at least with the
smoking of preferred medium-yield cigarettes, that the
psychostimulant-EEG pattern emerges during the first puff
and gradually builds during the first four puffs where it
reaches statistical significance (i.e., for delta and theta
reductions and alpha increments) and remains relatively
stable through to the end of puffing (Knott, 1988). Similar
dynamic profiles have been reported with 'preferred' low-
yield cigarettes but there appears to be distinct
qualitative and quantitative differences in the presence and
onset-time of significant EEG effects (Knott, 1989). Delta
reductions were not observed 'during' the puffing of low-
yield cigarettes and significant puff-by-puff reduction
effects, such as occurred with theta (see Fig. 7), did not

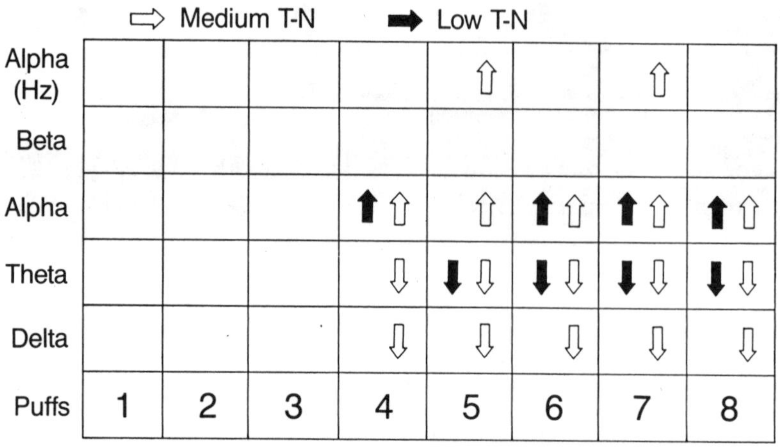

	1	2	3	4	5	6	7	8
Alpha (Hz)					⇧		⇧	
Beta								
Alpha				↑⇧	⇧	↑⇧	↑⇧	↑⇧
Theta				⇩	↓⇩	↓⇩	↓⇩	↓⇩
Delta				⇩	⇩	⇩	⇩	⇩
Puffs	1	2	3	4	5	6	7	8

⇨ Medium T-N   ➡ Low T-N

Fig. 7 Significant Puff-by-Puff EEG Changes from Medium and Low T-N Cigarettes
(Preferred, single cigarettes)

occur until the fifth puff (vs. fourth puff for medium-yield cigarettes).

Although pharmaco-EEG profiles have and continue to be assessed by single-channel EEG recording methodology, psycho-functional interpretation of these effects is markedly hampered by a restrictive approach which ignores potential regional brain differences in response to psychotropics. As psychotropic-induced shifts in neuroelectric activity have been reported to be far from homogeneous across brain sites (Coppola & Herrmann, 1987; Laurian et al., 1983; Pockberger et al., 1984; Sannita et al., 1983) an initial pilot study attempted to describe the scalp distribution of neuroelectric alterations induced by the smoking of standard, 'novel' cigarettes of various tar/nicotine yields (Knott, 1990). As evidenced in Fig. 8, sham smoking and the smoking of low, medium and high tar-nicotine yield cigarettes exerted differential effects on the distribution of EEG changes. With increasing yield, alpha appears to be augmented but this enhancement is much more prominent in the central-frontal regions than in

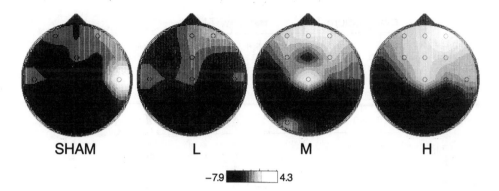

SHAM   L   M   H

−7.9 ▮▮▮▮ 4.3

Fig. 8   Effects of Low (L), Medium (M) and High (H)
Tar-Nicotine Cigarettes on
Topographic EEG (Alpha: %)

posterior brain regions.   Although it would be interesting
to speculate that this effect on the anterior 'executive-
control' regions, perhaps medicated by nicotine, is related
to reports of smoke-induced performance facilitation (Wesnes
& Warburton, 1983), future studies require that this be
examined with 'preferred' cigarettes, during non-performance
and performance conditions and preferrably with nicotine
manipulations.

REFERENCES

Church, R. (1989) In: Smoking and Human Behavior (A. Gale
    and T. Ney, Eds), John Wiley and Sons, New York, pp.
    115-140.
Conrin, J. (1986) Clin. Electroenceph. 20, 507-512.
Coppola, R. and Herrmann, W. (1987) Neuropsychobiol. 18,
    97-104.
Daniel, R. (1966) Psychophysiol. 2, 146-160.
Edwards, J. and Warburton, D. (1983) Pharm. Ther. 19,
    147-164.
Fink, M. (1980) Prog. Neuro-Psychopharm. 4, 495-502.
Golding, J. (1987) Pharm. Biochem. Behav. 29, 23-32.
Herrmann, W. (1982). Electroencephalography in Drug
    Research. Gustav Fischer, Stuttgart.

Herrmann, W. and Schaerer, E. (1986). In: Handbook of Electroencephalography and Clinical Neurophysiology, revised series, Vol. 2. (F. Lopes da Silva, W. Storm Van Leeuwen and A. Remond, Eds), Elsevier, Amsterdam, pp. 385-445.

Itil, T. (1982). In: Electroencephalography (E. Nierdermeyer and F. Lopes da Silva, Eds), Urban and Schwarzenberg, Munich, pp. 499-513.

Johnson, L. (1980). In: Techniques in Psychophysiology (I. Martin and P. Venables, Eds), John Wiley and Sons, Chichesten, pp. 329-357.

Knott, V. (1988). Neuropsychobiol. 19, 54-60.

Knott, V. (1989). Neuropsychobiol. 21, 216-222.

Knott, V. (1990). Neuropsychobiol., (in press).

Muller, H. and Muller, A. (1965). Int. J. Neuropsych. 1, 224-232.

Routenberg, A. (1968). Psych. Rev. 75, 51-80.

Saletu, B. (1987). In: Human Psychopharmacology: Measures and Methods, Vol. 1. (I. Hindmarch and P. Stonier, Eds), John Wiley and Sons, Chichester, pp. 173-200.

Sannita, W. (1983). In: Clinical and Experimental Neuropsychophysiology. (D. Papakostopoulos, S. Butler, I. Martin, Eds), (Croom Helm, London, pp. 350-369.

Sannita, W., Attonello, E., Rosadini, G. and Timitilli, C., (1983). Neuropsychobiol. 9, 66-72.

Wesnes, K. and Warburton, D. (1983). Pharmac. Ther. 21, 189-208.

NICOTINE: A UNIQUE PSYCHOACTIVE DRUG

E.F. Domino

Department of Pharmacology, University of Michigan, Ann Arbor, MI 48109-0626 USA.

SUMMARY: Nicotine is a unique psychoactive drug which produces arousal, relaxation, promotion of REM sleep and skeletal muscle relaxation in both animals and humans in tobacco smoking concentrations. Normal tobacco smokers, deprived of tobacco for 12 hr show a decrease in $alpha_1$, an increase in $alpha_2$ EEG power, and a decrease in the recovery cycle of the Hoffmann reflex after smoking cigarettes containing nicotine.

INTRODUCTION

The U.S. Surgeon General (1988) concluded that nicotine is a highly addictive substance and is the basis for the habit of tobacco smoking. While there is disagreement in relating tobacco smoking to cocaine or heroin use, there is little disagreement that nicotine has marked pharmacological actions. In tobacco smoking concentrations, nicotine is a remarkable psychoactive agent. A wide variety of complex stimulant and depressant effects are observed in animals and humans that involve the central and peripheral nervous, cardiovascular, endocrine, gastrointestinal, and skeletal motor systems. The electroencephalographic (EEG) activating effects of small doses of nicotine occur in intact as well as brainstem transected animals. These effects are directly on brainstem neuronal circuits. However, stimulation of peripheral afferents and release of catecholamines, and possibly other neurotransmitters and modulators, (serotonin, histamine) enhance the direct central effects of nicotine. The EEG activating effects of nicotine result in behavioral arousal. Nicotine increases the amount of REM sleep, an effect also seen with arginine vasopressin. Nicotine and tobacco smoking dramatically reduce phasic stretch reflexes such as the patellar reflex. Again, both central and peripheral mechanisms are involved. Conditioned avoidance behavior is selectively depressed by nicotine. Nicotine produces profound neuroendocrine changes, including elevated plasma catecholamines, glucose, glucagon, cortisol, insulin, and vasopressin. Considerable individual variability occurs in response to nicotine, suggesting important pharmacogenetic factors.

The present report summarizes the current status of nicotine and tobacco smoking in relationship to three major actions: (1) EEG effects, (2) behavioral arousal and sleep, and (3) effects on the patellar and Hoffmann reflex (the electrical equivalent of the ankle jerk).

EEG STUDIES IN ANIMALS

Studies on animals provide the basis for understanding the effects of nicotine on the EEG. Many researchers have shown that nicotine in subconvulsive doses can produce EEG desynchronization (Longo et al., 1954; Silvestrini, 1958; Stümpf, 1959; Dunlop, 1960; Floris et al., 1962; Silvette et al., 1962). Studies attempting to localize the central effects of nicotine have shown the hippocampus to be especially affected (Stümpf, 1959; Dunlop, 1960; Stümpf and Gogolak, 1967). Nicotine produces transient behavioral arousal and EEG activation in the cat (Knapp and Domino, 1962; Yamamoto and Domino, 1965). Much of the research on nicotine has revealed apparently contradictory actions. Yamamoto and Domino (1965) described nicotine's pharmacological actions as causing a brief wake up effect as well as a transient period of mild central nervous system depression. Phillis and York (1968) described nicotine as a cortical depressant, while Armitage and Hall (1968) described it as a cortical activator. Stadnicki and Schaeppi (1972) emphasized the biphasic effects of nicotine. They showed that nicotine causes initial cortical desynchronization followed by cortical synchronization.

The pertinence of animal research to human tobacco smoking has been examined with respect to nicotine dosage and method of administration. Bhattacharya and Goldstein (1970) studied the acute vs chronic effects of nicotine.

Chronic exposure resulted in EEG desynchronization. Furthermore, EEG activation has been shown upon both intravenous administration of nicotine and inhalation of tobacco smoke (Hall, 1969; Hudson, 1979; Sabelli and Giardina, 1972). In the cat, specific alpha frequency EEG effects were seen by Guha and Pradhan (1976). Nicotine initially induced EEG desynchronization and decreased the duration of both high and low alpha frequency bursts. Upon prolonged exposure to nicotine, the opposite was observed.

Ginzel (1988) studied the latency of electrocortical arousal caused by nicotine in cats. He concluded that the EEG activation caused by nicotine initially originates from the lungs. In addition to a peripheral effect, nicotine also has a direct central action in causing EEG activation (Knapp and Domino, 1962). Although the mechanisms by which nicotine produces its EEG effects are obviously complex, the EEG and behavioral effects in sleeping cats consist of three distinct phases. These include: (1) a brief period of EEG activation and arousal followed by (2) a period of sedation with EEG slow waves and, subsequently (3) a period of REM sleep.

EEG STUDIES IN MAN

Human research on tobacco smoking is almost as extensive as the non-human research, however much less invasive. Most of the current human research has been directed on the effects of tobacco smoking after certain periods of abstinence. Under such conditions, smoking has been shown to increase alpha and beta frequencies, decrease alpha power and decrease theta power (Herning et al., 1983; Pickworth et al., 1986; Pickworth et al., 1989). Knott (1988) noted an increase in alpha power when recorded from the right occipital region.

Lambiase and Serra (1957) were among the first to observe an increase in the frequency of the alpha rhythm and a decrease in the voltage in volunteers who smoked one tobacco cigarette. Other early studies attributed these alpha frequency increases to either the act of smoking or shifts in attention rather than the nicotine content (Hauser et al., 1958; Bickford, 1960; Murphree et al., 1967). Brown (1968) showed that the EEG of heavy smokers had small amounts of higher frequency alpha activity.

Abstinence from smoking has been shown to affect the EEG of tobacco smokers. Abstinence causes an increase in slow frequency alpha waves, a decrease in fast activity, and an increase of EEG amplitude. Resumption of smoking reverses these effects (Murphree et al., 1968; Ulett, 1969; Ulett and Itil,

1969; Itil et al., 1971; Philips, 1971; Knott and Venables, 1977). Knott and Venables (1979) showed that tobacco smoking either prior to or during ethyl alcohol consumption counteracted the alcohol induced slowing of alpha frequency which was evident in both nonsmokers and abstinent tobacco smokers. Lukas and Jasinski (1983) reported a nicotine induced dose related decrease in alpha power. Our recent studies (Domino and Matsuoka, 1990) are consistent with some of the above reports in which the mean differences in total power in the alpha₁ (8.0-10 Hz), alpha₂ (10.2-12.4 Hz), beta (13.6-3.0 Hz), delta (1.1-3.9 Hz), and theta (4.1-7.8 Hz) frequency bands were compared before and after smoking a cigarette with 0, 0.27, 2.0, and 2.16 mg nicotine content in 12 hr deprived tobacco smokers. The results are summarized in Fig. 1.

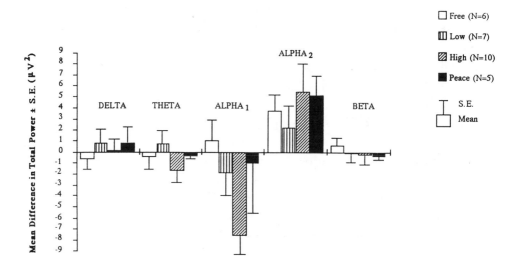

Fig. 1. Effects of smoking a cigarette of differing nicotine content on total cortical EEG power. The mean ± S.E. differences in total power before and after smoking four cigarettes of varying nicotine content are shown in bar graph format for five different EEG bands. N=number of subjects in each group.

EFFECTS ON THE PATELLAR AND HOFFMANN REFLEXES

Nicotine and tobacco smoke transiently reduce phasic stretch reflexes such as the patellar reflex (Domino and Von Baumgarten, 1969). Not all skeletal muscles are relaxed by nicotine. The trapezius muscle involved in the orienting reflex shows enhanced tension (Fagerström and Gotestam, 1977). Our own recent studies (Domino et al., 1990) have concentrated on the effects of smoking cigarettes of 0, 0.27, and 2.16 mg nicotine on the recovery cycle of the Hoffmann reflex in 12 hr deprived tobacco smokers. The Hoffmann reflex is the electrophysiologic equivalent of the ankle jerk. Its recovery cycle can easily be measured. The effects of smoking a high nicotine (2.16 mg) cigarette on the recovery cycle of the Hoffmann reflex are summarized in Fig. 2 (see Domino et al., 1990).

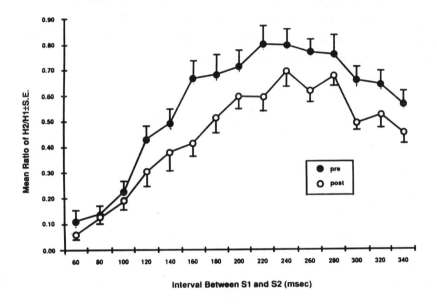

Fig. 2. Effects of smoking a high nicotine (2.16 mg) cigarette on the recovery cycle of the Hoffmann reflex. The mean ratio of the amplitude of the Hoffmann reflex post (H_2) divided by the amplitude pre (H_1) $\pm$ S.E. for 10 subjects is shown.

DISCUSSION

Two recent studies (Domino and Matsuoka, 1990; Domino et al., 1990) using normal tobacco smoking volunteers provide data which are consistent with previous reports in the literature. Smoking any of four different types of cigarettes (0, 0.27, 2.0 and 2.16 mg nicotine) increased total alpha EEG activity, although there were marked individual differences. Only after smoking nicotine containing cigarettes was there a mean decrease in alpha$_1$ power and an increase in alpha$_2$ power in the EEG, not only in the occipital areas but also more diffusely throughout the cerebral cortex. There were no consistent mean changes in delta, theta, or beta EEG activity. Individual differences were marked, irrespective of the nicotine content of the cigarette smoked. Two major trends were noted after tobacco smoking: 1) a more diffuse cortical distribution in alpha₂ activity and 2) an increase in alpha₂ activity related to the nicotine content of the cigarette smoked. Whether similar effects are observed in nondeprived tobacco smokers remains to be determined.

The Hoffmann reflex is the electrophysiological equivalent of a stretch reflex that allows one to easily measure its recovery cycle in man. It was postulated that the Hoffmann reflex recovery cycle would be prolonged by tobacco smoking. The Hoffmann reflex and its recovery cycle were obtained before and just after smoking one nonfiltered, nicotine free, low (0.27 mg) or high (2.16 mg) nicotine containing cigarette. Each cigarette was smoked on a different day. After smoking the high nicotine cigarettes, the subjects showed a clear cut reduction of the Hoffmann reflex, especially phase III-early inhibition. Individual differences were marked. Smoking low nicotine containing tobacco had only a small effect on the Hoffmann reflex and smoking nicotine free cigarettes had no effect at all. The data obtained are consistent with evidence in animals that nicotine stimulates Renshaw inhibitory neurons in the spinal cord (Ueki et al., 1961).

REFERENCES

Armitage, A.K., and Hall, G.H. (1968) Nature 219, 1179-1180.
Bhattacharya, I.C. and Goldstein, L. (1970) Neuropharmacol. 9, 109-118.
Bickford, R. (1960) Fed. Proc. 19, 619-625.
Brown, B.B. (1968) Neuropsychologia 6, 381-388.
Domino, E.F. and Matsuoka, S. (1990) FASEB J. 4, A994.
Domino, E.F. and Von Baumgarten, A.M. (1969) Clin. Pharmacol. Ther. 10, 72-79.
Domino, E.F., Kadoya, C., and Matsuoka, S. (1990) J. Clin. Pharmacol. 47, 167.
Dunlop, C.W., Stümpf, C., Maxwell, D.S., and Schindler, W. (1960) Am. J. Physiol. 198, 515-518.
Fagerström, K.O. and Gotestam, K.G. (1977) Addictive Behaviors 2, 203-206.
Floris, V., Morocutti, G., and Ayala, G.F. (1962) Boll. Soc. Ital. Biol. Sper. 38, 407-410.
Ginzel, K.H. (1988) In: The Pharmacology of Nicotine (M.J. Rand and K. Turau Eds), IRL Press, Washington, D.C., pp. 269-292.
Guha, D. and Pradhan, S.N. (1976) Neuropharmacol. 15, 225-232.
Hall, G.H. (1969) Brit. J. Pharmacol. 36, 200.
Hauser, H., Schwartz, B.E., Roth, G., and Bickford, R.G. (1958) EEG Clin. Neurophysiol. 10, 567.

Herning, R.I., Jones, R.T., and Bachman, J.(1983) Psychophysiol. 20, 507-512.

Hudson, R.D. (1979) Arch. Int. Pharmacodyn. 237, 191-212.

Itil, T.M., Ulett, G.A., Hsu, W., Klingenberg, H., and Ulett, J.A. (1971) Clin. Electroencephal. 2, 44-51.

Knapp, D.E. and Domino, E.F. (1962) Int. J. Pharmacol. 1, 333-351.

Knott, V.J. (1988) Neuropsychobiol. 19, 54-60.

Knott, V.J. and Venables, P.H. (1979) J. Studies on Alcohol 40 247-257.

Knott, V.J. and Venables, P.H. (1977) Psychophysiol. 14, 153-156.

Lambiase, M. and Serra, C. (1957) Acta Neurol. (Napoli) 12, 475-493.

Longo, V.G., Von Berger, G.P., and Bovet, D. (1954) J. Pharmacol. Exper. Ther. 111, 349-359.

Lukas, S.E. and Jasinski, D.R. (1983) Fed. Proc. 42, 1018.

Murphree, H.B., Pfeiffer, C.C., and Price, L.M. (1967) Ann. N.Y. Acad. Sci. 142, 245-260.

Murphree, H. and Schultz, R. (1968) Fed. Proc. 27, 220.

Philips, C. (1971) Psychophysiol. 8, 64-74.

Phillis, J.W. and York, D.H. (1968) Nature 219, 89-91.

Pickworth, W.B., Herning, R.I., and Henningfield, J.E. (1986) Pharmacol. Biochem. Behav. 25, 879-882.

Pickworth, W.B., Herning, R.I., and Henningfield, J.E. (1989) J. Pharmacol. Exper. Ther. 251, 976-982.

Sabelli, H.C. and Giardina, W.J. (1972) Biol. Psychiat. 4, 105-130.

Silvestrini, B. (1958) Arch. Int. Pharmacodyn. 116, 71-85.

Silvette, H., Hoff, E.C., Larson, P.S., and Haage, H.B. (1962) Pharmacol. Rev. 14, 137.

Stadnicki, S.W. and Schaeppi, V.H. (1972) Arch. Int. Pharmacodyn. Ther. 197, 72-85.

Stümpf, C. (1959) Naunyn Schmeideberg Arch. Exp. Path. 235, 421-436.

Stümpf, C. and Gogolak, G. (1967) Ann. N.Y. Acad. Sci. 142, 143-158.

Ueki, S., Koketsu, K., and Domino, E.F. (1961) Exper. Neurol. 3, 141-148.

Ulett, G.A. (1969) EEG Clin. Neurophysiol. 27, 658.

Ulett, J.A. and Itil, T.M. (1969) Science 164, 969-970.

U.S. Surgeon General (1988) The Health Consequences of Smoking: Nicotine Addiction. U.S. Department of Health and Human Services, Washington, D.C.

Yamamoto, K. and Domino, E.F. (1965) Int. J. Neuropharmacol. 4, 359-373.

THE COMPARATIVE PSYCHOPHARMACOLOGY OF NICOTINE

I. Hindmarch, J.S. Kerr, and N. Sherwood

Human Psychopharmacology Research Unit, The Robens Institute of Industrial and Environmental Health and Safety, University of Surrey, Guildford GU2 5XH, United Kingdom

SUMMARY: The psychopharmacological profile of nicotine was compared with those of 25 other substances including other social drugs and medicinal compounds. The sizes of effect of drugs on various items in a standardised test battery were calculated and ranked. It was shown that nicotine, in contrast to the other substances had small, specific, positive effects and a negative rating of behavioural toxicity.

How does nicotine compare with other psychoactive substances? Hindmarch et al. (1989) suggested that there might be some commonality of drug action, or concordance in the behavioural profile of psychoactivity of various drugs. These include substances which have direct CNS action (antidepressants, anxiolytics, sedatives, hypnotics), medicines with 'secondary' CNS effects (antihistamines, ephedrine, analgesics etc.) and those compounds used in a non-therapeutic situation (e.g. caffeine, nicotine, alcohol, cannabis). This paper examines the psychopharmacological profiles of various drugs and compares their behavioural toxicity to that of nicotine.

In order to assess the behavioural profile of different drugs, psychometric testing systems which meet scientific criteria are required. Such systems must be developed within an appropriate theoretical and methodological framework, and with due regard to the psychophysics of the human sensory, cognitive and response systems. Failure to take adequate account of the theoretical context of a test and a failure to consider the psychological aspects and limits of human performance invalidates many would-be psychometric tests, and does not permit generalisation of the findings and results obtained beyond the close confines of a particular study. These failures violate what must be the essential prerequisite for any study of comparative psychopharmacology: the ability to rate different drugs on the same measures in a scientific manner.

Hindmarch (1980) showed that many psychometric tests available at that time did not consider any of the basic theoretical or methodological aspects of measurement. Some of the tests were ingenious or applied creatively but were likely to be unreliable, not valid and insensitive to the important effects of psychoactive substances. It is not sufficient simply to create a test: a psychometric assessment or measure of drug effects in man requires that a test system be constructed and developed according to established theoretical, methodological, psychological and pragmatic standards. Tests which lack a history of reliable usage and validation against external norms are of no use to those wishing to investigate the psychoactive effects of pharmaceutical agents or self-administered, 'everyday' substances.

In contrast there are several psychometric tests which have been shown (Hindmarch, 1981, 1983, 1986) to be reliable and valid measures of the behavioural activity of a wide range of drugs including those compared here: commonly self-administered substances like nicotine, caffeine and alcohol, as well as medicines such as antihistamines, hypnotics, and antidepressants, and drugs of abuse such as amphetamine and the opiates. The tests, which are described below, assess many aspects of information processing, sensorimotor co-ordination, short term memory, reaction time and psychomotor functions related to skilled behaviour, e.g. driving. A judicious selection of different tests reflecting different aspects of CNS function can be made to form an effective battery of measures covering the range of possible effects which may be exhibited by psychotropic agents. The tests of psychological performance can be presented via microcomputers, as pencil and paper tasks, or on designated hardware with or without microprocessor control. The results of such assessments can be augmented with scores from subjective rating scales which, when properly constructed and implemented, can give reliable ratings of sedation and arousal (Hindmarch et al., 1980; Gudgeon & Hindmarch, 1983), mood (Hindmarch, 1979) and sleep (Hindmarch, 1984). The tests used in the experiments reviewed here are described below.

METHODS

This assessment of different psychoactive substances is based on the maximum acute drug effects found in studies conducted over the past 20 years using the same methodology and methods. All studies were double blind placebo controlled repeated measures designs. In a typical study 12 subjects were tested hourly for six hours after an initial assessment on the test battery and administration of the drug or placebo. After a one week washout period the subjects returned and were similarly tested under the alternative drug condition. The following tests were used to assess the extent of drug effect.

CRITICAL FLICKER FUSION: Critical flicker fusion threshold (CFF) provides an index of the state of arousal of the central nervous system, and allows accurate prediction of mental alertness and cognitive potential. (Hindmarch 1982, Parrott 1982). Subjects are required to discriminate flicker fusion in a set of four light emitting diodes held in foveal fixation at one metre. Individual thresholds are determined by the psychophysical method of limits on three ascending and three descending scales.

CHOICE REACTION TIME: Choice reaction time (CRT) is a measure of basic sensorimotor reaction to a critical stimulus and can be regarded as the basic performance skill inherant in and intrinsic to all overt activity (Hindmarch 1981). From a central starting position, subjects are required to extinguish one of six equidistant red lights illuminated at random by touching the appropriate response button next to the light. 20 such trials comprised one complete presentation of the task.

COMPENSATORY TRACKING TASK: The compensatory tracking task (CTT) has been forwarded as an efficient analogue of car driving in that a fine motor response is required as a result of the processing of visual information (Hindmarch 1983, 1986). The tracking task required subjects to use a joystick to keep a moving cursor in alignment with a moving target while simultaneously responding to visual stimuli (white lights), presented randomly in the bottom corners of the display, in the peripheral field of vision. This latter response approximates to a divided attention component in driving. The main response measure (r.m.s. error) is calculated as the mean deviation of the the joystick tracking the fixed programme, with a lower score indicating more accurate tracking. Responses to the peripheral stimuli were made by the subject pressing the keyboard space bar. The mean r.m.s. score and the mean reaction time (RT) were recorded at each test point.

SHORT-TERM MEMORY: High speed scanning and retrieval from short-term memory were assessed using a reaction time method (Sternberg 1966, 1975).The subject was required to judge whether a test digit was contained within a short memorised sequence of digits which were presented sequentially. The test digit appeared 1 second later and the time taken for the subject to react was recorded.

ANALYSIS

Quantitative comparison of the results of the individual studies was made using a meta-analytic technique to evaluate the strength of effect of an independent variable (drug effect) as

compared to a control (placebo) (Glass & Kliegl, 1983). The
technique was originally developed to evaluate treatment outcome
in response to psychotherapy (Prioleau et al., 1983; Wilson &
Rachman, 1983). Drug effect magnitudes, as seen in Tables I -
IV, are Cohen's 'd' scores derived from a priori t values
(Mullen & Rosenthal, 1985; Rosenthal & Rosnow, 1985), calculated
from the outputs of analyses of variance. Statistical
significance of drug effects compared to placebo was also
calculated by means of anovas and post hoc tests. The ranked
values for the size of effect of each drug on the various
dependent variables appear in Tables I - IV. Note that because
of missing values, not all drugs appear in all tables.

Table I. Ranked magnitudes of drug effects on critical flicker
fusion and total choice reaction time. The top and bottom groups
of drugs have been found to be significantly better and worse
respectively as compared to placebo at p<0.05 or better. The
doses (in mg unless stated otherwise) indicated for these drugs
are the minimum needed to produce an effect. The central group
of drugs have not been found to be significantly different to
placebo; the doses indicated for these drugs are the maximum at
which no difference from placebo was found in the original
studies. (a = minimum d values based on minimum estimated t).

Drug (dose)	d CFF	Drug (dose)	d TRT
sertraline (100)	1.719	sertraline (100)	0.684
pemoline (20)	1.391a	dexamphetamine (10)	0.393a
dexamphetamine (10)	1.391a	pemoline (20)	0.393a
astemizole (10)	1.052	dimethylxanthine (400)	0.393a
clobazam (30)	0.694	methylphenidate (10)	0.393a
dimethylxanthine (400)	0.646a	astemizole (10)	0.297
nicotine (2)	0.592	nicotine (2)	0.157
methylphenidate (10)	0.499a	nomifensine (100)	0.108
caffeine (400)	0.128	caffeine (400)	0.012
triazolam (0.25)	0.076	clobazam (30)	0.007

placebo	0.000	placebo	0.000
zopiclone (7.5)	0.006	zopiclone (7.5)	0.017
nomifensine (100)	0.152	triazolam (0.25)	0.025
mequitazine (5)	0.178	mequitazine (5)	0.061
triprolidine (10)	0.367	triprolidine (10)	0.149
alcohol(0.5g/kg)	0.370	amylobarbitone (100)	0.393a
nitrazepam (2.5)	0.408	codeine (120)	0.494
amylobarbitone (100)	0.459a	morphine (20)	0.862
codeine (120)	0.940	nitrazepam (2.5)	1.030
morphine (20)	1.272	chlorpheniramine (12)	1.199
lorazepam (1)	1.314	alcohol (0.5g/kg)	1.563
chlorpheniramine (12)	1.473	haloperidol (1)	1.834
amitriptyline (25)	2.194	amitriptyline (25)	2.300
haloperidol (1)	2.326	lorazepam (1)	2.442
mianserin (10)	3.205	mianserin (10)	3.286
chlorpromazine (50)	6.172	chlorpromazine (50)	6.172

Table II. Ranked magnitudes of drug effects on recognition reaction time (RRT) and motor reaction time (MRT). The top and bottom groups of drugs have been found to be significantly better and worse respectively as compared to placebo at $p < 0.05$ or better. The doses (in mg unless stated otherwise) indicated for these drugs are the minimum needed to produce an effect. The central group of drugs have not been found to be significantly different to placebo; the doses indicated for these drugs are the maximum at which no difference from placebo was found in the original studies.

Drug (dose)	d RRT	Drug (dose)	d MRT
nicotine (2)	0.561	nicotine (2)	0.853
astemizole (10)	0 .447	nomifensine (10)	0.103
mequitazine (5)	0.367		
		zopiclone (7.5)	0.101
nomifensine (100)	0.067	triazolam (0.25)	0.042
mianserin (10)	0.046	clobazam (30)	0.040

caffeine (400)	0.043	placebo	0.000
sertraline (100)	0.043	caffeine (400)	0.024
triazolam (0.25	0.020	triprolidine (10)	0.030
placebo	0.000	sertraline (100)	0.032
clobazam (30)	0.025	astemizole	0.037
zopiclone (7.5)	0.070		
		mequitazine (5)	0.308
codeine (120)	0.180	nitrazepam (2.5)	0.719
triprolidine (10)	0.310	alcohol (0.5g/kg)	0.744
morphine (20)	0.751	mianserin (10)	0.834
nitrazepam (2.5)	0.791	morphine (20)	0.897
alcohol (0.5g/kg)	0.804	codeine (120)	0.949
chlorphen (12)	0.827	chlorphen (12)	1.285
amitriptyline (25)	1.631	lorazepam (1)	1.468
lorazepam (1)	1.821	amitriptyline (25)	1.863

Table III. Ranked magnitudes of drug effects on tracker error (RMS) and tracker reaction time (RT). The top and bottom groups of drugs have been found to be significantly better and worse respectively as compared to placebo at p<0.05 or better. The doses (in mg unless stated otherwise) indicated for these drugs are the minimum needed to produce an effect. The central group of drugs have not been found to be significantly different to placebo; the doses indicated for these drugs are the maximum at which no difference from placebo was found in the original studies.

Drug (dose)	d RMS	Drug (dose)	d RTT
nicotine (2)	0.884	caffeine(400)	1.865
caffeine (400)	0.624	mequitazine (5)	1.217
astemizole (10)	0.345	nicotine (2)	0.467
nomifensine (100)	0.061	zopiclone (7.5)	0.134
clobazam (30)	0.035	nomifensine (100)	0.109
placebo	0.000	placebo	0.000
sertraline (100)	0.012	triazolam (0.25)	0.017

mequitazine (5)	0.043	sertraline (100)	0.044
zopiclone (7.5)	0.199	astemizole (10)	0.081
triazolam (0.25)	0.400	clobazam (30)	0.086
alcohol (0.5g/kg)	0.457	alcohol (0.5g/kg)	0.158
chlorphen (12)	0.636	amitriptyline (25)	0.393
amitriptyline (25)	1.447	chlorphen (12)	0.829
lorazepam (1)	1.879	mianserin (10)	1.103
mianserin (10)	1.929	lorazepam (1)	2.266

Table IV. Ranked magnitudes of drug effects on short term memory. The top and bottom groups of drugs have been found to be significantly better and worse respectively as compared to placebo at p<0.05 or better. The doses (in mg unless stated otherwise) indicated for these drugs are the minimum needed to produce an effect. The central group of drugs have not been found to be significantly different to placebo; the doses indicated for these drugs are the maximum at which no difference from placebo was found in the original studies.

Drug (dose)	d STM
nicotine (2)	0.158
caffeine (400)	0.158
nomifensine (100)	0.001
placebo	0.000
zopiclone (7.5)	0.020
clobazam (30)	0.066
sertraline (100)	0.098
mianserin (10)	0.359
amitriptyline (25)	0.655
lorazepam (1)	1.256
alcohol (0.5g/kg)	1.766
triazolam (0.25)	2.237
nitrazepam (2.5)	6.145

DISCUSSION

It is apparent from the data that nicotine compares "favourably" with the other substances in that it has small, positive effects on all the dependent variables; similar to those of caffeine, though with a slightly more stimulant action. This concords with the results of Hindmarch et al. (1990) who demonstrated that (in non-deprived smokers) nicotine improves performance on these measures, significantly so on measures of psychomotor function. A similar finding of increased locomotor activity with nicotine, in non-smokers, was reported by West & Jarvis (1986). The negative rating of caffeine and nicotine on behavioural toxicity (i.e. positive psychopharmacological benefit) cannot necessarily be ascribed as a cause of the popularity of self administration of these drugs: it is obvious that the other social drug, alcohol, even in relatively small doses produces a pronounced performance decrement, particularly on the recognition component of reaction time. This has important implications for skilled behaviour in everyday life e.g. driving.

In contrast to nicotine and caffeine, most of the substances included in this study have either a consistent sedating effect or no effect compared to placebo. It is interesting that drugs from the same therapeutic class can vary widely on the level of behavioural toxicity which they produce. Contrast for example clobazam and lorazepam. The two drugs are both used as anxiolytics, but clobazam (up to 30mg) raises the CFF threshold and has no effect on overall reaction time (Hindmarch, 1979; Hindmarch & Parrott, 1980) whereas lorazepam (over 1mg) has a pronounced sedating effect on both measures. There are no differences in the clinical effectiveness of the drugs however (Paes de Sousa et al., 1981; De Figueiredo et al., 1981) and this should be taken into account when calculating the psychopharmacological risk-benefit ratio of the two chemicals.

In conclusion, the psychopharmacological profile of nicotine is one of small, specific, positive effects in contrast to other agents which have large overall sedating (e.g. lorazepam) or stimulant (amphetamine) activity, and to substances which have little discernible effects compared to placebo (e.g. zopiclone). This supports Pomerleau & Pomerleau's (1984) assertion that nicotine may be used as a "psychopharmacological scalpel" in the investigations of psychopharmacologists.

REFERENCES

De Figueiredo, R., Franchini, A. & Martinho, A. (1981) Effectivness, tolerability and withdrawal phenomena in anxious patients following three weeks treatment with clobazam and lorazepam. Royal Society of Medicine ICSS, 43, 175-180

Glass, G.V. and Kliegl, R.M. (1983) An apology for research integration in the study of psychotherapy. Journal of Consulting and Clinical Psychology, 51 (1), 28

Hindmarch, I. (1979) Some aspects of the effects of clobazam, on on human performance. British Journal of Clinical Pharmacology, 7, 77S-82S

Hindmarch, I. (1980) Psychomotor function and psycho-active drugs. British Journal of Clinical Pharmacology, 10, 189-209

Hindmarch, I. (1981) Measuring the effect of psychoactive drugs on higher brain function. In G.D. Burrows & J. S. Werry (Eds.) Advances in Human Psychopharmacology II. Connecticut: JAI

Hindmarch, I. (1982) Critical Flicker Fusion Frequency (CFFF): The effects of psychotropic compounds. Pharmacopsychiatria, 15, Suppl.1, 44-48

Hindmarch, I. (1983) Measuring the side effects of psychoactive drugs: A pharmacodynamic profile of alprazolam. Alcohol & Alcoholism, 18, 361-367

Hindmarch, I. (1984) The Leeds Sleep Evaluation Questionnaire (LSEQ) as a measure of the subjective response to nocturnal treatment with benzodiazepines. In G. D. Burrows, T. R. Norman & K. P. Maguire (Eds) Biological Psychiatry: Recent Studies. London: Libbey

Hindmarch, I. (1986) The effects of psychoactive drugs on car handling and related psychomotor ability: A review. In J. F. O'Hanlon and J. J. de Gier (Eds.) Drugs and Driving. London: Taylor and Francis

Hindmarch, I. and Gudgeon, A.C. (1980) The effects of clobazam and lorazepam on aspects of psychomotor performance and car handling ability. British Journal of Clinical Pharmacology, 10 (2), 145

Hindmarch, I., Kerr, J. S. & Sherwood, N. (1989)

Psychopharmacological   aspects   of   psychoactive
substances.   In   D.   Warburton   (Ed.)   Addiction
Controversies: Proceedings   of   the   XXth Conference   on
Substance Abuse, Florence
Hindmarch,   I.,   Kerr,   J. S. & Sherwood, N. (1990) Effects   of
nicotine   gum   on psychomotor performance in   smokers   and
non-smokers. Psychopharmacology, 100, 535-541
Hindmarch, I. and Parrott, A.   C. (1980) The   effects   of
combined   sedative   and   anxiolytic   preparations   on
subjective   aspects   of   sleep   and   objective   measures   of
arousal   and   performance   in   the   morning   following
nocturnal   medication: I   Acute.   Arzneimittel-Forschung
(Drug Research), 30, 1025-1029
Hindmarch,   I.,   Parrott, A. C., Hickey, B. J. & Clyde, C.   A.
(1980) An   investigation   into the   effects   of   repeated
doses   of   temazepam on aspects of   sleep,   early   morning
behaviour   and   psychomotor   performance   in   normal
subjects.   Drugs   in   Experimental   and   Clinical   Research,
6, 399-403
Mullen, B. and   Rosenthal, R.   (1985)   Basic   meta-analysis:
Procedures   and   programs.   New   Jersey: Lawrence   Erlbaum
Associates
Paes   de   Sousa,   M.,   Figueiredo,   M.-L.,   Loueriro,   F. &
Hindmarch,   I.   (1981) Lorazepam and clobazam in   anxious
elderly   patients.   Royal Society of   Medicine   ICSS,   43,
119-123
Pomerleau,   O.   F.   & Pomerleau, C. S. (1984)   Neuroregulators
and   the   reinforcement   of   smoking: Towards   a
biobehavioural   explanation.   Neuroscience &   Biobehavioral
Review, 8, 503-513
Parrott,   A.   C. (1982)   Critical   flicker   fusion thresholds and
their   relationship   to   other   measures   of   alertness.
Pharmacopsychiatria, 15, (1), 39-44
Prioleau,   L.,   Murdock, M. & Brody, N. (1983) The   analysis of
psychotherapy versus placebo studies.   Behavioural & Brain
Sciences, 6, 275
Rosenthal,   R.   & Rosnow, R.   (1985)   Contrast analysis: Focused
comparisons   in   the   analysis   of   variance.
Cambridge: Cambridge University Press
Sternberg S (1966) High   speed   scanning   in   human   memory.
Science 153: 652-654
Sternberg S (1975) Memory scanning: new findings and   current
controversies.   Quarterly   Journal   of   Experimental
Psychology, 27, 1-32
West,   R.   J.   & Jarvis,   M. J. (1986) Effects of nicotine   on
finger   tapping   rate   in   non-smokers.   Pharmacology
Biochemistry & Behavior, 25, 727-731
Wilson, G.T. and   Rachman, S.J. (1983) Meta-analysis and   the
evaluation   of   psychotherapy   outcome: Limitations   and
liabilities.   Journal   of   Consulting   and   Clinical
Psychology, 51 (1), 54

SEPARATE AND COMBINED EFFECTS OF THE SOCIAL DRUGS ON PSYCHOMOTOR
PERFORMANCE

J.S. Kerr, N. Sherwood, and I. Hindmarch

Human Psychopharmacology Research Unit, The Robens Institute of
Industrial and Environmental Health and Safety, University of
Surrey, Guildford GU2 5XH, United Kingdom

SUMMARY: Ten female subjects (five smokers and five non-smokers)
performed a choice reaction time task (CRT), a compensatory
racking task (CTT) and were tested for their critical flicker
fusion threshold (CFF) after the administration of each
possible combination of nicotine (2mg gum or placebo), caffeine
(250mg capsule or placebo) and alcohol (30g or placebo). Motor
function was shown to be facilitated by nicotine or caffeine and
the debilitating effects of alcohol were frequently antagonised
by either drug. In spite of the differences in their
neuropharmacological actions, combinations of nicotine, caffeine
and alcohol may be compared through their effects on common
information processing mechanisms involved in psychomotor
performance.

Although social drugs are frequently taken in combination, few
studies have investigated the combined effects of nicotine,
caffeine and/or alcohol on performance. Demographic studies
(Istvan and Martarazzo 1984) suggest that the use of all three
substances co-varies in a complex fashion. It may be
hypothesised that the reason would be to maintain or induce a
preferred subjective state. Alternatively, people may use the
psychoactive properties of the drugs to modulate cognition and
arousal and so benefit from improved behavioural functioning.

One difficulty in attempts to compare the performance effects
of the social drugs remains that they have not been assessed on
the same standardised psychological measures within one
controlled study. The present study sought to alleviate this
problem and to compare and contrast the actions of social drugs

on performance in a controlled experiment using a series of tasks recognised as reliable indicators of drug effects. It has been possible (Hindmarch et al. 1989) to isolate the major variables of performance which make up an individuals psychomotor reaction to the administration of a psychoactive drug and to measure the resultant effects with a range of tests. Such laboratory tasks have become increasingly popular as a means of predicting real world performance while maintaining strict experimental control.

METHOD

Ten female volunteers aged between 21 and 50 years (mean 32 years) participated in the experiment. Five were smokers who had smoked 15 cigarettes or more per day for a minimum of five years prior to the start of the study. The remaining volunteers were non-smokers. All were in good health and underwent a full medical examination before they were accepted as study subjects. The study medication was administered orally as a drink (alcohol or placebo), together with a capsule (caffeine or placebo) and a piece of chewing gum (nicotine or placebo). Subjects were asked firstly to consume the drink and swallow the capsule within a five minute period. They then chewed the gum slowly and steadily for 20 minutes. Medication was given at the start of each daily session. The alcohol comprised 30g of 80% proof vodka in 200ml orange juice. In the placebo drink the vodka was replaced by water and the rim of the glass smeared with vodka to give an identical initial olfactory cue. The caffeine comprised 250mg hydrous caffeine in lactose capsule or placebo, where the caffeine was replaced by sucrose. The nicotine comprised 2mg or 0mg (placebo) nicotine polacrilex gum. Subjects were not required to abstain from smoking prior to each study day and so were tested at their preferred normal nicotine level.

Eight different treatments were given to subjects, one each morning of each experimental day. The order of presentation to each subject was varied randomly to ensure a counterbalanced design. The treatments were prepared daily by a research worker and administered by a study nurse who ensured that subjects complied with the treatment demands. A second research worker conducted the testing double-blind. Measures were taken of choice reaction time, compensatory tracking and critical flicker fusion.

RESULTS

One between group (smokers v non-smokers) and two within group (drug condition, trial number) analyses of variance have been used throughout with post-hoc comparisons of treatment means using Tukey's HSD test where appropriate.

CHOICE REACTION TIME: CRT data were analysed with regard to total reaction time (TRT) and split into recognition reaction time (RRT) and motor reaction time (MRT) components. No factor in the TRT or RRT ANOVA proved significant, but the drug factor in the MRT analysis reached significance, $F(7, 56)= 2.38$, $p>0.05$ with reaction times significantly faster than placebo with nicotine or caffeine alone or with the combination of nicotine and caffeine. Mean performance under each drug condition has been calculated and is shown in Figure 1.

COMPENSATORY TRACKING: Mean tracking performance under each drug condition is shown in Figure 2. The CTT ANOVA yielded a significant drug factor, $F(7, 56) = 2.66$, $p>0.05$. Tukey's HSD post-hoc pairwise tests conducted on the drug factor means showed that alcohol alone significantly impaired tracking accuracy in comparison to placebo, while all drug combinations which included nicotine, or caffeine alone, led to significantly improved tracking performance.

CRITICAL FLICKER FUSION: The group factor in the ANOVA proved

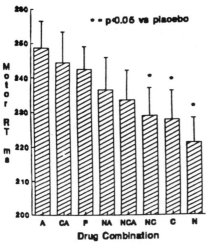

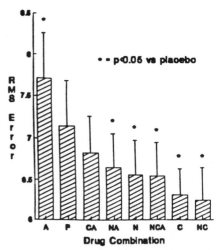

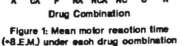

Figure 1: Mean motor reaction time
(+S.E.M.) under each drug combination

Figure 2: Mean R.M.S. error
(+S.E.M.) under each drug combination

significant, F(1, 8) = 6.85, p<0.05, with the smokers showing
overall lower mean thresholds (26.77 Hz) than non-smokers (29.88
Hz). Neither the drug factor nor related interactions reached
significance.

DISCUSSION

The effects of caffeine reported here support previous results
suggesting that the compound has a performance enhancing effect.
Of interest, there was no significant post drug rise in CNS
arousal as measured by CFF, suggesting that caffeine may have a
specific action on information processing.

The profile of nicotine results support those reported by
Hindmarch et al. (1990), where nicotine gum given in addition to
a subject's preferred self-administered nicotine level resulted
in significant improvements in performance on measures of motor
reaction time and tracking.

As hypothesised, alcohol disrupted performance on the
majority of tasks, although only significantly on the error
measure of the tracking task, despite the estimated peak blood

concentrations for all subjects being below the current United Kingdom legal limit for driving. A clear distinction can be drawn in behavioural terms between the effects of caffeine and nicotine which at the given doses appear to enhance performance on some of the tasks, and the debilitating effects of alcohol. The similarity between the effects of caffeine and that of nicotine is further supported by their apparent antagonistic action on the effects of alcohol, but in combination they appear to provide no greater a psychological effect than when administered alone.

The results suggest that the covariance in the use of all three social drugs may be influenced by their effects on performance and arousal. While the present study used standardised doses of each drug, it is apparent that in the real world an individual chooses their own regimen of administration for a desired effect. Alcohol appears not to benefit information processing. It may be presumed that the major motivation for consumption lies in the production of an altered subjective state. Caffeine in turn has a clear action in facilitating performance and cognition without gross changes in mood or affect and may be preferred because of this clean stimulant action. Nicotine would appear to fall someway between the two in that reports of subjective enjoyment from tobacco are common, yet studies continue to provide evidence of performance enhancement. As such nicotine appears to posess the capacity for the dual facilitation of performance and affect.

REFERENCES

Hindmarch, I., Kerr, J.S., Sherwood, N. (1989) Psychopharmacological aspects of psychoactive substances. In D. Warburton (ed) Addiction controversies: proceedings of the XXth conference on substance abuse, Florence
Hindmarch, I., Kerr, J.S., Sherwood, N. (1990) Effects of nicotine gum on psychomotor performance in smokers and non-smokers. Psychopharm 100: 535-541
Istvan, J., Matarazzo, J.D. (1984) Tobacco, alcohol and caffeine use: review of their relationships. Psychol Bull 95: 301-326
Leigh (1977)

THE INFLUENCE OF NICOTINE ON CNS AROUSAL DURING THE MENSTRUAL CYCLE

L. Dye, (1) N. Sherwood (2) and J.S. Kerr (2)

(1) Biopsychologie, Fakultat fur Psychologie, Ruhr-Universitat Bochum, Universitatsstr. 150, 4630 Bochum 1, FRG.

(2) Human Psychopharmacology Research Unit, The Robens Institute of Industrial and Environmental Health and Safety, University of Surrey, Guildford, GU2 5XH, UK.

SUMMARY: Critical flicker fusion Thresholds (CFFT), used as a measure of CNS arousal were recorded in 47 women (30 non-smokers, 9 light smokers and 8 heavy smokers) at regular intervals during the menstrual cycle. A significant linear trend was found within the pooled CFFT data, seen as a gradual increase across the menstrual cycle. The division of data into smoking groups indicated that this pattern was maintained in both non-smokers and light smokers but not among heavy smokers, who also showed significantly lower CFFTs ($p < 0.05$). A possible depressant effect of nicotine is discussed and the relationship of the present results to known changes in smoking patterns during the menstrual cycle considered.

For many women the menstrual cycle is characterised by a series of physiological and psychological changes which may be considered as stressful. In particular reports of increases in pain and negative affect (anxiety, irritability and depression) as well as increases in disturbing effects such as elation and emotional lability occur both before and during menstruation (Dalton 1977).

Steinberg and Cherek (1989) have shown that smoking increases significantly, both in terms of number of cigarettes smoked and number of puffs taken per cigarette during and before menstruation. The fact that smoking persists and even increases at these times has led some to propose that smoking may be a coping response to a range of menstrual symptoms (Sloss and

Frerichs 1983).

What remains to be shown is that these changes in smoking are the result of the psychopharmacological actions of nicotine rather than other aspects of smoking. The present study aimed to investigate the relationship between smoking during the menstrual cycle and CNS arousal.

METHOD

Forty-seven women aged between 19 and 49 years (mean 34 years) were studied. All were employees in a large retail outlet. Nine subjects smoked between 1 and 10 cigarettes a day, and eight subjects 15 cigarettes a day or more. The remaining subjects were all non-smokers.

Subjects were required to discriminate flicker fusion in a set of four light emitting diodes held in foveal fixation at one metre. Individual threasholds were determined on three ascending and three descending scales. Subjects were tested at the same time of day on a number of occasions, at least once and up to three times a week. This continued for up to seven weeks so as to obtain data on at least one complete cycle. Testing took place during work rest periods. No attempt was made to control subjects smoking and it was assumed that subjects were tested at their preferred nicotine level.

Subjects were also asked to keep a diary of symptoms and dates of menstruation. Exact day of cycle and cycle length were then calculated and standardised by the method described by Rossi and Rossi (1977).

RESULTS

A simple linear regression conducted upon the pooled CFFT data showed there to be a significant linear trend (P<0.004), with CNS arousal increasing across the menstrual cycle.

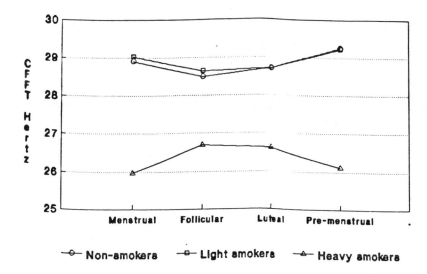

Figure 1: Mean CFFT at each phase of the menstrual cycle

Individual standardised cycles were then divided into the four 7 day phases comprising menstrual, follicular luteal and pre-menstrual stages of the cycle and subjects grouped according to smoking status. An unweighted means analysis of variance was then conducted upon the data. A significant group effect was found (P<0.05) with heavy smokers showing significantly lower CFFTs. No main effect of phase of cycle was found, but the smoking status by phase of cycle interaction was significant (P<0.025). An inspection of figure 1 suggests that the lower CFFTs had been maintained across all phases of the menstrual cycle in comparison to light and non-smokers who appeared to show an increase in CFFT during the pre-menstrual and menstrual phases.

DISCUSSION

The present data show clear variability in CNS arousal, as measured by CFFT, across the menstrual cycle. In particular an increasing level of arousal is in agreement with certain self-report questionnaires of arousal during the menstrual cycle

(Parlee 1980) and leads to the suggestion that some of the reports of menstrual distress may be the result of a heightened perception of menstrual activity.

Data from the heavy smoker group differed from light and non-smokers in two respects. Firstly the group showed significantly lower mean CFFTs and secondly these scores appeared not to vary across the menstrual cycle in the same fashion as non-smokers or light smokers.

A reduction in CFFT is commonly seen after the administration of drugs with sedative properties such as some benzodiazepines (Hindmarch 1982). It is hypothesised that for heavy smokers, high levels of systemic nicotine act to maintain a preferred level of sedation and that some events occuring during the menstrual cycle act as stimuli which may initiate or increase tobacco smoking in habitual smokers. The use of an objective measure of CNS arousal (CFFT) indicates that changes in smoking patterns across the menstrual cycle are not just the result of displacement activities in the face of psychological stress, but self-initiated coping responses that rely on the psychoactive properties of nicotine. These apparently sedative properties of nicotine have yet to be fully investigated.

REFERENCES

Dalton K. (1977) The Premenstrual Syndrome and Progesterone Therapy. London: Heinemann
Hindmarch, I. (1982) Critical flicker fusion frequency (CFFF): The effects of psychotropic compounds. Pharmacopsychiatrica, 15: 44-48
Parlee, M.B. (1980) Changes in mood and activation levels during the menstrual cycle in experimentally naive subjects. In A.J. Dan, E.A. Graham & C.P. Becker (Eds.) The Menstrual Cycle. New York: Springer
Rossi, A.S. and Rossi, T.E. (1977) Body time and social time: Mood patterns by menstrual cycle phase and day of the week. Social Science Research, 6: 273-308
Sloss, E.M. and Frerichs, R.R. (1983) Smoking and menstrual disorders. International Journal of Epidemiology, 12: 107-109
Steinberg, J.L. and Cherek, D.R. (1989) Menstrual cycle and cigarette smoking behaviour. Addictive Behaviours, 14: 173-179

EFFECTS OF NICOTINE GUM ON SHORT-TERM MEMORY

N. Sherwood, J.S. Kerr, and I. Hindmarch

Human Psychopharmacology Research Unit, The Robens Institute of Industrial and Environmental Health and Safety, University of Surrey, Guildford GU2 5XH, United Kingdom

SUMMARY: To investigate the effects of nicotine on memory function, 20 subjects (10 non-smokers and 10 smokers who had been allowed to smoke normally until testing) attended the laboratory at their "preferred nicotine level" and completed a short-term memory task (memory scanning) at set points for 4 hours after the administration of 2mg or 0mg (placebo) nicotine polacrilex gum. The results suggest that nicotine enhanced memory reaction time performance (P<0.01) when subjects were probed for information already present in short-term memory (correct positive responses) but had no effect on reaction time when the information was absent from memory (correct negative responses). It is suggested that nicotine facilitates the processing of stimulus information in short—term memory.

It has been suggested that the cholinergic system may be involved in certain aspects of human memory (Squire & Davies 1981) and that nicotine, as a cholinomimetic, may impinge upon this activity. Further speculation has centred on the possibility that the facilitation of memory by nicotine may contribute to the reinforcing nature of cigarette smoking.

Studies of the effects of nicotine on memory have commonly used habitual smokers as subjects (Houston 1978, Mangan and Golding 1983). Usually there has been a period of abstinence prior to the experiment and subjects have been in a state of partial or complete deprivation compared to their preferred everyday nicotine level when they reached the laboratories. As such it remains impossible to attribute the results as being due to deprivation in the control condition, the effects of an acute dose of nicotine after a period of abstinence, or a more general action related to the long-term use of cigarettes.

The methodology used in the present study follows that of Hindmarch et al. (1990). Their design allowed subjects to attend the laboratories at their preferred nicotine level. For smokers this meant that they were not required to abstain from smoking at any time beore testing and therefore could not initially be considered to be deprived. Nicotine gum was then administered double-blind, producing an absolute rise in systemic nicotine levels against which performance effects could be gauged.

METHOD

20 healthy female volunteers aged 21-45 years (mean 28 years) participated in the experiment. 10 were smokers who had smoked ten or more cigarettes a day for a minimum of five years prior to the start of the study. The remaining volunteers were all non-smokers, abstinent for a minimum of five years if ex-smokers. All were in good health and underwent a full medical examination before they were accepted as study volunteers.

Nicotine was administered orally as 2mg or 0mg (placebo) nicotine gum. Smoker subjects were not required to abstain from cigarettes prior to the experimental days and so attended the laboratories at their preferred nicotine level. All subjects were required to chew the gum slowly and steadily for 20 minutes to allow for buccal absorption of the nicotine. Hindmarch et al. (1990) recorded average increases in plasma nicotine of 4.6 ng/ml (smokers) and 3.7 ng/ml (non-smokers) over baseline evels 30 minutes after the administration of 2mg nicotine gum using a similar methodology.

The task was an adaptation of Sternberg's original test of high speed scanning and retrieval from short term memory (Sternberg 1966). Subjects memorised a set of 1, 3 or 5 digits (stimuli) presented sequentially in the centre of a screen for 1.2 seconds each, followed after a 1 second pause by a series of probe digits. The task was to judge as quickly and as accurately as possible whether the probe digit was or was not a member of the memorised set. Subjects responded by pressing either of two hand held buttons. Each memorised set was followed by 12 probe digits, and two presentations of each set size were made, in a random order, at each assessment (72 responses in total at each test point).

Subjects attended the laboratory on three days (Monday, Wednesday and Friday). The first day served as a practice session to familiarise subjects with the task and the surroundings. Although no medication was given on this day it was the same as experimental days in all other respects. Experimental sessions took place every other day to allow a minimum of 24 hours washout between treatments. The order of presentation of nicotine or placebo was varied randomly. The treatments were prepared daily by a research worker and administered by a study nurse. A second research worker then conducted the testing double blind. Subjects undertook test assessments on arrival (baseline test prior to treatment) and at 1, 2, 3 and 4 hours after administration of the gum.

RESULTS

Repeated measures analyses of covariance (with baseline data as the covariate) were calculated for the positive and negative responses. Significant main effects were found for both set size ($F = 63.02$; df = 2,38; $p < 0.001$) and drug condition ($F = 8.69$; df = 1,19; $p < 0.01$) in the positive response analysis. A significant set size term alone was found in the negative response analysis ($F = 41.15$; df = 2,38; $p < 0.001$). All other terms were non-significant. The main effects are summarised in figures 1 and 2.

An error count was conducted on the pooled subject data for each condition. Friedman's test was used to assess the differences between the drug conditions for both positive and negative responses. No significant differences were found between the drug and placebo condition ($X2 = 1.00$; N.S.)

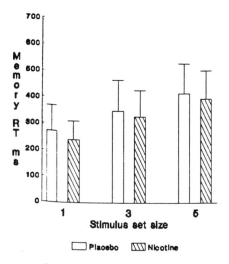

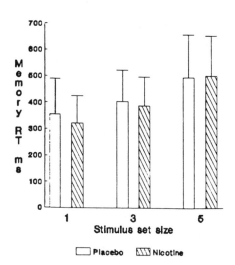

Figure 1: Mean memory reaction time
(+ S.D.) - probe present

Figure 2: Mean memory reaction time
(+ S.D.) - Probe absent

DISCUSSION

The memory scanning task used here proved sensitive in
uncovering the effects of nicotine on short-term memory. The
clear conclusion is that additional nicotine enhances memory
reaction time performance when subjects are probed for
information present in the memory store, but has no effect when
this information is absent. These results lend support to the
suggestion that nicotine may facilitate memory and cognitive
performance through its action on CNS activity (Wesnes &
Warburton 1983, Sahakian et al. 1989)

It has been proposed (Broadbent, 1984) that the decisions a
subject reaches in tasks such as memory scanning are the result
of evidence accumulating in two hypothetical parallel registers,
leading to either a "yes" or a "no" response once a criterion
level has been passed. The significant effects of nicotine on
memory reaction time responses in the present task may be due to
an increase in the absolute speed of scanning with no change in

the criterion level, but this should have led to a significant facilitation of correct negative responses as well as correct positive responses. Alternatively, nicotine may have lowered the criterion level, although this would have also lead to more errors in the nicotine condition. Although there was no significant error effect in the analysis, errors in fact accounted for only a small proportion of the total responses. It may be that if larger set sizes or more complex stimulus information were used, the concomitant increase in errors would reveal more on the underlying response strategy.

One final possibility is that nicotine influences the contents of STM, rather than the search or decision process per se. The rise in electrocortical arousal found after the administration of nicotine could heighten the availability of information in the cognitive processing system and increase the likelihood of a correct "match" when stimulus and probe information were congruent, while having no discernable effect when stimulus and probe were incongrous.

REFERENCES

Broadbent, D.E.(1984) The Maltese cross: a new simplistic model for memory. Behavioral and Brain Sciences, 7: 55-94

Hindmarch, I., Kerr, J.S., Sherwood, N. (1990) Effects of nicotine gum on psychomotor performance in smokers and non-smokers. Psychopharmacology, 100: 535-541

Houston, J.P., Schneider, N.G., Jarvik, M.E. (1978) Effects of smoking on free recall and organization. American Journal of Psychiatry, 135: 220-222

Mangan, G.L., Golding, J.F. (1983) The effects of smoking on memory consolidation. Journal of Psychology, 115: 65-77

Sahakian, B., Jones, G., Levy, R., Gray, J., Warburton, D. (1989) The effects of nicotine on attention, information processing and short-term memory in patients with dementia of the Alzheimer type. British Journal of Psychiatry 154: 797-800

Squire, L.R., Davies, N.P. (1981) The pharmacology of memory: a neurobiological perspective. Annual Review of Pharmacology and Toxicology, 22: 323-356

Sternberg, S. (1966) High speed scanning in human memory. Science 153: 652-654

Wesnes, K., Warburton, D. (1983) Smoking, nicotine and human performance. Pharmacology and Therapeutics 21: 189-208

RELATIONSHIPS BETWEEN PHOTIC DRIVING, NICOTINE AND MEMORY.

Gordon Mangan and Ian M. Colrain

Department of Psychology, University of Auckland, Auckland, New Zealand.

SUMMARY: Several studies by Soviet authors have reported an association between performance on a variety of memory tasks and the ability to entrain EEG to a periodic visual stimulus (photic driving). Little information is available, however, as to the exact parameters of the photic driving stimuli and of the memory tasks employed. The present study investigates the relationship between power at different driving frequencies and immediate recall of prose passages under nicotine and no-nicotine conditions. The driving measures were taken in response to 14 driving frequencies ranging from 5 to 40 Hz and measured at 16 EEG sites across the head. Results show a significant relationship between photic driving at high frequencies and recall, and a significant effect of nicotine on driving power and efficiency of recall. The implications of these data are discussed.

The present enquiry, which examines relationships between nicotine, memory and photic driving in human subjects, draws on three evidential sources. The first concerns studies which have investigated the effect of nicotine on recall. Two recent studies using human subjects (Peeke & Peeke, 1984; Warburton, Wesnes, Shergold & James, 1986) have indicated that immediate memory for auditory stimuli is enhanced following nicotine administration, the authors ascribing this phenomenon to improvement in attention and facilitation of information processing respectively. There have also been reports of a nicotine produced enhancement of delayed recall (Andersson, 1975; Mangan & Golding 1983, for example).

All of these data are consistent with reports indicating that structures involved in attentional and memoric processes such as the thalamus and hippocampus are major target areas for nicotine (Nelsen, Pelley & Goldstein, 1974; Clarke, Pert & Pert, 1984; Clarke, Schwartz, Paul, Pert & Pert, 1985; Adem, Synnergen, Botros, Ohman, Wingblad & Nordberg, 1987).

While there have been reports from studies using human subjects that initial learning or short term recall is reduced or not affected by nicotine (Andersson & Post, 1974; Andersson, 1975; Williams, 1980), differences in methodology relating to the learning tasks used and the amount of nicotine administered, and individual differences in tonic arousal level, interacting with the well established, bi-phasic dose response effects of nicotine, impose some constraints on the interpretation of these data.

The second area of relevance concerns the relationship between lability and memory. Photic driving, which describes the entrainment of the EEG by repetitive flash stimulation, is thought by Soviet psychophysiologists to index the nervous system property of lability when the frequency of stimulation is within the gamma band, that is, 35 to 80 Hz.

Lability was defined empirically by Wedenskii (1920) as the maximum number of impulses nervous tissue is able to reproduce in unit time in phase with the rhythm of stimulation. Several authors, most notably Nebylitsyn (1972), have considered lability to be directly proportional to the subject's capacity to process and encode information.

Studies performed by Golubeva (1972; 1973) reported relationships between high frequency driving and incidental but not intentional recall. Bokharova and Laktionov (1972), however, suggest that the lability/intentional recall link is expressed only when stimulus material is presented at a fast rate, inert subjects being then subject to proactive inhibition. This was supported by Mangan and Sturrock (1988) in their study of paired-associate recall and two-flash threshold, which is another measure of lability.

Results reported by Mundy-Castle (1953), Rozhdestvenskaya, Golubeva, and Yermolaeva-Tomina (1967), and others, demonstrate that photic driving is, to an extent, dependent on the subject's functional state, factors such as prevailing mood, age and vigilance status being influential. It is unclear, however, precisely what mechanisms underlie the reaction. Some authors suggest that the visual cortex is the only structure involved (Smirnov, 1956 for example); others, such as Trofimonov, Lyubimov and Laumova (1960), however, implicate additional structures such as superior colliculus and the lateral geniculate body. Elucidation of this problem awaits more definitive research.

The third area of interest is the effect of nicotine on EEG desynchronization and on photic driving efficiency or power. With some exceptions, it is clear that smoking doses of nicotine increase CNS arousal, measured by EEG desynchronization (Knott & Venables 1977; Michel & Battig, 1989). It has been suggested that this phenomenon could be secondary to the effects of nicotine on the mesencephalic reticular activating system (Edwards & Warburton, 1983). There are, however, reports of EEG depressant effects, and of mixed depressant and stimulant effects, although it is likely that these are interpretable in terms of the known bi-phasic dose response curves for the action of nicotine at synapse, and in terms of the starting arousal state of the organism.

In view of the nicotine / EEG desynchronization effect, and suggestions that increased cortical arousal improves photic driving (Golubeva, 1965; Hrbek, 1967) we might expect that nicotine would have a similar enhancing effect on photic driving. Evidence on this point, however, is generally not supportive. Vogel, Broverman and Klaiber (1977) report that while 3-hour deprived smokers have a greater incidence of driving response than non-smokers, fewer photic driving responses occurred in both groups after subjects had smoked a cigarette. Golding (1988) also reported no effect of nicotine on photic driving. In both cases, however, driving was at frequencies of 30 Hz or less, and it is possible that nicotine enhancement effects are not expressed through all frequency bands.

In the present study three hypotheses were tested: (1) there is a significant relationship between smoking doses of nicotine and recall; (2) there is a significant relationship between lability as indexed by photic driving in the gamma band and recall; (3) there is a significant relationship between photic driving power in the gamma band and nicotine.

METHOD

Subjects

Thirty three first year University students who were smokers, in normal health and had no family history of epilepsy volunteered for the study .

Equipment

Subjects wore an EEG cap which in addition to the standard 10-20 positions had electrodes, identified as Q1 and Q2, in the gaps between Pz and O1 and Pz O2 respectively. EEG was collected from 16 channels, 8 on each hemisphere, using a Nicolet IA98 21 channel EEG machine.The Nicolet output was sampled at 128 Hz by a Macintosh II computer running a National Instruments 16 channel 12 bit analog to digital converter board. Data collection and analysis were controlled with programs written in-house using the National Instruments LabVIEW and LabVIEW 2 software packages.

Photic Stimuli

Photic stimuli were produced using a Nicolet Ganzfeld. Light flashes of between 24 and 29 μ seconds with an intensity of 5000 lux were produced by a Xenon flash tube in the 18 inch sphere of the Ganzfeld. Subjects were exposed to a steady state light of 2 foot lamberts for 4 seconds. They were then exposed to 14 frequencies of flashing light, each for 4 seconds. The frequencies were 5, 6, and 7 Hz in the theta band, 9,10,11 and 12 Hz in the alpha band, 16, 18, and 20 Hz in the beta 1 band, 25 and 30 Hz in the beta 2 band and 35 and 40 HZ in the gamma band. In order to overcome the confounding of absolute illumination level with frequency, five Kodak Wratten filters of varying densities were used in an attempt to match the intensity of the light flashes with that of the steady state exposure.

Memory Task

Prose passages were taken from the Rivermead Behavioral Memory Test (Wilson, Cockburn & Baddeley, 1985). Each passage consists of a fictional news item containing 21 elements of information. Subject instructions and the news items were recorded on tape and then played to the subject at a rate of one element per second. At the end of the item, the subject was instructed to recall as much of the passage as possible. Scores ranged from 0 to 21.

Procedure

Four sessions were conducted with each subject. Data were collected under no-smoking, sham smoking , low nicotine and medium nicotine conditions. Subjects were asked to abstain from alcohol for 24 hours, and from nicotine and caffeinated drinks for two hours prior to each session. In each of the smoking sessions, subjects were fitted with an electro-cap and seated in front of the ganzfeld. They were then given a cigarette and instructed to light it, and to take three deep puffs in their own time. When this was completed they were given three 4-second driving epochs. After every third epoch, subjects took another puff. This procedure was continued until all frequencies had been presented twice, so that 12 puffs were taken overall.   The cigarette was extinguished by the experimenter, and the butt prepared and stored for later nicotine assay. The electro-cap was then removed and one of the Rivermead prose passage tapes was played. Immediately following the cessation of the passage the subject was instructed to repeat as much of the passage as he or she could remember into a tape recorder microphone.

In the no-smoking condition the procedure was similar, with rest breaks following every third driving epoch to substitute for puffs on the cigarette.

<u>Analysis</u>

The third and fourth seconds of the EEG recorded during each driving epoch were subjected to an FFT spectral analysis procedure which produced an output with a half Hz resolution. Thus power was measured at 5 Hz when driving at 5 hz, power at 6 Hz when driving at 6 Hz, 7 Hz at 7 Hz and so on. Responses of single frequencies at each electrode site were determined as were the means for five bands of interest: theta, alpha, beta 1, beta 2, and gamma. The correlations between the power values and Rivermead scores were then calculated.

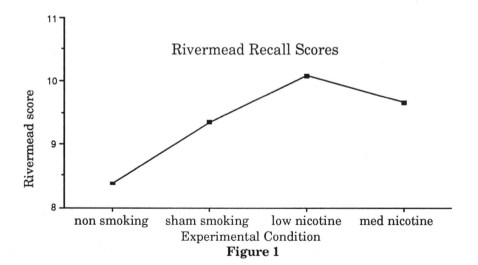

Figure 1

RESULTS

The first analysis was designed to assess the effects of smoking condition on recall of prose passages. Fig. 1 plots the recall means across the four smoking conditions.

A repeated measures ANOVA across all four experimental conditions revealed a significant effect for recall scores ($F(3,32) = 3.13$, $p = .03$). Post hoc comparisons revealed that the no-smoking condition was significantly different to the sham smoking ($t(32) = -1.93$, $p = .03$), low nicotine ($t(32) = -3.04$, $p = .002$) and medium nicotine ($t(32) = -2.22$, $p = .02$) conditions. The data indicate that all smoking conditions are related to an improvement in recall relative to the non-smoking condition, with the maximum improvement being found with the low nicotine (0.62 mg delivery) cigarette.

The next analysis was designed to assess the relationship between recall and photic driving. Table I shows Photic driving power/ Rivermead correlations based on data obtained during the no-smoking condition. The only significant relationship was between recall and driving in the gamma frequencies, that is between recall and lability. This relationship holds for both 35 and 40 Hz, the correlation at 35 Hz being +0.56, and at 40 Hz, +0.55.

Table I

Photic Driving Power / Rivermead correlations

Frequency Band	Correlation (r)	Significance
gamma (35 - 40 Hz)	0.64	$p = .0001$
beta 2 (25 - 30 Hz)	0.16	N.S
beta 1 (16 - 20 Hz)	0.09	N.S
alpha (9 - 12 Hz)	-0.04	N.S
theta (5 - 7 Hz)	-0.28	N.S

A Bonferroni correction was applied to the 16 electrode site correlations between recall and gamma driving. Sites C3 and C4 displayed the most significant correlations, .52 and .56 respectively (adjusted $p < .05$ for both). There was a tendency for the left hemisphere sites to show stronger driving / recall relationships.

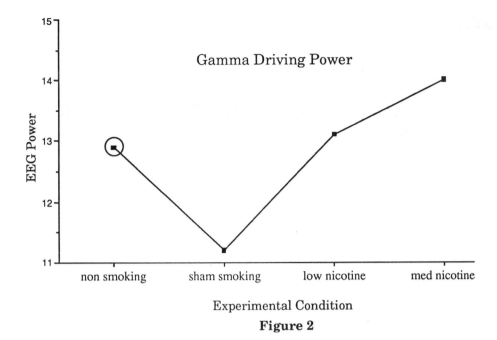

Figure 2

The next analysis sought to determine the effect of nicotine on the EEG response to photic driving. The effect of smoking conditions (sham, low nicotine, medium nicotine) on driving was only significant for responses to stimulation in the gamma band (F (2,32) = 4.11, p = .02).

As seen in Fig. 2, the medium nicotine condition is associated with the highest gamma photic driving power, and is significantly greater than the response in the sham smoking condition (t(32) = -3.24, p = .002). The low nicotine condition response falls between the other two but is not significantly different to either. Thus increasing nicotine dose is associated with increasing gamma photic driving power.

An unexpected result, however, is associated with the no-smoking condition, where the level of gamma photic driving is similar to that for the low nicotine condition, and is thus greater than the sham smoking response. Including the no-smoking data in an analysis of variance indicated that there was still a significant effect of experimental condition (F(3,32) = 3.09, p = .03). However, none of the post hoc comparisons involving the no-smoking data reached significance.

DISCUSSION

The first hypothesis was supported by our data. There are, however, two points worth noting. The first is that the sham smoking condition provoked better recall than the no-smoking condition, which could reflect some placebo or expectancy effect. On the other hand, it could indicate the existence of a physiological response, facilitatory to memory, which is associated with the act of smoking through a conditioning process. The second point of interest is that the low nicotine condition (0.62 mg delivery cigarette) was associated with better recall than the medium nicotine condition (1.12 mg delivery cigarette), although the difference between the conditions was not significant. This could indicate a titration phenomenon, that is, subjects extracted more nicotine from the lower nicotine delivery cigarettes through varying their smoking behaviour (puff intensity, puff volume, etc). Alternatively, it could reflect a bi-phasic nicotine recall relationship. This latter hypothesis could be tested using cigarettes with higher nicotine content.

The second hypothesis was also supported, by the significant positive relationship between gamma driving power and recall. Thus, subjects demonstrating the property of lability were seen to have an advantage in intentional recall of prose passages presented at a relatively fast rate.

The third hypothesis was also supported, but the support is qualified by the unexpected result relating to the no-smoking driving response. The elevated response was not seen in all subjects but was substantial in four of the 33. As stated earlier, little is known regarding the mechanisms underlying photic driving, but it is thought that functional states - mood, and vigilance status, e.g. - can influence the response. As data in the different conditions were collected on different days, usually at least a week apart, it is possible that the result is due to variation in one or more of the above factors, or others with similar impact on photic driving.

Overall, the finding of differential patterns of nicotine effect on gamma photic driving and memory, despite the substantial relationship between the two, indicates that the relationships between nicotine, memory and photic driving are not simple, and warrant continued, more intensive investigation.

Acknowledgements: The authors wish to thank Miss Ora Pellett and Mr Timothy Bates for their assistance with data collection and analysis. The present work was supported by research funding from Philip Morris U.S.A..

REFERENCES

Adem, A, Synnergen, B, Botros, M, Ohman, B, Wingblad, B, Nordberg, A. (1987) [^{3}H] Acetylcholine nicotinic recognition sites in human brain: characterization of agonist binding. Neuroscience Letters 83: 298-302.

Andersson,K. (1975) Effects of cigarette smoking on learning and retention. Psychopharmacology 41: 1 - 5.

Bokharova, S P, Laktionov, A N. (1972) Izuchenie interferentsii v kratkovremennoy pamyah v svyazi s tipologicheskimi osobennostyami nervnoy sistemy. Vop. Piskhol 1.

Clarke, P B S, Pert, C B, Pert, A. (1984) Autoradiographic distribution of nicotine receptors in rat brain. Brain Research 323: 390 - 395.

Clarke, P B S, Schwartz, R D, Paul, S M, Pert, C B, Pert, A. (1985) Nicotinic binding in rat brain: Autoradiographic comparisons of [^{3}H] Acetylcholine, [^{3}H] Nicotine and [^{125}I]-α-Bungaratoxin. The Journal of Neuroscience 5(5) 1307-1315.

Edwards, J A, Warburton, D M. (1983) Smoking, nicotine and electrocortical activity. Pharmacology Therapeutics 19:147-164.

Golding, J F. (1988) Effects of cigarette smoking on resting EEG, visual evoked potentials and photic driving. Pharmacology, Biochemistry and Behaviour 29: 23-32

Golubeva, E A. (1965) The reestablishment of brain biopotentials and typological properties of the nervous system. In: Typological features of higher nervous activity in man. (B M Teplov Ed) 4, Moscow, Izd. Akad. Pedagog. Nauk RSFSR.

Golubeva, E A. (1972) The driving reaction as a method of study in differential psychophysiology. In: Biological Bases of Individual Behaviour. (V D Nebylitsyn & J A Gray Eds) New York, Academic Press.

Golubeva, E A. (1973) The study of bioelectric correlates of memory in differential psychology. Soviet Psychology 11: 71-92.

Hrbek, A. (1967) Evolution of evoked potentials in children and their changes in oligophreny and in sleep. Photic driving as an indicator of the functional state of the brain. In: Mechanisms of orienting reaction in man. (I Ruttkay-Nedecky, L Ciganek, V Zikmund,& E Kellerova Eds) Publishing House of Slovak Academy of Sciences, Bratislava.

Knott V J, Venables, P H. (1977) EEG alpha correlates of non-smokers, smokers, smoking and smoking deprivation. Psychophysiology 14(2): 150-156.

Mangan, G L,Golding J F. (1983) The effects of smoking on memory consolidation. The Journal of Psychology 115: 65-67

Mangan G L, Sturrock R. (1988) Lability and Recall Personality and Individual Differences 9(2): 289-295.

Michel, Ch, Battig, K. (1989) Separate and combined psychophysiological effects of cigarette smoking and alcohol consumption. Psychopharmacology 97: 65 73

Mundy-Castle, A C. (1953) An analysis of central responses to photic stimulation in normal adults. Electroencephalography and Clinical Neurophysiology 5(1) 1-22.

Nebylitsyn, V D. (1966) Fundamental properties of the human nervous system. Moscow: Proveshcheniye. English Translation, G L Mangan (Ed), 1972, New York, Plenum.

Nelsen, J M, Pelley, K, Goldstein, L. (1973) Chronic nicotine treatment in rats: EEG amplitude and variability changes within and between structures. Research Communications in Chemical Pathology and Pharmacology 5 681-693.

Peeke, S C, Peeke, H V S. (1984) Attention, memory and cigarette smoking. Psychopharmacology 84: 205-216

Rozhdestvenskaya, V I, Golubeva, E A, Yermolaeva-Tomina, L B. (1967) The relation between functional states and typological properties of the nervous system. In: B M Teplov (Ed) Typological features of higher nervous activity in man. Vol. 5, Moscow, Prosveshcheniye.

Smirnov, G D. (1956) Recent advances in biology. Uspekhi sovrem. biol. 42: 3(6)

Trofimonov, L G, Lyubimov, N N, Laumova, T S. (1960) Cortical-subcortical relationship in the conditioning of defence and food reflexes in dog. In: The Moscow Colloquium on electroencephalography of higher nervous activity. (H H Jasper & G D Smirnov) The International Journal of electroencephalography and clinical neurophysiology. Suppl 13, 295-308.

Vogel, W, Broverman, D, Klaiber E L. (1977) Electroencephalographic responses to photic stimulation in habitual smokers and nonsmokers. Journal of Comparative and Physiological Psychology 91(2): 418-422

Warburton, D M, Wesnes, K, Shergold, K, James, M. (1986) Facilitation of learning and state dependency with nicotine. Psychopharmacology 89: 55-59.

Wedenskii, N. (1920) Accessory electrotonic alterations of excitability. Petersburg Bull. Acad. Imper. Sc. 14: Ser. VI, 332-359.

Wilson, B, Cockburn, J, Baddeley A. (1985) The Rivermead behavioural memory test manual. Thames Valley Test Company, England.

EFFECTS OF NICOTINE CIGARETTES ON MEMORY SEARCH RATE

R. West and S. Hack

Department of Psychology, Royal Holloway and Bedford New College,
London University, Egham, Surrey TW20 0EX, England

Two studies examined the effects of smoking a nicotine versus a
non-nicotine cigarette on performance on Sternberg's memory
search task. Subjects were shown short lists of digits followed
in each case by single "probe" digits. They had to indicate as
quickly as possible whether each "probe" was a member of the
"positive set" which they had previously seen. As positive set
size increases there is a linear increase in response times. In
the version of the task used for the present studies positive set
sizes of 2 and 5 were used. The search rate was significantly
faster after the nicotine cigarette than the non-nicotine
cigarette. No significant difference was found between the
results from occasional versus regular smokers and between the
effect of a cigarette before versus after a period of 24 hours'
abstinence. Smoking a cigarette containing nicotine can speed
search through items in short-term memory and this effect may not
be subject to significant chronic or acute tolerance under normal
smoking regimes.

There has been a great deal of research into the acute effects
of cigarette smoking on perceptuomotor performance and memory
(USDHHS, 1988). However, there is still considerable uncertainty
about the nature and extent of any effects. The issue is
complicated by the possibility of acute tolerance if the effects
of a cigarette are tested without smokers having first abstained
for a while. If smokers are tested after a period of abstinence
then any effects of smoking a cigarette might amount only to
relief of a withdrawal decrement. There is also the possibility
of chronic tolerance. In theory non-smokers could be used to get
around this problem, but they usually cannot inhale tobacco smoke
because they have not adapted to its irritancy.

There has recently been a move to test the effects of alternative methods of nicotine delivery. These include nicotine tablets, nicotine chewing gum, nasal nicotine drops, and subcutaneous nicotine injections. However, the rate of absorption is typically much slower than from inhaled cigarette smoke. This may or may not affect their pharmacological action.

It is possible to address many of the problems of tolerance by testing subjects before and after a period of abstinence and also using both regular and occasional smokers. This latter group appear to have adapted to the cigarette smoke sufficiently that they can inhale it and obtain a substantial nicotine dose but would not be expected to suffer withdrawal effects.

The main study described in this paper examined the effects of smoking a tobacco cigarette versus a non-tobacco cigarette on memory search rates in regular and occasional smokers, before and after a period of abstinence. This was preceded by a preliminary study in which a group of occasional smokers were tested on a single occasion after a period of abstinence.

In choosing a task on which to test the effects of cigarette smoking, we wanted one which had been tested with other drugs so that any effects could be interpreted in relation to their similarity to or difference from these drugs. We also wanted to use a task which had a ready interpretation in terms of the cognitive processes which were involved.

The task we chose was Sternberg's memory search task (Sternberg, 1966). This involves presenting subjects with a short list of digits. This list is known as the "positive set". The digits not in the list are known as the "negative set". The subjects are then shown a series of "probe" digits one at a time and they have to indicate as quickly as possible in each case whether or not it was a member of the original list. This involves them searching their memory of the original list until they find a match, or fail to find one. In his original series of experiments, Sternberg showed that the average time taken to respond to the probes was a simple linear ascending function of the size of the positive set. He concluded that subjects were undertaking an exhaustive serial search through their mental representation of the positive set when each probe was presented. This conclusion has been challenged over the years because of further work carried out with the task (Baddeley, 1985), but the notion of some kind of memory search still appears to be the most satisfactory interpretation.

The memory search task has been extensively used in tests of the effects of drugs on performance and has been shown in one form or another to be sensitive to the effects of a range of stimulant and sedative drugs (e.g. Subhan & Hindmarch, 1985). In general, stimulant drugs appear to speed up the search rate and sedative drugs slow it down. It is a task that might be expected to be affected by nicotine on two counts. First, nicotine appears to have a broad CNS and autonomic stimulant action. Secondly, the role of cholinergic mechanisms in memory suggest that a cholinergic agonist such as nicotine would facilitate retrieval.

Hindmarch et al. (1990) recently reported finding no effect of nicotine chewing gum on performance on the Sternberg task. Unfortunately, the version of the task used included only one positive set size (4 digits) making it impossible to assess any effect on the memory search rate per se. It is necessary to use at least two different set sizes for the slope of the response time/set size function to be estimated.

The methods and results of the studies reported in this paper are described in detail in West & Hack (1990).

STUDY 1

The first study was a preliminary attempt to assess whether a nicotine cigarette would speed up memory search by comparison with a non-nicotine cigarette in a sample of occasional smokers after a period of abstinence. The presumption was that any effect would be unlikely to be merely relief of a withdrawal decrement because no withdrawal syndrome would be present.

SUBJECTS AND METHODS; Ten undergraduate smokers took part in the experiment. For inclusion they had to be smoking fewer than five cigarettes per day and not smoking every day. The subjects were asked to abstain on the day of testing, prior to attending the laboratory. Testing took place in the afternoon.

Table I: Study 1 overall design. CO=CO measure, Stern=Sternberg task, Nic cig=smoke nicotine cigarette, non-nic=smoke non-nicotine cigarette.

Group 1:	N=5			
CO + Stern	Nic cig	Stern + CO	Non-nic	Stern + CO
Group 2:	N=5			
CO + Stern	Non-nic	Stern + CO	Nic cig	Stern + CO

Table I shows the design of the study. The allocation of subjects to groups was random. The nicotine cigarette was Benson & Hedges King Size (FTC ratings: Tar 18mg, Nicotine 1.5mg, CO 19mg). The non-nicotine cigarette used was a brand called "Free" derived from wheat, cocoa and citrus plants. The cigarettes were smoked on a fixed regimen of one puff every 10 seconds until the butt was reached. Approximately 10 minutes was allowed for smoking the cigarette including preparation, lighting up and stubbing out.

The Sternberg task was implemented on a BBC Model B microcomputer using a modified version of a program kindly provided by Professor Ian Hindmarch of Leeds University, UK. The task involved presenting subjects with a short list of digits (the positive set) for three seconds. Subjects were then presented with a series of probe digits one after the other and they had to indicate in each case whether the probe digit was or was not a member of the positive set. They responded by pressing one key on the computer keyboard if the digit was a member of the positive set and another key if it was not. They used the index finger of their right hand in the former case and the index of finger of their left hand in the latter case. Their fingers were resting on or just above their respective keys at all times. The subjects were instructed to respond as quickly as possible but

it was stressed that they should make as few errors as possible.
Response times were recorded in milliseconds. The Sternberg task
and measurement of the expired-air CO took about 10 minutes to
complete in all.

In each of the three testing sessions (baseline, post-test 1
and post-test-2), the subjects were allowed one practice trial
with positive set size 2 and one with positive set size 5. Each
trial consisted of presentation of a positive set followed by 16
probe digits, eight randomly chosen from the positive set and
eight randomly chosen from the negative set. After the practice
trial, the subjects underwent four trials with positive set size
2 and four with positive set size 5. The trials were randomly
interspersed.

The subjects' mean response times for correct responses were
calculated separately for set sizes 2 and 5. The search rate was

calculated as $\dfrac{3}{T5-T2}$

T5 is the mean response time (secs) for positive set size 5
and T2 (secs) is the response time for positive set size 2. Only
subjects who made one or no errors were included. Three subjects
made a large number of errors and also had very short response
times (under 400ms) and their data had to be discarded.

RESULTS

The subjects' mean CO increase following the nicotine cigarette
was 7.6ppm. There was an increase in search rate following the
nicotine cigarette from an average of 18 items per second to 37.5
items per second. By contrast the search rate following the non-
nicotine cigarette was 25 items per second which was the same as
it was beforehand. The increase following the nicotine cigarette
was evident in six of the seven subjects (one subject showing no
change). This was significant by binomial test (p<.05 one-
tailed). The results were sufficiently encouraging to carry out

a larger trial which would examine the issue in more detail.

STUDY 2

This study elaborated on the first in several ways. First, we
tested both regular and occasional smokers so that the we could
compare smokers who were likely to have experienced different
degrees of adaptation to the effects of nicotine. Secondly, we
tested subjects before and after a period of abstinence (24hrs)
to assess the extent to which acute tolerance may be present.
Thirdly, we included a number of additional measures to check on
some of the assumptions in the first study.

SUBJECTS AND METHODS; The subjects were all college students.
Fourteen were occasional smokers and 15 were regular smokers.
Occasional smokers were defined as smoking fewer than 20
cigarettes per week and not smoking for at least one day every
week. The occasional smokers smoked an average of 1.4 cigaretttes
per day whereas the regular smokers averaged 14.6 per day. The
occasional smokers also scored as minimally dependent (1.5 out
of 9) on the Smoking Motivation Questionnaire dependence scale
(see West & Russell, 1985).
    The occasional and regular smokers were tested before and
after a period of abstinence. Each testing session took a similar
format to that used in Study 1. For the pre-abstinence session,
no restrictions were stipulated for smoking prior to the session.
All the regular smokers smoked on the day in question. The
occasional and regular smokers were randomly allocated to receive
either the nicotine cigarette first or the non-nicotine cigarette
first. Those who smoked the nicotine cigarette first in the pre-
abstinence session, also smoked the nicotine cigarette first in
the post-abstinence session.
    Table II shows the main features of the design of Study 2. The
same nicotine and non-nicotine cigarettes were used as with Study

1, but on this occasion, the subjects were permitted to smoke their cigarettes ad lib.

Table II: Design of Study 2. CO=CO measure, Stern=Sternberg task, HR=heart rate measure.

Pre-abstinence session	Post-abstinence session
CO + mood quest	CO + mood and craving quest
HR + Stern	HR + Stern
Cigarette 1	**Cigarette 1**
HR + Stern + CO	HR + Stern + CO
Smoking effects quest	Smoking effects quest
Cigarette 2	**Cigarette 2**
HR + Stern + CO	HR + Stern + CO
Smoking effects quest	Smoking effects quest

Four measures were added to the Sternberg task and expired-air CO used in Study 1 (See West & Hack, 1990 for details):

1. After each cigarette, subjects completed a smoking effects questionnaire in which they rated on a three-point scale the extent to which the cigarette had led to certain subjective effects including dizziness, tremor, and palpitations.

2. Before and after each cigarette their resting heart rate was recorded over a one minute period.

3. At the start of the pre-abstinence and post-abstinence sessions, all subjects completed a mood questionnaire. This included five-point ratings of: depression, irritability, restlessness, hunger and poor concentration (see West & Russell, 1985).

4. At the beginning of the post-abstinence session, the subjects completed a craving questionnaire containing six-point ratings of difficulty not smoking, time spent with urges to smoke and strength of urges to smoke (West & Russell, 1985).

RESULTS

The occasional smokers experienced no discernible withdrawal symptoms following their period of abstinence, confirming that they were not physically dependent. The regular smokers by contrast showed a significant increase in irritability compared with the occasional smokers (F=8.8, p<.01) and hunger (F=4.6, p<.05). They also reported much greater difficulty not smoking and urges to smoke during the 24 hours' abstinence than did the occasional smokers (F>11.0, p<.005 for all three items). Thus the regular smokers did show evidence of a withdrawal syndrome.

The start-of-session CO dropped from 16.9 ppm in the regular smokers before abstinence to 7.4 ppm after abstinence. In the occasional smokers it remained virtually unchanged (4.5 ppm versus 4.6 ppm).

The mean CO boost from the nicotine cigarette was 2.9 ppm in the occasional smokers before abstinence, and 2.4 ppm after abstinence. For the regular smokers the corresponding figures were significantly higher at 5.5 ppm and 6.6 ppm (F=7.4 and F=10.6 respectively, p<.02). Thus the regular smokers obtained more CO and therefore almost certainly more nicotine from the nicotine cigarettes than did the occasional smokers. However, the CO intake from the cigarette post-abstinence was not significantly different in either group from pre-abstinence.

The nicotine cigarette caused significantly more dizziness, tremor, and palpitations than did the non-nicotine cigarette both before and after abstinence among both the occasional and regular smokers (F>5 in all cases, p<.05). The regular smokers reported greater dizziness, tremor and palpitations from the post-abstinence nicotine cigarette than the pre-abstinence one (F>5 in all cases, p<.05). This was not the case with the occasional smokers.

The increase in memory search rate was significantly greater after the nicotine cigarette than the non-nicotine cigarette among the occasional and regular smokers combined and before and after abstinence (F=3.26, P<.05). There was no significant effect of abstinence and no significant difference between the occasional and regular smokers. Figure 1 shows the result graphically. When the nicotine cigarette was smoked first there was an increase in search rate after the first cigarette, and when the nicotine cigarette was smoked second there was an increase in search rate after the second cigarette. This increase in search rate resulted from a decrease in the response time for positive set size 5, and not an increase in response time for positive set size 2 (See West & Hack, 1990 for details).

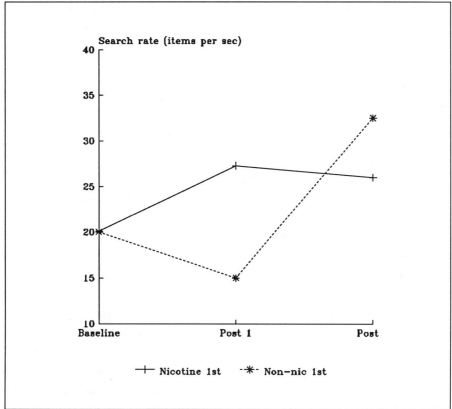

Figure 1: Memory search rates before and after nicotine and non-nicotine cigarettes

DISCUSSION

The results of the two studies indicate that smoking a nicotine cigarette can result in an increase in memory search rate. This does not appear to differ according to whether smokers abstain from smoking before smoking the test cigarette or whether they are occasional or regular smokers. Thus regular smokers who were shown to experience cigarette withdrawal symptoms experienced the effect to an indistinguishable degree before and after a period of abstinence, as did occasional smokers. This was despite the fact that the regular smokers experienced other effects of smoking (dizziness, palpitations and tremor) to a greater extent after the period of abstinence than before.

The time period between the first and second post-test was approximately 10 minutes which probably explains why the increase in search time which occurred when the nicotine cigarette was smoked first did not then fall again after the non-nicotine cigarette (Figure 1).

The Sternberg task offers much greater scope for exploring the processes which might be affected by a drug than we have taken advantage of here. For example, it is possible to perform manipulations which examine the response selection and motor program processes and the stimulus acquisition components (Sternberg, 1966), Hindmarch's program does allow these factors to be manipulated. It would be worth expoloring these in more detail. It would probably need a separate sessions because the increase in number of conditions would extend the testing session to a point where systemic nicotine levels would have fallen considerably by the end. Also, there would be potential problems of an interaction with the effects of nicotine on ability to sustain attention.

Part of the problem with the literature on the effects of cigarettes on performance is the failure to replicate findings. Before discussing in detail the relevance of the present findings for smoking motivation and issues relating to tolerance, we should await a replication. If the findings are replicated it will reinforce the important notion that the extent of acute and chronic tolerance to the effects of nicotine depends critically on the specific variable being assessed. In parallel with an attempt to replicate the findings using cigarettes, it would be worthwhile using alternative forms of nicotine delivery. If the effect is not greatly subject to acute tolerance as our results suggest, then it should occur with the relatively slow absorption from subcutaneous nicotine or a nicotine patch.

REFERENCES

Baddeley, A. (1985) The Psychology of Memory, Harper & Row, London.
Hindmarch, I., Kerr, J.S., and Sherwood, N. (1990) Psychopharm. 100, 535-541.
Sternberg, S. (1966) Science 153, 652-654.
Subhan, Z. and Hindmarch, I. (1985) Eur. J. Clin. Pharmacol. 28, 567-571.
USDHHS (1988) The Health Consequences of Smoking: Nicotine Addiction, A Report of the Surgeon general. Office on Smoking and Health, Rockville M.D..
West, R., and Hack, S. (1990) Manuscript submitted for publication.
West, R., and Russell, M.A.H. (1985) Psychopharm. 87, 334-336.

NICOTINE POLACRILEX GUM AND SUSTAINED ATTENTION

A.C. Parrott, D. Craig, M. Haines and G. Winder

Department of Psychology, Polytechnic of East London, London
E15 4LZ, UK.

SUMMARY: The effects of nicotine chewing gum, and cigarette
smoking, upon aspects of attention were investigated in three
studies, using young nicotine-deprived smokers. In the first
study, nicotine led to a significant linear dose-related
improvement in rapid visual information processing (RVIP),
target detection. In the second study, nicotine again led to
improvements in RVIP target detection. A linear dose-response
function was evident at the first post-test, while a curvilinear
(Inverted-U) function was present at the second post-test
period. In both studies, RVIP response time demonstrated
curvilinear functions, with faster performance under 2mg and 4mg
gum, but unchanged response times under placebo and cigarette.
In the second study, three further attentional tasks were also
undertaken. Letter cancellation response time was significantly
improved by nicotine, in a linear dose-response manner. Neither
Stroop task performance, nor performance on a width of attention
task, were however affected. Thus in study 2, nicotine led to
significant improvements in both measures of sustained
attention, while indices of width of attention and
distractability were not affected. In study 3, 4mg nicotine gum
was compared with placebo gum. Regular smokers (+15
cigarettes/day), showed a non-significant tendency for improved
RVIP target detection following the nicotine gum, whereas non-
regular smokers (-5 cigarettes/day) showed similar RVIP
performance levels after placebo and 4mg nicotine gum.

INTRODUCTION AND METHODOLOGIES

The effects of nicotine gum were investigated in three studies.
Each study was double blind and placebo controlled with regard
to gum administration. Sham smoking was used when appropriate.
Drugs were administered in counterbalanced or latin square
order. The subject groups comprised 16 smokers (age range: 18-
35), different in each study, who had abstained from smoking for
12 hours prior to testing. In studies 1 and 2, all subjects
smoked +15 cigarettes/day. The first study was designed to
assess the effects of nicotine gum (2mg, 4mg), placebo gum, and

cigarette smoking (own brand), upon rapid visual information processing (RVIP) performance. The aim of this study (Parrott and Winder, 1989) was to assess whether the nicotine gum would affect performance on a task with known sensitivity to smoking/nicotine (Wesnes and Warburton, 1983; Hasenfrantz et al, 1989). In the second study, four attentional tasks were assessed: RVIP, letter cancellation, the Stroop task, and a width of attention task. The aim of this second study (Parrott and Craig, unpublished) was to investigate the effects of nicotine upon different aspects of attention. In the third study, 8 non-regular smokers (-5 cigarettes/day), were compared with 8 regular smokers (+15 cigarettes/day) following 4mg nicotine gum, or placebo gum. The aim of this last study (Parrott and Haines, unpublished), was to investigate the influence of level of nicotine deprivation, upon RVIP performance. Resting heart rate, Profile of Mood State Questionnaire, and subjective response to the gum, were also assessed in these studies, but are not reported here.

RESULTS: STUDY 1

Nicotine increased RVIP target detection in a linear dose-related manner ($p < .01$). Performance showed the traditional vigilance decline under placebo, it improved after cigarette smoking, while intermediate detection rates were evident in the nicotine gum conditions (Table 1). A different performance pattern was shown for RVIP response time. Placebo and cigarette led to unchanged response time, while performance levels were comparatively improved in the two nicotine gum conditions. The second-order (Inverted-U, Yerkes-Dodson, or curvilinear) function was significant ($p < .01$; Table 1). Commission error remained low in frequency, and was not affected by drug condition, in any of the three studies.

Table 1. Results from the first study.

Assessment measure	Placebo gum	2mg nicotine gum	4mg nicotine gum	Cigarette smoking	Statistical effects
RVIP Response Time (msec)					
pre-post 1 diff	+2	−14	−29*	−2	D+ C*
pre-post 2 diff	+2	−11	−22	−3	
RVIP Target detection					
pre-post 1 diff	+0.8	+1.2	+0.6	+3.7	
pre-post 2 diff	−2.2	+0.2	+1.9*	+2.2*	D* L**

	one-tail	two-tail
+	p<.10	p<.05
*	p<.05	p<.025
**	p<.01	p<.005
***	p<.001	p<.0005

D= ANOVA Drug effect
L= Linear dose effect
C= Curvilinear dose effect
Placebo/drug differences comprise:
Dunnet test comparisons

post 1 = first 5 minutes of RVIP; post 2 = second 5 minutes of RVIP task

RESULTS: STUDY 2

The RVIP task findings were broadly similar to those found in the first study, although a number of differences were evident. Response time showed a curvilinear function, with faster responses in the 2mg and 4mg gum conditions, but unchanged times for the placebo and cigarette conditions. This second-order polynomial function was borderline in significance ($p < .10$, two-tail; $p < .05$, one-tail; Table 2). RVIP target detection again decreased under placebo, while they were improved in the nicotine conditions. At the first post-test, the pattern of findings was similar to study 1, with the greatest performance improvement in the cigarette condition, and intermediate values in the nicotine gum conditions. At the second post-test, a different pattern was evident, with highest performance in the 2mg gum condition (Table 2), and a significantly faster under smoking, and the linear dose-response function was significant (Table 2). Neither width of attention nor Stroop task performance, were however affected by nicotine.

Table 2. Results from the second study.

Assessment measure	Placebo gum	2mg nicotine gum	4mg nicotine gum	Cigarette smoking	Statistical effects	
RVIP Response Time (msec)						
pre-post 1 diff	+8	-10	-13	-3		
pre-post 2 diff	-3	-17	-21	+3	C*	
RVIP Target detection						
pre-post 1 diff	-1.1	+1.5	+0.9	+2.7*	D+	L*
pre-post 2 diff	-1.1	+4.4**	+1.1	+1.1	D*	C*
Letter Cancellation Response Time (msec)						
pre-post diff	+27	-15	+14	-54**	D**	L*

Legend: as Table 1.

RESULTS: STUDY 3

Each subject self-assessed their level of cigarette craving before the RVIP task. As expected, the deprived regular smokers reported significantly higher levels of craving, than the deprived non-regular smokers (p.<.05). On the RVIP task, target detections in the regular smokers increased under nicotine gum (+3.3 targets), but decreased under placebo gum (-1.1 targets). In non-regular smokers, target detection rates were similar following nicotine gum (+1.9 targets), and placebo gum (+1.5 targets). The ANOVA subject group x drug interaction was not however significant (p=0.13 two-tail; p=.065 one tail).

DISCUSSION AND CONCLUSIONS

This series of studies has shown that nicotine polacrilex gum can have significant effects upon rapid visual information processing. The RVIP task has previously been shown to be sensitive to cigarette smoking (Wesnes and Warburton, 1983; Hasenfrantz et al, 1989), and nicotine tablet (Wesnes and Warburton, 1983, 1984). Synder and Henningfield (1989) also found significant effects on a computerised performance test battery, with 2mg and 4mg gum. The present findings do however contrast with those of Michel et al (1988), who found no effect of 4mg gum upon a variant of the RVIP task. Some aspects of

Michel's study may have reduced its sensitivity: the parallel groups design, smaller group size, brief pre-experiment task practice, and short period of nicotine deprivation (+2 hours). The present studies, by including a cigarette smoking condition, also provided a 'high' dose condition, against which the fairly subtle low-dose effects of the nicotine gum could be estimated.

The RVIP dose-response relationships comprised a mixture of monotonic and curvilinear functions (Tables 1,2). The literature on the performance effects of nicotine, contains many examples of monotonic and curvilinear functions (Mangan and Golding, 1984). Wesnes and Warburton found increasing RVIP performance with increasing dose in some studies, but optimum performance at low-mind doses in others (Wesnes and Warburton, 1983). Williams (1980) noted significant monotonic <u>and</u> curvilinear functions within the same letter cancellation data. While empirical demonstrations of the inverted-U function are quite rare, nicotine may be an ideal stimulant for examining the nature of this function. The significant increase in letter cancellation performance, confirmed that nicotine improves this measure of sustained attention (Parrott and Roberts, current symposium; Williams, 1980). Neither Stroop task performance, an index of distractability, nor performance on a width of attention task were affected by nicotine. Surgeon General (1988) briefly reviewed the few distractability/attentional-width studies then available, concluding that current evidence was 'equivocal'. Further studies are obviously needed into these rarely-studied aspects of attention. The findings from the third study were inconclusive, but suggested that nicotine only improves the performance of nicotine deprived subjects. Thus while deprived regular smokers showed similar performance under placebo and nicotine gum. Subjects may need to be in a state of nicotine deprivation to demonstrate performance change (Parrott and Roberts, this symposium).

Acknowledgments: We would like to thank Professor Urbain Sawe of Pharmacia Leo for the gum supplies, Dr. Keith Wesness and Pauline Simpson for the RVIP program, Dr. Anne Richards for the width of attention program, and Dr. Andy Burton for the Stroop program.

REFERENCES

Hasenfrantz M, Michel C, Nil R, Batting K (1989). Can smoking increase attention in rapid information processing during noise? Electrocortical, physiological and behavioural effects. Psychopharmacology, 98, 75-80.

Mangan GL, Golding JF (1984). The psychopharmacology of smoking. Cambridge UK, Cambridge University Press.

Michel Ch, Hasenfrantz M, Nil R, Batting K (1988). Cardiovascular, electrocortical and behavioural effects of nicotine chewing gum. Klin, Wochenscr, 66, (suppl.) 72-79.

Parrott AC, Roberts G (1990). Nicotine deprivation and nicotine reinstatement: effects upon a brief sustained attention task. Proceedings of the Current Symposium.

Parrott AC, Winder G (1989). Nicotine chewing gum (2mg, 4mg) and cigarette smoking: comparative effects upon vigilance and heart rate. Psychopharmacology, 97, 257-261.

Snyder FR, Henningfield JE (1989). Effects of nicotine administration following 12h of tobacco deprivation: assessment on computerised performance tasks. Psychopharmacology, 97, 17-22.

Surgeon General (1988). Nicotine addiction: the health consequences of smoking, Washington, US Government Printing Office.

Wesnes K, Warburton DM (1983). Smoking nicotine and human performance. Pharmacology and Therapeutics, 21, 189-208.

Williams GD (1980). Effect of cigarette smoking on immediate memory and performance. British Journal of Psychology, 71, 83-90.

THE BEHAVIORAL EFFECTS OF NICOTINE IN LABORATORY ANIMALS AND IN
THE HUMAN SMOKER

K. Bättig

Behavioral Biology Lab, Swiss Federal Institute of Technology,
Turnerstr. 1, ETH-Zentrum, CH-8092 Zurich, Switzerland

SUMMARY: Taken together, the findings presented at this symposium
and the recent scientific literature of the past years are more
of descriptive than explanatory value. Nicotine appears as a mild
stimulant, which, at least in smoking, does not disrupt normal
behavior, as opposed to amphetamine and cocaine. The tolerance
effects are in part modest, but in part also poorly investigated.
The possible withdrawal effects of smoking abstinence have so far
mainly been illustrated by subjective reports rather than by
objectively measuring indices of different performances. A major
unresolved question concerns thus the reasons underlying smoking
dependence in man. Although animal models for self-application of
nicotine have successfully been developed, they have revealed
aspects which at present resist a satisfactory explanation, such
as the poor dose effect relations, the low resistance to
extinction and the absence of reports on withdrawal effects. In
the ligth of such heterogeneous findings, multifactorial concepts
of smoking dependence are more likely to develop further in the
near future than are satisfactory and comprehensive unifactorial
concepts, as recently summarized in a book edited by ney and gale
(1989), highlighting the biobehavioral consequences of smoking,
the aspects of dependence, the aspects of neuronal and behavioral
functioning, the aspects of emotion, anxiety and stress, and last
but not least, those of sensory and motor habit formation in
smoking.

Behavioral research on the nicotinic effects in the human
smoker and in laboratory animals has undergone a considerable
methodological diversification over the last years. Several
specific questions related to these two topics were also
illustrated through the communications of this symposium.

A. THE STIMULANT EFFECTS OF NICOTINE

The list of task performances demonstrated to be improved by smoking has increased in recent years. It appears by now that the substance acts as a mild stimulant, thereby increasing general arousal and facilitating manifold cognitive and psychomotor performances and skills including the formation and recall of memories. How do such effects compare to those of other psychotropic substances? In a meta-analysis, Hindmarch compared in his contribution the effects obtained in earlier studies with a large series of substances to those of nicotine. Nicotine appeared to significantly improve critical flicker fusion, choice and recognition reaction times, motor reaction time, tracking, tracking reaction time, and short-term memory. Although significant, these effects were generally small in magnitude, particularly in comparison to amphetamine and its derivatives.

It has been well established through research at many laboratories that such improvements go in parallel with characteristic changes in the profile of EEG activity, as demonstrated by Knott in his contribution. Psychostimulants appear to augment rather than to depress EEG activity in the alpha band. With nicotine, this effect is characteristic after smoking a habituated cigarette in the normal, habituated fashion. Higher yields, otherwise unhabituated cigarettes or requiring subjects to inhale more deeply than usual can lead to an overdosage of nicotine without activity increases in the alpha band. With continuous recording and analysis of EEG across the successive puffs, it can be shown that the shifts in the power spectrum are mostly already visible after the fourth to fifth puff. The topographic analysis further revealed that the changes are more prominent in the frontal than in posterior brain regions. As this brain region can also be considered as the "executive control" area of the cortex, the speculation of the author that this distribution of activity might be critical for the performance improvements after nicotine certainly merits closer attention.

Do stimulants of the nicotine or caffeine type increase the speed of processing information in the brain? Several laboratories have obtained evidence that this may indeed be the case, as evidenced by a shortening of the latencies of stimulus-evoked brain potentials (Edwards et al., 1985). Another argument for this assumption could also be seen in demonstrations that nicotine increases the maximum frequencies at which the brain processes information. One hint in this direction is presented by increases in the visual flicker fusion freqeuncy. Another one has been explored by Mangan and Colrain, who used the paradigm of photic driving of the EEG. They obtained evidence that smoking increased the power of photic driving, which is an enhancement of the EEG by repetitive flashes selectively at the highest frequencies of the EEG spectrum. This increase in high frequency photic driving was paralleled by improvements in memory recall.

B. Facilitation of stress coping?

Numerous reports have demonstrated that stress loads increase the self-reported consumption of cigarettes (USDHHS, 1988). Reports on the objective effects of smoking in stress situations are, however, more equivocal (Bättig & Nil, 1988). In human studies, changes have been seen mostly for the vegetative responses to the stress loads. These were mostly dampened if the subjects were presented with stressful stimuli that did not allow escape or avoidance, such as noise bursts, electrical pain stimuli, or highly emotional video scenes. However, when the subjects were required to perform stressful tasks such as demanding video games, the vegetative responses to the task and to the cigarette appeared to be additive.

This suggests that interactions of nicotine and stress might be rather complex. An illustration of this complexity was also communicated from our laboratory at this symposium by Driscoll et al. They treated pregnant rats of two strains with the

stress of saline injections and compared the subsequent
maternal behavior and the behavior of the offspring with that
of animals who received nicotine injections. Effects in the
direction of stress induced by the saline injections and relief
of stress by nicotine were seen only in the RLA strain of rats.
These animals are more anxious and emotional than animals of
the RHA strain, which showed no changes in behavior in response
to the injection stress or to nicotine.

C. The individuality of the response to nicotine

It is well known that cigarette smokers differ considerably
not only with respect to the number of cigarettes smoked per
day and the strength of the preferred personal brands but even
more so with respect to the intensity of smoke absorption (Nil
& Bättig, 1989). This enormous variation was also documented at
this symposium by the several communications on nicotine
absorption through smoking (Höfer & Bättig; Kolonen et al.;
Sepkovic et al.; Heller et al.).

Recently, the possible background of such differences has
also been investigated on the molecular level. The past years
have demonstrated that the neuronal receptors for agonistic and
antagonistic nicotinic compounds are highly complex, consisting
of high- and low-affinity and of both competitive and
noncompetitive binding sites (Nordberg et al., 1989). Strain
differences for nicotinic effects have been demonstrated
earlier in rats (Garg, 1969; Bättig et al., 1976) and in mice
(Hatchell & Collins, 1977).

In this light, the results presented by Collins and his
group merit particular interest. They adopted the technique of
assessing nicotine and α-bungarotoxin binding in discrete brain
areas of a large series of inbred mouse strains. These results
were compared with the nicotinic effects on motor behavior,
vegetative response and the incidence of convulsions in the
same strains. It appeared not only that the results of the

binding assays differed widely across the strains, but also that these differences were closely related to the differences in the action profiles of nicotine across the strains.

Interindividual variation in nicotinic responses is therefore, at least in part, genetically determined. Beyond this there are by now sufficient arguments to assume that different factors affecting neural development may additionally contribute to shape individual susceptibility to nicotine. As an early example of such studies, Lichtensteiger et al. (1983) showed that prenatal nicotine can affect the development of the ascending catecholaminergic pathways. These systems are not only important in mediating the effects of nicotine. Their developmental maturation can be manipulated by prenatal nicotine exposures as well as by a series of other prenatal challenges such as heavy metals, anoxia, alcohol and stress. Other neurochemical systems involved in the manifestation of nicotinic action appear now to be similarly sensitive to manipulation in the early phases of development. Apparently this also holds for the cholinergic system, which develops later than the catecholaminergic projections. Along this line, the communication of Zhang and coworkers is particularly interesting. They found that early postnatal nicotine affected later spontaneous motor behavior and the development of subunits of the nicotinic receptors to a similar extent. This finding is extended by the communication of Yanai and colleagues, who observed parallel changes in muscarinic receptor density in the hippocampus and in the efficiency of spatial learning after early postnatal nicotine treatment in mice. This finding is of particular interest in the light of the numerous studies which have related hippocampal micro-anatomy and different manipulations of the hippocampus to a whole series of behavioral performances (Lipp et al., 1988).

D. Tolerance

The question as to whether and which effects of nicotine
might be subject to tolerance could be of critical importance
for the explanation of the dependence of smokers on nicotine.
However, the available evidence remains, at least in part,
quite equivocal.

The question has been raised in particular for the stimulant
and performance-improving effects of nicotine in smokers. Do
such gains represent an improvement above the normal level or
do they merely represent escape from or avoidance of the
performance-deteriorating effects of nicotine withdrawal?
Parrott and coworkers presented data in their communications
suggesting that, at least for tasks requiring sustained
attention, nicotine can do no more than restoring withdrawal
deficits to baseline levels. The finding is in contrast to an
extended study presented at this symposium by West and Hack.
These authors compared regular smokers with occasional smokers.
Smoking abstinence produced increases in cigarette craving and
irritability in the regular but not in the occasional smokers.
Behavioral performance was tested using the classical Sternberg
serial memory test. This measure was not impaired by smoking
withdrawal, but it was similarly improved in both groups of
smokers by smoking nicotine-containing cigarettes. Thus the
substance was followed by a significant increase in the rate of
memory search.

Although such results suggest that tolerance and withdrawal
effects might be task dependent, they nevertheless call for a
consideration of other functions which are affected by
nicotine. A rather robust effect of nicotine can be seen in the
acceleration of heart rate. Laboratory studies showed that this
effect is modestly subject to acute intraday tolerance, which
dissipates overnight. In a field study, we monitored heart
rate, activity and cigarette consumption across 24-hour periods
in a group of 10 smokers and 10 nonsmokers, all females doing
housework and taking care of one to three preschool children.

Figure 1 shows the main results, consisting in a rather stable elevation of heart rate in the smokers across the entire day and in a modest and gradual increase in the rate of smoking throughout the day. Motor activity, however, did not differ between the two groups. Smokers appear therefore to smoke so as to maintain a consistently elevated heart rate, which returns to normal levels with smoking abstinence.

Similar on/off effects of nicotine being only modestly affected by chronic tolerance is also seen for weight regulation in rats and man and for spontaneous motor activity in the rat. In the rat, nicotine produces after an initial depression an increase in activity, which in chronic experiments done in our laboratory was seen to persist with daily nicotine treatments without any sizable change in the magnitude of the changes for months (Welzl et al., 1988). A similar picture can be seen for weight regulation (USDHHS, 1988), and it has been proposed that nicotine lowers the set point of body weight regulation as opposed to amphetamine, which produces a transient anorexia and has little effect on body weight in chronic experiments.

E. Reward

The question as to which of the elements of the action of nicotine might constitute the rewards and thereby the motivation to smoke remains a critical one. Research into this question has been guided recently to a great extent by the neurophysiological concept of reward as proposed by Wise (1988). This concept is based on the neurochemical analysis of intracranial self-stimulation in the rat. The electrical stimulation of active sites of this system immediately evokes an increase of locomotion, an orienting response of the head and an apparent need to repeat such stimulation. A great number of studies carried out at different laboratories has produced evidence that the occurrence of these reponses

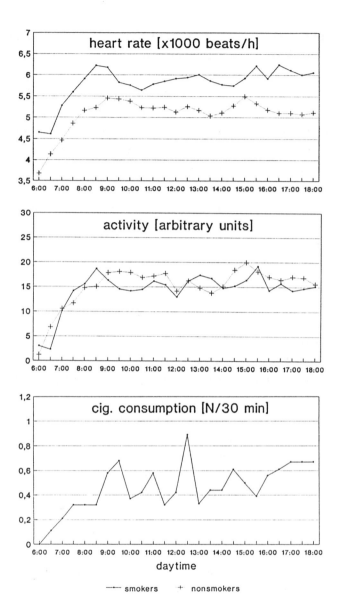

Fig. 1. Heart rate, motor activity and cigarette consumption averaged across two workdays in 10 smokers and in 10 nonsmokers.

depends critically on the functional intactness of the mesolimbic dopaminergic projection. Furthermore, it has been seen that the injection of opiates into the tegmental areas containing dopaminergic cell bodies increases firing of the ascending axons and that this is mediated through the presence of opiate receptors at these neurons. Moreover, such injections produce the same triad of symptoms which is typical for electrical stimulation. Amphetamine and cocaine, on the other hand, were found to produce similar effects when injected into the nucleus accumbens, the critical target area of the mesolimbic dopaminergic projections. Other studies have produced evidence that natural rewards such as food or water activate the same system. Based on such findings, Wise's concept, elaborated in more detail in a recent book edited by Liemann and Cooper (1989), proposes that drugs which induce self-application may do so by directly or indirectly activating the mesolimbic deopaminergic system. If this were the case, activation of the dopaminergic reward system would be the critical element for inducing self-application and all other actions of the drug would be side effects.

A critical argument against this hypothesis could be seen in the fact that relatively high doses of nicotine have been needed so far to activate the mesolimbic dopaminergic projection. This argument may lose some of its impact through the demonstration by Balfour et al. that chronic pretreatment with nicotine sizably increases the susceptibility of the dopaminergic system to the effects of the substance.

On the other hand, Corrigall, in his contribution, has not only reviewed arguments which might support the dopaminergic hypothesis for the case of nicotine, but he has also presented findings which suggest a more differentiated view. Although the self-injection of nicotine by rats can be obtained rather reliably, the behavior depends, in contrast to other self-applied drugs, on the use of particular reward schedules. It is considerably less dependent on the dosage of the substance than in the case of cocaine. Furthermore, decreases rather than compensatory increases in self-application responses were obtained after treatment with dopamine antagonists, whereas the same doses of these antagonists increased cocaine self-administration.

Such findings indicate that a possible dopaminergic link for nicotine self-application needs further experiments and that the profile of such a link would probably be a quite different one than for other drugs of dependence. In this context, the particularities of nicotine self-application by laboratory animals as well as of other paradigms to evaluate rewarding effects of drugs merit closer attention. Attempts to induce rats to self-administer nicotine remained rather unsuccessful for a long time until Goldberg and Spealman (1982) discovered in the squirrel monkey that relatively stable rates of administration can be obtained when each injection is followed by a time-out of several minutes, after which the next injection can be dependent on the first lever press or on a specific reward schedule for responding. Toward a better understanding of this phenomenon, approaches manipulating different aspects of the paradigm appear to be highly desirable. The contribution of Yanagita, elaborating on such aspects in the monkey, highlights several particularities of the behavior. It appears from these results that, in contrast to the opiates, nicotine pretreatment does not enhance nicotine-seeking behavior, that the behavior is much less resistant to extinction than cocaine self-application, that dose regulation by adequate responding is considerably less efficient than with other drugs of dependence and that elaborate and prolonged training is needed to stabilize the self-application of nicotine.

Another paradigm to assess the reward value of drugs used widely in recent years is that of the conditioned place preference, which tests whether an undrugged animal will prefer a place where it was trained to experience the action of the drug against places associated with the experience of no drug or saline, as recently reviewed by Carr et al. (1989).

According to this review, the commonly known drugs of dependence induce conditioned place preference rather unequivocally, whereas this seems not to be the case for nicotine, for which more recently even conditioned place aversion has been reported (Jorenby et al., 1990).

Taken together, there remain thus several missing links toward seeing nicotine as a "quasi heroin-amphetamine-cocaine-like" substance, as favored recently in an extensive review on nicotine dependence (USDHHS, 1988). In contrast to the self-application paradigm in rats and other laboratory animals, smokers develop nicotine self-application rather easily, and its resistance to extinction is very high. Furthermore, increases in motor activity are a main symptom in rats habituated to nicotine but not in humans, as recently reviewed (USDHHS, 1988) and also shown above in Figure 1.

REFERENCES

Bättig, K., Driscoll, P., Schlatter, J., and Uster, H.J. (1976) Pharmacol. Biochem. Behav. 4, 435-439

Bättig, K., and Nil, R. (1988) Activ. Nerv. Sup. 30, 103-105

Carr, G.D., Fibiger, H.C., and Phillips, A.G. (1989) In: The Neuropharmacological Basis of Reward (J.M. Liebman and S.J. Cooper, Eds), Clarendon Press, Oxford, pp. 264-319

Edwards, J.A., Wesnes, K., Warburton, D.M., and Gale, A. (1985) Addict. Behav. 10, 113-126

Garg, M. (1969) Psychopharmacologia 14, 432-438

Goldberg, S.R., and Spealman, R.D. (1982) Fed. Proc. 41, 216-220

Hatchell, P.C., and Collins, A.C. (1977) Pharmacol. Biochem. Behav. 6, 25-30

Jorenby, D.E., Steinpreis, R.E., Sherman, J.E., and Baker, T.B. (1990) Psychopharmacology 101, 533-538

Lichtensteiger, W., Schlumpf, M., and Davis, M.D. (1983) Monogr. Neur. Sci. 9, 213-224

Liebman, J.M., and Cooper, S.J. (Eds). (1989) The Neuropharmacological Basis of Reward, Clarendon Press, Oxford

Lipp, H.-P., Schwegler, H., Heimrich, B., and Driscoll, P. (1988) J. Neurosci. 8, 1905-1921

Ney, T., and Gale, A. (Eds). (1989) Smoking and Human Behavior, John Wiley & Sons, Chichester

Nil, R., and Bättig, K. (1989) In: Smoking and Human Behavior (T. Ney and A. Gale, Eds), John Wiley & Sons, Chichester, pp. 199-221

Nordberg, A., Fuxe, D., Holmstedt, B., and Sundwall, A. (Eds). (1989) Nicotinic Receptors in the CNS: Their Role in Synaptic Transmission, Elsevier, Amsterdam

U.S. Department of Health and Human Services. (1988) The Health Consequences of Smoking: Nicotine Addiction. A Report of the Surgeon General, DHHS, Rockville, MD

Welzl, H., Alessandri, B., Oettinger, R., and Bättig, K. (1988) Psychopharmacology 96, 317-323

Wise, R.A. (1988) J. Abnorm. Psychol. 97, 118-132

VII.
Nicotine and Human Disease

NICOTINE AND CARDIOVASCULAR DISEASE

Neal L. Benowitz

University of California, San Francisco, San Francisco General Hospital Medical Center, San Francisco, California 94110

SUMMARY: The evidence that nicotine may cause or aggravate cardiovascular disease is reviewed. Nicotine may contribute to accelerated atherogenesis by inducing hyperlipidemia, injuring endothelial cells and/or promoting thrombosis, although the evidence is not conclusive. Nicotine is likely to contribute to acute ischemic events in people who already have coronary heart disease via adverse effects on systemic hemodynamics, by promoting thrombosis, constricting coronary arteries and/or facilitating arrhythmogenesis. Pharmacodynamic studies suggest that nicotine inhaled in cigarette smoke may have different cardiovascular effects than that absorbed more slowly, such as from nicotine gum or transdermal nicotine. The safety of chronic nicotine exposure, such as with medicinal use of nicotine, cannot be predicted and requires empiric evaluation.

INTRODUCTION

Cigarette smoking is a substantial risk factor for cardiovascular diseases, which are a major cause of disability and death around the world. Causal relationships have been demonstrated between cigarette smoking and accelerated atherosclerosis, acute myocardial infarction, unstable angina pectoris, sudden cardiac death, peripheral arterial occlusive disease, atherosclerotic aortic aneurysm and stroke (U.S. Department of Health and Human Services, 1989). Although epidemiologic studies of smoking and stable angina pectoris have been conflicting, smoking clearly reduces exercise tolerance in people with stable angina pectoris (Danfield et al., 1984).

While the evidence for a causal link between cigarette smoking and coronary heart disease is well established, the pathophysiology is not. Of the thousands of chemicals in cigarette smoke, a few are suspected to contribute to smoking-induced vascular disease. These include carbon monoxide, nicotine, polycyclic aromatic hydrocarbons, and tobacco glycoproteins. This paper will focus on nicotine.

The issue of whether and, if so, to what extent nicotine contributes to cardiovascular disease is important for several reasons. From a basic science perspective, what is learned about nicotine may expand our understanding of the pathogenesis of cardiovascular disease related to other toxic exposures. Nicotine is now available as a medication for use in smoking cessation treatment and could conceivably find other medical applications. Knowledge of the long-term toxicity of nicotine is essential for rational decisions about such therapy. Smokeless tobacco delivers levels of nicotine similar to those of cigarette smoking (Benowitz et al., 1989), without other combustion products. Insofar as nicotine increases the risk of cardiovascular disease, this would be a concern as well for users of smokeless tobacco. Finally, in a search for a safer cigarette the idea of a low tar-high nicotine cigarette has been proposed (Russell, 1976). In that people regulate the intake of nicotine from their cigarettes, nicotine enhancement would be expected to reduce the amount of smoke inhaled, thereby reducing the exposure to tars and some noxious gases that contribute to lung cancer and chronic obstructive lung disease. A risk-benefit analysis of such a proposition requires an assessment of the risks of nicotine.

CARDIOVASCULAR PHARMACOLOGY OF NICOTINE

In considering the potential of nicotine to contribute to or aggravate cardiovascular disease, it is necessary to review its actions in healthy people. Smoking a cigarette or an infusion of nicotine activates the sympathetic nervous system (Cryer et al., 1976). In healthy people, this results in an increase in heart rate and blood pressure, cardiac stroke volume and output, and coronary blood flow (Nicod et al., 1984). The peripheral vascular changes include cutaneous vasoconstriction, associated with a decrease in skin temperature, systemic venoconstriction, and increased muscle blood flow (Freund & Ward, 1960; Eckstein & Horsley, 1960; Rottenstein et al., 1960). Increased circulating concentrations of norepinephrine and epinephrine indicate neural adrenergic stimulation and adrenal medullary stimulation. The release of vasopressin may mediate some of the vasoconstriction, as evidenced by findings that pretreatment with a vasopressin antagonist reduces the nicotine-induced constriction of blood vessels of the skin (Waeber et al., 1984). Circulating free fatty acids, glycerol, and lactate concentrations increase. Cardiovascular and metabolic effects are prevented by combined alpha and beta adrenergic blockade, indicating that cardiovascular effects of cigarette smoking are mediated by activation of the sympathetic nervous system (Cryer et al., 1976).

Nicotine inhibits the release of prostacyclin from the rings of rabbit or rat aorta and human veins (Sonnenfeld & Wennmalm, 1980; Wennmalm, 1980). The suggested mechanism is the inhibition of prostacyclin synthesis through the inhibition of cyclo-oxygenase. Decreased urinary excretion of the prostacyclin metabolite 6-keto-prostaglandin $F_{1\alpha}$ ($PGF_{1\alpha}$) has been reported in smokers after they have smoked nicotine-containing but not nicotine-free cigarettes (Nadler et al., 1983), but a more recent study measuring the urinary excretion of 2,3-dinor-6-keto-$PGF_{1\alpha}$, another metabolite of prostacyclin, by specific techniques of gas chromatography-mass spectrometry has suggested that smoking led to an increased rather than reduced biosynthesis of prostacyclin (Nowak et al., 1987).

Pharmacodynamic considerations may also be important in understanding the long-term effects of nicotine. Of major importance is that the magnitude of effects of nicotine may vary according to rate of administration, duration of exposure and the extent of tolerance (Porchet et al., 1987, 1988). With respect to tolerance, nicotine infusion studies in humans indicate that heart rate increases at relatively low blood levels of nicotine, but then heart rate reaches a maximum despite further increases in nicotine levels. Other pharmacodynamic studies indicate that in smokers heart rate increases most with the first few cigarettes of the day, but subsequently does not vary with the degree of nicotine intake. Although there is considerable development of tolerance during a single exposure, heart rate acceleration persists as long as moderate concentrations of nicotine persist (Benowitz et al., 1984).

PATHOGENESIS OF ATHEROSCLEROTIC VASCULAR DISEASE

As noted above, cigarette smoking is a major risk factor for accelerated development of atherosclerosis. To understand how nicotine might aggravate atherosclerosis, it is useful to examine mechanisms of atherogenesis. The natural history of atherosclerosis involves a progression from fatty streaks (lipid deposition in macrophages) on the inner surface of the arterial wall to fibrous plaques to calcified lesions that may ulcerate or hemorrhage and provide a site for thrombosis and occlusion (Ross, 1986). A number of factors appear to promote the atherosclerotic process. Hyperlipidemia, particularly increased concentrations of low density lipoproteins (LDL) that contain apolipoproteins B and E, promotes lipid deposition in macrophages and the development of fatty streaks. The vascular endothelium normally serves as a protective barrier between the blood and the arterial wall. Injury to the endothelial cells lining blood vessels is believed to contribute to the initiation of atherosclerosis. Endothelial injury may occur as a consequence of hemodynamic stress, exposure to toxins or immunological insults. Platelets adhere to vascular lesions, aggregate and secrete substances such as thromboxane A_2 that promote vasoconstriction and further platelet aggregation, and factors such as platelet-derived growth factor (PDGF) that promote the growth of smooth muscle cells. Endothelial cells secrete factors such as prostacyclin and endothelium-derived relaxing factor that have opposite effects such as inhibition of platelet aggregation and/or vasodilation.

In addition to cigarette smoking, other major risk factors for atherosclerosis are hyperlipidemia and hypertension. The latter is suspected to act by inducing hemodynamic trauma. Diabetes is also a strong risk factor, thought to act primarily by inducing abnormal lipid profiles, but may also do so by enhancing thrombosis. Sedentary lifestyle and/or chronic psychological stress have also been described as risk factors.

There are no good models for cigarette smoking or nicotine-induced atherosclerosis in animals. In a few studies, treatment with high-dose nicotine in the presence of an atherogenic diet has been shown to accelerate atherogenesis (Stefanovich et al., 1969), but these models are of questionable relevance to human atherosclerosis. We are left primarily to examine data on the pharmacology of nicotine and those physiological actions that are believed to contribute to athero-sclerosis.

NICOTINE AND LIPIDS

Cigarette smokers have on average higher levels of LDL and lower levels of HDL cholesterol, although the magnitude of difference between smokers and non-smokers is relatively small (usually about 10%) (Craig et al., 1989). However, a small difference could account for substantial additional numbers of patients with atherosclerotic vascular disease. Nicotine, via release of catecholamines, induces lipolysis and increases plasma free fatty acid concentrations. Increased free fatty acid concentrations could result in enhanced synthesis of LDL (Brischetto et al., 1983). Results of studies of effects of nicotine on lipids in animals are conflicting. Injections of nicotine or feeding of nicotine has been described to increase total cholesterol in dogs and rabbits on high cholesterol diets (Booyse et al., 1981; Kershbaum et al., 1965). Nicotine feeding to squirrel monkeys for two years has shown increased plasma LDL (Cluette-Brown et al., 1986). The mechanism in monkeys included both accelerated synthesis of LDL via lipolysis of HDL and VLDL and impaired clearance of LDL (Cluette-Brown et al., 1986; Hojnacki et al., 1986). Of importance in interpreting the animal studies is that high doses of nicotine have been administered, often by an oral route, without measurement of blood levels of nicotine. Although the relevance of these studies to human smoking has not been established, the data from animal studies is consistent with data in human cigarette smokers, supporting the possibility that nicotine is involved in elevating lipids in humans.

In people, one of the main effects of smoking is to reduce HDL cholesterol. Lower HDL cholesterol levels are associated with a higher risk of coronary heart disease. Initiation of cigarette smoking is associated with reductions in HDL con-centrations, and cessation of cigarette smoking results in increased HDL levels (Stubbe et al., 1982; Dwyer et al., 1988). That cigarette smoking acutely reduces HDL (Gnasso et al., 1986), and HDL increases within two weeks of cessation of smoking (Stubbe et al., 1982) suggests a short-term pharmacologic action of cigarette smoking, such as could be mediated by nicotine.

A study of serum lipids during two weeks of nicotine exposure via nicotine chewing gum in healthy nonsmokers found no changes in lipid profiles (Quensel et al., 1989). However, nicotine chewing gum results in relatively low levels of nicotine that rise and fall gradually, and gum is not chewed overnight. The high levels developed from more rapid administration such as by cigarette smoking or the sustained levels of nicotine that occur with transdermal nicotine administration could conceivably have different effects, and these effects need to be investigated. Of some interest in this regard is a recent study of users of smokeless tobacco indicating a higher prevalence of hypercholesterolemia (when normalized for age and education) compared with nonusers of tobacco (Tucker, 1989). The prevalence of hypercholesterolemia in smokeless tobacco users was similar to that of cigarette smokers.

Based on the above considerations, one must conclude that there is a possible link between nicotine and the more atherogenic lipid profile observed in cigarette smokers.

NICOTINE AND HYPERTENSION

Although smoking a cigarette acutely increases blood pressure, habitual cigarette smoking is not associated with an increased prevalence of hypertension (Green et al., 1986). In fact, in survey studies cigarette smokers tend to have lower blood pressures than nonsmokers, even with control for body weight. However, these measurements of blood pressure are made during at least short-term abstinence from smoking (during a physical examination). Cigarette smoking transiently increases blood pressure many times throughout the day (Cellina et al., 1975). Thus, while cigarette smoking does not seem to accelerate atherosclerosis by inducing chronic hypertension, it is possible that transient but frequently repeated blood pressure elevations could contribute to endothelial injury.

Progression of chronic hypertension to accelerated or malignant hypertension is more common in cigarette smokers (Isles et al., 1979). Nicotine could contribute to this progression by aggravating vasoconstriction, either via sympathetic activation or inhibition of prostaglandin synthesis. Animal studies indicate that nicotine may reduce renal blood flow (Downey et al., 1981) which, in a patient with marginal renal blood flow due to hypertensive vascular disease, could produce renal ischemia and aggravate hypertension. Thus, there is concern about nicotine replacement therapies in patients with severe hypertension.

Tobacco, due to effects of nicotine, may, however, interact with particular hypertensive diseases. For example, a patient with pheochromocytoma developed paroxysmal hypertension and angina pectoris following the use of oral snuff (McPhaul et al., 1984). Within 10 minutes, blood pressure increased from 110/70 mmHg to 300/103 mmHg, with a heart rate increase from 70 to 110. Rechallenge with snuff after surgical removal of the pheochromocytoma revealed only a mild blood pressure increase. Another patient with previously controlled essential hypertension presented with a blood pressure of 210/115 mmHg prior to surgery (Wells & Rustick, 1986). A mass of snuff was found in the patient's cheek. The snuff was removed and blood pressure returned to 150/85 mmHg within 15 minutes.

NICOTINE AND ENDOTHELIAL INJURY

Degenerative changes in the endothelium of umbilical arteries have been observed in newborn children of smoking mothers (Asmussen & Kjeldsen, 1975). Nicotine concentrations similar to those of cigarette smokers were shown to modulate the structural and functional characteristics of cultured vascular smooth muscle and endothelial cells (Csonka et al., 1985; Thyberg, 1986). Nicotine-induced structural damage and increased mitotic activity in aortic endothelial cells of nicotine-treated animals have been reported (Booyse et al., 1981; Zimmerman & McGeachie, 1987). Nicotine given i.v. or orally produced dose-dependent increases in circulating anuclear carcasses of endothelial cells (Hladovec, 1978). Intravenous nicotine administered to dogs to produce blood levels similar to those of smokers markedly enhanced the uptake of labeled fibrinogen into the arterial wall (Allen et al., 1989). While the mechanism of the nicotine effects has not been determined, enhanced deposition of fibrinogen and lipids within the walls of blood cells could contribute to atherogenesis.

Oral nicotine dosed to rats to achieve blood levels comparable to human smokers produced greater myointimal thickening of the aorta following experimental injury (denudation of the endothelium with a balloon catheter) (Krupski et al., 1987). The excessive myointimal thickening in nicotine-treated animals is consistent with a response to persistent injury to endothelial cells. In support of the relevance of animal or in vitro studies to effects of nicotine in humans, Davis and colleagues reported an increase in the number of endothelial cells found in venous blood (reflecting endothelial injury) and a decrease in the platelet aggregate ratios (reflecting platelet aggregation) in nonsmokers who smoked tobacco but not non-tobacco cigarettes (Davis et al., 1985).

The above findings suggest that nicotine may produce endothelial injury and otherwise modify the function of the arterial wall so as to contribute to the development of atherosclerosis.

NICOTINE AND THROMBOSIS

The blood of smokers coagulates more readily than the blood of nonsmokers. Smoking may affect both platelet function as well as coagulation factors. Effects of cigarette smoking on platelet aggregation have been demonstrated by finding reduced platelet survival time (Mustard & Murphy, 1963), increased circulating platelet aggregates and release of platelet specific proteins (platelet factor 4 and beta thromboglobulin) following cigarette smoking (Schmidt & Rasmussen, 1984). Bleeding time is shorter in smokers (Gerrard et al., 1989). Increased sensitivity of platelets in vitro to various aggregating chemicals has been reported in some studies (Belch et al., 1984), although some recent studies have not confirmed such an effect (Siess et al., 1982).

While early studies suggested that nicotine might directly enhance platelet aggregation (Saba & Mason, 1975), recent studies have shown more complex responses (Pfueller et al., 1988). At higher concentrations of nicotine (1-10 mM), nicotine inhibits aggregation responses to some stimuli, while enhancing responses to others. No effects were seen at concentrations relevant to human smoking (about 0.2 µM). Studies of nicotine effects in vivo are also conflicting. Folts has shown that both cigarette smoking and intravenous nicotine enhance cyclic reduction in coronary artery blood flow in stenosed canine coronary arteries (Folts & Bonebrake, 1982). This phenomenon had been shown in previous studies to be a consequence of platelet thrombi formation, and the effect was blocked by aspirin. The effects of nicotine in this system were blocked by phentolamine, suggesting that catecholamines mediate the nicotine response. It should be noted, however, that the dose of nicotine (80 µg/kg given over 3-5 min) is expected to result in higher nicotine levels and perhaps a greater intensity short-term response (i.e., release of catecholamines) than occurs in cigarette smokers. Studies of nicotine administered chronically to rodents via miniosmotic pumps have shown inhibition of ADP-induced aggregation (Becker et al., 1988; Terres et al., 1989). In that this inhibitory effect was blocked by propranolol, a catecholamine-mediated mechanism was suggested. In these rodent studies there was no evidence of enhancement of platelet aggregation.

. The importance of nicotine in causing platelet aggregation in humans was suggested by a study showing a correlation between blood concentrations of nicotine and enhanced platelet aggregation responses to cigarette smoking (Renaud et al., 1984). It has likewise been reported that there was a greater increase in circulating platelet aggregates in humans after smoking tobacco compared to nontobacco cigarettes (Davis et al., 1985). However, nicotine chewing gum has not been associated with any acute change in platelet aggregation responses (Johnston et al., 1984).

Nicotine could affect platelets by releasing epinephrine, which is known to enhance platelet reactivity, or by inhibiting prostacyclin, an antiaggregatory hormone secreted by endothelial cells. Data of Nowak et al. (1987) indicating that cigarette smokers excrete higher amounts of thromboxane A$_2$ and prostacyclin metabolites in their urine are instructive. These data suggest that platelet aggregation is more active with cigarette smokers, but that this effect may be more related to a generalized increase in platelet activation due to underlying vascular disease rather than to acute effects of cigarette smoking.

Other biochemical changes in cigarette smokers that may be related to an enhanced thrombotic potential include increased plasma fibrinogen, hematocrit and blood viscosity (Yarnell et al., 1987; Ernst & Matrai, 1987). The mechanism for smoking-induced elevations in fibrinogen is unknown, but the observation that the normalization of fibrinogen levels after smoking cessation takes eight weeks or longer argues against an acute effect from nicotine (Ernst & Matrai, 1987). The effect of smoking on hematocrit appears most likely due to the relative hypoxia induced by chronic carbon monoxide exposure (Smith & Landaw, 1978).

In summary, whereas chronic cigarette smoking appears to be associated with platelet activation and decreased platelet survival as well as an increased thrombotic tendency in general, the role of nicotine in this effect is questionable. Nicotine does not appear to directly affect platelets. Animal studies with acute dosing of nicotine indicate that nicotine can induce platelet aggregation, presumably by systemic release of catecholamines. Chronic dosing of nicotine in animals does not appear to enhance aggregation. Enhanced platelet activation in human smokers may be a consequence of underlying vascular disease.

NICOTINE AND ATHEROGENESIS

To summarize the above discussion, nicotine may contribute to or aggravate several risk factors for accelerated atherosclerosis. There are plausible mechanisms and some evidence to indicate that nicotine contributes to hyperlipidemia and to endothelial injury. Nicotine does not appear to produce chronic hypertension, although other hemodynamic effects such as transient increases in blood pressure or persistent heart rate acceleration could contribute to endothelial injury. High doses of nicotine, presumably mediated by release of catecholamines, may enhance platelet aggregation, but most of the evidence indicates that chronic nicotine exposure is not the cause of platelet activation. On balance, I would conclude that while there is not conclusive evidence that chronic nicotine exposure contributes to premature atherosclerosis, there is enough evidence to be concerned about the possibility and to warrant further research.

PATHOGENESIS OF ACUTE CORONARY ISCHEMIC EVENTS

Acute cardiac ischemia occurs when the myocardial demand for oxygen cannot be met by coronary blood flow. In stable angina pectoris, exercise or emotion increases heart rate and blood pressure, thereby increasing myocardial oxygen demand, which cannot be met owing to atherosclerotic coronary stenosis. In unstable angina and acute myocardial infarction, a reduction in coronary blood flow is the primary precipitant. Coronary blood flow may be reduced as a consequence of thrombosis, often on an ulcerated plaque and/or of coronary vasoconstriction. Coronary vasoconstriction may result from local release of vasoconstrictors such as thromboxane A_2 from platelets that have aggregated on ulcerated plaques, from diminished or absent release of vasodilators such as prostacyclin or endothelial-derived relaxing factors from injured endothelial cells and/or from autonomic neural influences on coronary tone. Even a moderate degree of constriction of a coronary artery at a site already partially occluded by an atherosclerotic plaque and/or thrombus can result in severe stenosis or occlusion with resultant acute ischemia. Sudden death occurs when acute ischemia precipitates lethal arrhythmias.

Nicotine may contribute to acute cardiac ischemia via its systemic hemodynamic effects, enhancement of thrombosis, reduction of coronary blood flow and/or arrhythmogenesis. The data supporting these mechanisms will be examined.

NICOTINE AND SYSTEMIC HEMODYNAMICS

As noted previously, cigarette smoking, acting via nicotine, increases blood pressure (about 10 mmHg systolic and 5 mmHg diastolic) transiently and increases heart rate (on average 7 BPM) in a sustained pattern throughout the entire day (Benowitz et al., 1984). Cardiac output increases as a result of increased heart rate, enhanced cardiac contractility, and enhanced cardiac filling, the latter a result of systemic venoconstriction from nicotine. Myocardial work and oxygen consumption, which are determined by heart rate, myocardial contractility and afterload (arterial blood pressure), increase.

The cardiac hemodynamic effects of nicotine per se, consumed as nicotine chewing gum, have been studied in healthy people and in patients with coronary heart disease. In healthy people, chewing nicotine gum increases myocardial contractility as measured by echocardiography and increases myocardial oxygen consumption (by an average of 22%) as measured from changes in coronary blood flow and arterial-coronary sinus oxygen differences (Bayer et al., 1985; Kaijser &

Berglund, 1985). Of note in the latter study is that the increase in myocardial oxygen consumption produced by nicotine was greater than that associated with the same increase in the heart rate-blood pressure product (an index of myocardial work) produced by cardiac pacing. This discrepancy suggests that, in addition to its systemic hemodynamic effects, direct effects of nicotine on contractility (Greenspan et al., 1969) or myocardial metabolism contribute to the increased myocardial oxygen demand. In patients with coronary heart disease, chewing nicotine gum resulted in reduced myocardial contractility in regions of the heart shown on previous exercise testing to be at high risk for ischemia (Kaijser & Berglund, 1985). Reduced regional myocardial contractility occurred due to ischemia, indicating that coronary blood flow was not able to adequately meet the nicotine-enhanced myocardial oxygen demand. Thus, nicotine definitely produces systemic hemodynamic changes that can adversely affect a person with coronary heart disease.

A role of nicotine in enhancing thrombosis, as discussed previously, is possible but has not yet been clearly established.

NICOTINE AND CORONARY BLOOD FLOW

Several studies have demonstrated that cigarette smoking increases coronary vascular resistance. In healthy people, smoking is associated with an increase in coronary blood flow that occurs in response to the increased metabolic demand (Nicod et al., 1984; Klein et al., 1984). In the presence of coronary heart disease, however, cigarette smoking either does not affect or decreases coronary blood flow. In studies of regional myocardial perfusion using positron emission tomography with a rubidium-82 flow marker in patients with stable angina pectoris, cigarette smoking reduced perfusion in ischemic regions of the heart (Deanfield et al., 1986). The absolute decrease in coronary blood flow occurred at lower levels of myocardial oxygen consumption compared with exercise. These data suggest that cigarette smoking produces coronary vasoconstriction in the specific regions of the heart that are vulnerable to ischemia.

Nicotine chewing gum in healthy people increases coronary blood flow (Kaijser & Berglund, 1985). However, the extent of increase in coronary blood flow was less than observed when a similar increase in myocardial oxygen consumption was induced by atrial pacing. This suggests that nicotine limits the increase in coronary blood flow by vasoconstriction even in a nonischemic heart.

Cigarette smoking has been reported to be a risk factor for myocardial infarction with normal coronary arteries, presumed to be related at least in part to coronary artery spasm, and for recurrence of angina due to coronary spasm after withdrawal of anti-anginal medications (McKenna et al., 1980; Freedman et al., 1982). Coronary spasm has been documented by angiography during cigarette smoking (Maouad et al., 1984). It is noteworthy that silent ischemia, assessed as ST segment depression on ambulatory electrocardiography, which may reflect spontaneous coronary vasoconstriction, occurs more frequently and is of greater duration when people smoke cigarettes compared to when they do not (Barry et al., 1989).

Several mechanisms have been suggested to explain nicotine-induced reduction of coronary blood flow. Nicotine activates the sympathetic nervous system, and alpha adrenergic stimulation constricts coronary arteries (Feigl, 1987). The observation that cigarette smoking increases coronary blood flow in patients with coronary artery disease in the presence of phentolamine, an alpha blocker, but not in the absence of this drug supports the idea of an adrenergic mechanism (Winniford et al., 1986). Nicotine has been shown to inhibit the release of prostacyclin from rings of rabbit or rat aorta as well as human veins (Sonnenfeld & Wennmalm, 1980; Wennmalm, 1980). Prostacyclin, released by endothelial cells, dilates coronary arteries. The local release of prostacyclin is thought to be an important homeostatic response to local ischemia. Although recent studies in human smokers indicate an increase rather than the predicted decrease in the urinary excretion of prostacyclin metabolites (reflecting total systemic release of prostacyclin) (Nowak et al., 1987), it is possible that nicotine inhibits prostacyclin release in specific circulations such as the coronary circulation and thereby constricts coronary arteries. Enhanced aggregation of platelets (if this is, in fact, an effect of nicotine) could result in enhanced release of thromboxane A_2, which could also constrict coronary arteries.

NICOTINE AND ARRHYTHMOGENESIS

The increased risk of sudden cardiac death in smokers might result from ischemia combined with the arrhythmogenic effects of nicotine. Nicotine increases circulating concentrations of catecholamines. Cigarette smoking, most likely mediated by sympathomimetic actions of nicotine, facilitates atrioventricular nodal conduction in humans (Peters et al., 1987). Two case reports of arrhythmias after chewing nicotine gum are noteworthy. In one case, atrial fibrillation with a ventricular response rate of 150 developed in a man who chewed 30 pieces of 2 mg nicotine gum per day (Stewart & Catterall, 1985). Another man, who had a prior history of paroxysmal atrial fibrillation, developed a recurrence 5 minutes after chewing the day's first piece of nicotine gum (Rigotti & Eagle, 1986). On the other hand, smoking has not been shown to increase the prevalence or severity of ventricular ectopy or the inducibility of arrhythmia by electrical stimulation in patients with ischemic heart disease (Peters et al., 1987; Meyers et al., 1988). It is possible that the sympathomimetic actions of nicotine may render arrhythmias that occur in patients with ischemic heart disease more life-threatening. Of note in this regard is that cigarette smoking decreases the ventricular fibrillation threshold in dogs (Bellet et al., 1972), although this effect could also be contributed to by both nicotine and carbon monoxide.

NICOTINE, PIPE SMOKING AND CORONARY HEART DISEASE

Pipe smokers in general absorb as much nicotine, but, because they inhale less, absorb less carbon monoxide and other combustion products than do cigarette smokers (Wald et al., 1981). The epidemiology of coronary heart disease in pipe as compared to cigarette smokers has been of interest as a possible test of the role of nicotine in causing coronary heart disease. In several studies, the risk of smoking pipes has been reported to be lower than that from smoking cigarettes (Wald et al., 1981). However, a recent large Swedish study found an equivalent dose-related increase in rates for ischemic heart disease for pure cigarette versus pure pipe smokers (Carstensen et al., 1987). It was speculated that Swedish smokers may inhale more pipe smoke than do pipe smokers in other countries, although the evidence for this is not clear. People who use smokeless tobacco also have levels of nicotine similar to those of cigarette smokers (Benowitz et al., 1989). Data from large-scale epidemiological studies of smokeless tobacco use and coronary heart disease are not yet available, but when they are available will be useful in examining the role of nicotine.

It must be remembered, however, that even if it is confirmed that pipe smoking, smokeless tobacco use and/or prolonged exposure to nicotine for medicinal purposes does not enhance coronary heart disease, this does not exclude a role of nicotine in cigarette smoke. Cigarette smoking differs from these other routes of exposure in that nicotine is delivered to the blood stream and the nervous system more quickly and in higher concentrations following inhalation than is achieved by other routes. The result is a greater degree of neural stimulation with a greater intensity of cardiovascular effects that could have an adverse impact on the circulatory system.

CONCLUSION

Epidemiologic evidence indicates a causal relationship between cigarette smoking and coronary heart disease. Because cigarette smoking results in exposure to so many different chemicals, it is impossible to ascertain the contribution of nicotine per se from epidemiologic data. Studies of the pharmacology and toxicology of nicotine in animals, and experimental studies of the effects of cigarette smoking and nicotine administration in people, suggest that nicotine may contribute to accelerated atherogenesis and most likely contributes to acute coronary ischemic events (Table 1). Despite the potential hazards, short-term administration of nicotine, such as in nicotine replacement therapy as an adjunct to smoking cessation therapy, poses little cardiovascular risk to healthy individuals. Patients with coronary heart disease are likely to suffer increased risk when taking nicotine compared to not taking nicotine, but certainly less than they would when smoking cigarettes, which exposes them both to higher levels of nicotine as well as carbon monoxide and other potential toxins. The safety of chronic nicotine exposure, such as would occur during nicotine treatment of a chronic medical disease, cannot at this time be predicted and requires empiric evaluation.

Acknowledgements. Research supported in part by USPHS grants DA02277, CA32389, and DA01696.

Table 1. Summary of Evidence that Nicotine Contributes to Coronary Heart Disease: Pathophysiologic Factors

Atherosclerosis		Acute Ischemic Events	
Hyperlipidemia	+	Systemic Hemodynamics	+++
Hypertension	o	Thrombosis	+
Endothelial Injury	+	Coronary Vasoconstriction	++
Thrombosis	+	Arrhythmogenesis	+

o   No evidence
+   Possible
++   Probable
+++ Definite

REFERENCES

Allen, D.R., Browse, N.L., and Rutt, D.L. (1989) Atherosclerosis. 77, 83-88.
Asmussen, I., and Kjeldsen, K. (1975) Circ. Res. 36, 579-589.
Barry, J., Mead, K., Nabel, E.G., Rocco, M.B., Campbell, S., Fenton, T., Mudge, G.H., Jr., and Selwyn, A.P. (1989) JAMA. 261, 398-402.
Bayer, F., Bohn, I., and Strauer, B.E. (1985) Therapiewoche. 35, 1968-1974.
Becker, B.F., Terres, W., Kratzer, M., and Gerlach, E. (1988) Klin Wochenschr. 66(Suppl XI), 28-36.
Belch, J.J.F., McArdle, B.M., Burns, P., Lowe, G.D.O., and Forbes, C.D. (1984) Thromb Haemost. 51, 6-8.
Bellet, S., DeGuzman, N.T., Kostis, J.B., Roman, L., and Fleischmann, D. (1972) Am. Heart J. 83, 67-76.
Benowitz, N.L., Jacob III, P., and Yu, L. (1989) Ann. Int. Med. 111, 112-116.
Benowitz, N.L., Kuyt, F., and Jacob, P., III. (1984) Clin. Pharmacol. Ther. 36, 74-81.
Booyse, F.M., Osikowicz, G., and Quarfoot, A.J. (1981) Am. J. Pathol. 102, 229-238.
Brischetto, C.S., Connor, W.E., Connor, S.L., and Matarazzo, J.D. (1983) Am. J. Cardiol. 52, 675-680.
Carstensen, J.M., Pershagen, G., and Eklund, G. (1987) J. Epidem. Comm. Hlth. 41, 166-172.
Cellina, G.U., Honour, A.J., and Littler, W.A. (1975) Am. Heart J. 89, 18-25.
Cluette-Brown, J., Mulligan, J., Doyle, K., Hagan, S., Osmolski, T., and Hojnacki, J. (1986) Proc. Soc. Exp. Biol. Med. 182, 409-413.
Craig, W.Y., Palomaki, G.E., and Haddow, J.E. (1989) Br. Med. J. 298, 784-788.

Cryer, P.E., Haymond, M.W., Santiago, J.V., and Shah, S.D. (1976) N. Engl. J. Med. 295, 573-577.

Csonka, E., Somogyi, A., Augustin, J., Haberbosch, W., Schettler, G., and Jellinek, H. (1985) Virchows Archiv A. Pathological Anatomy and Histology. 407, 441-447.

Danfield, J., Wright, C., Krikler, S., Ribeiro, P., and Fox, K. (1984) N. Eng. J. Med. 310, 951-954.

Davis, J.W., Shelton, L., Eigenberg, D.A., Hignite, C.E., and Watanabe, I.S. (1985) Clin. Pharmacol. Ther. 37, 529-533.

Deanfield, J.E., Shea, M.J., Wilson, R.A., Horlock, P., de Landsheere, C.M., and Selwyn, A.P. (1986) Am. J. Cardiol. 57, 1005-1009.

Downey, H.F., Crystal, G.J., and Bashour, F.A. (1981) J. Pharmacol. Exp. Ther. 216, 363-367.

Dwyer, J.H., Rieger-Ndakorerwa, G.E., Semmer, N.K., Fuchs, R., and Lippert, P. (1988) JAMA. 259, 2857-2862.

Eckstein, J.W., and Horsley, A.W. (1960) Ann. N.Y. Acad. Sci. 90, 133-137.

Ernst, E., and Matrai, A. (1987) Atherosclerosis. 64, 75-77.

Feigl, E.O. (1987) Circulation. 76, 737-745.

Folts, J.D., and Bonebrake, F.C. (1982) Circulation. 65, 465-469.

Freedman, S.B., Richmond, D.R., and Kelly, D.T. (1982) Am. J. Cardiol. 50, 711-715.

Freund, J., and Ward, C. (1960) Ann. N.Y. Acad. Sci. 90, 85-101.

Gerrard, J.M., Taback, S., Singhroy, S., Docherty, J.C., Kostolansky, I., McNicol, A., Kobrinsky, N.L., McKenzie, J.K., and Rowe, R. (1989) Circulation. 79, 29-38.

Gnasso, A., Haberbosch, W., Schettler, G., Schmitz, G., and Augustin, J. (1986) Proc. Soc. Exp. Biol. Med. 182, 414-418.

Green, M.S., Jucha, E., and Luz, Y. (1986) Am. Heart J. 111, 932-940.

Greenspan, K., Edmands, R.E., Knoebel, S.B., and Fisch, C. (1969) Arch. Intern. Med. 123, 707-712.

Hladovec, J. (1978) Experientia. 34, 1585-1586.

Hojnacki, J., Mulligan, J., Cluette-Brown, J., Igoe, F., and Osmolski, T. (1986) Proc. Soc. Exp. Biol. Med. 182, 414-418.

Isles, C., Brown, J.J., Cummings, A.M. et al. (1979) Br. Med. J. 1, 579-581.

Johnston, R.V., Belch, J.J.F., McArdle, B., and Forbes, C.D. (1984) Thromb. Res. 35, 99-104.

Kaijser, L., and Berglund, B. (1985) Clin. Physiol. 5, 541-552.

Kershbaum, A., Bellet, S., and Khorsandian, R. (1965) Am. Heart J. 69, 206-210.

Klein, L.W., Ambrose, J., Pichard, A., Holt, J., Gorlin, R., and Teichholz, L.E. (1984) J. Amer. Coll. Cardiol. 3, 879-886.

Krupski, W.C., Olive, G.C., Weber, C.A., and Rapp, J.H. (1987) Surgery. 102, 409-415.

Maouad, J., Fernandez, F., Barrillon, A., Gerbaux, A., and Gay, J. (1984) Am. J. Cardiol. 53, 354-355.

McKenna, J., Chew, C.Y.C., and Oakley, C.M. (1980) Br. Heart J. 43, 493-498.

McPhaul, M., Punzi, H.A., Sandy, A., Borganelli, M., Rude, R., and Kaplan, N.M. (1984) JAMA. 252, 2860-2862.

Meyers, M.G., Benowitz, N.L., Dubbin, J.D., Haynes, R.B., and Sole, M.J. (1988) Chest. 93, 14-19.

Mustard, J.F., and Murphy, E.A. (1963) Br. Med. J. 1, 846-849.

Nadler, J.L., Velasco, J.S., and Horton, R. (1983) Lancet. 1, 1248-1250.

Nicod, P., Rehr, R., Winniford, M.D., Campbell, W.B., Firth, B.G., and Hillis, L.D. (1984) J. Amer. Coll. Cardiol. 4, 964-971.

Nowak, J., Murray, J.J., Oates, J.A., and FitzGerald, G.A. (1987) Circulation. 76, 6-14.

Peters, R.W., Benowitz, N.L., Valenti, S., Modin, G., and Fisher, M.L. (1987) Am. J. Cardiol. 60, 1078-1082.

Pfueller, S.L., Burns, P., Mak, K., and Firkin, B.G. (1988) Haemostasis. 18, 163-169.

Porchet, H.C., Benowitz, N.L., and Sheiner, L.B. (1988) J. Pharmacol. Exp. Ther. 244, 231-236.

Porchet, H.C., Benowitz, N.L., Sheiner, L.B., and Copeland, J.R. (1987) J. Clin. Invest. 80, 1466-1471.

Quensel, M., Agardh, C.-D., and Nilsson-Ehle, P. (1989) Scand. J. Clin. Lab. Invest. 49, 149-153.

Renaud, S., Blache, D., Dumont, E., Thevenon, C., and Wissendanger, T. (1984) Clin. Pharmacol. Ther. 36, 389-395.

Rigotti, N.A., and Eagle, K.A. (1986) JAMA. 255, 1018.

Ross, R. (1986) N. Engl. J. Med. 314, 488-500.

Rottenstein, H., Peirce, G., Russ, E., Felder, D., and Montgomery, H. (1960) Ann. N.Y. Acad. Sci. 90, 102-113

Russell, M.A.H. (1976) Brit. Med. J. 1, 1430-1433.

Saba, S.R., and Mason, R.G. (1975) Thromb. Res. 7, 819-824.

Schmidt, K.G., and Rasmussen, J.W. (1984) Thromb Haemostas (Stuttgart). 51, 279-282.

Siess, W., Lorenz, R., Roth, P., and Weber, P.C. (1982) Circulation. 66, 44-48.

Smith, J.R., and Landaw, S.A. (1978) N. Engl. J. Med. 298, 6-10.

Sonnenfeld, T., and Wennmalm, A. (1980) Br. J. Pharmacol. 71, 609-613.

Stefanovich, V., Gore, I., Kajiyama, G., and Iwanaga, Y. (1969) Exp. Molec. Path. 11, 71-81.

Stewart, P.M., and Catterall, J.R. (1985) Br. Heart J. 54, 222-223.

Stubbe, I., Eskilsson, J., and Nilsson-Ehle, P. (1982) Br. Med. J. 284, 1511-1513.

Terres, W., Becker, B.F., Schrodl, W., and Gerlach, E. (1989) J. Cardiovas. Pharmacol. 13, 233-237.

Thyberg, J. (1986) Virchows Archiv. 52, 25-32.

Tucker, L.A. (1989) Am. J. Public Health. 79, 1048-1050.

U.S. Department of Health and Human Services. (1989). Reducing the Health Consequences of Smoking: 25 Years of Progress. A Report of the Surgeon General. Washington, D.C., Government Printing Office (DHHS Publication No. 88-8406).

Waeber, B., Schaller, M., Nussberger, J., Bussien, J., Hofbauer, K.G., and Brunner, H.R. (1984) Am. J. Physiol. 249(Heart Circ Physiol 16), H895-H901.

Wald, N.J., Idle, M., Boreham, J., Bailey, A., and Van Vunakis, H. (1981) Lancet. 2, 775-777.

Wells, D.G., and Rustick, J.M. (1986) Anesthesiology. 65, 339.

Wennmalm, A. (1980) Br. J. Pharmacol. 69, 545-549.

Winniford, M.D., Wheelan, K.R., Kremers, M.S., Ugolini, V., Van Den Berg, E.J., Niggemann, E.H., Jansen, D.E., and Hillis, L.D. (1986) Circulation. 73, 662-667.

Yarnell, Y.W.G., Sweetnam, P.M., Rogers, S., Elwood, P.C., Bainton, D., Baker, I.A., Eastham, R., O'Brien, J.R., and Etherington, M.D. (1987) J. Clin. Pathol. 40, 909-913.

Zimmerman M., and McGeachie, J. (1987) Atherosclerosis. 63, 33-41.

NICOTINE AND SMOKING IN PATIENTS WITH ULCERATIVE COLITIS

R. V. Heatley, G. F. Cope & J. Kelleher

Department of Medicine, St. James's University Hospital, Leeds, U.K.

SUMMARY: Few patients with ulcerative colitis smoke. We have demonstrated that smoking affects colonic mucus production and human immune mechanisms in vitro and this may have relevance to the pathogenesis of the disease. There is anecdotal evidence to suggest that nicotine may be the active moiety associated with smoking responsible for these effects, but we have been unable to confirm this in in vitro test systems.

The chance observation in 1982 that very few patients with ulcerative colitis were current cigarette smokers (Heatley, Thomas, Prokipchuk, Gauldie, Sieniewicz & Bienenstock, 1982) has encouraged much research into this association and the publication of a number of anecdotal reports. The new evidence has substantiated and expanded the original findings and has provided a new approach to investigate the pathogenesis of ulcerative colitis.

The original observation came from an investigation into the occurrence of lung disorders in patients with inflammatory bowel disease. Despite a high incidence of pulmonary abnormalities, only 11% of patients with ulcerative colitis were current cigarette smokers. In contrast, 36% of patients with Crohn's disease smoked, a figure similar to the national average.

Shortly after, Harries et al. (1982) carried out a specific questionnaire survey and found that only 8% of ulcerative colitis patients were current cigarette smokers compared with 42% of Crohn's disease patients and 44% of age- and sex-matched controls.

Following these reports, supporting anecdotal evidence linking non-smoking with ulcerative colitis was published. De Castella (1982) described the case-history of a 33-year old woman smoker who abruptly stopped and shortly afterwards developed the classic symptoms of ulcerative colitis.

Initially, nicotine was proposed as the active constituent of cigarette smoke responsible for this effect. This followed a report by Roberts & Diggle (1982) of an individual for whom remission of symptoms could be induced equally by resuming cigarette smoking or by 16mg of nicotine per day delivered from nicotine chewing gum. Marijuana smoking has also been reported to be beneficial in another single case report (Kaminski & Haase, 1990).

The idea that nicotine chewing gum might, however, provide a useful mode of therapy was given no encouragement by the results of an eight-week uncontrolled clinical trial reported by Perera et al. (1984). Here, non-smoking patients with ulcerative colitis given conventional therapy plus nicotine chewing gum, starting at 2mg once a day and progressing to 4mg five times a day, showed no objective improvement over patients given conventional therapy alone and there was also poor patient tolerance of the treatment.

Since these original observations, international interest has been intense. There is no doubt that the risk of ulcerative colitis is very much lower among current cigarette smokers than among those who have never smoked, the estimated relative risk (RR) being 0.25 (p<0.001). There is also no doubt that such a reduction in risk is not evident in former smokers (Cope, Heatley, Kelleher & Lee, 1987).

Although many theories have been advanced, little experimental evidence has been forthcoming to explain these findings. Based on the knowledge that colonic mucus is known to be quantitatively and qualitatively abnormal in ulcerative colitis, and that cigarette smoking alters respiratory mucus by systemic as well as local effects, we have examined colonic mucus production in vitro in patients with ulcerative colitis. The results show that patients with ulcerative colitis have significantly reduced colonic mucus production compared with controls (Table I). When smoking habits were

TABLE I

Total in vitro glycoprotein production (DPM x 10^3/mg biopsy
protein) in patients with ulcerative colitis and
controls in relation to smoking habit.

Values are means (standard errors).

	Non-smokers	Smokers	All patients
Controls (n = 121)	151.7 (27.3)	146.4 (16.3)	152.6 (11.1)
Ulcerative Colitis (n = 82)	112.1 (10.2)+	153.4 (35.6)	117.6 (10.1)*

* $p < 0.05$ compared with controls
+ $p < 0.05$ compared with non-smoking controls

taken into account, the majority of patients who were non-smokers had reduced mucus production compared with non-smoking controls. Those patients who did smoke had a level of colonic mucus production that was equivalent to smoking controls. Smoking appeared to have no effect on colonic mucus production in controls, since both smokers and non-smokers had the same level of mucus production (Cope, Heatley & Kelleher, 1986a; Cope, Heatley & Kelleher, 1986b). It would appear from these results that smoking does not affect mucus production under normal circumstances, but in a situation where this is abnormal, as in ulcerative colitis, then smoking appears to have a restorative effect on production of colonic mucus.

The effects of drugs on in vitro mucus synthesis and secretion has also been evaluated. Those examined included nicotine and its major metabolites, nicotine-1-oxide and cotinine, hexamethonium (a competitive nicotinic receptor antagonist), catecholamines, adrenaline and noradrenaline; and prostaglandins F_{1a} and I_2. Also included were acetylcholine ± eserine as a positive control, and sodium fluoride as a metabolic inhibitor. Neither nicotine nor its metabolites had any significant effects on in vitro mucus production (Figs. 1 and 2).

Altered immunoglobulin production has also been associated with ulcerative colitis, and smoking has been shown to influence peripheral immunoglobulin levels. We measured pre-formed and newly-synthesised IgA, IgG and IgM by colonic biopsies in organ culture (medians, ng/mg biopsy protein), from patients with ulcerative colitis and control subjects (Cope, Purkins, Trejdosiewicz, Heatley & Kelleher, 1989). Separation of patients by current smoking habit showed that smoking had little effect on the pre-formed levels of all three immunoglobulins; however, smoking has a profound influence on the synthesis of immunoglobulins after culture. Thus, IgA synthesis was reduced in both controls (6.4) and inflammatory bowel disease (IBD) patients (7.0) who smoked, compared with non-smokers (15.8 and 38.3 respectively). IgG production increased in smokers with IBD (25.9 compared with 5.3), while IgM production was greatly decreased in IBD smokers (0.4 compared with 2.2), but increased in controls who smoked (2.2 compared with 0.4).

These studies suggest that smoking has profound influence on colonic mucus and colonic mucosal immunoglobulin production, rendering patients' humoral and physical colonic mucosal defences relatively suppressed. A number of explanations for the protective effects of cigarette smoking on ulcerative colitis have been postulated. One possibility is that the increased levels of circulating catecholamines, which are released by the action of nicotine on the adrenal medulla, may rectify a possible genetic/constitutional deficiency of biogenic monoamines which exists in ulcerative colitis (Bowers, Allen & Hickey, 1983). An alternative theory implicated the antagonistic effect of nicotine on the over-production, or over-sensitivity, of prostaglandins that have been associated with ulcerative colitis (McGarry, 1983). An alternative hypothesis was that smoking in some way influences the pathological appearance, possibly by an effect on the immune system, leading to features of Crohn's disease rather than ulcerative colitis in patients who smoke (Holdstock, Savage, Harman & Wright, 1984). An effect on the immune system was further implicated when it was pointed out that smoking produces complex changes in immune function, including reduced natural killer cell activity in peripheral blood lymphocytes and decreased immunoglobulin concentrations in serum and saliva (Somerville, Logan, Edmond & Langman, 1984).

It would appear that after numerous and widespread epidemiological studies, the negative association between cigarette smoking and ulcerative colitis is now established. Although a number of explanations have been put forward to suggest how smoking could exert a protective effect on ulcerative colitis, the mechanism is still open for debate. Evidence is accumulating that cigarette smoking has a profound effect on several parameters of mucosal defence. Smoking is also considered to have an antagonistic effect on the pharmacological effects of prostaglandins, as well as increasing the rate of mucus production by the colon in vitro. These, and other, factors need to be explored more extensively in order to explain this phenomenon and hopefully this approach will help to elucidate the pathogenesis of inflammatory bowel disease.

REFERENCES

Bowers, E.J., Allen, I.E. and Hickey, R.J. (1983) New England Journal of Medicine 308, 1476-1477.
Cope, G.F., Heatley, R.V., Kelleher, J. and Lee, P.N. (1987) Human Toxicology 6, 189-193.
Cope, G.F., Heatley, R.V. and Kelleher, J. (1986a) Gut 27, A618-619.
Cope, G.F., Heatley, R.V. and Kelleher, J. (1986b) British Medical Journal 293, 481.
Cope, G.F., Purkins, L., Trejdosiewicz, L.K., Heatley, R.V. and Kelleher, J. (1989) Gut 30, A749.
De Castella, H. (1982) British Medical Journal 284, 1706.
Harries A.D., Baird, A. and Rhodes, J. (1982) British Medical Journal 284: 706.

Heatley, R.V., Thomas, P., Prokipchuk, E.J., Gauldie. J., Sieniewicz, D.J. and Bienenstock, J. (1982) Quarterly Journal of Medicine 203, 241-250.

Holdstock, G., Savage, D., Harman, M. and Wright, R. (1984) British Medical Journal 288, 362.

Kaminski, M.V. and Haase, T. (1990) Annals of Internal Medicine 112, 471.

McGarry, J.M. (1983) Lancet ii, 1498.

Perera, D.R., Janeway, C.M., Feld, A., Ylvisaker, J.T., Belic, L. and Jick, H. (1984) British Medical Journal 288, 1533.

Roberts, C.J. and Diggle, R. (1982) British Medical Journal 285, 440.

Somerville, K.W., Logan, R.F.A., Edmond, M. and Langman, M.J.S. (1984) British Medical Journal 289, 954-956.

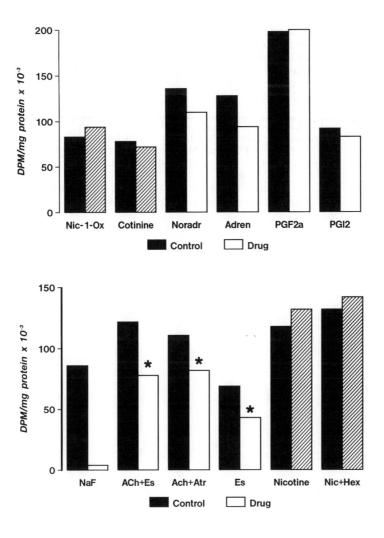

Fig.1 and 2: The effects of exogenously applied drugs on total [³H] glucosamine incorporation into newly synthesised mucus glycoproteins (Nic-1-Ox, Nicotine-1-Oxide; Noradr, Noradrenaline; Adren, Adrenaline; Pg, Prostaglandin; NaF, Sodium fluoride; Ach, Acetylcholine; Es, Eserine; Atr, Atropine; Nic, Nicotine; Hex, Hexamethonium)

CIGARETTE SMOKING INFLUENCES EICOSANOID PRODUCTION

BY THE COLONIC MUCOSA IN A DOSE DEPENDENT MANNER

G.F. COPE[*], R.V. HEATLEY, J. KELLEHER

Department of Medicine, St James's Hospital, Leeds.
*Present address Wolfson Res Labs, Queen Elizabeth
Hospital, Birmingham, UK.

SUMMARY In vitro production of prostaglandin E_2 and
leukotriene B_4 by colonic biopsies from patients with
ulcerative colitis and controls were found to be dependent on
current smoking habit in a dose dependent manner, being
increased in light smokers but reduced in moderate and heavy
smokers.

INTRODUCTION

Cigarette smoking is an important factor in the
pathogenesis of inflammatory bowel disease (IBD), with smoking
being positively associated with Crohn's disease (CD); and
negatively associated with ulcerative colitis (UC) (Calkins
1989). Smoking also effects the clinical course of the
disease, with smokers having a greater degree of recurrence
of CD than non-smokers (Sutherland et al 1990), while patients
with active UC who recommenced moderate smoking (median
20/day) improved, however those who smoked less (median

group, which contained 30 smokers. The control group comprised patients with the irritable bowel syndrome or diverticular disease and were found to have a normal colonic mucosa. Within three hours of removal the biopsies were cultured for 48 hours at 37°C in a humified mixture of 5% CO_2 in air. Following which PGE_2 and LTB_4 were measured in the homogenised biopsy tissue and in the culture medium by radioimmunoassay, and the total amount expressed per mg of protein in the biopsy (ng/mg protein). Statistical analysis was by Kruskal-Wallis analysis.

RESULTS

TABLE 1   Eicosanoid production in (ng/mg protein) different patient groups relative to current smoking (No cigs/day) against never smokers [median (range)].

	PGE_2	LTB_4
ALL PATIENTS (n=100)	34.7(0.0-1846.1)	3.7 (0.0-17.0)
Never (56)	44.3 (0.0-1846.1)	4.2 (0.0-17.0)
1-9 (13)	210.5 (14.8-1037.2)	4.8 (0.0- 8.8)
10-19 (23)	7.6 (2.9- 248.9)	3.7 (0.0- 6.1)
20+ (8)	22.8*(7.3- 35.2)	2.3 (0.0- 5.4)
CONTROLS (n=55)	28.8 (0.0- 792.0)	3.3 (0.0-10.9)
Never (25)	8.8 (0.0- 465.0)	4.0 (0.0-10.9)
1-9 (7)	54.4 (17.1- 792.0)	4.8 (0.0- 6.5)
10-19 (17)	27.4 (2.9- 248.9)	3.7 (0.0- 6.1)
20+ (6)	24.2 (11.5- 35.2)	2.3 (0.0- 5.4)
UC (n=31)	59.4#(0.0-1037.2)	4.2##(0.0-17.0)
Never (25)	59.4 (0.0- 619.0)	4.8 (0.0-17.0)
1-9 (3)	341.6*(236.7-1037.2)	7.7 (5.6- 8.8)
10+ (3)	22.4 (7.8- 27.6)	3.3 (2.9- 3.7)

# $p<0.05$; ## <0.01 relative to controls
* $p<0.05$ relative to never smokers

10/day) did not (Rudra et al 1989). Nicotine may be the active constituent of tobacco smoke that protects against UC, as nicotine chewing gum (16 mg/day) has been shown to suppress disease activity in a current smoker, however, a small uncontrolled clinical trial with nicotine gum failed, largely due to poor tolerance of the treatment (Perera et al 1984). Inflammatory bowel disease is characterised by chronic and acute inflammation of the intestinal mucosa, with resultant chronic diarrhoea. The mediators responsible for the enhanced inflammatory response include the eicosanoids, particularly prostaglandin E_2 (PGE_2) and leukotriene B_4 (LTB_4), which are elevated, especially in the active phase of the disease (Stenson 1988). Cigarette smoking has been shown to inhibit PGE_2 synthesis by the gastric mucosa, while alveolar macrophages from smokers have reduced capacity to synthesis both PGE_2 and LTB_4. The aim of this study was to determine the effect of smoking habit on in vitro eicosanoid production by the intestinal mucosa in patients with UC and controls.

PATIENTS AND METHODS

Patients attending a colonoscopy clinic were questioned about their smoking habits, together with other relevant social and clinical details. Pinch biopsies from the colonic mucosa were obtained from 100 patients, including 45 with IBD (31 with UC, 8 with CD, and 6 non-specific inflammation), of whom 14 were current smokers; and 55 patients in the control

Table I shows that the amount of PGE_2 and LTB_4 synthesised by UC patients were significantly greater than by control patients ($p<0.05$), this was largely due to patients with active disease. Analysis relative to current smoking habit showed the same trend whether in all patients, controls, or in patients with UC, in that smoking appeared to have a biphasic effect on eicosanoid production. Prostaglandin E_2 production was enhanced in light smokers (<10 cigarette per day) compared with never smokers, the difference reaching significance in patients with UC ($p<0.05$). However, in moderate smokers (10-19/day) PGE_2 production was reduced, again reaching significance in UC patients ($p<0.05$). In those patients who were heavier smokers (20+/day) the production was reduced still further, reaching significance when all patients were considered ($p<0.05$). The same pattern was seen for LTB_4 production with respect to cigarette consumption, however the differences did not reach significance except in heavier smokers in the all patient group ($p<0.05$).

DISCUSSION

Cigarette smoking clearly has a profound effect on the pathogenesis and clinical course of IBD. The eicosanoids, particularly PGE_2 and LTB_4, are important mediators in the perpetuation of the inflammatory response in this condition. This present study supports the finding that elevated levels of these compounds are found in ulcerative colitis, particularly in its active phase. Eicosanoid production relative to smoking habit shows that PGE_2 and LTB_4 production are elevated in individuals who smoke a relative few number of cigarettes a day (<10), but in heavier smokers (>10) eicosanoid production is inhibited. This study provides experimental data to support the epidemiological finding that patients with colitis who recommence smoking after a period of abstention improve, but only if they smoke on average 20 cigarettes a day, those who smoke less do not improve. The causes of IBD remain unknown, yet the eicosanoids are clearly important in the inflammatory reaction. Smoking habit appears to be a critical factor in the development and progression of the disease. Further research into how Crohn's disease and ulcerative colitis patients differ in their eicosanoid response to cigarette smoking may provide invaluable information towards our outstanding of these diseases.

REFERENCES

Calkins BM. A meta-analysis of the role of smoking in inflammatory bowel disease. Dig Dis Sci 1989; 12: 1841-54.
Rudra T, Motley R, Rhodes J. Does smoking improve colitis? Scand J Gastroenterol 1989; 24: suppl 170: 61-3.
Stenson WF. Arachidonic acid metabolites in inflammatory bowel disease. In: Lewis A (Ed) Advances in Inflammation Research vol 12, Raven Press 1988: 215-22.
Sutherland LR, Ramcharan S, Bryant H, Fick G. Effect of cigarette smoking on recurrence of Crohn's disease. Gastronenterology 1990; 99: 1123-8.

ABSTENTION FROM SMOKING MAY INCREASE THE RISK OF ULCERATIVE
COLITIS BY ENHANCING EICOSANOID PRODUCTION

G.F. COPE[*], R.V. HEATLEY, J. KELLEHER

Department of Medicine, St James's Hospital, Leeds.
*Present address Wolfson Res Labs, Queen Elizabeth Hospital,
Birmingham, UK.

SUMMARY Production of prostaglandin E_2, and to a lesser
extent leukotriene B_4, by the colonic mucosa in vitro was
shown to be increased in recent ex-smokers (<10 yrs), but
increased in patients who have refrained from smoking for more
than 10 years.

INTRODUCTION

Former smokers are at greater risk of developing ulcerative

colitis (UC) than either never smokers or current smokers,

with current smokers having a 60% reduction in risk of the

disease, while former smokers have a 2.1 fold increase in risk

compared to never smokers (Boyko 1988a). The onset of

symptoms following smoking cessation varies from one month to

several years, with a mean time lapse of 5.7 years (Motley et

al 1987). Smoking before the onset of the disease may also

affect the clinical course of the condition, in that former

smokers have a severer form of the disease, require more hospitalisation, and more frequently undergo colectomy than never smokers (Boyko 1988b).

The eicosanoids, particularly prostaglandin E_2 (PGE_2) and leukotriene B_4 (LTB_4), are elevated in the peripheral circulation and at the mucosal surface of patients with ulcerative colitis, particularly when the disease is in its active phase (Nielsen 1988). Smoking has been shown to reduce eicosanoid synthesis, by, amongst others, alveolar macrophages and the gastric mucosa. Nicotine may be the important factor in cigarette smoke responsible for this inhibition, as it has been shown to reduce prostaglandin synthesis by the endocardium (Alster 1986), however, other studies have failed to show any effect of nicotine on prostaglandin synthesis. The aim of the present study was to determine the effect of smoking history on in vitro eicosanoid production by colonic tissue from patients with ulcerative colitis and controls.

PATIENTS AND METHODS

Patients attending a colonoscopy clinic were questioned about their current and previous smoking habit, plus other social and clinical details. Pinch biopsies from the colonic mucosa were removed from 112 patients, 56 ex-smokers and 56 never smokers. These included 39 with UC, consisting of 25 never smokers, 8 recent abstainers (1 to 10 years) and 6 long-term abstainers (>10 yrs). The control group, contained 25 never smokers, 17 (≤10 years) and 23 (>10 years) ex-smokers. The biopsies were cultured for 48 hours at 37°C in a humified mixture of 5% CO_2 in air. Following culture PGE_2 and LTB_4 were measured in the homogenised biopsy tissue and in the culture medium by radioimmunoassay, and the total amount expressed per mg of protein in the biopsy (ng/mg protein).

RESULTS

TABLE 1   Eicosanoid production (ng/mg protein) in all patients, controls and patients with ulcerative colitis relative to smoking history with never smokers, ex-smokers I = 1-10 yrs and ex-smokers II= >10 yrs [median (range)].

	PGE_2	LTB_4
ALL PATIENTS		
never (n=56)	44.3 (0.0-1846.1)	4.2 (0.0-17.0)
ex-smokers I (27)	83.6 (10.4-3842.8)	4.2 (0.0-12.1)
·ex-smokers II (29)	21.2 (0.0- 530.6)	2.7* (0.0-13.2)
CONTROLS		
never (25)	28.8 (0.0- 456.6)	4.0 (0.0-10.9)
ex-smokers I (17)	73.8 (10.4- 265.3)	3.8 (0.0- 6.7)
ex-smokers II (23)	19.7 (0.0- 227.9)	2.7 (0.0-13.2)
UC		
never (25)	59.4 (0.0- 619.0)	4.8 (0.0-17.0)
ex-smokers I (8)	131.6 (11.0-3842.8)	4.2 (0.0-12.1)
ex-smokers II (6)	32.7 (15.3- 530.6)	4.2 (0.8- 6.0)

* $p<0.05$ relative to never smokers

The results (Table 1) showed that abstention from smoking has a biphasic effect on prostaglandin production, which was seen to a greater or lesser extent in the all patient group, controls and patients with UC. The pattern was that recent abstainers (≤10 years) had increased PGE_2 production compared with never smokers, while long-term abstainers (>10 years) had reduced production. In general PGE_2 production in recent abstainers was double that in never smokers, while in long

term abstainers the production halved. However, due to the
wide distribution of data and the relatively small number of'
patients the difference only approached significance in the
all patient group (p<0.1). Leukotriene B_4 production
responded in a difference manner. Recent abstainers in the
all patient group and the control group had LTB_4 production
that was relatively unchanged, but in long-term abstainers
production was decreased, reaching significance in the all
patient group (p>0.05). In patients with UC, there was
relatively little difference between never smokers and ex-
smokers, being only marginally lower in long-term abstainers.

DISCUSSION

The finding that ex-smoking is a risk factor in the
development of ulcerative colitis, and that the onset of the
disease may be after several years of smoking abstention is an
intriguing factor in the aetiology of the disease. The
present study shows that ex-smoking has an effect on in vitro
prostaglandin production, and to a lesser extent leukotriene
production, that appears to be time dependent. Recent
abstainers, irrespective of their disease status, have
increased prostaglandin production compared with never
smokers, but in long-term ex-smokers PG synthesis is
decreased. The effect of smoking abstention on leukotriene
production was negligible, except for a reduced production
detected in long-term abstainers the all patient group, and in

the control group, compared with never smokers.

These data, in conjunction with the effect of current smoking on prostaglandin production, suggests that inhibition of eicosanoid production occurs as long as a relatively high level of tobacco consumption is maintained, but if this is reduced or stopped altogether then there is a rebound effect when production increases, and it is at this stage that, in susceptible individuals, ulcerative colitis develops. As smoking cessation continues however, prostaglandin production declines towards, and ultimately falls below, the never smoking values. Clearly, smoking has a profound effect on colonic mucosal defence physiology which is still manifest many years after smoking is discontinued.

REFERENCES

Boyko EJ. A critical review of the association between cigarette smoking and the risk of ulcerative colitis. In: MacDermott RP (ed). Inflammatory Bowel Disease: current status and future approach. Elsevier Publishers 1988: 671-76.

Motley RJ, Rhodes J, Ford GA, Wilkinson SP, Chesner IM, Asquith P, Hellier MD, Mayberry JF. Time relationship between cessation of smoking and onset of ulcerative colitis. Digestion 1987; 37: 125-7.

Bokyo EJ, Perera DR, Koepsell TD, Keane EM, Inui TS. Effects of cigarette smoking on the clinical course of ulcerative colitis. Scand J Gastroenterol 1988; 23: 1147-52.

Alster P, Brandt R, Koul BL, Nowak J, Sonnenfeld T. Effect of nicotine on prostacyclin formation in human endocardium in vitro. Gen Pharmacol 1986; 17: 441-4.

Nielsen OH. In vitro studies on the significance of arachidonate metabolism and other oxidative processes in the inflammatory response of human neutrophils and macrophages. Scand J Gastroenterol 1988; 23 (suppl 150): 1-22.

SMOKING AND PARKINSON'S DISEASE

Richard Godwin-Austen

Nottingham NG7 2UH
Department of Neurology, University Hospital, Queens Medical Centre,

SUMMARY

Evidence is presented for a negative correlation between smoking and
Parkinson's disease. The reduced risk of developing Parkinson's disease
amounts to about half the risk in non-smokers and is demonstrable even
in those who have ceased smoking up to 20 years before the onset of
the disease. How this protective effect is mediated is unknown and in
particular it is uncertain whether or not it is related to nicotine.

Parkinson's disease is a progressive disabling disorder of the nervous
system of unknown causation. In the last 30 years the neurochemistry
and neuropharmacology of the disease has been worked out so that there
is now effective treatment for the majority of patients, and the excess
mortality previously associated with Parkinson's disease has now been
substantially reduced. But the cause of the disorder is unknown and
any possible clues justify careful investigation.

As part of a large case-control study to investigate the medical, social,
behavioural and family history of patients with Parkinson's disease,
smoking history was also investigated. This investigation was set up
to try to discover whether there were any epidemiological clues as to

the cause of the disease. The results of this survey were published
in 1982 (Godwin-Austen et al, 1982) but there had already been reports
that Parkinson's disease is less common in smokers than non-smokers
(Kahn 1966, Nefzger et al, 1962, Kessler 1972, Baumann et al, 1980).
We aimed to establish that smoking was indeed negatively correlated
with Parkinson's disease and that this disocciation was not attributable
either to a confounding factor less common in smokers or an artefact
attributable to cessation of smoking at the onset of the disease.

The investigation was organised in such a way as to identify patients
with Parkinson's disease and a corresponding age and sex matched control
from the same family doctor's practice. The practices identified were
uniformly distributed throughout England, Scotland and Wales. Patients
and controls were interviewed so that a comprehensive questionnaire
could be completed. This questionnaire included information about the
disease (severity and manifestations) smoking habits and a wide range
of questions relating to medical and social history and environment.

From 350 case/control pairs an analysis was carried out to assess the
association between a factor (such as smoking) and the disease. The
ratio of those pairs where the factor was present in the case with
Parkinson's disease but absent from the control to those pairs where
the reverse was true gave an estimate of relative risk.

In this series there were 45 case/control pairs in which the case was
a current smoker and the control a non-smoker, whereas the reverse
situation was present in 87 pairs. This difference is significant (p
$<$ 0.001) confirming the strong negative correlation between smoking
and Parkinson's disease (relative risk 0.52). In order to investigate
whether this result arose from patients with Parkinson's Disease giving
up smoking we compared smoking habits when the symptoms of the disease
first developed and 20 years earlier with that of the controls at
the same times. A history of smoking even 20 years earlier was
associated with a reduced risk of developing Parkinson's Disease
(relative risk 0.55; p $<$ 0.01). And there was a trend towards a higher

risk of Parkinson's disease the earlier the individual had given up smoking. (See Table)

		Relative Risk	
Never smoked		1.0	
Given up smoking			
1.	10 yrs before P.D.	0.84)	
2.	1-10 yrs before P.D.	0.62)	0.71
3.	After onset P.D.	0.58)	
Still smoking		0.40	

TABLE   Relative Risk of Parkinson's Disease in Smokers

Investigation of aspects of the smoking habit revealed only that there was a lower risk of Parkinson's disease in those who smoked more heavily. This observation has subsequently been confirmed by Ward et al (1983) in their study of 16 twin pairs. They showed that there is a statistically significant negative correlation between heavier smoking and the development of Parkinson's disease. [Parkinson's disease twin, 193 mean pack years, non-parkinsonian twin, 330 mean pack years (p < 0.01)]

The association between Parkinson's disease and the tar, nicotine content and carbon monoxide of the tobacco product failed to demonstrate any consistent pattern. Similarly when we looked at the manifestations of Parkinson's disease in terms of severity, age of onset, tremor or speech difficulty, smoking was associated with a similar reduction of risk. Thus smoking does not seem to alter the symptoms and signs of the disease nor to protect against one particular form of the condition.

Degenerative vascular disease afflicts people of similar age to those
with Parkinson's disease and cerebral arteriosclerosis has been suggested
as a cause for Parkinson's disease. Furthermore it may be difficult
clinically to distinguish between Parkinson's disease and vascular
disease affecting the brain. Our results showed that a history of
stroke, heart disease or hypertension did not influence the relative
risk of Parkinson's disease in relation to smoking (Godwin-Austen 1982).
Parkinson's disease patients had a lower incidence of vascular disease
than the controls. Similarly there was discernible effect from other
factors likely to be relevant (concussive head injury, family history
of Parkinson's disease, alcohol three or more times a week).

CONCLUSION

Our results confirmed the negative correlation between smoking and
Parkinson's but we were unable to identify a specific negative
correlation with nicotine (using data linking specific products with
their nicotine content). The relative risk in our series was 52%, and

the reduction of risk of developing Parkinson's disease applied to some
degree even in those who had stopped smoking up to 20 years previously.
There is some evidence the quantity smoked directly increases this
negative correlation.

Although patients who smoke may succumb to smoking-related diseases
at an earlier age than those who do not smoke, we could not demonstrate
a differential effect of smoking on the age at onset of the Parkinson's
disease. Our data does not therefore support the possibility that the
reduced prevalence of Parkinson's disease in smokers is attributable
to selective mortality.

Kessler and Diamond (1971) suggested that nicotine might have a
therapeutic role in Parkinson's disease. Our data would not support
this hypothesis since not only does smoking long before the onset of
the disease appear to exert an affect; but also the rate of progression

of Parkinson's disease is not differentially affected in those still smoking compared with those who have given up.

There is a positive correlation between smoking and degenerative vascular disease in contrast to the opposite correlation with Parkinson's disease. We therefore conclude that arteriosclerosis is not involved in the causation of Parkinson's disease.

These results may indicate a behavioural effect so that one of the earliest features of the pre-clinical phase of Parkinson's disease is a disinclination to smoke. Alternatively nicotine may have a protective effect on dopamine pathways (Owman et al 1989).

REFERENCES

Baumann R J, Jameson H D, McKean H E, Haack D G, Weisberg L M
    Neurology (Minneap) 1980; 30: 839–43
Godwin-Austen R B, Lee P N, Marmot M G and Stern G M
    J. Neurol., Neurosurg and Psychiat 1982; 45: 577–81
Kahn H A
    In: "Epidemiological Approaches to the Study of Cancer and other
    Chronic Diseases".
    Monograph No. 19. Washington D C
    National Cancer Institute, U S Govt. Printing Office 1966: 1–125
Kessler I I
    Am. J. Epidemiol 1972; 96: 242–54
Nefzger M D, Quadfasel F A, Karl V C
    Am. J. Epidemiol 1967; 88: 149–58
Owman Ch, Fuxe K, Jansen A M and Kahrstrom J
    Progress in Brain Research 1989; 79: 267–276
Ward C D, Duvoisin R C, Ince S E, Nutt J D, Eldridge R and Calne D B
    Neurology (Cleveland) 1983; 33: 815–24

THE EFFECTS OF NICOTINE ON ATTENTION, INFORMATION PROCESSING, AND WORKING MEMORY IN PATIENTS WITH DEMENTIA OF THE ALZHEIMER TYPE

B.J. Sahakian[1] and G.M.M. Jones[1,2]

[1] Section of Old Age Psychiatry, Department of Psychiatry, Institute of Psychiatry, De Crespigny Park, London, SE5 8AF, U.K.
[2] Department of Psychology, Institute of Psychiatry, De Crespigny Park, London, SE5 8AF, U.K.

SUMMARY: The reduction in nicotine receptors in the neocortex and hippocampus of patients dying with dementia of the Alzheimer type (DAT) suggests that stimulating the remaining receptors with nicotine might enable an alternative approach to cholinergic treatment of patients with mild/moderate DAT. A preliminary study of subcutaneous nicotine in groups of young normals and patients with DAT (all n=7) is described. Tests of rapid visual information processing (RIP) (non spatial working memory), short term spatial working memory and cortical arousal (critical flicker fusion test (CFF) were modified to allow the baseline performance of the DAT patients to approximate that of normal subjects. Nicotine (0.4, 0.6, 0.8 mg s.c.) produced dose-dependent improvements in the DAT patients in the accuracy and speed of responding in the RIP task, reduced the threshold for OFF, but had no beneficial effect on the test of spatial short term memory. These preliminary observations are extended in two ways: by studying the effects of acute nicotine on larger populations of subjects, including equivalent numbers of smokers and non-smokers; and by investigating the effects of chronic nicotine, administered in the form of chewing gum, on similar measure, but also including an evaluation of effects on indices of everday living.

In this chapter, we would like to present our recent work which looks at the effects of nicotine on some of the deficits we have seen in patients in the predominantly moderate stage of dementia of the Alzheimer type (DAT), namely on attention and working memory. In these psychopharmacological studies, we are looking at a type of attention termed rapid visual information processing.

By way of introduction to these studies, we will just briefly

mention the theoretical framework in which they are based, namely,
the cholinergic hypothesis of ageing and dementia (see Sahakian,
1988). This hypothesis attributes impairment of memory and
cognition to reduced central cholinergic function. Critical
evidence for this hypothesis is that the degree of cognitive
impairement in patients with DAT is positively correlated with the
decrease in choline acetyltransferase activity and a reduction in
acetylcholine synthesis in brain tissue measured in post mortem and
biopsy studies, respectively (Perry et al. 1978; Francia et al,
1985). The evidence for changes in cholinergic receptors of the
muscarinic type has been controversial. However, it has only proved
possible to carry out comparable studies on nicotinic receptors in
the last few years with the advent of new techniques. Using
receptor autoradiographic techniques, Clarke and colleagues (1984;
1986) mapped the density of nicotine receptors in several areas of
rat brain. The regions with high densities include certain thalamic
nuclei, the dentate gyrus and presubiculum of the hippocampal
formation, the ventral tegmental area and substantia nigra, pars
compacta of the midbrain, and the neocortex. Within the neocortex,
the muscarinic M2 receptor subtype and nicotinic receptors are
particularly dense in certain regions including primary visual
cortex, unlike the muscarinic M1 receptor subtype which is widely
distributed throughout most cortical areas (Mash et al., 1988;
Zilles 1990; Zilles et al., 1990).

In patients dying with DAT, the work of Perry and colleagues
(1986, 1987) has shown a huge reduction in nicotine receptors in
neocortex and hippocampus. However, the significance of this
result, both in understanding the cognitive deficits in DAT and
their possible remediation is not yet clear. Not only does nicotine
act post-synaptically but recent evidence shows that it increases
the release of acetylcholine (Richard et al. 1989).

Therefore, a novel, but logical treatment strategy, based on the
cholinergic hypothesis, would be to treat DAT patients with
nicotine, which might be expected to boost cholinergic function
both pre- and post-synaptically.

In view of the distinctive distribution of nicotine receptors, particularly in regions such as the hippocampus, sensory regions of cortex and in the midbrain (Clarke et al. 1984; 1986; Zilles, 1990; Zilles et al., 1990), where they may also modulate mesolimbic dopamine function, we chose tasks which would be especially sensitive to functions subserved by these regions. Thus, we included a test of short term spatial memory, based on animal studies which indicate a role of the hippocampus. We also included tests of working memory, attention and rapid visual information processing, as well as critical flicker fusion (CFF), an index of temporal resolution in the visual system that serves as an indirect measure of cortical arousal.

Finally, wherever possible, we took reaction time measures for these tasks. In fact, we will be showing that DAT patients exhibit large deficits in this type of attentional function, in addition to their primary memory impairment, and that these attentional deficits can be ameliorated by treatment with nicotine.

We will be reporting on the results of a preliminary study performed in collaboration with Professor Raymond Levy, Jeffrey Gray and David Warburton using subcutaneous injections of nicotine in 7 young and 7 old control subjects and in 7 patients with DAT.

Because nicotine had never been given by sub-cutaneous injection, the blood levels and half-life had to be established before we could design our testing protocol.

Russell and colleagues (1990) have shown that following sub-cutaneous adminstration, plasma nicotine levels rise shortly after injection, reach a peak about 15 minutes following injection, and then begin to decline. We therefore decided to work within a 40 minute test session in order to fit the tests into this window of maximal physiological activity of nicotine.

We will discuss the preliminary nicotine injection study. We gave two groups of normal subjects (young and old) and a DAT patient group, subcutaneous injections of nicotine. There were seven persons in each of the three groups. The groups included male and female subjects with approximately equal numbers of smokers and non-smokers. The three groups were matched for premorbid verbal IQ using the National Adult Reading Test (NART). The young adult subjects were students and community volunteers, the elderly controls were community volunteers and spouses of patients, and all the patients were out-patients who had been assessed at the Maudsley Hospital Memory Clinic.

We will not go into the details of the diagnostic procedure, but the full details of both exclusion and inclusion criteria have been published in paper by Philpot and Levy (1987). Briefly, patients were diagnosed using McKhann et al. (1984) criteria. All patients in the study were in the mild and moderate stages of the disease and fell in stages 1 and 2 of the Clinical Dementia Rating Scale of Hughes and colleagues (1982).

Subjects and patients attended seven test sessions in the following order. An undrugged baseline, a placebo session, three sessions with nicotine (0.4, 0.6 and 0.8 mg), and follow-up placebo and baseline sessions. Data from the two baseline and two placebo sessions were averaged. The amount of nicotine at the high dose is approximately equivalent to that in one cigarette, so these doses are not considered to be high. Both placebo injections were normal saline.

Two computerised tests, the digit span sub-test of the WAIS, the critical flicker fusion test, and the finger tapping computer test (a test of motor speed) were administered at each test session.

The first of our computerised tasks was primarily a test of attention and information processing, with a small working memory component.

It is a substantially modified and simplified form of the task that Wesnes and Warburton (1984) used in their studies. Briefly, in our task, subjects were asked to detect only consecutive, ascending odd or even sequences of digits (ie, 2,4,6; 3,5,7; 4,6,8; and 5,7,9).

Digits were presented on the computer screen at a rate of 100 per minute and the task lasted 7 minutes. Whenever a target sequence was detected subjects were asked to press a button. DAT patients were severely impaired on this task compard with normal subjects. In order to titrate performance in the three groups to an approximately equivalent level, which was not at ceiling and could show improvements, our task was individually graded in difficulty. This meant that generally normal young subjects were asked to detect all four sequences; normal old subjects were asked to detect between two and four sequences, and DAT patients were asked to detect only one sequence.

The measures obtained from this task are reaction time, correct detection (termed hits), incorrect responses (termed false alarms), and failure to detect or omissions. From these measures, indices of stimulus sensitivity and response bias can be calculated. In signal detection analysis, an increase in sensitivity without a change in bias would be the best indicator that nicotine is acting on attention and not some other process.

There were significant group differences in the accuracy of detection or sensitivity index on this rapid visual information processing task. The normal aged group showed a slight decrease in sensitivity in both the baseline and placebo condition as compared with the normal young adults. The DAT group was impaired when compared with both groups. With nicotine, both the normal young group and the DAT group showed a significant dose-dependent improvement in sensitivity. The effect was largest in the case of the DAT group.

Reaction times in all conditions were found to be significantly slower in patients with DAT as compared with the controls. Nicotine caused a significant and marked quickening of reaction times in patients with DAT.

The accuracy or % correct by 2 minute interval for the DAT group is sustained throughout the duration of the testing in the nicotine condition. Therefore, nicotine appeared to affect both response vigour as well as stimulus sensitivity. There was no significant change in the Bias index. It is interesting to note there is no specific speed/accuracy trade off, since both measures improve with nictine.

Furthermore, this is not an effect of non-specific arousal, since Hazenfratz and colleagues (1989) have recently shown that although white noise increased both heart rate and plasma cortisol, it had no effect on detection. or hit rate in a very similar paradigm. While smoking increased performance on this task, noise had no effect.

The other computer task tested attention and short term memory. In its simplest form, it was a modified version of the delayed response test used in rats by Sahgal (1987) and Dunnett (1985). In our version of the task for human subjects, they were first required to attend to, and register by touching a touch-sensitive screen, a sequence of stimuli, "yellow happy faces", presented in different locations on the VDU.

After touching the model sequence presented, subjects had to recall the sequences in the correct location/order following a 0, 4 or 16 second delay, by again touching the same windows in which the model sequence was presented. So this task had two components, an attentional one, in which the subjects had to prove they were paying attention by touching the happy faces as they appeared on the screen; and a memory one, where they again had to touch the location/order correctly. Like the first computer task, this one was individually graded in difficulty to titrate performance in the three groups to approximately equivalent levels in the memory phase of the task. Thus both control groups were shown model sequences of up to six happy faces, and the patients were generally shown sequences of beteewn 1 and 4 happy faces.

Only the patient group made attentional errors and these were reduced by the high dose of nicotine.

As you would expect, DAT patients were significantly impaired on the short term memory component of this computer task, while the control groups performed well even at the most difficult level of sequencing. Nicotine did not significantly alter the performance on any of the three groups on the memory component of this task. Furthermore, digit span traditionally regarded as an index of short term memory capacity, was unaffected by nicotine.

Nicotine increased detection of a flashing or flickering light as evidenced by the decreased difference between ascending and descending threshold values, in the CFF test, especially in the DAT group.

Therefore, our first results indicated that nicotine was

exerting its effects on attention and information processing rather than on memory processes, and it seemed that nicotine might be acting on cortical mechanisms involved in visual perception and attention.

We have now extended our study by testing a large sample, and there are now about 24 patients or subjects in each of the three groups. These results are to be reported on shortly. In addition, we will be reporting on the results of a study of nicotine given chronically in the form of a chewing gum in 9 patients with DAT.

In conclusion, the striking positive effects of acute nicotine in our preliminary study are most encouraging because of the previous resistance of the cognitive decline in DAT to pharmacological treatment. They suggest that such an approach may still prove to be viable for DAT patients who are mildly or moderately affected by the disease.

ACKNOWLEDGEMENTS: We thank Margaret Derrick for typing. Dr. B.J. Sahakian thanks the Wellcome Trust and the Eleanor Peel Foundation for support.

REFERENCES

Clarke, P., Hamill, G., Nadi, N., Jacobowitz, D., Pert, A. (1986) ^{3}H-Nicotine and 125 I-alpha-bungarotoxin labelled nicotine receptors in the interpeduncular nucleus of rats. II. Effects of habenular deafferentation. J.Comp.Neurol. 251: 407-413.

Clarke, P.S., Pert, C.B., Pert, A. (1984) Autoradiographic distribution of nicotine receptors in rat brain, Brain Research, 232, 390-395.

Dunnett, S.B. (1985) Comparative effects of cholinergic drugs and lesions of nucleus basalis or fimbria-fornix on delayed matching in rats. Psychopharmacology, 87, 357-363.

Francis, P.I., Palmer, A.M., Sims, N.R., Bowen, D., Davison, A., Esiri, M., Neary, D., Snowden, J., Wilcock, G. (1985) Neurochemical studies of early-onset Alzheimer's disease: possible influence on treatment. New Engalnd Journal of Medicine, 313, 7-11.

Hasenfratz, M., Michel, C., Nil, R., Bättig, K. (1989). Can smoking increase attention in rapid information processing during noise? Electrocortical, physiological and behavioral effects. Psychopharmacolgy, 98, 75-80.

Hughes, C.P., Berg, L., Danziger, W.L., Cohen, L.A., Martin, R.L. (1982) A new clinical scale for the staging of dementia. British Journal of Psychiatry, 140, 566-572.

Mash, D.C., White, F.W., and Mesulam, M.-M. (1988) Distribution of muscarinic receptor subtypes within architectonic subregions of the primate cerebral cortex. The Journal of Comparative Neurology, 278, 265-274.

McKhann, G., Drachman, D., Folstein, M., Katzman, R., Price, D. and Stadlan, E.M. (1984). Clinical diagnosis of Alzheimer's disease: report of the NINCDS/ADRDA work group under the auspices of Department of Health and Human Services Task force on Alzheimer's Disease. Neurology, 34, 939-944.

Perry, E.K., Perry, R.H., Smith, C.J. Dick, D.J., Candy, J.M., Edwardson, J.A., Faibrun, A., Blessed, G. (1987) Nicotinic receptor abnormalities in Alzheimer's and Parkinson's diseases. Journal of Neurology, Neurosurgery, and Psychiatry, 50, 806-809.

Perry, E., Perry, R., Smith, C., Purohit, J., Bonham, P., Dick, D., Candy, J., Edwardson, J.A., Faiburn, A. (1986). Cholinergic receptors in cognitive disorders. Can.J.Neurol.Sci., 13, 521-527.

Perry, E.K., Tomlinson, B.E., Blessed, G., Bergmann, K., Gibson, P., and Perry, R. (1978) Correlation of cholinergic abnormalities with senile plaques and mental test scores in senile dementia British Medical Journal, 11, 1457-1459.

Philpot, M.P., Levy, R. (1987) A memory clinic for the early diagnosis of dementia. International Journal of Geriatric Psychiatry, 2, 195-200

Richard, J., Aranjo, D.M., Quirion, R. (1989) Modulation of cortical acetylcholine release by cholinergic agents. Society for Neuroscience Abstracts, Vol. 15, Part 2, 472.15, 1197.

Russell, M., Jarvis, M., Jones G., Feyerabend, C. (1990) Nonsmokers show acute tolerance to subcutaneous nicotine. Psychopharmacology, 102, 56-58.

Sahakian, B.J. (1988) Cholinergic drugs and human cognitive performance. In: Handbood of Psychopharmacology, Volume 20, Psychopharmacology of the Aging Nervous System. (edited by Iversen, L.L., Iversen, S.D., Snyder, S.H.) London: Plenum Press.

Sahgal, A. (1987) Contrasting effects of vasopressin, desglycinamide-vasopressin and amphetamine on a delayed matching to position task in rats. Psychopharmacology, 93, 243-249.

Wesnes, K., Warburton, D.M. (1984) Effects of scopolamine and nicotine on human rapid information processing performance. Psychopharmacology, 82, 147-150.

Zilles, K. (1990) Codistribution of receptors in the human cerebral cortex. In: Receptors in the Human Nervous System (edited by Mendelsohn, F.A.O. and Paxinos, G.) London: Academic Press.

Zilles, K., Schröder, H., Schröder, U., Horvath, E., Werner, L., Luiten, P.G.M., Maelicke, A., Strosberg, A.D. (1990) Distribution of cholinergic receptors in the rat and human neocortex. Manuscript in submission.

NICOTINIC RECEPTOR LOSS IN BRAIN OF ALZHEIMER PATIENTS AS
REVEALED BY IN VITRO RECEPTOR BINDING AND IN VIVO POSITRON
EMISSION TOMOGRAPHY TECHNIQUES.

Agneta Nordberg

Department of Pharmacology, Uppsala University, Uppsala and
Department of Geriatric Medicine, Karolinska Institute, Stockholm,
Sweden.

SUMMARY: Heterogenous nicotinic receptor subtypes exist in brain.
In vitro receptor binding of ^{3}H-nicotine to tissue homogenates
from cortical autopsy brain tissue from patients with
Alazheimer´s disease indicates a marked loss in the high affinity
nicotinic binding sites concomittant with a change in the
proportion of high and low affinity nicotinic binding sites. When
^{11}C-nicotine is given intravenously to Alzheimer patients in
order to in vivo visualize the nicotinic receptors a lower uptake
of ^{11}C radioacitivity to the brain is observed compared to
healthy volunteers. A significant larger difference between the
uptake of the two enantiomers (S)(-) and (R)(+) ^{11}C-nicotine is
observed in Alzheimer brains compared to healthy controls. This
observation might be of diagnostic importance.

INTRODUCTION

Alzheimer´s disease, senile dementia of Alzheimer type (AD/SDAT)
is the most common form of dementia in the westword. It is
characterized as a progressive neurodegenerative disorder with
global detoriation of cognitive functions and preservation of
many other important brain functions. An early clinical diagnosis
is prevented by the lack of early diagnostic markers. The final

diagnosis can only be verified at autopsy by histopathological investigations. Neurochemical studies in autopsy brain tissue from Alzheimer patients have revealed a loss in multi-transmitter systems (Hardy et al.1985). The cholinergic system in brain which correlates best with cognitive function has shown the most consistent changes in AD/SDAT (Perry 1986). Among the afflicted cholinergic parameters in AD/SDAT brains are the nicotinic receptors. The aim of this chapter is to illustrate how a deficit in nicotinic receptors in AD/SDAT brains explored using in vitro ligand binding studies can be verified in vivo using positron emission tomography (PET) technique.

Human brain nicotinic receptors

The presence of nicotinic receptors in human autosy brain tissue has been characterized in vitro by receptor binding studies using ligands with various affinity (Volpe et al. ,1979; Nordberg and Winblad,1981; Larsson et al.,1987; Shimohama et al.,1985; Araujo et al.,1988). The nicotinic agonist binding sites in human brain can be rationalized in term of superhigh, high and low affinity binding sites by competion studies between ^{3}H-nicotine and unlabelled nicotine (Nordberg et al., 1988 a,b,c, Nordberg et al. 1989b). The high affinity nicotinic bindingsites can be measured using nanomolar concentrations of ^{3}H-nicotine or ^{3}H-acetylcholine (^{3}H-ACh) .The regional distribution of high affinity nicotinic binding sites in human brain is illustrated in Figure 1. The number of high affinity nicotinic receptors is high in the thalamus, caudate nucleus, putamen, substantia nigra, intermediate in the hypothalamus, cortical areas, cerebellum and low in the hippocampus, pons and globus pallidus. The distribution of nicotinic receptors does not totally follow the distribution of ACh throughout the brain since areas such as the cerebellum which nearly is devoid of ACh shows a fairly high number of nicotinic receptors.

A similar distribution of nicotinic receptors has also been

observed by in vitro autoradiography using large cryosections of the human brain and ^{3}H-nicotine (Adem et al., 1989). Within a brain region a regional distribution of nicotinic receptors in discrete brain nucleus can be observed as has been done for the human thalamus (Adem et al, 1988). Studies in human brain is facilitated by the fact that the nicotinic receptors are rather stable in autopsy brain tissue at least when the postmortem intervals are less than 24 h (Nordberg et al. 1989b).

The nicotinic receptors in human brain are influenced by aging. Thus the ^{3}H-nicotine binding sites decrease with aging in the cortex (Flynn and Mash, 1986; Nordberg et al., 1988a) and hippocampus (Perry et. al., 1987) while increase in the thalamus (Nordberg et al, 1988a).

NICOTINIC RECEPTORS IN HUMAN BRAIN

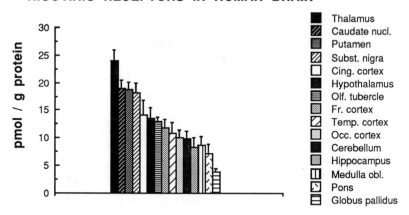

Figure 1. Regional distribution of nicotinic receptors in human brain. ^{3}H-nicotine (5 nM) was incubated with P2 fractions prepared from different regions of the brain. Correction for unspecific binding was made in parallel experiments in the presence of nicotine 10^{-3} M. Mean value ± S.E.M, n=4.

Recently, attempts have been made to visualize nicotinic receptors in vivo in human brain by ^{11}C-nicotine and positron emission tomography (PET) (Nybäck et al. 1989, Nordberg et al. 1990 a,b). The studies were preceeded by experiments in monkeys (Mazie`re et al., 1986; Nordberg et al., 1989a). When ^{11}C-nicotine is injcected intravenously to healthy volunteers the ^{11}C radiooactivity is rapidly taken up by the brain and distributed to various regions in a pattern which closely agrees with the distribution of nicotinic binding sites measured by in vitro receptor binding. A similar regional distribution of ^{3}H-radioactivity was recently observed ex vivo in rats intravenously injected with ^{3}H-nicotine (Broussolle et al.1989). Nicotine has been found to stimulate the glucose utilization in rat brain areas rich in nicotinic receptors suggesting that the nicotinic receptors are coupled to energy metabolism (London 1990). Chronic treatment with nicotine to rats induces an increase in the number of high affinity nicotinic binding sites in brain measured in vitro (Larsson et al. 1986, Romanelli et al. 1988, Nordberg et al.1989c). An increased number of high affinity nicotinic receptors has also been measured in autopsy brain tissue from smokers (Benwell et al. 1989). Similarly, in vivo studies by PET indicate an increased uptake of ^{11}C-nicotine to brain of smokers compared to nonsmokers (Nybäck et al. 1990). These observations motivated a comparison of in vitro and in vivo measurements of nicotinic receptors also in pathological states such as Alzheimer´s disease.

Loss of nicotinic receptors in autopsy Alzheimer brain tissue

Several research groups have now confirmed a marked reduction in nicotinic receptors in autopsy (Whitehouse et al., 1986, Nordberg and Winblad, 1986; Flynn and Mash, 1986; Perry et al. 1987, Arajou et al., 1988) or biopsy (Giacobini et al., 1989) AD/SDAT brain tissue. The marked loss in nicotinic receptors measured in the frontal and temporal cortex of autopsy AD/SDAT brain tissue by ^{3}H-nicotine is illustrated in Figure 2. The nicotinic receptors have been observed to be mainly diminished in cortical areas (Whitehouse et al., 1986; Nordberg and Winblad ,1986) but also in the hippocampus (Perry et al. 1987) and putamen (Shimohama et al., 1986). The number of high affinity nicotinic receptors and the activity of choline acetyltransferase decrease to a similar extent suggesting that the nicotinic receptors in contrast to the muscarinic receptors are located on cholinergic axons that degenerate in AD/SDAT. Analysis of subtypes of nicotinic receptors by competition experiments reveal a shift in the properties of high and low affinity sites with an increased proportion of low affinity sites (Nordberg et al. 1988b). The observation does not exclude that there is a loss in both subtypes of nicotinic receptors concomitant with a change in the proportion of sites.

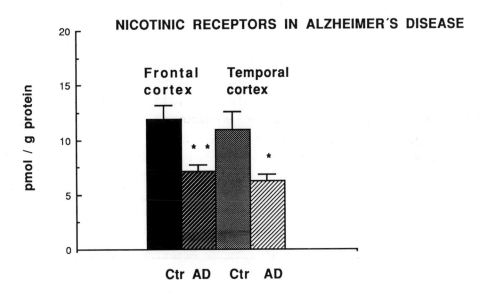

Figure 2. High affinity nicotinic receptors in frontal and temporal cortex of AD/SDAT brains. ^{3}H-nicotine (5 nM) was incubated with P2 fractions of cortical tissue. Corrections for unspecific binding was performed in parallel experiments in the presence of nicotine 10^{-3} M. Mean value ± S.E.M , n=6; ** p< 0.01; * p< 0.5.

Nicotinic receptors in AD/SDAT brains visualized by PET

Studies in postmortem brain tissue have shown a loss in nicotinic
receptors in AD/SDAT. These studies have to be confirmed in vivo.
Attempts have therefore been made to study the in vivo binding of
(-)(S) nicotine labelled by ^{11}C in brain of AD/SDAT patients
using PET (Nordberg et al. 1990a,b). Receptor studies in autopsy
brains only give information at the final end of the disease
while in vivo PET studies have the advantage that they can be
performed early in the course of the disease and might be used as
a diagnostic tool.

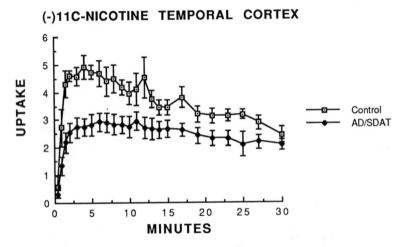

(-)11C-NICOTINE TEMPORAL CORTEX

Figure 3. Uptake and time course of ^{11}C-radioactivity in the
temporal cortex of six AD/SDAT patients and four healthy
volunteers following intravenous injection of (S)(-)-N-[11]C-
methyl-nicotine. The uptake is expressed in nCi/ cm^3/ dose bw^{-1}.
Mean value ± S.E.M .

The data presented here were obtained in studied performed in
AD/SDAT patients assessed in the Geriatric Clinic, Huddinge
hospital. The patients fulfilled the clincal diagnosis of
AD/SDAT. The AD/SDAT patients (mean age 68±2 years) had a

duration of the disease of 1- 10 years and the healthy volunteers (mean age 68±2 years) had no history of psychiatric or neurological disease. None of the AD/SDAT patients nor the healthy volunteers were smokers. The AD/SDAT patients as well as the healthy volunteers underwent PET investigations a the PET centre, Uppsala University, Uppsala. Since both enantiomers of nicotine bind to nicotinic receptors in brain tissue in vitro (Copeland et al. 1990) the (-)(S) form as well as the less biological active (+)(R) form of ^{11}C-nicotine were used in the study. When the two enantiomers of ^{11}C-nicotine were injected intravenously to healthy volunteers or AD/SDAT patients they were rapidly distributed from the arterial blood and peaked in brain within 2-5 minutes. The uptake of (-)(S) and (+)(R) ^{11}C-nicotine to the brain was very similar in healthy volunteers while in AD/SDAT patients a significantly markedly lower uptake of (+)(R) ^{11}C-nicotine was observed compared to (-)(S) ^{11}C-nicotine. The uptake of (-)(S) ^{11}C-nicotine was in general lower in cortical areas as the temporal cortex (Figure 3).

The uptake of (+)(R) ^{11}C-nicotine was also significantly lower in AD/SDAT patients than in healthy volunteers. Figure 4 illustrates the lower uptake of (+)(R) ^{11}C-nicotine in the frontal cortex of AD/SDAT patients compared to healthy volunteers.

Earlier studies in AD/SDAT patients using PET have focused on parameters such as cerebral blood flow, oxygen consumption and glucose utilization which are all reduced (Frackowiak et al. 1981; Ferris et al., 1980). PET might however give further valuable information in neurogenerative disorders such as AD/SDAT by visualizing neuroreceptors. An important question is however whether the time course of ^{11}C-nicotine solely mimicks the cerebral blood flow or is a tracer for nicotinic receptors. Studies using ^{11}C-butanol as a blood flow marker reveal a different course for ^{11}C-butanol in brain in comparison to (-)(S) and (+)(R) ^{11}C-nicotine in both control and AD/SDAT patients (Nordberg et al.1990b). These studies support the assumption that

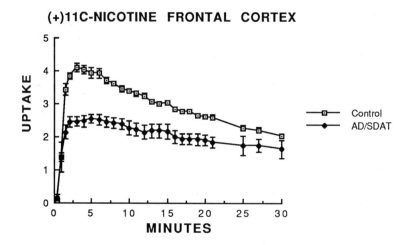

Figure 4 Uptake and time course of ^{11}C-radioactivity in the frontal cortex of five AD/SDAT patients and five healthy volunteers following intravenous injections of (R)(+)-N-^{11}C-methyl-nicotine. The uptake is expressed in nCi/ cm^3/ dose bw^{-1}, Mean value ± S.E.M.

the uptake and time course of ^{11}C-nicotine might reflect specific nicotinic receptor binding sites in the brain. Kinetic analysis of the binding data obtained from PET studies are hampered by the rapid on/off rate of nicotine to its receptor which is known from in vitro receptor binding studies (Larsson and Nordberg, 1985). Further studies are needed to evaluate whether subtypes of nicotinic receptors in brain can be visualized by PET. Although the ^{11}C-compounds are given in tracer amounts the specific radioactivity might not be high enough to make a distinction between receptor subtypes. Furthermore, a distinction of subtypes might be easier in Alzheimer than in control brains. The properties of the (R) form compared to the (S) form of the ^{11}C-nicotine have to be further evaluated by PET.

IN CONCLUSIONS: The (R)(+) and (S)(-) enantiomers of ^{11}C-nicotine and PET might be valuable tools in order to visualize nicotinic receptors in vivo in man. Hopefully, the technique will have implicity both for study of dependence processes induced by nicotine as well as for neurodegenerative diseases such as Alzheimer's disease. In the latter case the PET technique might be used both as a diagnostic tool and for evaluation of new therapeutic strategies.

ACKNOWLEDGEMENTS
This study was supported by grants from the the Swedish Medical Research Council, the Swedish Tobacco Company, Loo and Hans Osterman's foundation and Stohne's foundation.

REFERENCES

Adem, A., Nordberg, A., Singh Jossan, S., Sara, V., and Gillberg, P.G., (1989) Neurosci. Lett. 101, 247-252
Adem, A., Singh Jossan, S.,d´,Argy, R., Brandt, I., Winblad, B., and Nordberg. A., (1989) J. Neural Transm. 73, 77-83
Araujo, D.M., Lapchak, P.A., Robitaille, Y., Gauthier, S., and Quirion, R., (1988) J. Neurochem. 50, 1914-1923
Benwell, M.E.M., Balfour, D.J.K., and Andersson, J.M. (1989) J. Neurochem. 50, 1243-1247
Broussolle, E.P., Wong, D.F., Fanelli, R.J., and London, E.D. (1989) Life Sci. 44, 1123-1132
Copeland, J.R., Adem, A., Jacob 111, P., and Nordberg, A. (1990) Naunyn Schmiedebergs Arch Pharmacol., submitted
Flynn,D.D. and Mash., D.C., (1986) J. Neurochem. 47, 1948-1954
Ferris, S.H., de Leon, M.J., Wolf, A.P., Farkas, T., Christman, D.R., Reisberg, B., Fowler, J.S., MacGregor, R., Goldman, A., George, A.E., and Rampal, S. (1980) Neurobiol. Aging 1, 127-131
Frackowiak, R.S.J., Pozzilli, C., Legg, N.J., Du Boulay, G.H., Marshall, J., Lenzi, G.L., and Jones, T. (1981) Brain 104, 753-778
Giacobini, E., DeSarno, P., Clark, B., and McIlhany, M. (1989) Progr. Brain Res. 79, 335-343
Hardy, J.A., Adolfsson, R., Alafuzoff, I., Bucht, G., Marcusson, J., Nyberg, P., Perdahl,E., Wester,P., and Winblad, B. (1985) Neurochem. Int. 7, 545-563
Larsson, C. and Nordberg, A. (1985) J. Neurochem. 45, 24-31
Larsson, C., Nilsson,L., Halen,A., and Nordberg., A. (1986) Drug and Alcohol Dependence 17, 37-45
Larsson, C., Lundberg, P.Å., Halen, A., Adem, A., and Nordberg,A. (1987) J. Neural Transm. 69, 3-18
London, E.D. (1990) In: The Biology of Nicotine Dependence, Ciba Foundation Symposium 152) John Wiley, Chichester, pp. 131-146
Maziere. M., Comar, D., Marazano, C., and Berger, G. (1976) Eur. J. Nucl. Med. 1, 255-258

Nordberg, A. and Winblad, B. (1981) Life Sci. 29, 1937-1944
Nordberg, A. and Winblad, B. (1986) Neurosci. Lett. 72, 115-119
Nordberg, A., Adem, A., Nilsson, L., and Winblad, B. (1988a) In:
New Trends in Aging Research,Fidia Research Series, vol. 15
(Tomlinson, B., Pepeu, G., Wischik, C.M.,eds) Liviana Press,
Italy, pp. 27-36
Nordberg, A., Adem, A., Hardy, J., and Winblad, B. (1988b)
Neurosci. Lett. 86, 317-321
Nordberg, A., Adem, A., Nilsson, L., Romanelli, L., and Zhang, X.
(1988c) In: Nicotinic Acetylcholine Receptors in the nervous
System, NATO Asi Series vol. H25 (Clementi, F., Gotti, C.,
Sher, E., eds) Springer Verlag , Berlin, pp. 305-315
Nordberg, A., Hartvig, P., Lundqvist, H., Antoni, G., Ulin,
J.,and Långström, B. (1989a) J Neural Transm. (P-D Sect) 1:
195-205
Nordberg, A., Hartvig, P., Lilja, A., Viitanen, M., Amberla, K.,
Lundqvist, H., Andersson, Y., Ulin, J., Winblad, B., and
Långström, B. (1990a) J. Neural Transm. (P-D Sect) in press
Nordberg, A., Hartvig, P., Lundqvist, H., Lilja, A., Viitanen,
M., Amberla, K., Ulin, J., Winblad, B., and Långström, B.
(1990b) In: Current Research in Alzheimer Therapy: Early
Diagnosis, (Giacobini, E., Becker, R.E., Eds), Taylor &
Francis, New York, in press
Nordberg, A., Nilsson-Håkansson, L., Adem, A., Hardy, J.,
Alafuzoff, I., Lai, Z., Herrera-Marschitz, M., and Winblad, B.
(1989b) Progr. Brain Res. 79, 353-362
Nordberg, A., Romanelli,.L., Sundwall, A., Bianchi, C., and
Beani, L. (1990c) Br. J. Pharmacol. 98 71-78
Nybäck, H., Nordberg, A., Långström, B., Halldin, C., Åhlin, P.,
Swan, C.G., and Sedvall, G. (1989) Progr. Brain Res. 79, 313-
319
Perry, E.K. (1986) Br. Med. Bull. 42, 63-69
Perry, E.K., Perry, R.H., Smith, C.J., Dick, D.J., Candy,
J.M.,Edwardson,J.A.,Fairbairn, A.,and Blessed,G.(1987)
J.Neurol. Neurosurg. Psychiatry 50, 806-809
Romanelli, L., Öhman, B., Adem, A, and Nordberg, A., (1988) Eur.
J. Pharmacol. 148, 289-291.
Shimohama, S., Taniguchi, T., Fujiwara, M., and Kameyama, M.
(1985) J. Neurochem. 45, 604-610
Shimohama, S., Taniguchi, T., Fujiwara, M., and Kameyama., M.
(1986) J. Neurochem. 46, 288-293
Volpe, B.T., Francis, A., Gazzanigra, S., and Schechter, N.
(1979) Exp. Neurol. 66, 737-744
Whitehouse, P.J., Martino, A.M., Antuono, P.G., Lowenstein, P.R.,
Coyle, J.T., Price, D.L., and Kellar, K.J. (1986) Brain Res.
371, 146-151

Author Index

Subject Index